Theory and Applications of Partial Differential Equations

MATHEMATICAL CONCEPTS AND METHODS IN SCIENCE AND ENGINEERING

Recent volumes in this series:

33 **PRINCIPLES OF ENGINEERING MECHANICS**
Volume 2: Dynamics–The Analysis of Motion • *Millard F. Beatty, Jr.*

34 **STRUCTURAL OPTIMIZATION**
Volume 1: Optimality Criteria • *Edited by M. Save and W. Prager*

35 **OPTIMAL CONTROL APPLICATIONS IN ELECTRIC POWER SYSTEMS**
• *G. S. Christensen, M. E. El-Hawary, and S. A. Soliman*

36 **GENERALIZED CONCAVITY**
• *Mordecai Avriel, Walter E. Diewert, Siegfried Schaible, and Israel Zang*

37 **MULTICRITERIA OPTIMIZATION IN ENGINEERING AND IN THE SCIENCES**
• *Edited by Wolfram Stadler*

38 **OPTIMAL LONG-TERM OPERATION OF ELECTRIC POWER SYSTEMS**
• *G. S. Christensen and S. A. Soliman*

39 **INTRODUCTION TO CONTINUUM MECHANICS FOR ENGINEERS**
• *Ray M. Bowen*

40 **STRUCTURAL OPTIMIZATION**
Volume 2: Mathematical Programming • *Edited by M. Save and W. Prager*

41 **OPTIMAL CONTROL OF DISTRIBUTED NUCLEAR REACTORS**
• *G. S. Christensen, S. A. Soliman, and R. Nieva*

42 **NUMERICAL SOLUTION OF INTEGRAL EQUATIONS**
• *Edited by Michael A. Golberg*

43 **APPLIED OPTIMAL CONTROL THEORY OF DISTRIBUTED SYSTEMS**
• *K. A. Lurie*

44 **APPLIED MATHEMATICS IN AEROSPACE SCIENCE AND ENGINEERING**
• *Edited by Angelo Miele and Attilio Salvetti*

45 **NONLINEAR EFFECTS IN FLUIDS AND SOLIDS**
• *Edited by Michael M. Carroll and Michael A. Hayes*

46 **THEORY AND APPLICATIONS OF PARTIAL DIFFERENTIAL EQUATIONS**
• *Piero Bassanini and Alan R. Elcrat*

A Continuation Order Plan is available for this series. A continuation order will bring delivery of each new volume immediately upon publication. Volumes are billed only upon actual shipment. For further information please contact the publisher.

Theory and Applications of Partial Differential Equations

Piero Bassanini

University of Rome "La Sapienza"
Rome, Italy

and

Alan R. Elcrat

Wichita State University
Wichita, Kansas

Plenum Press • New York and London

Library of Congress Cataloging-in-Publication Data

Bassanini, Piero.
Theory and applications of partial differential equations / Piero Bassanini and Alan R. Elcrat.
p. cm. -- (Mathematical concepts and methods in science and engineering ; 46)
Includes bibliographical references (p. -) and index.
ISBN 0-306-45640-0
1. Differential equations, Partial. I. Elcrat, Alan R. II. Title. III. Series.
QA377.B295 1997
515'.353--dc21 97-38643
CIP

ISBN 0-306-45640-0

A Division of Plenum Publishing Corporation
233 Spring Street, New York, N.Y. 10013

http://www.plenum.com

10 9 8 7 6 5 4 3 2 1

Printed in the United States of America

Preface

This book is a product of the experience of the authors in teaching partial differential equations to students of mathematics, physics, and engineering over a period of 20 years. Our goal in writing it has been to introduce the subject with precise and rigorous analysis on the one hand, and interesting and significant applications on the other.

The starting level of the book is at the first-year graduate level in a U.S. university. Previous experience with partial differential equations is not required, but the use of classical analysis to find solutions of specific problems is not emphasized. From that perspective our treatment is decidedly theoretical. We have avoided abstraction and full generality in many situations, however. Our plan has been to introduce fundamental ideas in relatively simple situations and to show their impact on relevant applications. The student is then, we feel, well prepared to fight through more specialized treatises.

There are parts of the exposition that require Lebesgue integration, distributions and Fourier transforms, and Sobolev spaces. We have included a long appendix, Chapter 8, giving precise statements of all results used. This may be thought of as an introduction to these topics. The reader who is not familiar with these subjects may refer to parts of Chapter 8 as needed or become somewhat familiar with them as prerequisite and treat Chapter 8 as Chapter 0. On the other hand, this book is entirely self-contained with regard to partial differential equations. We have refrained from referring to more advanced treatises in order to complete an exposition. The exercises are an important part of the exposition: They serve both to illustrate and to extend the theory, as well as to train the student by having her or him fill in the details omitted in the text.

The book has six main chapters. Chapter 2 concerns the wave equation, Chapter 3 the heat equation, and Chapter 4 potential theory and the Laplace equation. Chapter 5 deals with second-order elliptic equations, Chapter 6 abstract evolution equations, and Chapter 7 hyperbolic conservation laws. Together with appropriate parts of Chapter 8, the book covers material for two one-semester courses in partial differential equations, an introductory course (Chapters 1–4) and an advanced course (Chapters 5–7).

A book on this subject is certain to have a large overlap with existing expositions. There are, however, certain novel features, as follows.

Chapter 2 includes a section on linear hyperbolic systems with constant coefficients in two variables and an application to transmission-reflection problems in layered media. Chapter 3 contains a variety of problems from applications aimed at stimulating the interest of the reader. Chapter 4 presents Perron's method of subharmonic functions for the Dirichlet problem as well as a fairly complete discussion of the boundary integral equations method for the solution of the Dirichlet and Neumann problems. This includes certain interesting variants that have recently arisen especially in connection with applications to aerodynamics. In addition, a complete proof is given of Wiener's criterion for regularity of boundary points for the Laplace equation. The part of the theory of capacities required for this is developed in a manner bridging the gap between physics and modern potential theory. Furthermore, a theorem of Widman, which shows that first derivatives of harmonic functions are continuous up to the boundary for $C^{1+\alpha}$ boundary and data, is proven. This result has previously appeared only in the research literature. In Chapter 5 we present a proof of the celebrated De Giorgi–Nash–Moser theorem for linear elliptic divergence structure operators, and we give a self-contained exposition of the proof of existence for the nonparametric Plateau problem using Perron's method. Also, results on surfaces with constant mean curvature and capillary surfaces are proven. These results have hitherto been available only in research papers and monographs. Chapter 6 is mainly devoted to the Hilbert space theory of parabolic evolution equations and the study of the initial value problem for a quasilinear "viscous" conservation law together with some a priori estimates in the "vanishing viscosity limit." This enables us to give a proof of Kruzhkov's theorem concerning "entropy solutions" of the scalar conservation law in Chapter 7. In this chapter we have also given a new exposition of the proof of local existence of Lipschitz continuous solutions of quasilinear hyperbolic systems in one space variable along with a companion continuation theorem. (We have avoided systems of conservation laws in several space variables because more sophisticated techniques are required.) A proof of Glimm's theorem on existence of weak solutions concludes Chapter 7. The exposition of this difficult theorem has been given in sufficient detail to put it at the level of this book.

Finally, each chapter includes applications that have not appeared outside of the research literature.

We are indebted to Professor Victor Isakov and to Dr. Carlo Sinestrari for reading parts of the manuscript and suggesting valuable improvements. The figures were drawn by Megan Elcrat.

P. Bassanini
Rome, Italy

A. R. Elcrat
Wichita, Kansas

Contents

1

Introduction to Partial Differential Equations

We begin with some terminology and notation. We use the word *domain* for an open connected set in $\mathbb{R}^n$. If Ω is a domain in $\mathbb{R}^n$ with sufficiently regular boundary (precise hypotheses will be given later), then the *Gauss lemma*

$$\int_\Omega f_{x_i}\, d\mathbf{x} = \int_{\partial\Omega} f n_i\, dS$$

for $f \in C^1(\bar{\Omega})$ holds, where $\mathbf{x} = (x_1, \ldots, x_n)$ and $\mathbf{n} = (n_1, \ldots, n_n)$ is the exterior normal. This immediately implies the *divergence theorem*

$$\int_\Omega \operatorname{div} \mathbf{F}\, dx = \int_{\partial\Omega} \mathbf{F} \cdot \mathbf{n}\, dS,$$

where $\mathbf{F}$ is a vector-valued function whose components are in $C^1(\bar{\Omega})$.

The idea of a *well-posed problem* plays a central role. We say that a differential equation problem is well posed if:

a. A solution exists.
b. This solution in unique.
c. Solutions depend continuously on the data.

The meaning of (a) and (b) must be made specific in the problem at hand by giving the class of functions for which a solution is to be found, and then (c) requires a choice of how the closeness of data and solutions is to be measured (usually with norms in appropriate Banach spaces). When the problem is *linear*, uniqueness and continuous dependence can be rephrased in the following form:

b. The homogeneous problem has only the zero solution.
c. If the data approach zero, the solution approaches zero.

The study of partial differential equations is usually built around the equations of classical mathematical physics, in particular the *wave*, *heat*, and *Laplace* equations. In addition, many modern applications are concerned with *hyperbolic systems of conservation laws*. In this introduction we will sketch derivations of some typical problems from first principles (e.g., Ref. 1). We will also say a few things about classification and characteristics of second-order partial differential equations (e.g., Refs. 2, 3).

1. Population Diffusion

Let $u = u(\mathbf{x}, t)$ be the population density at a point $\mathbf{x}$ of a domain Ω in $\mathbb{R}^2$, and at time t, so that the population in a domain $D \subset \Omega$ is given by (the integer part of) $\int_D u(\mathbf{x}, t)\, d\mathbf{x}$. We suppose that this density varies only because of the birth–death rate density $F(\mathbf{x}, t)$ and the flux vector $\mathbf{W}(\mathbf{x}, t)$ across the boundary. Then

$$\frac{d}{dt}\int_D u(\mathbf{x}, t)\, d\mathbf{x} = -\int_{\partial\Omega} \mathbf{W} \cdot \mathbf{n}\, ds + \int_D F(\mathbf{x}, t)\, d\mathbf{x}$$

for any D. This can be written, using the divergence theorem, as

$$\int_D (u_t + \operatorname{div} \mathbf{W} - F)\, d\mathbf{x} = 0,$$

and, because D is arbitrary,

$$u_t + \operatorname{div} \mathbf{W} = F.$$

To specify the problem, a "constitutive relation" between $\mathbf{W}$ and u must be assigned. We will take the simple relation

$$\mathbf{W} = -k \operatorname{grad} u. \tag{F}$$

This says that the population flux is proportional to the density gradient, directed from high to low density ("town to country") for $k = a^2 > 0$ and from low to high density ("country to town") for $k = -a^2 < 0$. We assume that a is constant. Then if $k > 0$, we obtain the equation

$$u_t = a^2 \Delta u + F, \tag{H}$$

and, if $k < 0$,

$$u_t = -a^2 \Delta u + F. \tag{BH}$$

If the density u is stationary (independent of time) and $F = 0$, it follows from (H) or (BH) that u satisfies the *Laplace equation*

$$\Delta u = 0.$$

Equation (H) also arises from the theory of heat conduction (hence its name, the *heat equation*), where u is the temperature and $\mathbf{W}$ is the heat flux. The constitutive relation (F) then is termed the *Fourier's law*, and we take $k > 0$ because heat flows from hot to cold regions. We note that (BH), the *backward heat equation*, arises from (H) formally by reversing the direction of time. In both of these models u is a nonnegative quantity. This fact may or may not be taken into account when dealing with the equations.

If (H) or (BH) holds in a bounded domain Ω, we might take

$$u = 0 \qquad \text{on } \partial\Omega$$

(in the context of population dynamics, Ω is surrounded by a "desert"), or

$$\frac{\partial u}{\partial n} = 0 \qquad \text{on } \partial\Omega$$

(the "frontier" is closed), or

$$\frac{\partial u}{\partial n} = -\alpha u \qquad \text{on } \partial\Omega$$

(flux is regulated according to the population density at the frontier). For the one-dimensional heat equation, $u_t = a^2 u_{xx} + F$, the normal derivative of u is replaced by $\pm u_x$. If u is the temperature, the relation $\partial u/\partial n + \alpha u = g$ (given) follows from *Newton's law of cooling*, which says that the heat flux across a boundary is proportional to the temperature difference $u - u_0$, with $u_0 = g/\alpha$.

If, in addition, the initial density is given,

$$u(\mathbf{x}, 0) = \varphi(\mathbf{x}) \qquad \text{in } \Omega,$$

we obtain a problem that is a candidate to be well posed. Another possibility is that the frontier is so far away that we may as well assume that $\Omega = \mathbb{R}^2$. This leads to the pure *initial-value problem* for (H) or (BH). The question of the well-posedness of this problem will be dealt with in Chapter 3.

2. Vibrating String Equation

Consider small, transverse (say, vertical) vibrations of a string with constant linear density ρ and vertical displacement $u(x, t)$. Suppose that the string in its unperturbed state lies along the segment $[0, l]$ of the x-axis and has tension $\mathbf{T}_0 = T_0(x)\mathbf{i}$, where $\mathbf{i}, \mathbf{j}$ are the unit vectors of the x- and u-axes. Then the total lengthening of the string in the course of its motion will be

$$\Delta l = \int_0^l (\sqrt{1 + u_x^2} - 1)\, dx \simeq \frac{1}{2} \int_0^l u_x^2 dx,$$

because by the hypothesis of small oscillations $| u_x | \ll 1$. Hence, the local "strain" $dl/dx = \frac{1}{2} u_x^2$ is of second order in $| u_x |$, and up to first order the string is "inextendible."

The tension in the string is

$$\mathbf{T} = T(x, t)\mathbf{t}, \tag{T}$$

where $\mathbf{t}$ is the tangent to the string configuration $u = u(x, t)$, and by Hooke's law,

$$T(x, t) - T_0(x) = O(u_x^2),$$

so that to first order in $| u_x |$ the tension $T(x, t)$ coincides with $T_0(x)$ and is independent of time. Neglecting second-order terms, (T) becomes

$$\mathbf{T} \simeq T_0(x)\mathbf{i} + T_0(x)u_x\mathbf{j}.$$

The assumption of no horizontal motion requires $T_0(x) = T_0$ constant (horizontal forces acting on any string segment must balance out); hence, the tension is independent of x, too. The vertical force due to tension on an arbitrary segment (x_1, x_2) is then

$$T_0\big[u_x(x_2, t) - u_x(x_1, t)\big] = T_0 \int_{x_1}^{x_2} u_{xx}(x, t)\, dx,$$

and by Newton's law this equals $\int_{x_1}^{x_2} \rho u_{tt}\, dx$, so that $\rho u_{tt} = T_0 u_{xx}$, or

$$u_{tt} = c^2 u_{xx}, \tag{V}$$

where $c^2 = T_0/\rho$. This is the (one-dimensional) *wave equation*. Similar reasoning applied to a vibrating membrane leads to the same equation with the second x-derivative replaced by the Laplacian (Ref. 4). If there is an external vertical force $f(x, t)$ per unit mass (e.g., due to gravity), this must be added on the right-hand side of (V).

Note that second-order terms cannot be neglected when writing the energy equation for the string, as the potential energy is proportional to the lengthening

$$V = \frac{T_0}{2}\int_0^l u_x^2 dx \equiv T_0 \Delta l;$$

the total energy is then $T + V$, with

$$T = \frac{\rho}{2}\int_0^l u_t^2 dx$$

the kinetic energy. Typical boundary conditions at $x = 0$, $x = l$ are

$$u = 0$$

(string with fixed endpoints), or

$$u_x = 0$$

(string with free endpoints, where the vertical tension $T_0 u_x$ vanishes; for a membrane, u_x is replaced by the normal derivative, $\partial u/\partial n$: see Ref. 1). If we add the initial data at $t = 0$,

$$u(x, 0) = u_0(x), \qquad u_t(x, 0) = u_1(x),$$

(the wave equation is of second order in t), we obtain an initial–boundary value problem. This problem will be shown in Chapter 2 to be well posed.

3. Equations for Isentropic Flow of a Perfect Gas

The equations of gas dynamics play a fundamental role in applications. We will use them as an example of a system of conservation laws.

From the Eulerian viewpoint, in which one "watches" the flow velocity $\mathbf{v}(\mathbf{x}, t) = (v_1, v_2, v_3)$ and fluid density $\rho(\mathbf{x}, t)$ at a fixed (arbitrary) point

$\mathbf{x} = (x_1, x_2, x_3)$ of the flow domain as a function of time t, the equations for a compressible inviscid fluid can be written in the conservative (divergence) form

$$\frac{\partial \rho}{\partial t} + \sum_{j=1}^{3} \frac{\partial}{\partial x_j}(\rho v_j) = 0, \frac{\partial}{\partial t}(\rho v_i) + \sum_{j=1}^{3} \frac{\partial}{\partial x_j}(\rho v_i v_j + p\delta_{ij}) = 0, \qquad i = 1, 2, 3$$

(the *Euler equations*), where δ_{ij} is the Kronecker delta. These equations can be derived from conservation of mass and momentum. For isentropic flows the pressure p is a function of ρ alone, which, for an ideal gas, with constant specific heats, is given by

$$p = p_0(\rho/\rho_0)^\gamma, \tag{P}$$

where γ, p_0, ρ_0 are positive constants, with $\gamma > 1$.

We assume for simplicity that the flow is one-dimensional with (scalar) velocity $v_1 = u(x, t)$ along the x-axis ($x = x_1$). The Euler equations then reduce to the 2×2 system of quasilinear equations in divergence form:

$$\begin{aligned} \rho_t + (\rho u)_x &= 0, \\ (\rho u)_t + (\rho u^2 + p)_x &= 0, \end{aligned} \tag{I}$$

where p is given by (P). The quantity

$$c = \sqrt{\partial p/\partial \rho} = \sqrt{\gamma p_0/\rho_0^\gamma}\sqrt{\rho^{\gamma-1}}$$

is the (local) speed of sound. If $\gamma = 3$, c is proportional to ρ and system (I) reduces to the scalar conservation law

$$U_t + (\tfrac{1}{2}U^2)_x = 0 \tag{S}$$

for each of the quantities $U = u + c$, $U = u - c$. If $\gamma < 3$, letting $\mathbf{u} = (u_1, u_2) = (\rho, u)$, (I) is of the form

$$\mathbf{u}_t + \mathbb{A}(\mathbf{u})\mathbf{u}_x = 0 \tag{II}$$

for an appropriate 2×2 matrix $\mathbb{A}(\mathbf{u})$. This matrix has a full set of (left and right) eigenvectors for each $\mathbf{u}$ (Exercise 3.1). We will study the initial value problem for (S) and (II) in Chapter 7.

4. Classification and Characteristics

A second-order partial differential equation in N variables $(y_1, \ldots, y_N) := \mathbf{y}$ and one unknown function $u = u(\mathbf{y})$, of the form

$$\sum_{i,j=1}^{N} a_{ij} u_{y_i y_j} + F(\mathbf{y}, u, \text{grad } u) = 0, \tag{G}$$

where $\mathbb{A} = [a_{ij}]$ is a constant $N \cdot N$ symmetric matrix, can be classified into types on the basis of the two invariants of $\mathbb{A}$, the *index of inertia* $\mathscr{I}$ (the number of negative eigenvalues) and the *defect* $\mathscr{D}$ (the number of zero eigenvalues). This classification depends only on the *principal part* $\sum_{ij} a_{ij} u_{y_i y_j}$ of the equation. Precisely, the equation is said to be:

i. of *elliptic* type in $\mathbb{R}^N$ if $\mathscr{D} = 0$, $\mathscr{I} = 0$ ($\mathbb{A}$ is positive definite), or $\mathscr{I} = N$ ($\mathbb{A}$ is negative definite);
ii. of *hyperbolic* type in $\mathbb{R}^N$ if $\mathscr{D} = 0$, $\mathscr{I} = 1$, or $N - 1$ ($\mathbb{A}$ is indefinite);
iii. of *parabolic* type in $\mathbb{R}^N$ if $0 < \mathscr{D} < N$ ($\mathbb{A}$ is singular); the case of interest is $\mathscr{I} = 0$, $\mathscr{D} = 1$.

In physics the 'hyperbolic' or 'parabolic' variable (the variable corresponding to the negative or null eigenvalue) is usually a time variable t, say $y_N = t$, $\mathbf{y} = (\mathbf{x}, t)$, with $\mathbf{x} \in \mathbb{R}^{N-1}$ a space variable. The equation is in *canonic form* if $\mathbb{A}$ is diagonal: this can always be achieved by a linear transformation. If F is linear in the variables u, u_{y_i} with constant coefficients, then the corresponding linear canonic forms are

$$\begin{aligned} \Delta u + au &= f && \text{(elliptic)}, \\ \Delta u - c^{-2} u_{tt} + au &= f && \text{(hyperbolic)}, \\ \Delta u + a u_t &= f && \text{(parabolic)}, \end{aligned}$$

where a is a real constant, f is a given function of $\mathbf{y}$, and Δ is the Laplace operator with respect to $\mathbf{y}$ (in the elliptic case) or $\mathbf{x}$. This classification is complete only for the case $N = 2$ (and for constant coefficients). The situation increases in complexity when higher-order equations and systems and variable coefficients are considered. We will instead consider classes of equations that are defined by examples arising in significant applications.

A *characteristic manifold* of (G) is a manifold defined by the implicit equation $\varphi(\mathbf{y}) = \text{const}$, where φ is a C^1 solution of the *characteristic equation*

$$\sum_{i,j=1}^{N} a_{ij} \varphi_{y_i} \varphi_{y_j} = 0,$$

with $\sum_{i=1}^{N}(\varphi_{y_i})^2 > 0$. For instance, the wave equation in three space variables

$$u_{tt} = c^2 \Delta u \equiv c^2(u_{x_1x_1} + u_{x_2x_2} + u_{x_3x_3}) \tag{W}$$

is hyperbolic in $\mathbb{R}^4$, and the characteristic equation is

$$\varphi_t^2 = c^2 \operatorname{grad}^2 \varphi.$$

If $\varphi(\mathbf{x}, t)$ is of the form $\varphi(\mathbf{x}, t) = t - X(\mathbf{x})$, then the time-dependent manifolds $X(\mathbf{x}) = t + \text{const}$ in $\mathbb{R}^3$ are called *wave fronts*.

Exercises

2.1. Write the energy equation for the string. *Hint:* $(d/dt)(T + V) = T_0[u_x u_t|_0^l$.

3.1. Find the matrix $\mathbb{A}(\mathbf{u})$ and show that it has a full set of (left and right) eigenvectors for each $\mathbf{u}$, corresponding to the real eigenvalues $u \pm c$. *Hint:* Right eigenvectors are $(\pm\rho, c)$.

4.1. Show that if $Au := \sum_{i,j=1}^{n} a_{ij}u_{x_ix_j}$ is an elliptic operator in $\mathbb{R}^n$, $n = N - 1$, then the equations $Au = 0$, $u_t = Au$ are parabolic in $\mathbb{R}^N$ and $u_{tt} = Au$ is hyperbolic in $\mathbb{R}^N$.

4.2. Show that an elliptic equation has no characteristic manifolds.

4.3. Show that for parabolic equations in the linear canonic form (LC) the characteristic manifolds are given by $t = \text{const}$.

4.4. Show that the wave fronts can be determined as solutions of the "eikonal equation" $\operatorname{grad}^2 X = c^{-2}$. In particular, verify that the wave equation $u_{tt} = c^2\Delta u$ admits spherical wave fronts $|\,\mathbf{x} - \mathbf{x}_0\,| = \pm c(t - t_0)$, corresponding to characteristic cones in space-time.

4.5. Show that a regular (locally) invertible change of variables $\xi = \xi(\mathbf{y})$ transforms (G) into an equation of the same form with principal part defined by the new matrix $\tilde{\mathbb{A}} = \mathbb{J}\mathbb{A}\mathbb{J}^T$, where $\mathbb{J}$ is the Jacobian matrix of the transformation. Hence, deduce that the classification is (locally) invariant with respect to such changes of variables.

4.6. For $N = 2$, define $\Delta := a_{12}^2 - a_{11}a_{22}$. Show that equation (G) is elliptic if $\Delta < 0$, hyperbolic if $\Delta > 0$, and parabolic if $\Delta = 0$, and that the characteristics are straight lines satisfying $dx_2/dx_1 = (a_{12} \pm \sqrt{\Delta})/a_{11}$ (for $\Delta \geq 0$).

4.7. Suppose we seek solutions of the wave equation in the form $u = e^{\nu S}$, where $\nu > 0$ is a parameter. Substituting in (W), we find the equation for S:

$$\nu^{-1}(S_{tt} - c^2\Delta S) + S_t^2 - c^2\operatorname{grad}^2 S = 0.$$

Hence, $S = \text{const}$ approaches a characteristic manifold in the limit $\nu \to \infty$ (this corresponds to the high-frequency limit of geometrical optics). On the other hand, if $\nu = i\lambda$ is purely imaginary, then $e^{i\lambda S}$ satisfies the wave equation if and only if S satisfies the wave equation and $S = \text{const}$ is a characteristic manifold.

4.8. The plane wave $u = e^{i(\omega t - \xi \cdot \mathbf{x})}$ is a solution of the wave equation in $N-1$ space variables if and only if the 'dispersion relation' $|\xi| = \omega/c$ holds, and the (hyper) planes of constant phase $\omega t - \xi \cdot \mathbf{x} = \text{const}$ are wave fronts in $\mathbb{R}^{N-1}$. $\omega > 0$ is the frequency and $\xi \in \mathbb{R}^{N-1}$ the wave number.

References

1. TIHONOV, A. N., and SAMARSKII, A. A., *Equazioni della Fisica Matematica*, Mir, Moscow, Russia, 1981.
2. HELLWIG, G., *Partial Differential Equations*, Teubner, Stuttgart, Germany, 1977.
3. COURANT, R., and HILBERT D., *Methods of Mathematical Physics*, Vol. 2, Interscience, New York, New York, 1965.
4. BITSADZE, A. V., *Partial Differential Equations*, World Scientific, Singapore, 1994.

2

Wave Equation

The ideas in this chapter are built around a study of the one-dimensional wave equation

$$u_{tt} - c^2 u_{xx} = 0, \tag{V}$$

also called the vibrating string equation. This equation and its higher-dimensional version

$$u_{tt} - c^2 \Delta u = 0 \tag{W}$$

also governs many other phenomena, e.g., sound waves, electromagnetic waves, and is of great interest in its own right. It is also, however, representative of the large class of hyperbolic partial differential equations, and it is useful to meet fundamental properties of this class in this relatively simple case.

1. Initial Value Problem

At the beginning we will assume that solutions of (V) have two continuous derivatives. The transformation of variables

$$\xi = x - ct, \qquad \eta = x + ct \tag{T}$$

is very useful in dealing with (V).

Theorem 1.1. If D is a convex domain in the (x, t)-plane, any solution $u \in C^2(D)$ of (V) can be written in the form

$$u(x, t) = \omega_1(x - ct) + \omega_2(x + ct),$$

where ω_1, ω_2 are arbitrary functions of class C^2.

Proof. If (T) is applied, we obtain for $U(\xi, \eta) = u(x, t)$, the equation

$$U_{\xi\eta} = 0$$

in the domain $D^* = \mathrm{T}(D)$. Because D is convex, D^* is also, and $U_\xi = g(\xi)$. As $U_{\eta\xi} = U_{\xi\eta}$, we also obtain $U_\eta = f(\eta)$. f, g are C^1 functions and U is actually defined on the smallest rectangle containing D^*. By performing the inverse transformation we see that u is defined on the smallest parallelogram, bounded by lines $x \pm ct = \text{const}$, that contains D. For an arbitrary point $(\xi_0, \eta_0) \in D^*$ we can write

$$U(\xi, \eta) = U(\xi_0, \eta_0) + \int_{\xi_0}^{\xi} g(s)ds + \int_{\eta_0}^{\eta} f(s)ds := \omega_1(\xi) + \omega_2(\eta).$$

Transforming back to D, the theorem follows. □

We may think of the expression given in Theorem 1.1 as a "general solution" of (V).

Example 1.1. The hypothesis of convexity of D in Theorem 1.1 is required, as the example in Fig. 1, given in the (ξ, η)-plane, shows.

The two components given by $\omega_1(x - ct)$ and $\omega_2(x + ct)$ may be thought of as "traveling waves"; each translates with speed c without distortion, ω_1 to the right, ω_2 to the left. Thought of in another way, ω_1 is constant along the lines $x - ct = \text{const}$, and ω_2 is constant when $x + ct = \text{const}$. These two families of curves are called the *characteristic lines* of (V), the first are denoted by C_+ (positive slope, $1/c$), the second by C_- (negative slope, $-1/c$). We see that if the values of ω_1 and ω_2 are known along a curve Γ that is nowhere tangent to a C_+ or C_-, the corresponding solution of (V) is known in a curvilinear triangle made up of Γ, the C_- from the left endpoint of Γ and the C_+ from the right endpoint (Fig. 2).

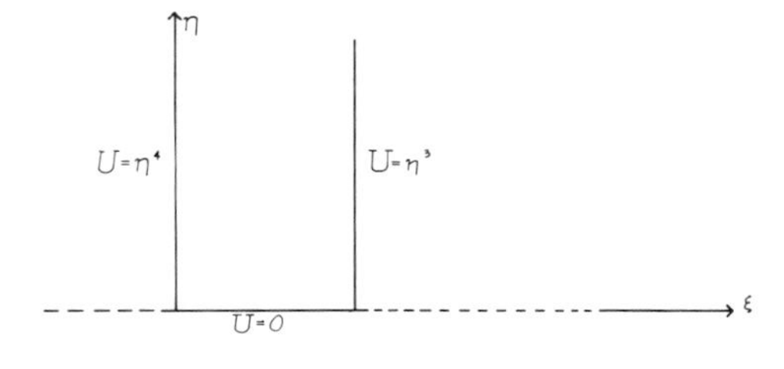

Fig. 1. An example showing convexity is needed.

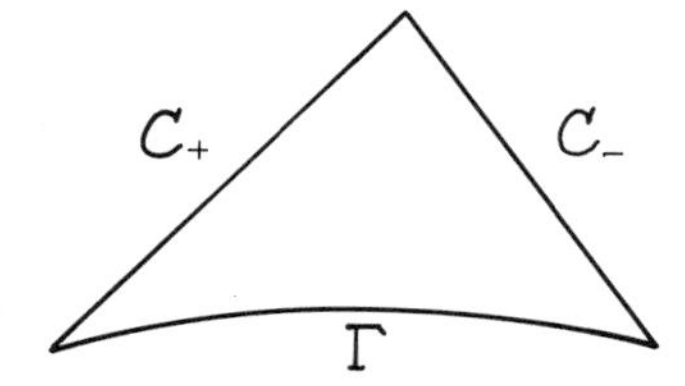

Fig. 2. Determination of the solution in a curvilinear triangle.

We introduce the notation

$$W_\psi(x, t) = \frac{1}{2c}\int_{x-ct}^{x+ct} \psi(s)\, ds$$

for t times the average of a function ψ defined on $\mathbb{R}$ on the interval $(x - ct, x + ct)$. The first basic problem that we solve for (V) is the "Cauchy problem,"

$$u_{tt} - c^2 u_{xx} = 0, \quad x \in \mathbb{R}, t \in \mathbb{R}, \quad u(x, 0) = u_0(x), \quad u_t(x, 0) = u_1(x), \quad x \in \mathbb{R}. \tag{CP}$$

Theorem 1.2. Suppose that $u_0 \in C^2(\mathbb{R})$, $u_1 \in C^1(\mathbb{R})$. Then (CP) has a unique solution $u \in C_2(\mathbb{R}^2)$ given by

$$u(x, t) = W_{u_1}(x, t) + \frac{\partial}{\partial t} W_{u_0}(x, t) \equiv \frac{1}{2}[u_0(x - ct) + u_0(x + ct)] + \frac{1}{2c}\int_{x-ct}^{x+ct} u_1(s)ds. \tag{D}$$

This expression is known as the *d'Alembert solution* of the wave equation.

Proof. The fact that (D) is a solution is immediate. If u is a solution, write $u(x, t) = \omega_1(x - ct) + \omega_2(x + ct)$ (Theorem 1.1). Then $u_0 = \omega_1(x) + \omega_2(x)$, $u_1(x) = c[\omega_1'(x) - \omega_2'(x)]$. Differentiating the first of these equations, and solving the resulting two linear equations for ω_1', ω_2' leads directly to (D). □

The inequality

$$|u(x, t)| \leq \sup_{\mathbb{R}}|u_0(x)| + |t|\sup_{\mathbb{R}}|u_1(x)|$$

follows immediately from (D). From this we can derive continuous dependence of solutions of (CP) on the initial data, in each finite strip $\{x \in \mathbb{R}, |t| \leq T\}$ for $T > 0$.

We remark here that a solution of the inhomogeneous Cauchy problem,

$$u_{tt} - c^2 u_{xx} = f(x, t),$$
$$u(x, 0) = u_0(x), \qquad u_t(x, 0) = u_1(x),$$

can be obtained by superimposing (D) with the particular solution

$$u_p(x, t) = \frac{1}{2c} \int_0^t d\tau \int_{x-c(t-\tau)}^{x+c(t-\tau)} f(s, \tau) ds,$$

which is a solution of $u_{tt} - c^2 u_{xx} = f$ with zero initial conditions. This can be verified directly, or derived from the so-called *Duhamel principle* (Exercise 1.2). The above double integral extends over the triangle $\mathcal{T}$, with upper vertex (x, t), bounded by segments of characteristic lines through this vertex and by the interval $[x - ct, x + ct]$ on the initial line.

We make a series of remarks, which follow from (D) by inspection. Each is elementary, but points to a fundamental property of (CP).

i. The class of initial conditions $(u_0, u_1) \in C^2 \times C^1$ is *persistent* in the sense that if the solution is "stopped" at time t the values of u and u_t obtained are in this class and can be used as initial data for a new (CP) starting at t.
ii. The *domain of dependence* $\mathcal{D}(P)$ of the point $P = (x, t)$, defined as the subset of $\mathbb{R}$ on whose complement a change in the data of (CP) does not affect $u(x, t)$, is the interval $[x - ct, x + ct]$ on the initial line. This means that the solution $u(P)$ depends only on the values of the initial data in $\mathcal{D}$. For the inhomogeneous (CP), $u(x, t)$ depends only on the values of f in $\mathcal{T}$.
iii. The *domain of determinacy* $\Delta(I)$ of an interval $I = [a, b]$ on the initial line [the set of points P such that $\mathcal{D}(P)$ is contained in I] is the upper(and lower) triangle determined by I and the characteristic lines through the endpoints of I. By definition, this is the domain of $\mathbb{R}^2$ where the solution u is uniquely determined by the values of the initial data in I.
iv. The *domain of influence* $\mathcal{J}(I)$ [the set of points P such that $\mathcal{D}(P)$ intersects I] is the unbounded domain bounded by the other characteristics through the endpoints (Fig. 3).
v. If u_0 and u_1 vanish outside I, then u vanishes outside $\mathcal{J}$ in $\mathbb{R}^2$ and

$$\lim_{t \to \infty} u(x, t) = \frac{1}{2c} \int_a^b u_1(s) ds$$

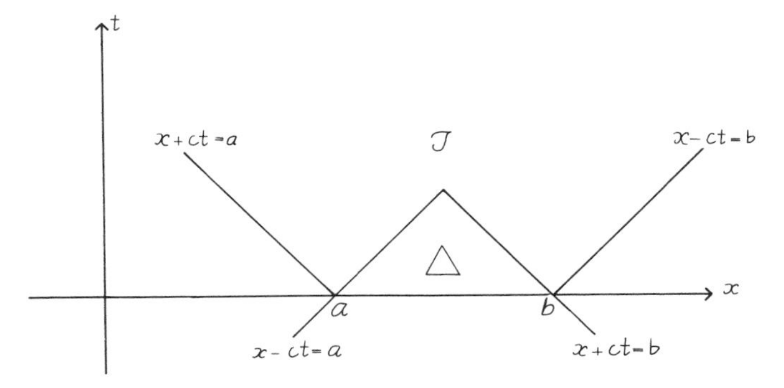

Fig. 3. The domains of influence and determination.

for every x. This shows that the support of $u(x, t)$ with respect to the x-variable "expands in time with speed c," and the "signals" do not decay as $t \to \infty$.

The concept of solution can be extended to include problems with discontinuous initial data. Suppose that u_0, u_1 have a discrete set of discontinuities and $u_0 \in C^2$, $u_1 \in C^1$ on the intervals with these discontinuity points as endpoints. Then the function defined by the formula in (D) is clearly a solution in the subsets of $\mathbb{R}^2$ bounded by characteristics emanating from the discontinuity points. A suitable concept of *generalized solution* is obtained by multiplying the equation by a test function φ and shifting derivatives to φ using integration by parts (i.e., the Gauss lemma).

Definition 1.1. A function u is a *weak solution* of (V) if

$$\int_{\mathbb{R}^2} u(\varphi_{tt} - c^2\varphi_{xx})dxdt = 0$$

for all $\varphi \in C_0^\infty(\mathbb{R}^2)$.

We have left the class of functions u is allowed to vary in unspecified here. This logical gap seems reasonable because all of this is most simply and elegantly understood in the context of the theory of distributions as explained in Chapter 8. The reader may think in terms of the class of "piecewise C^2 functions" for the moment. Then it is a straightforward exercise in integration by parts to show that the solution given by (D) is a weak solution (see Ref. 1).

Example 1.2. Let

$$u_0 = \begin{cases} h, & |x| < a, \\ 0 & |x| \geq a, \end{cases} \qquad u_1 = 0.$$

We can define the solution in an (x, t)-diagram (Fig. 4).

The reader can easily construct a sequence of time snapshots of the solution using this figure, and is strongly encouraged to do so.

2. Initial–Boundary Value Problems

Suppose that our string is only infinite on one side and extends from zero to infinity. We have

$$u_{tt} - c^2 u_{xx} = 0, \qquad 0 < x < \infty, t > 0, \tag{1}$$

$$u(x, 0) = u_0(x), \qquad u_t(x, 0) = u_1(x), \qquad 0 \leq x < \infty, \tag{2}$$

and we need an additional condition at $x = 0$. For simplicity we consider first the fixed endpoint condition,

$$u(0, t) = 0, \qquad t \geq 0. \tag{3}$$

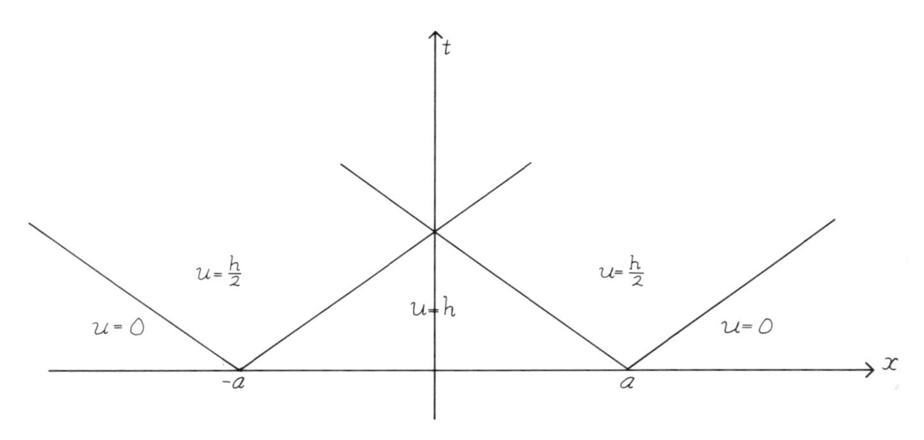

Fig. 4. Support of a discontinuous IVP.

A solution can be obtained using the ideas of the previous section. In fact, for $x - ct > 0$ the d'Alembert formula gives the solution [see the definition of $\Delta(I)$ in Section 1], and we can write

$$u(x, t) = \omega_1(x - ct) + \omega_2(x + ct),$$

where

$$\omega_1(x - ct) = \frac{1}{2}u_0(x - ct) - \frac{1}{2c}\int_0^{x-ct} u_1(s)ds$$

for $x > ct$, and

$$\omega_2(x + ct) = \frac{1}{2}u_0(x + ct) + \frac{1}{2c}\int_0^{x+ct} u_1(s)ds.$$

Then $u(0, t) = \omega_1(-ct) + \omega_2(ct)$, and (3) requires, formally, that $\omega_1(-ct) = -\omega_2(ct)$, i.e., $\omega_1(s) = -\omega_2(-s)$. Because $x + ct > 0$ always, $\omega_2(s)$ need only be defined for $s > 0$, and this relation can be used to extend the definition of ω_1 to negative values of s. We have

$$\omega_1(x - ct) = -\frac{1}{2}u_0(ct - x) - \frac{1}{2c}\int_0^{ct-x} u_1(s)ds, \qquad 0 < x < ct,$$

and

$$u(x, t) = \begin{cases} \frac{1}{2}[u_0(x - ct) + u_0(x + ct)] + \dfrac{1}{2c}\displaystyle\int_{x-ct}^{x+ct} u_1(s)ds, & x > ct, \\ \frac{1}{2}[u_0(ct - x) - u_0(ct - x)] + \dfrac{1}{2c}\displaystyle\int_{ct-x}^{ct+x} u_1(s)ds, & 0 < x < ct. \end{cases} \tag{4}$$

We can easily interpret this solution geometrically and physically. For a point (x_0, t_0) with $x_0 < ct_0$, draw the C_+ and C_- characteristics backwards in time. The C_- will hit the initial line [at $P = (ct_0 + x_0, 0)$], but the C_+ will hit the t-axis at $(0, t_0 - x_0/c)$ (Fig. 5).

If this C_+ is reflected into a C_- receding to the initial line, it will hit at $Q = (ct_0 - x_0, 0)$. The solution is given by $(1/2)[u_0(P) - u_0(Q)]$ plus the result of integrating $(1/2c)u_1$ from Q to P. In (CP), the effect of data, or "disturbances," propagates along characteristic lines. Here it is the same except that *reflection* at $x = 0$ inverts the amplitude. In fact, equation (4) follows from the d'Alembert formula (D) (Section 1) by extending the initial data (2) to

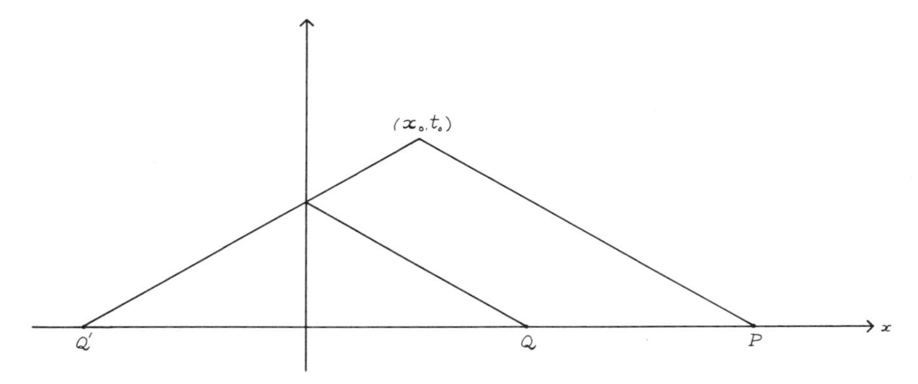

Fig. 5. Reflection of characteristics.

negative x as odd functions, so $u_0(Q') = -u_0(Q)$ and the integral of u_1 over $Q'Q$ cancels out.

Remarks

2.1. If we do not assume that $u_0(0) = 0$, there will be a discontinuity in the solution that propagates along $x = ct$. Suppose $u_0 \in C^2([0, +\infty))$, $u_1 \in C^1([0, +\infty))$. The full compatibility conditions in order to have a C^2 solution u,

$$u_0(0) = u_1(0) = u_0''(0) = 0,$$

can be derived as a consequence of the extension process (see Ref. 2 and Exercise 2.1).

2.2. If the boundary condition is $u_x(0, t) = h(t)$, the reflection law at $x = 0$ becomes $\omega_1(-ct) + \omega_2(ct) = h(t)$. The solution proceeds as before.

2.3. If a vertical spring is attached to the string at $x = 0$, the boundary condition

$$u_x(0, t) - ku(0, t) = 0 \tag{5}$$

$(k > 0)$ is obtained. If the solution for $0 < x < ct$ is written in the form $u(x, t) = \omega_2(x + ct) + \phi(t - x/c)$, and we assume $u_1 = 0$, the function ϕ can be obtained as the solution of a first-order differential equation.

2.4. Suppose that the endpoint of the string moves to the right with constant speed u. We can then formulate the following problem:

$$\begin{aligned} &u_{tt} - c^2 u_{xx} = 0, && Ut < x < \infty,\ t > 0, \\ &u(x, 0) = u_0(x), && u_t(x, 0) = 0, \qquad 0 < x < \infty, \end{aligned}$$

and

$$u(Ut, t) = 0.$$

If $U \geq c$, $x = Ut$ lies below or coincides with $x = ct$ and there is no solution, unless $u_0 = 0$. If $0 \leq U < c$, we have the reflection law $\omega_1[(U - c)t] + \frac{1}{2}u_0[(U + c)t] = 0$ along $x = Ut$, and this can be written as

$$\omega_1(s) = -\tfrac{1}{2}u_0(-as),$$

where $a = (c + U)/(c - U)$. Because $a > 0$, this extends the definition of ω_1 to negative s, so that u can be defined for $Ut < x < ct$. We have

$$u(x, t) = \begin{cases} \frac{1}{2}u_0(x - ct) + \frac{1}{2}u_0(x + ct), & 0 \leq t < x/c, \\ -\frac{1}{2}u_0[a(ct - x)] + \frac{1}{2}u_0(x + ct), & x/c < t \leq x/U. \end{cases}$$

The next problem that we consider involves an *interface* between two media located at $x = 0$ (Ref. 3). For example, consider a string made of two pieces with different densities tied together at $x = 0$, so that $u_{tt} - c^2(x)u_{xx} = 0$, $x \neq 0$, where

$$c^2 = \begin{cases} c_1^2, & x < 0, \\ c_2^2, & x > 0. \end{cases}$$

We want to consider an "incoming wave," i.e., a solution of the form $\omega_1(x - ct)$, which we will write as $u = \mathscr{I}(t - x/c)$, which has a sharp front, i.e., $\mathscr{I}(s) = 0$ for $s < 0$. Then $u = 0$ for $t < x/c$ and the waveform given by $\mathscr{I}$ translates to the right with speed c. Alternatively, $\mathscr{I}(t - x/c)$ is the solution of (CP) with initial data

$$u(x, 0) = \begin{cases} \mathscr{I}(-x/c), & x < 0, \\ 0, & x > 0, \end{cases} \tag{6}$$

and

$$u_t(x, 0) = \begin{cases} \mathscr{I}'(-x/c), & x < 0, \\ 0, & x > 0. \end{cases} \tag{7}$$

Suppose that this incoming wave strikes the interface at $x = 0$ at time $t = 0$. We assume the string is at rest to the right when $t = 0$. In (6) and (7), we take $c = c_1$, and we seek the solution in the form

$$u = \begin{cases} \mathscr{I}(t - x/c_1) + \mathscr{R}(t + x/c_1), & x < 0, \\ \mathscr{T}(t - x/c_2) + r(t + x/c_2), & x > 0. \end{cases}$$

The results of Section 1 show that $u = 0$ for $x > c_2 t$, and, then, that $r(s) \equiv 0$ and $\mathscr{T}(s) = 0$ for $s < 0$. Similarly, $u = \mathscr{I}(t - x/c_1)$ for $x < -c_1 t$, hence $\mathscr{R}(s) = 0$ for $s < 0$. We must then determine

$$\mathscr{T}(t - x/c_2), \qquad 0 < x < c_2 t,$$

called the *transmitted wave*, and

$$\mathscr{R}(t + x/c_1), \qquad -c_1 t < x < 0,$$

called the *reflected wave*. In order to do this we need to impose conditions on how the strings interact at the interface. We assume continuity,

$$u(0-, t) = u(0+, t), \qquad t > 0,$$

and that the vertical force given by $T_0 u_x$ is continuous, i.e.,

$$u_x(0-, t) = u_x(0+, t).$$

We then have

$$\mathscr{I}(t) + \mathscr{R}(t) = \mathscr{T}(t)$$

and

$$\mathscr{I}'(t) - \mathscr{R}'(t) = \frac{c_1}{c_2} \mathscr{T}'(t).$$

Differentiating the first equation and solving simultaneously for $\mathscr{R}'(t)$ and $\mathscr{T}'(t)$ we obtain

$$\mathscr{T}(t) = \frac{2c_2}{c_1 + c_2} \mathscr{I}(t) := T\mathscr{I}(t)$$

and

$$\mathscr{R}(t) = \frac{c_2 - c_1}{c_2 + c_1} \mathscr{I}(t) := R\mathscr{I}(t).$$

The expressions T and R are called the transmission and reflection coefficients.

We can also apply the above methods to a finite string. As an illustration, we consider the problem

$$\begin{aligned} u_{tt} - c^2 u_{xx} = 0, \qquad & 0 < x < l, t > 0, \\ u(x, 0) = u_t(x, 0) = 0, \qquad & 0 < x < l, \\ u(0, t) = \mu(t), \qquad u(l, t) = 0, \qquad & t > 0. \end{aligned}$$

It will be convenient to extend $\mu(t)$ by $\mu(t) = 0$ for $t \leq 0$. Then, using Exercise 1.1, we have

$$u(x, t) + u(P_2) = \mu(t_1),$$

where t_1 is the time at which the C_+ characteristic through (x, t) intersects the t-axis, and P_2 is as shown in the diagram (Fig. 6).

Continuing recursively, we can find $u(P_2)$ in terms of $u(P_3)$, etc. We have

$$u(x, t) = \mu(t_1) - u(P_2) = \mu(t_1) - \mu(t_2) + u(P_3) = \cdots = \sum_{i=1}^{\infty}(-1)^{i+1}\mu(t_i),$$

where

$$t_1 = t - x/c, \qquad t_2 = t_1 - 2(l - x)/c,$$

and

$$t_{2n+1} = t_{2n} - 2x/c, \qquad t_{2n+2} = t_{2n+1} - 2(l - x)/c,$$

for $n = 1, 2, \ldots$. The sum is actually finite as μ vanishes for $t < 0$.

Suppose we now consider the finite string problem

$$\begin{aligned} u_{tt} - c^2 u_{xx} = 0, \qquad & 0 < x < l, t > 0, \\ u(x, 0) = u_0(x), \qquad u_t(x, 0) = u_1(x), \qquad & 0 < x < l, \\ u(0, t) = u(l, t) = 0, \qquad & t > 0. \end{aligned}$$

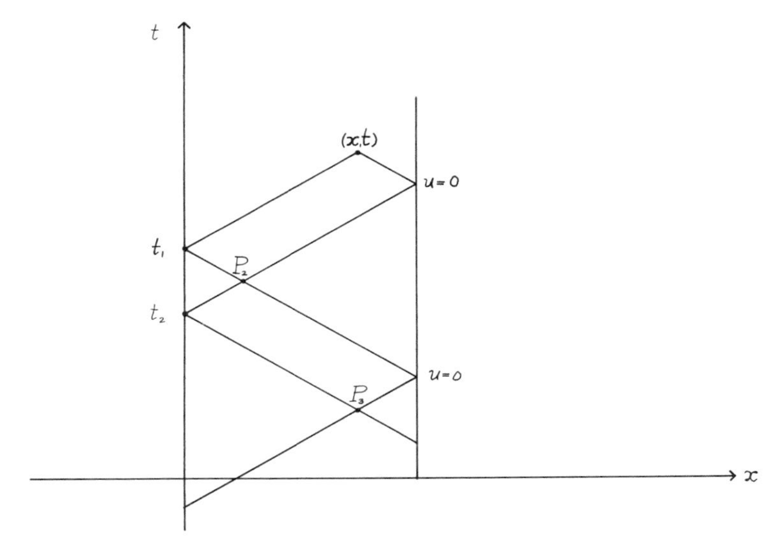

Fig. 6. Characteristic solution of the problem on a finite interval.

There is another procedure that is well suited to this problem. Assume a solution of (V) of the form $X(x)T(t)$. Then

$$X''/X = T''/c^2T = \text{const.}$$

We may assume the constant is negative real, say $-\lambda$. (The reader will be able to show easily that other possibilities lead nowhere for this problem.) Then the resulting constant-coefficient differential equation for T has

$$T = C\cos(c\sqrt{\lambda}t) + D\sin(c\sqrt{\lambda}t)$$

as general solution, whereas X and λ are determined as solutions of the eigenvalue problem

$$X'' + \lambda X = 0, \qquad 0 < x < l, \qquad X(0) = X(l) = 0.$$

The equation for X has the general solution $X = A\cos(\sqrt{\lambda}x) + B\sin(\sqrt{\lambda}x)$, and by imposing $X(0) = X(l) = 0$ we find $A = 0$, $\sin\sqrt{\lambda}l = 0$. Thus, the nontrivial solutions (eigenfunctions) are $X_n = \sin(n\pi x/l)$, $n = 1, 2, \ldots$, in correspondence to $\lambda = \lambda_n = (n\pi/l)^2$ (eigenvalues). We then obtain a sequence of solutions

$$u_n = [a_n \cos(n\pi ct/l) + b_n \sin(n\pi ct/l)] \sin(n\pi x/l)$$

of (V), vanishing at $x = 0, l$. Physically, each of these is a "standing wave," with frequencies

$$\omega_n = n\omega_1, \qquad \omega_1 = \pi c/l; \tag{8}$$

ω_1 is called the *fundamental frequency*. It remains to satisfy the initial conditions, and we do this "superimposing" (adding) the solutions u_n, and representing $u(x, t)$ as the eigenfunction expansion

$$u(x, t) = \sum_{n=1}^{\infty} (a_n \cos \omega_n t + b_n \sin \omega_n t) \sin(n\pi x/l).$$

Then for $t = 0$,

$$u(x, 0) = u_0(x) = \sum_{n=1}^{\infty} a_n \sin\left(\frac{n\pi x}{l}\right), \quad u_t(x, 0) = u_1(x) = \sum_{n=1}^{\infty} b_n \frac{n\pi c}{l} \sin\left(\frac{n\pi x}{l}\right).$$

The initial conditions are then satisfied, formally, by requiring that $\{a_n\}$ and $\{n\pi c b_n/l\}$ are the coefficients in Fourier sine series expansions of $u_0(x)$ and $u_1(x)$. There remains the question of showing that the series for $u(x, t)$ is convergent in an appropriate sense when hypotheses are made on properties of u_0, u_1. We leave this question as an exercise.

We conclude this section with a Fourier series solution of a forced vibration problem. Consider

$$\begin{aligned} u_{tt} - c^2 u_{xx} &= f(x, t), \qquad && 0 < x < l,\ t > 0, \\ u(x, 0) = u_t(x, 0) &= 0, && 0 < x < l, \\ u(0, t) = u(l, t) &= 0, && t > 0. \end{aligned}$$

We expand f in a Fourier sine series,

$$f(x, t) = \sum_{n=1}^{\infty} f_n(t) \sin(n\pi x/l), \qquad f_n(t) = \frac{2}{l}\int_0^l f(\xi, t) \sin(n\pi\xi/l)d\xi,$$

and seek u in the form

$$u(x, t) = \sum_{n=1}^{\infty} u_n(t) \sin(n\pi x/l),$$

where the functions $u_n(t)$ are unknown. We find that

$$u_n'' + \omega_n^2 u_n = f_n$$

and

$$u_n(0) = u_n'(0) = 0.$$

The function $u_n(t)$ can then be written as

$$u_n(t) = \frac{1}{\omega_n} \int_0^t \sin[\omega_n(t - \tau)] f_n(\tau) d\tau.$$

If the function $f_n(t)$ is sinusoidal, say $f_n(t) = a_n \cos(k\Omega t)$, the phenomenon described as *resonance* in elementary differential equations arises, i.e.,

$$u_n(t) = \begin{cases} (2k\Omega)^{-1} a_n t \sin(k\Omega t), & k\Omega \equiv n\omega_1, \\ (n^2\omega_1^2 - k^2\Omega^2)^{-1} a_n[\cos(k\Omega t) - \cos(n\omega_1 t)], & k\Omega \neq n\omega_1. \end{cases}$$

If $k\Omega = \omega_n \equiv n\omega_1$ ($k\Omega$ is the "forcing frequency," ω_n an "eigenfrequency" of the string), the amplitude grows linearly with t ("secular term"), and $u_n(t)$ becomes unbounded. Suppose ω_1^2/Ω^2 is an irrational number (so ω_1/Ω is, too). Then if a general function $f_n(t)$ of period $2\pi/\Omega$ is expanded in a Fourier series, each term has the potential of creating a term with a *small denominator* $n^2\omega_1^2 - k^2\Omega^2$, so that convergence of the series for u may be impaired. In fact, a theorem of Dirichlet states that if $\sigma = \omega_1^2/\Omega^2$ is an irrational number, there are infinitely many pairs of integers (m, j) such that $|m\sigma - j| < 1/m$ (for the proof, see Ref. 4, p. 448).

3. Reflection Problem

We consider here the equations governing the propagation of electromagnetic waves, usually known as Maxwell's equations, in the simplest case in which the electric and magnetic fields, denoted by E and H, vary in only one direction. These equations can be written

$$H_x + \varepsilon E_t = 0, \qquad E_x + \mu H_t = 0, \tag{ME}$$

where ε and μ are material properties called *permittivity* and *permeability*. In particular, we will consider a system of plane layers in which ε is constant in each layer and μ is constant everywhere. If ε and μ are constant, we can formally differentiate both equations in (ME) to obtain wave equations for both H and E with $c^2 = 1/\varepsilon\mu$. The problems that we consider can then be formulated as boundary value problems for the wave equation in the spirit of the reflection–transmission problem that was studied in the last section, but we will maintain (ME) as a first-order system for E and H because this is in many ways more convenient and instructive.

We will consider m plane layers, defined by $a_k < x < a_{k+1}$ $k = 1, \ldots, m$, lying between semi-infinite layers defined by $x < a_1 = 0$ and $x > a := a_{m+1}$. Also, $\varepsilon = \varepsilon_i$ for $a_{i-1} < x < a_i$, $i = 1, \ldots, n$, $n = m + 2$, where $a_0 := -\infty$, $a_n := +\infty$. The problem that we consider is to find the response of this system to an "incoming wave" moving to the right in the first semi-infinite medium, $x < 0$. The functions $E(x, t)$ and $H(x, t)$ are assumed to be continuous at the interfaces. (A more precise description of the problem is that a linearly polarized TEM wave travels to the right in the first medium at normal incidence to the interfaces. E and H are tangential components of the electric and magnetic fields at the interfaces, and we are imposing the physically natural condition that tangential components are continuous. See Exercise 3.1.)

It is convenient to introduce the normalized variables

$$\tau = t/\sqrt{\varepsilon_2 \mu}, \qquad u(x, \tau) = \sqrt{\varepsilon_2/\mu}E(x, t), \qquad v(x, \tau) = H(x, t);$$

(ME) then become

$$u_x + v_\tau = 0, \qquad v_x + h_i^2 u_\tau = 0; \qquad a_{i-1} < x < a_i, \qquad \tau \in \mathbb{R}, \tag{ME$'$}$$

where $h_i := \sqrt{\varepsilon_i/\varepsilon_2}$. Denoting by (u_i, v_i) the solution (u, v) in the ith layer, we can write a general solution of this system as combinations of traveling waves

$$u_i(x, \tau) = (1/h_i)[U_i(\tau - h_i x) + V_i(\tau + h_i x)], \quad v_i(x, \tau) = U_i(\tau - h_i x) - V_i(\tau + h_i x) \tag{GS}$$

for $a_{i-1} < x < a_i$, $\tau \in \mathbb{R}$, $i = 1, \ldots, n$. See Exercise 3.2. We note that

$$U_i(\tau - h_i x) = \tfrac{1}{2}[h_i u_i(x, \tau) + v_i(x, \tau)] = \tfrac{1}{2}[\sqrt{\varepsilon_i/\mu}E(x, t) + H(x, t)]$$

so that we may think of

$$U_1(\tau) = \tfrac{1}{2}[\sqrt{\varepsilon_1/\mu}E(0, t) + H(0, t)] := \mathscr{I}(\tau)$$

as the (given) *incoming wave*. Then

$$V_1(\tau) := \mathscr{R}(\tau)$$

is the (unknown) *reflected wave* translating to the left for $x < 0$, and

$$U_n(\tau) := \mathscr{T}(\tau)$$

gives the (unknown) *transmitted wave* propagating in $x > a_{n-1}$. We tacitly assume that $V_n \equiv 0$, hence U_n will be the only wave left for $x > a_{n-1}$. We may state the so-called transmission problem (see Fig. 7):

(TP) For an arbitrary bounded incident wave, find a bounded solution (u, v) of (ME′) for $(x, \tau) \in \mathbb{R}^2$, matching continuously at *all* interfaces $x = a_k$, $k = 1, \ldots, m + 1$.

Continuous matching implies

$$u_{k+1}(a_k, \tau) = u_k(a_k, \tau), \qquad v_{k+1}(a_k, \tau) = v_k(a_k, \tau), \qquad k = 1, \ldots, m + 1. \tag{10}$$

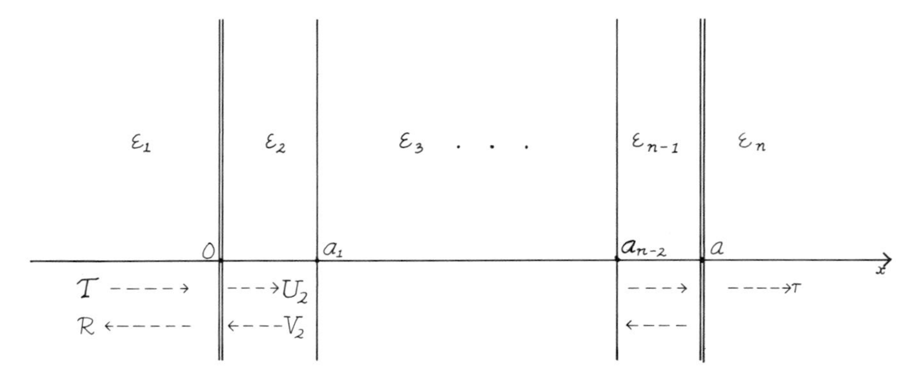

Fig. 7. The layer transmission problem.

From (GS) and $V_n = 0$, we find

$$h_1 u(0, \tau) + v(0, \tau) = 2\mathscr{I}(\tau),$$

where $u = u_1 = u_2,\ v = v_1 = v_2$, and

$$h_n u(a, \tau) - v(a, \tau) = 0,$$

where $u = u_n = u_{n-1},\ v = v_n = v_{n-1}$. Moreover,

$$\mathscr{R}(\tau) = \tfrac{1}{2}[h_1 u(0, \tau) - v(0, \tau)] \equiv \mathscr{I}(\tau) - v(0, \tau) \tag{11}$$

for $u = u_1 = u_2,\ v = v_1 = v_2$, and

$$\mathscr{T}(\tau) = \tfrac{1}{2}[h_n u(a, \tau + h_n a) + v(a, \tau + h_n a)] \tag{12}$$

for $u = u_n = u_{n-1},\ v = v_n = v_{n-1}$. Equations (11), (12) yield $\mathscr{R}(\tau),\ \mathscr{T}(\tau)$ as soon as the solution $u(x, \tau)$ inside the slab $D_a = \{0 < x < a, \tau \in \mathbb{R}\}$ is known. Therefore, (TP) is equivalent to:

(BVP) For an arbitrary bounded function $\mathscr{I}(\tau),\ \tau \in \mathbb{R}$, find a bounded solution (u, v) of (ME′) for $(x, \tau) \in D_a$, satisfying the *impedance* boundary conditions at $x = 0$ and $x = a$,

$$h_1 u(0, \tau) + v(0, \tau) = 2\mathscr{I}(\tau), \qquad h_n u(a, \tau) - v(a, \tau) = 0 \qquad (\tau \in \mathbb{R}), \tag{I}$$

and matching continuously (for $m \geq 2$) at the *inner* interfaces $x = a_j$, $j = 2, \ldots, m$.

See Exercise 3.11 and Refs. 5, 6. We are here considering weak solutions, in the sense of Section 1, and the functions $U_i,\ V_i$ in (GS) may be discontinuous as a consequence of discontinuities in $\mathscr{I}(\tau)$. In fact, the solution will be seen to have the same regularity as $\mathscr{I}(\tau)$.

If $\mathscr{I}(\tau)$ vanishes for $\tau \leq \tau_0$, (TP) can be made into an initial value problem as in the transmission–reflection problem in the previous section. We are interested here, however, in cases in which $\mathscr{I}(\tau)$ is periodic, e.g., $\mathscr{I}(\tau) = \sin \omega\tau$ or $\mathscr{I}(\tau) = \operatorname{sgn}(\sin \omega\tau)$, sinusoids or "square waves" with period $p = 2\pi/\omega$. In particular, we want to find conditions under which a solution exists, is unique, and is periodic with the same period as $\mathscr{I}$. Our problem then is (BVP), a pure boundary value problem, without initial conditions.

We first consider the case of a single layer ($n = 3$). The unknowns in the layer $0 < x < a$ are then $u = u_2,\ v = v_2$ and (ME′) reduces to

$$u_x + v_\tau = 0, \qquad v_x + u_\tau = 0; \qquad 0 < x < a,\ \tau \in \mathbb{R}. \tag{13}$$

Equivalently, the unknowns are (U_2, V_2), and from (I) and (GS) we have

$$U_2(\tau) - r_1 V_2(\tau) = \frac{2}{1+h_1}\mathscr{I}(\tau), \qquad r_2 U_2(\tau - a) + V_2(\tau + a) = 0, \tag{14}$$

where

$$r_1 = \frac{1-h_1}{1+h_1}, \qquad r_2 = \frac{h_3 - 1}{h_3 + 1}$$

are reflection coefficients at $x = 0, a$, respectively. (We recall that $h_2 = 1$.) If $r_1, r_2 \neq 0$ (i.e., $\varepsilon_1 \neq \varepsilon_2,\ \varepsilon_2 \neq \varepsilon_3$), we can write $U_2(\tau) = -r_2^{-1} V_2(\tau + 2a)$, and $V_2(\tau)$ is a solution of

$$V(\tau + 2a) + r_1 r_2 V(\tau) = F(\tau) := -\frac{2r_2}{1+h_1}\mathscr{I}(\tau).$$

This difference equation has a unique bounded solution for any bounded F, provided $|r_1 r_2| \neq 1$. The definition of h_i immediately implies that $|r_1 r_2| < 1$ here, and the solution is given by

$$V(\tau) = \sum_{n=0}^{\infty} (-r_1 r_2)^n F(\tau - 2a - 2na). \tag{S}$$

See Exercise 3.3. The reflected wave is then given by $\mathscr{R}(\tau) = \mathscr{I}(\tau) - v_2(0, \tau)$ [see equation (11)], and from (GS), (14), (S), and manipulations, we find

$$\mathscr{R}(\tau) = -r_1 \mathscr{I}(\tau) - r_2(1 - r_1^2) \sum_{n=0}^{\infty} (-r_1 r_2)^n \mathscr{I}(\tau - 2a - 2na).$$

This solution is a superposition of the first reflected wave and waves arising from multiple reflections at the slab walls. We see that if $\mathscr{I}(\tau)$ is periodic, the solution is periodic, with the same period. With this formula we can easily solve several "reflection reduction" problems (Exercises 3.5, 3.6).

We remark here that if $\mathscr{I}(\tau) \equiv 0$ for $\tau \leq \tau_0$, the series reduces to a finite number of terms for each fixed τ, the restriction $|r_1 r_2| \neq 1$ (and the boundedness assumptions) can be dropped, and the solution coincides with the solution of an initial–boundary value problem.

For m layers the continuous matching equations (10) become

$$U_{k+1}(\tau - h_{k+1}a_k) + V_{k+1}(\tau + h_{k+1}a_k) = \frac{k_{k+1}}{h_k}[U_k(\tau - h_k a_k) + V_k(\tau + h_k a_k)],$$
$$U_{k+1}(\tau - h_{k+1}a_k) - V_{k+1}(\tau + h_{k+1}a_k) = U_k(\tau - h_k a_k) - V_k(\tau + h_k a_k),$$

$k = 1, \ldots, m+1$. These may be recast as $2n-2$ equations for the determination of the $2n-2$ unknown functions $U_2, \ldots, U_n, V_1, \ldots, V_{n-1}$ in terms of U_1. The resulting equations have a unique solution, under an appropriate hypothesis on the reflection coefficients

$$r_k = \frac{h_{k+1} - h_k}{h_{k+1} + h_k}.$$

If $m = 2$, this (sufficient) condition is

$$|r_1 r_2| + |r_1 r_3| + |r_2 r_3| < 1,$$

and the reflected wave is given by

$$\begin{aligned}\mathscr{R}(\tau) = -r_1 \mathscr{I}(\tau) - (1 - r_1^2) \sum_{n_1=0}^{\infty} \sum_{n_2=0}^{\infty} \sum_{n_3=0}^{\infty} \frac{(n_1 + n_2 + n_3)!}{n_1! n_2! n_3!} \\ \times (-r_3 r_1)^{n_1} (-r_1 r_2)^{n_2} (-r_2 r_3)^{n_3} \\ \cdot \{r_2 \mathscr{I}[\tau - (1 + n_{12}) d_2 - n_{13}\, d_3] + r_3 \mathscr{I}[\tau - (1 + n_{12}) d_2 - (1 + n_{13}) d_3]\},\end{aligned}$$

where $d_j := 2h_j(a_j - a_{j-1})$ and $n_{ij} := n_i + n_j$. The demonstration is left as an exercise.

4. Linear Hyperbolic Systems with Constant Coefficients in Two Variables

The previous considerations can be easily extended to linear systems of first-order partial differential equations in two independent variables, i.e.,

$$\mathbf{u}_t + \mathbb{A}\mathbf{u}_x = 0, \tag{15}$$

where $\mathbb{A}$ is a (constant) $m \times m$ matrix and $\mathbf{u} = (u_1, \ldots, u_m) = \mathbf{u}(x, t)$. We suppose that $\mathbb{A}$ is *strictly hyperbolic*, that is, $\mathbb{A}$ has m real, distinct eigenvalues

$\lambda_1 < \cdots < \lambda_m$. Let $\mathbf{r}_k, \mathbf{l}_k$, $k = 1, \ldots, m$, be linearly independent right and left eigenvectors,

$$\mathbb{A}\mathbf{r}_k = \lambda_k \mathbf{r}_k, \qquad \mathbf{l}_k \mathbb{A} \equiv \mathbb{A}^T \mathbf{l}_k = \lambda_k \mathbf{l}_k, \qquad k = 1, \ldots, m.$$

We may normalize these so that

$$|\mathbf{r}_k| = 1, \qquad \mathbf{l}_j \cdot \mathbf{r}_k = \delta_{jk}, \qquad j, k = 1, \ldots, m.$$

By left multiplying (15) by $\mathbf{l}_j$ we find

$$\mathbf{l}_j \cdot \mathbf{u}_t + \mathbf{l}_j \cdot \mathbb{A}\mathbf{u}_x = (\mathbf{l}_j \cdot \mathbf{u})_t + \lambda_j (\mathbf{l}_j \cdot \mathbf{u})_x = 0.$$

let $z_j := \mathbf{l}_j \cdot \mathbf{u}$. Then

$$\frac{\partial z_j}{\partial t} + \lambda_j \frac{\partial z_j}{\partial x} = 0, \qquad j = 1, \ldots, m,$$

and $z_j(x, t) = Z_j(x - \lambda_j t)$, where $Z_j(x)$ are arbitrary C^1 functions.

We can then write (Exercise 4.1)

$$\mathbf{u}(x, t) = \sum_{j=1}^{m} z_j \mathbf{r}_j = \sum_{j=1}^{m} Z_j(x - \lambda_j t)\mathbf{r}_j \tag{16}$$

as the general solution of (15). We immediately have a theorem of existence and uniqueness of the initial value problem for (15).

Theorem 4.1. If $\mathbf{u}_0 \in C^1(\mathbb{R})$, there is a unique solution $\mathbf{u} \in C^1(\mathbb{R}^2)$ of (15) satisfying $\mathbf{u}(x, 0) = \mathbf{u}_0(x)$.

Proof. From (16) we find $\mathbf{u}_0(x) = \sum_{j=1}^{m} Z_j(x)\mathbf{r}_j$, hence $Z_j(x) = \mathbf{l}_j \cdot \mathbf{u}_0(x)$. Then

$$\mathbf{u}(x, t) = \sum_{j=1}^{m} \mathbf{l}_j \cdot \mathbf{u}_0(x - \lambda_j t)\mathbf{r}_j$$

is a solution. Because the problem is linear and $\mathbf{u}$ vanishes if $\mathbf{u}_0 = 0$, uniqueness follows. □

The solution is a linear combination of terms (traveling waves) each of which is constant along "characteristic lines" $x - \lambda_j t = \text{const}$, and there is an obvious analogy with the one-dimensional wave equation. In fact, the wave equation can be written as a system of the form (15) (Exercise 4.3).

Suppose that we want to determine a solution of (15) in a half strip

$$S = \{(x, t): \ 0 < x < l, \ t > 0\}.$$

In addition to initial data $\mathbf{u}(x, 0)$, $0 < x < l$, we must give some boundary data at $x = 0, l$. The representation (16) shows how this can be done. For example, consider the following problem:

a. $\mathbf{u}_t + \mathbb{A}\mathbf{u}_x = 0$, $(x, t) \in S$,
b. $\mathbf{u}(x, 0) = \mathbf{u}_0(x)$, $0 \leq x \leq l$,
c. $\mathbf{l}_j \cdot \mathbf{u}(0, t) = f_j(t)$, $t \geq 0$, $j \in J_+$,
d. $\mathbf{l}_j \cdot \mathbf{u}(l, t) = g_j(t)$, $t \geq 0$, $j \in J_-$,

where, for $j \in 1, \ldots, m$,

$$J_+ = \{j : \lambda_j > 0\}, \qquad J_- = \{j : \lambda_j < 0\}.$$

Theorem 4.2. If $\mathbf{u}_0 \in C^1[0, l]$, $f_j \in C^1[0, \infty)$ for $j \in J_+$, $g_j \in C^1[0, \infty)$ for $j \in J_-$, and the compatibility conditions

$$f_j(0) = \mathbf{l}_j \cdot \mathbf{u}_0(0), \qquad f_j'(0) = -\lambda_j \mathbf{l}_j \cdot \mathbf{u}_0'(0), \qquad j \in J_+,$$
$$g_j(0) = \mathbf{l}_j \cdot \mathbf{u}_0(l), \qquad g_j'(0) = -\lambda_j \mathbf{l}_j \cdot \mathbf{u}_0'(l), \qquad j \in J_-$$

hold, then there is a unique solution $\mathbf{u} \in C^1(\bar{S})$ of (a)–(d).

The proof is more or less obvious if characteristic lines are drawn backwards from a point (x, t) to the boundary of S.

By analogy with the wave equation we may define the *energy* for a solution of (a)–(d),

$$\mathscr{E}(t) := \frac{1}{2}\int_0^l \sum_{j=1}^m (\mathbf{l}_j \cdot \mathbf{u})^2 dx \equiv \frac{1}{2}\int_0^l \sum_{j=1}^m z_j^2(x, t) dx.$$

Under our assumptions, $\mathscr{E}(t)$ is a C^1 function for $t \geq 0$, and

$$\mathscr{E}'(t) = \int_0^l z_j \sum_{j=1}^m \frac{\partial z_j}{\partial t}\, dx = -\int_0^l \sum_{j=1}^m \lambda_j z_j \frac{\partial z_j}{\partial x}\, dx = -\frac{1}{2}\sum_{j=1}^m \lambda_j [z_j^2(l, t) - z_j^2(0, t)].$$

If $f_j(t) = 0$ for $j \in J_+$ and $g_j(t) = 0$ for $j \in J_-$, then

$$\mathscr{E}'(t) = -\frac{1}{2}\sum_{j\in J_+} \lambda_j z_j^2(l, t) + \frac{1}{2}\sum_{j\in J_-} \lambda_j z_j^2(0, t) \leq 0$$

for $t \geq 0$. We see that the energy decays if the boundary conditions are homogeneous. In particular, if $\mathbf{u}_0 = 0$, then $\mathscr{E}(t) = 0$, and uniqueness for (a)–(d) follows. Theorem 4.2 implies the rule that *the number of boundary conditions to be assigned at a boundary point must be equal to the number of characteristic lines "entering" the domain at that point.* See Exercise 4.4.

The boundary conditions in (c), (d) can be generalized to

c′. $\mathbf{l}_j \cdot \mathbf{u}(0, t) = \sum_{k\in J_-} L_{jk}\mathbf{l}_k \cdot \mathbf{u}(0, t) + f_j(t),\ t \geq 0,\ j \in J_+,$

d′. $\mathbf{l}_j \cdot \mathbf{u}(l, t) = \sum_{k\in J_+} R_{jk}\mathbf{l}_k \cdot \mathbf{u}(l, t) + g_j(t),\ t \geq 0,\ j \in J_-.$

A theorem analogous to Theorem 4.2 can be given (Exercise 4.5). It is interesting to ask what hypotheses on L_{jk} and R_{jk} imply that energy decays if $f_j(t)$, $g_j(t)$ are all zero. Some special cases are considered in the exercises.

5. Wave Equation in Two and Three Dimensions

Consider the wave equation in $\mathbb{R}^3$,

$$u_{tt} - c^2 \Delta u = 0, \tag{W}$$

where $u = u(x_1, x_2, x_3, t) = u(\mathbf{x}, t)$. We will first consider a method based on "spherical means."

Definition 5.1. For $f \in C^2(\mathbb{R}^3)$ the *spherical mean* of f is given for $\mathbf{x} \in \mathbb{R}^3$, $r > 0$ by

$$Mf(\mathbf{x}, r) = \frac{1}{4\pi r^2}\int_{|\mathbf{y}-\mathbf{x}|=r} f(\mathbf{y})dS_y,$$

i.e., the average of f over the sphere of radius r centered at $\mathbf{x}$.

If we denote by $d\Omega$ the element of surface area over the unit sphere $\Omega = \{|\nu| = 1\}$, and write $\nu = (\mathbf{y} - \mathbf{x})/|\mathbf{y} - \mathbf{x}|$, i.e., $\mathbf{y} = \mathbf{x} + r\nu$, then

$$M \equiv Mf(\mathbf{x}, r) = \frac{1}{4\pi}\int_{|\nu|=1} f(\mathbf{x} + r\nu)d\Omega.$$

Theorem 5.1.

a. If $f \in C(\mathbb{R}^3)$, then $Mf(\mathbf{x}, 0) := \lim_{r\to 0} Mf(\mathbf{x}, r) = f(\mathbf{x})$.
b. if $f \in C^1(\mathbb{R}^3)$, then $(\partial/\partial r)Mf(\mathbf{x}, 0) := \lim_{r\to 0}(\partial/\partial r)Mf(\mathbf{x}, r) = 0$.
c. If $f \in C^2(\mathbb{R}^3)$, then $(\partial^2 M/\partial r^2) + (2/r)(\partial M/\partial r) = \Delta_x M$ for $r > 0$.

Proof. It is immediate that $f \in C^k(\mathbb{R}^3)$ implies that $M \in C^k(\mathbb{R}^3 \times (0, \infty))$, and that (a) holds. We have

$$\frac{\partial M}{\partial r} = \frac{1}{4\pi}\int_{|\nu|=1} \frac{\partial}{\partial r} f(\mathbf{x} + r\nu)d\Omega = \frac{1}{4\pi}\int_{|\nu|=1} \nu \cdot \text{grad}_x f(\mathbf{x} + r\nu)d\Omega,$$

and then the divergence theorem implies

$$\lim_{r\to 0}\frac{\partial M}{\partial r} = \frac{1}{4\pi}\text{grad}\, f(\mathbf{x}) \cdot \int_{|\nu|=1} \nu d\Omega = \frac{1}{4\pi}\text{grad}\, f(\mathbf{x}) \cdot \int_{|\nu|\le 1} \text{grad}(1)d\mathbf{x} = 0,$$

establishing (b). Noting that if $f \in C^2$,

$$\begin{aligned}\frac{\partial M}{\partial r} &= \frac{1}{4\pi r^2}\int_{|\mathbf{y}-\mathbf{x}|=r} \nu \cdot \text{grad}_y f(\mathbf{y})dS = \frac{1}{4\pi r^2}\int_{|\mathbf{y}-\mathbf{x}|\le r} \Delta_y f(\mathbf{y})d\mathbf{y} \\ &= \frac{1}{4\pi r^2}\int_0^r d\rho \int_{|\mathbf{y}-\mathbf{x}|=\rho} \Delta_y f(\mathbf{y})dS = \frac{1}{4\pi r^2}\int_0^r d\rho\rho^2 \int_{|\nu|=1} \Delta_x f(\mathbf{x} + \rho\nu)d\Omega,\end{aligned}$$

and $\Delta_x M = M\Delta_x$, we have

$$\begin{aligned}\frac{\partial^2 M}{\partial r^2} &= \frac{1}{4\pi}\int_{|\nu|=1} \Delta_x f(\mathbf{x} + r\nu)d\Omega - \frac{2}{4\pi r^3}\int_0^r d\rho\rho^2 \int_{|\nu|=1} \Delta_x f(\mathbf{x} + \rho\nu)d\Omega \\ &= M\Delta_x f - \frac{2}{r}\frac{\partial M}{\partial r} = \Delta_x Mf - \frac{2}{r}\frac{\partial M}{\partial r},\end{aligned}$$

and we have (c). □

It is useful to note that $Mf(\mathbf{x}, r)$ can be extended to $r < 0$ as an even function (replace ν by $-\nu$), and with $Mf(\mathbf{x}, 0) = f(\mathbf{x})$ it is continuous as a function of r. In fact, if $f \in C(\mathbb{R}^3)$, $Mf \in C(\mathbb{R}^4)$. Also, if $f \in C^2(\mathbb{R}^3)$, $Mf \in C^2(\mathbb{R}^4)$. To see this it suffices to recall that $M_r f(\mathbf{x}, 0) = 0$, and that

$$\begin{aligned}
\lim_{r\to 0\pm} \frac{\partial^2 M}{\partial r^2} &= \lim_{r\to 0} \frac{\partial}{\partial r}\left[\frac{1}{4\pi r^2}\int_0^r d\rho\rho^2 \int_{|\nu|=1} \Delta_x f(\mathbf{x}+\rho\nu)d\Omega\right] \\
&= \lim_{r\to 0}\left[-\frac{2}{3}\frac{3}{4\pi r^3}\int_{|\mathbf{y}-\mathbf{x}|\le r} \Delta_y f(\mathbf{y})d\mathbf{y} + \frac{1}{4\pi r^2}\int_{|\mathbf{y}-\mathbf{x}|=r} \Delta_y f(\mathbf{y})dS_y\right] \\
&= \tfrac{1}{3}\Delta_x f(\mathbf{x})
\end{aligned}$$

exists. It follows that the extended M satisfies $M_{rr} + 2M_r/r = \Delta_x M$ as well. We can derive from this the following important result.

Theorem 5.2. If $f \in C^2(\mathbb{R}^3)$, then

$$\Delta_x[rMf(\mathbf{x}, r)] = \frac{\partial^2}{\partial r^2}[rMf(\mathbf{x}, r)], \qquad (\mathbf{x}, r) \in \mathbb{R}^4.$$

Proof. We need only remark that $(rM)_{rr} = rM_{rr} + 2M_r$. □

If we write $r = ct$, this implies that, for $f \in C^2(\mathbb{R}^3)$, $rMf(\mathbf{x}, r) = ctMf(\mathbf{x}, ct)$ is in $C^2(\mathbb{R}^4)$ and a solution of (W).

Theorem 5.3. If $u_0 \in C^3(\mathbb{R}^3)$, $u_1 \in C^2(\mathbb{R}^3)$, then $u(\mathbf{x}, t)$ defined by

$$\begin{aligned}
u(\mathbf{x}, t) &= tMu_1(\mathbf{x}, ct) + \frac{\partial}{\partial t}[tMu_0(\mathbf{x}, ct)] \\
&= \frac{t}{4\pi}\int_{|\nu|=1} u_1(\mathbf{x}+ct\nu)d\Omega + \frac{\partial}{\partial t}\left[\frac{t}{4\pi}\int_{|\nu|=1} u_0(\mathbf{x}+ct\nu)d\Omega\right]
\end{aligned}$$

is in $C^2(\mathbb{R}^4)$ and is a solution of

$$\begin{aligned}
&u_{tt} - c^2\Delta u = 0, && (\mathbf{x}, t) \in \mathbb{R}^4, \\
&u(\mathbf{x}, 0) = u_0(\mathbf{x}), && u_t(\mathbf{x}, 0) = u_1(\mathbf{x}), \qquad \mathbf{x} \in \mathbb{R}^3.
\end{aligned}$$

Proof. The previous results imply that $tMu_0(\mathbf{x}, ct) \in C^2(\mathbb{R}^4)$ and is a solution of (W). Similarly, $u_0 \in C^3(\mathbb{R}^3)$ implies $(tMu_0)_t \in C^2(\mathbb{R}^4)$ is a solution of (W). Also, as $t \to 0$,

$$tMu_1(\mathbf{x}, ct) \to 0, \qquad \frac{\partial}{\partial t}[tMu_0(\mathbf{x}, ct)] = Mu_0 + t(Mu_0)_t \to u_0(\mathbf{x}),$$

and

$$u_t = Mu_1 + t(Mu_1)_t + (tMu_0)_{tt} = Mu_1 + t(Mu_1)_t + c^2 t\Delta(Mu_0) \to u_1. \qquad \square$$

We can also study the dependence of u on u_0, u_1 using this formula, Differentiating under the integral in $t(Mu_0)_t$ we obtain

$$t\frac{\partial}{\partial t}(Mu_0) = \frac{tc}{4\pi}\int_{|\nu|=1}\sum_{k=1}^{3}\nu_k \frac{\partial}{\partial x_k}u_0(\mathbf{x} + ct\nu)d\Omega,$$

so that

$$|u(\mathbf{x}, t)| \leq |t| \sup_{\mathbb{R}^3}|u_1(\mathbf{x})| + \sup_{\mathbb{R}^3}|u_0(\mathbf{x})| + 3c|t| \max_k \sup_{\mathbb{R}^3}\left|\frac{\partial u_0(\mathbf{x})}{\partial x_k}\right| \tag{CD}$$

for $(\mathbf{x}, t) \in \mathbb{R}^4$. This inequality immediately implies that u can be bounded, in a fixed interval $|t| \leq T$, by quantities involving u_0, u_1, and $\partial u_0(\mathbf{x})/\partial x_k$. Hence, contrary to what happens for the vibrating string equation, the gradient of u_0 must also be "controlled." There is something unappealing in this result. In particular, if we freeze the motion at some time T and attempt to use $u(\mathbf{x}, T)$, $u_t(\mathbf{x}, T)$ as new initial data, we have no guarantee that $u(\mathbf{x}, T)$ is in $C^3(\mathbb{R}^3)$, and there is a possibility of "loss of derivatives." In what follows, we will see that this loss is indeed a real possibility and give a formulation of (CP) that avoids this difficulty.

The above results can be used to solve the initial value problem for the two-dimensional wave equation, by means of Hadamard's *method of descent* (Refs. 1, 7).

Theorem 5.4. If $u_0(\mathbf{X}) \in C^3(\mathbb{R}^2)$, $u_1(\mathbf{X}) \in C^2(\mathbb{R}^2)$, $\mathbf{X} = (x_1, x_2)$, then

$$\begin{aligned} u(\mathbf{X}, t) = {} & \frac{\operatorname{sgn}(t)}{2\pi c}\int_{|\mathbf{Y}|\leq ct}\frac{u_1(\mathbf{X} + \mathbf{Y})}{(c^2t^2 - |\mathbf{Y}|^2)^{1/2}}\,d\mathbf{Y} \\ & + \frac{\partial}{\partial t}\left[\frac{\operatorname{sgn}(t)}{2\pi c}\int_{|\mathbf{Y}|\leq ct}\frac{u_0(\mathbf{X} + \mathbf{Y})}{(c^2t^2 - |\mathbf{Y}|^2)^{1/2}}\,dY\right], \end{aligned} \tag{17}$$

where $\mathbf{Y} = (y_1, y_2)$ is a solution [in $C^2(\mathbb{R}^3)$] of

$$u_{tt} = c^2 \Delta u \equiv c^2(u_{x_1 x_1} + u_{x_2 x_2}), \qquad (\mathbf{X}, t) \in \mathbb{R}^3,$$
$$u(\mathbf{X}, 0) = u_0(\mathbf{X}), \qquad u_t(\mathbf{X}, 0) = u_1(\mathbf{X}), \qquad \mathbf{X} = (x_1, x_2) \in \mathbb{R}^2.$$

Proof. The proof is inherited from Theorem 5.3, using the special initial data $u(\mathbf{x}, 0) = u_0(\mathbf{X})$ and $u_t(\mathbf{x}, 0) = u_1(\mathbf{X})$ independent of x_3. We observe that the function u given in Theorem 5.3,

$$u = \frac{1}{4\pi c^2 t} \int_{|\mathbf{y}|=ct} u_1(\mathbf{X} + \mathbf{Y}) dS_y + \frac{\partial}{\partial t}\left[\frac{1}{4\pi c^2 t} \int_{|\mathbf{y}|=1} u_0(\mathbf{X} + \mathbf{Y}) dS_y\right], \tag{17a}$$

with $\mathbf{y} = (\mathbf{Y}, y_3)$, is independent of x_3 and hence is a C^2 solution of the initial value problem in $\mathbb{R}^2$. Thus, all that is required is to verify that (17a) can be written in the form (17) by transforming the integral over the sphere $|\mathbf{y}| = ct$ in $\mathbb{R}^3$ into twice the integral over the meridian disk $|\mathbf{Y}| \leq ct$ in $\mathbb{R}^2$. In this way, we "descend" from three to two dimensions. The details are left as an exercise. □

We can define the *domain of dependence* of a point $(\mathbf{x}_0, t_0)$ as the subset of $\mathbb{R}^n$ ($n = 2$ or 3) on whose complement a change in the data of (CP) does not affect $u(\mathbf{x}_0, t_0)$ (see Section 1). For the three-dimensional wave equation, it is easily shown using the result of Theorem 5.3 that the domain of dependence of (CP) is the sphere $|\mathbf{x} - \mathbf{x}_0| = c|t_0|$. Hence, for $n = 3$ *the domain of dependence coincides with its boundary (Huygens' principle)*. It is interesting to contrast this with the two-dimensional wave equation ($n = 2$); there, as shown by Theorem 5.4, the domain of dependence is the disk centered at $\mathbf{X}_0$ with radius $c|t_0|$, $|\mathbf{Y} - \mathbf{X}_0| \leq c|t_0|$.

The validity of Huygens' principle is strictly related to the "sharpness" of transmission of signals for phenomena governed by (W) (in three dimensions): It implies that a sharply localized initial state is observed later at a different place as an effect that is equally sharply delimited (Ref. 7). If the support of u_0 and u_1 is bounded, say contained in a ball around the origin of radius R, an elementary consequence of Theorem 5.3 is that $u(\mathbf{x}, t) \equiv 0$ for fixed $\mathbf{x}$ and t sufficiently large. Conversely, for fixed (sufficiently large) t the support of u is bounded in space by two (spherical) surfaces, a *leading wave front* where the disturbance begins and a *rear wave front* where the disturbance vanishes, respectively. These wave fronts can be constructed by taking envelopes of spherical surfaces (Huygens' construction, see Refs. 2, 7–9).

This should be contrasted with the results in the one- and two-dimensional cases, which imply that the rear wave front is missing and signals are permanent, giving rise to a "reverberation" (Exercise 5.4): Disturbances propagate with finite speed but never die out completely, like the surface waves resulting from a stone dropped into a pond (Ref. 1). In general, it can be proven that Huygens' principle holds for the wave equation in $\mathbb{R}^n$ with n odd ≥ 3 only. This fact and further considerations (see Exercise 5.16 and Ref. 7) show that the three-dimensional space in which we live is characterized by the fact that signals can be received sharply and without distortion.

All of this shows a marked dependence on the number of space dimensions. In contrast, the proof of uniqueness of solutions of (CP) (based on an "energy method") works in the same way for all dimensions.

Theorem 5.5. The solution of (CP) is unique.

Proof. We will show, more precisely, that u_0, u_1 vanishing on $|\mathbf{x} - \mathbf{x}_0| \leq r$ implies that $u(\mathbf{x}, t)$ vanishes on $|\mathbf{x} - \mathbf{x}_0| \leq r - c|t|$. We will consider $0 \leq t \leq r/c$. Negative t are dealt with by reflection. If $B(t) = \{|\mathbf{x} - \mathbf{x}_0| \leq r - ct\}$, $\partial B(t) = \{|\mathbf{x} - \mathbf{x}_0| = r - ct\}$, let

$$\begin{aligned}
\mathscr{E}(t) &= \frac{1}{2}\int_{B(t)} (u_t^2 + c^2|\text{grad } u|^2)d\mathbf{x} \\
&= \frac{1}{2}\int_0^{r-ct} d\rho \int_{|\mathbf{x}-\mathbf{x}_0|=\rho} (u_t^2 + c^2|\text{grad } u|^2)dS.
\end{aligned}$$

Then

$$\begin{aligned}
\mathscr{E}'(t) &= \int_{B(t)} \left(u_t u_{tt} + c^2 \sum_{k=1}^{d} u_{x_k} u_{x_k t}\right) d\mathbf{x} - \frac{c}{2}\int_{\partial B(t)} (u_t^2 + c^2|\text{grad } u|^2)dS \\
&= \int_{B(t)} c^2 \sum_{k=1}^{d} (u_t u_{x_k x_k} + u_{x_k} u_{x_k t}) d\mathbf{x} - \frac{c}{2}\int_{\partial B(t)} (u_t^2 + c^2|\text{grad } u|^2)dS \\
&= c^2 \int_{B(t)} \sum_{k=1}^{d} (u_t u_{x_k})_{x_k}\, d\mathbf{x} - \frac{c}{2}\int_{\partial B(t)} (u_t^2 + c^2|\text{grad } u|^2)dS \\
&= \frac{c}{2}\int_{\partial B(t)} (2cu_t \text{grad } u \cdot \mathbf{n} - u_t^2 - c^2|\text{grad } u|^2)dS,
\end{aligned}$$

where $\mathbf{n}$ is the exterior normal vector to $\partial B(t)$ in $\mathbb{R}^d$, and $d = 3$ here. Applying the inequality $2\mathbf{a} \cdot \mathbf{b} \leq |\mathbf{a}|^2 + |\mathbf{b}|^2$ to the term $2c \text{ grad } u \cdot \mathbf{n} u_t$ we find that $d\mathscr{E}(t)/dt \leq 0$. Because $\mathscr{E}(0) = 0$, we deduce that $\mathscr{E}(t) \equiv 0$ on $B(t)$. □

We can define the *domain of determinacy* $\Delta(I)$ and the *domain of influence* $\mathscr{I}(I)$ for a set I on the initial manifold (say, a ball or a disk, see Exercise 5.6).

The definitions are similar to those in Section 1 for the vibrating string equation. If u_0 and u_1 vanish outside I, then u vanishes outside $\mathscr{J}$ in $\mathbb{R}^2$ or $\mathbb{R}^3$ and *the support* of $u(\mathbf{x}, t)$ with respect to the $\mathbf{x}$-variable *expands in time with finite speed* c, independently of the number of dimensions.

We now present an example illustrating the loss of derivatives in (CP). Let $u_0 = 0$ and

$$u_1 = \begin{cases} (c^2 - |\mathbf{x}|^2)^{5/2}, & |x| \leq c, \\ 0 & |x| > c. \end{cases} \tag{18}$$

From the formula of spherical means the solution satisfies

$$u(0, t) = \frac{1}{4\pi c^2 t} \int_{|\mathbf{y}|=ct} u_1(\mathbf{y}) dS = \frac{4\pi c^2 t^2 (c^2 - c^2 t^2)^{5/2}}{4\pi c^2 t} = tc^5 (1 - t^2)^{5/2}$$

for $t \leq 1$, and $u(0, t) = 0$ for $t > 1$. Then $u_{tt}(0, t)$ develops a singularity at $t = 1$. We say that *focusing* occurs. (This term comes from the corresponding physical phenomenon of formation of caustics in optics.) Because the initial data are spherically symmetric, it can be argued that the full solution will be too, $u = u(|\mathbf{x}|, t)$ (Exercise 5.5).

This example shows that the class of initial data $(u, u_t) \in C^3 \times C^2$ is not "persistent," i.e., is not preserved for $t > 0$. We now consider a class of data that is persistent. This discussion requires the use of Sobolev spaces and Fourier transforms; the reader unfamiliar with these ideas is referred to Chapter 8.

We consider the wave equation $u_{tt} - c^2 \Delta u = 0$ in d space dimensions, $d \geq 1$, with Cauchy data

$$u(\mathbf{x}, 0) = u_0(\mathbf{x}), \qquad u_t(\mathbf{x}, 0) = u_1(\mathbf{x}), \qquad \mathbf{x} \in \mathbb{R}^d.$$

The Fourier transform applied to this problem leads to $\hat{u}_{tt} + c^2 |\xi|^2 \hat{u} = 0$, and

$$\hat{u}(\xi, t) = \hat{u}_0(\xi) \cos(c|\xi|t) + \frac{\hat{u}_1(\xi)}{c|\xi|} \sin(c|\xi|t), \tag{19}$$

where $\xi = (\xi_1, \ldots, \xi_d)$ and $\tilde{u}$ denotes the Fourier transform. Denoting the norm in $H^s(\mathbb{R}^d)$ by $\| \cdot \|_s$, we have

$$\|u(\cdot, t)\|_s^2 = \int_{\mathbb{R}^d} (1 + |\xi|^2)^s |\hat{u}(\xi, t)|^2 d\xi.$$

Theorem 5.6. There is a constant $C > 0$ such that

$$\|u(\cdot, t)\|_s^2 \le C[\|u_0\|_s^2 + (1 + c^2t^2)\|u_1\|_{s-1}^2]. \tag{20}$$

Proof. Using (19) and the inequality $(a + b)^2 \le 2a^2 + 2b^2$, we have

$$\|u(\cdot, t)\|_s^2 \le 2\int_{\mathbb{R}^d}(1 + |\xi|^2)^s|\hat{u}_0(\xi)|^2 d\xi + 2\int_{\mathbb{R}^d}\frac{(1 + |\xi|^2)^s}{c^2|\xi|^2}\sin^2(c|\xi|t)|\hat{u}_1(\xi)|^2 d\xi.$$

As $(1 + |\xi|^2)(\sin(c|\xi|t)/|\xi|)^2 \le 1 + c^2t^2$, we obtain

$$\|u(\cdot, t)\|_s^2 \le 2\|u_0\|_s^2 + \frac{2}{c^2}(1 + c^2t^2)\|u_1\|_{s-1}^2,$$

and the theorem follows. □

Similarly, one can prove the further inequality

$$\|u_t(\cdot, t)\|_{s-1}^2 \le C[\|u_0\|_s^2 + \|u_1\|_{s-1}^2] \tag{21}$$

(Exercise 5.7). From (20), (21) we see that $(u_0, u_1) \in H^s(\mathbb{R}^d) \times H^{s-1}(\mathbb{R}^d)$ is a *persistent class* for the wave equation for any $s \ge 0$ and any $d \ge 1$. If $d = 1$, the d'Alembert formula shows that there is no loss of derivatives, and $C^s(\mathbb{R}) \times C^{s-1}(\mathbb{R})$ is also a persistent class.

The inverse Fourier transform can be envisaged as a representation of a function as a superposition of plane waves. Suppose we seek plane wave solutions of the wave equation in d space dimensions $u_{tt} - c^2\Delta u = 0$ in the form $u(\mathbf{x}, t) = e^{i\mathbf{x}\cdot\xi}T(t)$, for a fixed $\xi \in \mathbb{R}^d$ (the "wave number"). An easy calculation yields $T''(t) = -c^2|\xi|^2T$, hence

$$u(\mathbf{x}, t) = [A(\xi)e^{i\omega t} + B(\xi)e^{-i\omega t}]e^{i\mathbf{x}\cdot\xi},$$

where $A(\xi)$, $B(\xi)$ are arbitrary amplitude functions and the angular frequency ω is given by the *dispersion relation* $\omega = c|\xi|$. A more general (formal) solution is given by the superposition of plane traveling waves

$$u(\mathbf{x}, t) = \int_{\mathbb{R}^d}[A(\xi)e^{i(\mathbf{x}\cdot\mathbf{k}+ct)|\xi|} + B(\xi)e^{i(\mathbf{x}\cdot\mathbf{k}-ct)|\xi|}]d\xi, \tag{IF}$$

where $\mathbf{k} = \xi/|\xi|$. This has the form of an inverse Fourier transform, and by enforcing initial data at $t = 0$ we get the formula (19).

The following result is often useful in dealing with inhomogeneous problems.

Theorem 5.7 (Duhamel's principle). Suppose that L is a linear constant-coefficient differential operator in the space variables, $v = v(x, t; \tau)$ is a solution of the Cauchy problem

$$\frac{\partial^m v}{\partial t^m} = Lv, \qquad t > 0,$$

$$v(\mathbf{x}, 0; \tau) = \frac{\partial v}{\partial t}(\mathbf{x}, 0; \tau) = \cdots = \frac{\partial^{m-2} v}{\partial t^{m-2}}(\mathbf{x}, 0; \tau) = 0,$$

$$\frac{\partial^{m-1} v}{\partial t^{m-1}}(\mathbf{x}, 0; \tau) = f(\mathbf{x}, \tau),$$

where $\tau \geq 0,\ m \geq 1$. Let $u(\mathbf{x}, t)$ be the Duhamel integral,

$$u(\mathbf{x}, t) = \int_0^t v(x, t - \tau; \tau)d\tau.$$

Then

$$\frac{\partial^m u}{\partial t^m} = Lu + f(\mathbf{x}, t), \qquad t > 0,$$

$$u(\mathbf{x}, 0) = \cdots = \frac{\partial^{m-1} u}{\partial t^{m-1}}(\mathbf{x}, 0) = 0.$$

Proof. The initial conditions follow from successive differentiation of u and evaluation at $t = 0$, for example,

$$u_t(\mathbf{x}, t) = v(\mathbf{x}, 0; t) + \int_0^t v_t(x, t - \tau; \tau)d\tau,$$

so that $u_t(0, t) = 0$. Also,

$$\begin{aligned}\frac{\partial^m u}{\partial t^m}(\mathbf{x}, t) &= \frac{\partial^{m-1} v}{\partial t^{m-1}}(\mathbf{x}, 0; t) + \int_0^t \frac{\partial^m v}{\partial t^m}(\mathbf{x}, t - \tau; \tau)d\tau \\ &= f(\mathbf{x}, t) + \int_0^t Lv(\mathbf{x}, t - \tau; \tau)d\tau \\ &= f(\mathbf{x}, t) + L\int_0^t v(\mathbf{x}, t - \tau; \tau)d\tau = f(\mathbf{x}, t) = Lu.\end{aligned}$$

□

We have tacitly assumed that the function v is such that passages to the limit and interchange of differentiation and integration are permissible. Rather than formulating general hypotheses, we will depend on properties of functions involved in specific applications. In particular, we can solve the inhomogeneous wave equation in three dimensions using this idea.

Theorem 5.8. If $f(\mathbf{x}, t) \in C^2(\mathbb{R}^4)$, then

$$u(\mathbf{x}, t) = \frac{1}{4\pi c^2}\int_{|\mathbf{y}-\mathbf{x}|\leq c|t|} f\left(\mathbf{y}, t \mp \frac{|\mathbf{y}-\mathbf{x}|}{c}\right)\frac{d\mathbf{y}}{|\mathbf{y}-\mathbf{x}|} \tag{RP}$$

($-$ for $t > 0$, $+$ for $t < 0$) is of class $C^2(\mathbb{R}^4)$ and is a solution of

$$\begin{aligned} &u_{tt} - c^2\Delta u = f(\mathbf{x}, t), \qquad &&\mathbf{x} \in \mathbb{R}^3, t > 0(t < 0), \\ &u(\mathbf{x}, 0) = u_t(\mathbf{x}, 0) = 0, &&\mathbf{x} \in \mathbb{R}^3. \end{aligned} \tag{IVP}$$

Proof. Consider first $t > 0$. We solve the Cauchy problem

$$\begin{aligned} &v_{tt} - c^2\Delta v = 0, \qquad &&t > 0, \\ &v(\mathbf{x}, 0; \tau) = 0, &&v_t(\mathbf{x}, 0; \tau) = f(\mathbf{x}, \tau), \end{aligned}$$

for $v(\mathbf{x}, t; \tau)$, $\tau \geq 0$, and obtain

$$v(\mathbf{x}, t; \tau) = \frac{t}{4\pi}\int_{|\nu|=1} f(\mathbf{x} + ct\nu, \tau)d\Omega.$$

Then

$$u(\mathbf{x}, t) = \int_0^t v(\mathbf{x}, t - \tau; \tau)d\tau = \int_0^t d\tau \frac{t-\tau}{4\pi}\int_{|\nu|=1} f[\mathbf{x} + c(t-\tau)\nu, \tau]d\Omega$$

is a solution of (IVP) as $f \in C^2$ implies Duhamel's principle can be applied. If $r = c(t - \tau)$, this can be written as

$$\begin{aligned} u(\mathbf{x}, t) &= \frac{1}{4\pi c^2}\int_0^{ct} drr \int_{|\nu|=1} f(\mathbf{x} + r\nu, t - r/c)d\Omega \\ &= \frac{1}{4\pi c^2}\int_0^{ct} dr \frac{1}{r}\int_{|\mathbf{y}-\mathbf{x}|=r} f\left(\mathbf{y}, t - \frac{|\mathbf{y}-\mathbf{x}|}{c}\right)dS_y \\ &= \frac{1}{4\pi c^2}\int_{|\mathbf{y}-\mathbf{x}|\leq ct} f\left(\mathbf{y}, t - \frac{|\mathbf{y}-\mathbf{x}|}{c}\right)\frac{d\mathbf{y}}{|\mathbf{y}-\mathbf{x}|}. \end{aligned}$$

The case $t < 0$ can be proven similarly, by letting $r = c(\tau - t)$. □

The formula given in (RP) is known as the *retarded potential.* We can see from this representation that the solution at $(\mathbf{x}, t)$ depends only on the values of f on the "characteristic cone" $|\mathbf{y}-\mathbf{x}| = c(|t|-\tau)$, $0 \le \tau \le |t|$. This is to be contrasted with the situation in one and two dimensions as will be shown in the exercises.

Corollary 5.1. For the two-dimensional Cauchy problem

$$\begin{aligned} u_{tt} &= c^2 \Delta u + f(\mathbf{X}, t), \qquad && (\mathbf{X}, t) \in \mathbb{R}^3, \\ u(\mathbf{X}, 0) &= u_t(\mathbf{X}, 0) = 0, && \mathbf{X} = (x_1, x_2) \in \mathbb{R}^2, \end{aligned}$$

where $f \in C^2(\mathbb{R}^3)$, a solution [in $C^2(\mathbb{R}^3)$] is given by

$$u(\mathbf{X}, t) = \frac{1}{2\pi c^2} \int_0^{c|t|} dr \int_{|\mathbf{Y}-\mathbf{X}| \le r} f\left(\mathbf{Y}, t \mp \frac{r}{c}\right) \frac{d\mathbf{Y}}{(r^2 - |\mathbf{Y}-\mathbf{X}|^2)^{1/2}},$$

where $-$, $+$ corresponds to t positive and negative, respectively.

The proof is left as an exercise.

The retarded potential can be used to prove the following result on "finite transients in three dimensions."

Theorem 5.9. Suppose that $f(\mathbf{x}, t) = F(\mathbf{x}) e^{i\omega t}$, where F has compact support G, in (IVP). Then the solution $u(\mathbf{x}, t) = U(\mathbf{x}) e^{i\omega t}$ for $t > T(\mathbf{x}) = c^{-1} \sup_{\mathbf{y} \in G} |\mathbf{y} - \mathbf{x}|$.

Proof. In general,

$$u(\mathbf{x}, t) = \frac{1}{4\pi c^2} \int_{\substack{|\mathbf{y}-\mathbf{x}| \le c|t| \\ \mathbf{y} \in G}} F(\mathbf{y}) \exp\left[i\omega\left(t - \frac{|\mathbf{y}-\mathbf{x}|}{c}\right)\right] \frac{d\mathbf{y}}{|\mathbf{y}-\mathbf{x}|}.$$

If $t > T(\mathbf{x})$, we can write

$$u(\mathbf{x}, t) = \frac{e^{i\omega t}}{4\pi c^2} \int_G F(\mathbf{y}) e^{-i\omega|\mathbf{y}-\mathbf{x}|/c} \frac{d\mathbf{y}}{|\mathbf{y}-\mathbf{x}|} := e^{i\omega t} U(\mathbf{x}). \qquad \square$$

For the corresponding problem in one dimension, because

$$\begin{aligned} u(x, t) &= \frac{1}{2c} \int_0^t d\tau \int_{x-c(t-\tau)}^{x+c(t-\tau)} f(y, \tau) dy = \frac{1}{2c^2} \int_0^{ct} dr \int_{|y-x| \le r} f(y, t - r/c) dy \\ &= \frac{e^{i\omega t}}{2c^2} \int_0^{ct} dr e^{-i\omega r/c} \int_{|y-x| \le r} F(y) dy := e^{i\omega t} U(\mathbf{x}, t), \end{aligned}$$

we see that transients do not vanish at any finite time.

Certain problems of electromagnetism and optics lead to the consideration of *traveling wave solutions* of the wave equation of the form $u = u(x, y, z - ct)$ or $u = u(x, y, z - ct, z + ct)$, where u may be required to be "localized" (e.g., of compact support) with respect to x and y. We leave this interesting question as an exercise (Exercise 5.14).

Exercises

1.1. If P, Q, R, S are corners of a parallelogram bounded by characteristic lines, show that any solution u of (V) satisfies

$$u(P) + u(R) = u(Q) + u(S)$$

(Fig. 8).

1.2. (Duhamel's principle). Suppose that $v = v(x, t; \tau)$ is a solution of the Cauchy problem

$$v_{tt} = Lv, \qquad v(x, 0; \tau) = 0, \qquad v_t(x, 0; \tau) = f(x, \tau),$$

where L is a linear constant-coefficient differential operator in the space variable. Show, formally, that the "Duhamel integral"

$$u(x, t) = \int_0^t v(x, t - \tau; \tau) d\tau$$

solves $u_{tt} = Lu + f$, $u(x, 0) = u_t(x, 0) = 0$. Verify in the case of $Lu = u_{xx}$ that for $f \in C^1$ these formalities can be justified.

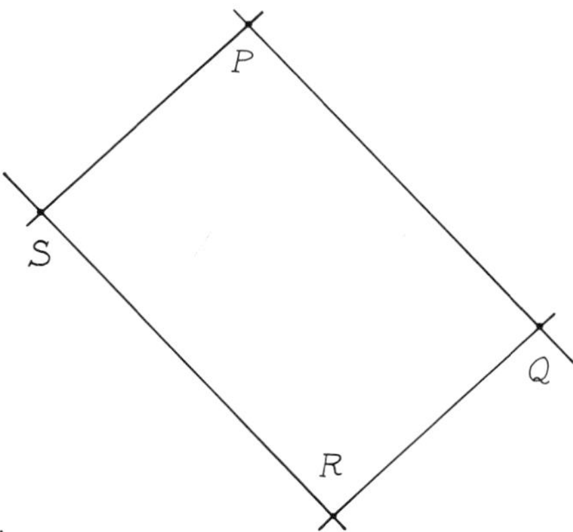

Fig. 8. A characteristic quadrilateral.

1.3. Show that (D) defines a weak solution if u_0 and/or u_1 have isolated discontinuities.

1.4. Show that traveling wave solutions, of the form $u = f(x \pm at)$, of the equation

$$u_{tt} = c^2 u_{xx} + bu, \qquad b \neq 0,$$

do not exist ($f \equiv 0$) for $a = c$, and are real or complex exponentials for $a \neq c$.

1.5. Prove that for any continuous piecewise C^2 solution the discontinuity curves must be contained in characteristic lines.

1.6. Prove that any continuous piecewise C^2 solution of the Cauchy problem is uniquely given by (D).

1.7. Use the formula in Exercise 1.1 to show that the Dirichlet problem for the vibrating string equation

$$u_{tt} = u_{xx} \quad \text{in } Q = (0, 1) \times (0, 1), \qquad u \text{ given on } \partial Q$$

is ill-posed. *Hint*: Show that a solution in general does not exist.

2.1. Derive the compatibility conditions in Remark 2.1 as a consequence of the extension process.

2.2. Show that for the boundary condition $u_x(0, t) = 0$ ["free endpoint," see (5) with $k = 0$], the solution can be obtained from (D) of Section 1 by extending the data (2) as even functions of x to all of $\mathbb{R}$. Show that the compatibility conditions in order to have a C^2 solution in this case are $u_0'(0) = u_1'(0) = 0$.

2.3. Discuss the compatibility conditions at $x = Ut$ for a string with moving endpoint, in the case $0 < U < c$.

2.4. Do the details in Remark 2.3.

2.5. Construct an (x, t)-diagram analogous to that for Example 1.2 for

$$u_0 = \begin{cases} h & 0 < a < x < a + b, \\ 0 & \text{otherwise,} \end{cases} \qquad u_1 = 0, \qquad u(0, t) = 0.$$

2.6. Investigate what happens in Remark 2.4 when $U < 0$.

2.7. Find the solution of $u_{xx} - u_{tt} = 0$ for $|x| < t$, $0 < t < \infty$ satisfying $u(x, -x) = g(x)$, $u(x, x) = f(x)$.

2.8. Find the solution of (V) in $((0, x_0) \cup (x_0, l)) \times (0, \infty)$ with $u(x, 0) = u_t(x, 0) = 0, 0 < x < l$, $u(0, t) = u(l, t) = 0$, and the "interface conditions"

$$u(x_0-, t) = u(x_0+, t), \qquad u_x(x_0+, t) - u_x(x_0-, t) = A \sin \omega t$$

for $t > 0$.

2.9. Show that the Fourier series solution of the finite string problem is uniformly convergent to a C^2 solution u of (V) under the sufficient hypotheses: $u_0 \in H^3(0, l)$, $u_1 \in H^2(0, l)$, $u_0(0) = u_0(l) = u_1(0) = u_1(l) = 0$, $u''(0) = u''(l) = 0$.

2.10. Show that the solution u has *finite energy*

$$\int_0^l (u_t^2 + u_x^2)dx < \infty$$

(see Chapter 1), constant in t, if and only if the series $\sum_{n=1}^{\infty} n^2(a_n^2 + b_n^2)$ converges. Show that this implies $u_0 \in H^1(0, l)$, $u_1 \in L^2(0, l)$, and that u is a weak solution with $u(x, t) \in H^1(0, l)$ for every t.

3.1. The electric and magnetic fields $\mathbf{E}$ and $\mathbf{H}$ satisfy the Maxwell equations

$$\text{curl } \mathbf{H} = \varepsilon \frac{\partial \mathbf{E}}{\partial t}, \qquad \text{curl } \mathbf{E} = -\mu \frac{\partial \mathbf{H}}{\partial t}.$$

Let $\mathbf{E} = E(x, t)\mathbf{j}$, $\mathbf{H} = H(x, t)\mathbf{k}$, where $\mathbf{i}, \mathbf{j}, \mathbf{k}$ are unit vectors along the x, y, z Cartesian axes, respectively. Then (E, H) is a TEM wave. Show that E and H satisfy (ME). *Hint*: Multiply the first equations scalarly by $\mathbf{j}$ and the second by $\mathbf{k}$.

3.2. Show that the general solution (GS) holds. *Hint*: Dividing the second equation (ME′) by h_i and adding the first yields

$$\frac{\partial}{\partial \tau}(h_i u_i \pm v_i) \pm \frac{1}{h_i}\frac{\partial}{\partial x}(h_i u_i \pm v_i) = 0.$$

Hence, $h_i u_i + v_i$ is constant along the lines $C_+ : \tau - h_i x = \text{const}$, $h_i u_i - v_i$ is constant along the lines $C_- : \tau + h_i x = \text{const}$ (characteristic lines).

3.3. Prove (S) by considering the corresponding homogeneous equation, and showing that its general solution is

$$V_0(\tau) = |r_1 r_2|^{\tau/2a} P(\tau),$$

where $P(\tau)$ is (arbitrary) periodic with $P(\tau + 2a) + \text{sgn}(r_1 r_2)P(\tau) = 0$. *Hint*: Define $P(\tau) := |r_1 r_2|^{-\tau/2a} V_0(\tau)$. To find V, proceed recursively.

3.4. Show that for $\mathscr{I}(\tau) \equiv 0$ the homogeneous boundary value problem (13), (I) (with $n = 3$) has the unbounded solutions

$$u(x, \tau) = r_1 |r_1 r_2|^{(\tau - x)/2a} P(\tau - x) + |r_1 r_2|^{(\tau + x)/2a} P(\tau + x),$$
$$v(x, \tau) = r_1 |r_1 r_2|^{(\tau - x)/2a} P(\tau - x) - |r_1 r_2|^{(\tau + x)/2a} P(\tau + x),$$

where P is (arbitrary) periodic with $P(\tau + 2a) + \mathrm{sgn}(r_1 r_2)P(\tau) = 0$. [This can be used to show that uniqueness requires boundedness, hence inclusion of this in the formulation of (BVP).]

3.5. If $\mathscr{I}(\tau)$ has period $p = 2a$ (a half wave layer) and $r_1 + r_2 = 0$ (i.e., $\varepsilon_1 = \varepsilon_3$), then $\mathscr{R}(\tau) \equiv 0$.

3.6. If $\mathscr{I}(\tau + p) = -\mathscr{I}(\tau)$, $p = 2a$ (a quarter wave layer), and $r_1 = r_2$, then $\mathscr{R}(\tau) \equiv 0$. (This condition, $\varepsilon_2 = \sqrt{\varepsilon_1 \varepsilon_3}$, is approximately satisfied by a soap film lying on a glass plate.)

3.7. Prove that if $\mathscr{I}(\tau)$ has compact support in $[\tau_0, \tau_1]$, the support of $\mathscr{R}(\tau)$ is in $[\tau_0, +\infty)$ and $\mathscr{R}(\tau) \to 0$ as $\tau \to +\infty$. (The solution decays as $\tau \to +\infty$. This corresponds to the physical fact that the wave amplitude is reduced at each reflection.)

3.8. Prove that the solution of (BVP) depends continuously on $\mathscr{I}(\tau)$.

3.9. Establish the reflection formula for two layers. Show that $\mathscr{R}(\tau) \equiv 0$ if $\mathscr{I}(\tau)$ is d_3-periodic, $d_2 = nd_3$ for some positive integer n, and $\varepsilon_1 = \varepsilon_4$. Find other conditions for absence of the reflected wave.

3.10. In the case of one layer a pulse of width less than the layer thickness can be used to retrieve the parameters a, h_1, h_3 in an "inverse problem." Verify this.

3.11. Show that the boundary conditions (I) can be written in terms of E, H in the form

$$E(0, t) + \alpha H(0, t) = I(t), \qquad E(a, t) - \beta H(a, t) = 0, \qquad \text{(I}')$$

where $\alpha = \sqrt{\mu/\varepsilon_1}$, $\beta = \sqrt{\mu/\varepsilon_3}$, $I(t) = 2\sqrt{\mu/\varepsilon_1}\mathscr{I}(\tau)$ (α and β are "impedances"). Define the energy, corresponding to a solution (E, H) of (ME) and (I$'$) in D_a, as

$$\mathscr{E}(t) = \frac{1}{2}\int_0^a [\varepsilon E^2(x, t) + \mu H^2(x, t)]dx$$

(with $\varepsilon = \varepsilon_2$). If $E,\ H \in C^1(\bar{D}_a)$, show that these boundary conditions are dissipative, i.e., that $\mathscr{E}'(t) \le 0$ for $I(t) \equiv 0$. *Hint*: $\mathscr{E}'(t) \le -\beta H^2(a, t) - \alpha H^2(0, t)$.

4.1. Prove (16). *Hint*: Any vector $\nu \in \mathbb{R}^m$ can be represented as $\nu = \sum_j \boldsymbol{\nu} \cdot \mathbf{l}_j \mathbf{r}_j = \sum_j \boldsymbol{\nu} \cdot \mathbf{r}_j \mathbf{l}_j$.

4.2. Show that if $\mathbf{u}_0(x)$ is assigned only for $0 \le x \le l$, the solution of the initial value problem is uniquely determined (only) in the region bounded by $t = 0$ and the lines $x = \lambda_m t$, $x - l = \lambda_1 t$.

4.3. Show that the substitution $u_1 = u_x + u_t$, $u_2 = u_x - u_t$ transforms $u_{xx} - u_{tt} = 0$ into a system of the form (15).

4.4. Verify the rule on the (system obtained from the) wave equation.

4.5. State and prove an existence–uniqueness theorem for (a), (b), (c′), (d′). (What should the compatibility conditions be?)

4.6. Consider two wave equations

$$\rho_1 y_{tt} - T_1 y_{xx} = 0, \qquad 0 < x < 1, t > 0,$$
$$\rho_2 y_{tt} - T_2 y_{xx} = 0, \qquad 1 < x < 2, t > 0,$$

with fixed end conditions $y(0, t) = y(2, t) = 0$, and the coupling conditions

$$T_1 y_x(1-, t) = T_2 y_x(1+, t)$$

(continuity of vertical force) and

$$y_1(1-, t) - y_t(1+, t) = -kT_1 y_x(1-, t),$$

where $k > 0$. (This second condition is a dissipative condition introduced to stabilize the system.) Show directly that $\mathscr{E}'(t) \leq 0$ where

$$\mathscr{E}(t) = \frac{1}{2}\int_0^1 (\rho_1 y_t^2 + T_1 y_x^2)dx + \frac{1}{2}\int_1^2 (\rho_2 y_t^2 + T_2 y_x^2)dx.$$

Write this as a first-order system introducing $y = y(x, t),\ z = z(2 - x, t)$,

$$u_1 = \tfrac{1}{2}(y_t - c_1 y_x), \qquad u_2 = \tfrac{1}{2}(z_t - c_2 z_x), \qquad u_3 = \tfrac{1}{2}(y_t + c_1 y_x),$$
$$u_4 = \tfrac{1}{2}(z_t + c_2 z_x),$$

where $c_i = \sqrt{T_i/\rho_i}$, then write the boundary conditions in the form (c′), (d′).

4.7. In (a), (b), (c′), (d′) let $f_j(t) = 0$ for $j \in J_+$, $g_j(t) = 0$ for $j \in J_-$, and $R_{jk} = 0$ for $j \in J_-$, $k \in J_+$. If $\mathbb{L} = \{L_{jk}\}_{j\in J_+, k\in J_-}$ is the matrix defining (c′), give a sufficient condition that $\mathscr{E}'(t) \leq 0$ in terms of $\mathbb{L}$. *Hint*:

$$\sum_{j=1}^{m} \lambda_j z_j^2(0, t) = \sum_{j\in J_-} \lambda_j z_j^2(0, t) + \sum_{j\in J_+} \lambda_j \left[\sum_{k\in J_-} L_{jk} z_k(0, t)\right]^2$$

can be written as a quadratic form acting on the vector $\{z_j(0, t)\}_{j\in J_-}$.

4.8 (Ref. 10). Suppose that $c_1 = c_2 = c$ in Exercise 4.6. If $u_- = (u_1, u_2)$, $u_+ = (u_3, u_4)$, write the boundary conditions in the form $u_-(0, t) = -u_+(0, t)$, $u_+(1, t) = \mathbb{D}u_-(1, t)$, showing that

$$\mathbb{D} = a^{-1}\begin{bmatrix} (T_1 - T_2)/c + kT_1T_2/c^2 & 2T_2/c \\ 2T_1/c & (T_2 - T_1)/c + kT_1T_2/c^2 \end{bmatrix},$$

where $a = (T_1 + T_2)/c + kT_1T_2/c^2$. Verify that $\lambda = 1$ and

$$\lambda = -a^{-1}\left(\frac{T_1 + T_2}{c} - k\frac{T_1T_2}{c^2}\right)$$

are the eigenvalues of $\mathbb{D}$. Then, using the method of characteristics, show that the solution of the system has both conserved and decaying components.

4.9. If $y_2(2, t) = 0$ is replaced by $y_{2x}(2, t) = 0$ and the boundary conditions are written in the form $u_+(1, t) = \mathbb{D}u_-(1, t)$, show that the eigenvalues of $\mathbb{D}$ satisfy $|\lambda| < 1$. Deduce that solutions of the problem decay exponentially.

5.1. Complete the proof of Theorem 5.4. *Hint*: The integral over the sphere $y_3 = \pm f(y_1, y_2) = \pm(c^2t^2 - y_1^2 - y_2^2)^{1/2}$ of a function independent of y_3 can be written as twice an integral over the meridian disk, $y_1^2 + y_2^2 \leq c^2t^2$, with

$$dS = (1 + f_{y_1}^2 + f_{y_2}^2)^{1/2}\, d\xi_1 d\xi_2 = c|t|(c^2t^2 - y_1^2 - y_2^2)^{-1/2}\, dy_1 dy_2.$$

5.2. Derive an inequality analogous to (CD) under the hypotheses of Theorem 5.4.

5.3. Derive the d'Alembert solution by applying the method of descent to (17). Does Huygens' principle hold here?

5.4. Recall the result on permanence of signal for the one-dimensional wave equation. In a similar way we can consider the two-dimensional case. Let $\delta(\mathbf{X}) = \min|\mathbf{X} - \mathbf{Y}|/c$ where the minimum is over the (bounded) support of the data. Show that $u(\mathbf{X}, t) \neq 0$ for $t > \delta(\mathbf{X})$. Show also that $u(\mathbf{X}, t) = O(1/t)$ for fixed $\mathbf{X}$ and $t \to +\infty$.

5.5. Show that Theorem 5.3 implies that if $\mathbf{x}$ is allowed to vary, $\max|u(\mathbf{x}, t)| = O(1/t)$ as $t \to \infty$, where the maximum is taken over $\mathbf{x} \in \mathbb{R}^3$. *Hint*: All that is needed is to show that the area of the part of the sphere of radius $c|t|$ centered at $\mathbf{x}$, and contained in the ball $\{|\mathbf{x}| = R\}$, is bounded independently of t.

5.6. Find $\Delta(I)$ and $\mathcal{J}(I)$ for a ball I in $\mathbb{R}^3$.

5.7. The solution in the example illustrating the focusing phenomenon can be written by seeking spherically symmetric solutions $u = u(r, t)$, $r = |\mathbf{x}|$. It follows that ru satisfies the one-dimensional wave equation, and the "general solution" is given by the superposition of two "spherical waves"

$$u(r, t) = \frac{1}{r}[f(ct + r) + g(ct - r)].$$

Using the initial conditions (18) and the uniqueness theorem, show that the solution is given by

$$u(r, t) = \frac{1}{14cr}[c^2 - (r - ct)^2]^{7/2}H[c - (r - ct)$$
$$- [c^2 - (r + ct)^2]^{7/2}H[c - (r + ct)]$$

5.8. Prove inequality (21).
5.9. Use the method of descent to prove Corollary 5.1.
5.10. Use Theorems 5.4 and 5.8 to show that the estimate

$$|u(\mathbf{x}, t)| \leq |t|\sup_{\mathbb{R}^3}|u_1| + \sup_{\mathbb{R}^3}|u_0| + 3c|t| \max_k \sup_{\mathbb{R}^3}\left|\frac{\partial u_0}{\partial x_k}\right|$$
$$+ \frac{t^2}{2}\sup_{\mathbb{R}^4}|f|$$

holds for solutions of

$$u_{tt} - c^2\Delta u = f(\mathbf{x}, t), \qquad (\mathbf{x}, t) \in \mathbb{R}^4, \qquad u(\mathbf{x}, 0) = u_0(\mathbf{x}),$$
$$u_t(\mathbf{x}, 0) = u_1(\mathbf{x}), \qquad \mathbf{x} \in \mathbb{R}^3.$$

Find u (i.e., u_0, u_1, f) such that the estimate becomes an equality.
5.11. Carry out the calculations in the one-dimensional transient example if F is the characteristic function of [0, 1].
5.12. Give an example that shows nonvanishing of transients in two dimensions.
5.13. Derive the d'Alembert formula from (IF) for $d = 1$.
5.14 (Localized traveling wave solutions).
(i) Traveling wave solutions $u = u(x, z - ct) \in C^2(\mathbb{R}^2)$ of the wave equation in two space variables,

$$u_{xx} + u_{zz} - \frac{1}{c^2}u_{tt} = 0, \qquad \mathbf{X} = (x, z) \in \mathbb{R}^2, \; t \in \mathbb{R},$$

are *linear* in x, and if we add the "localization" properties

$$u \text{ bounded}, \qquad u(x, z - ct) \to 0 \; |x| \to \infty(\forall z, t),$$

they are identically *zero*. *Hint*: Putting $\sigma = z + ct$, $\tau = z - ct$ yields $u = u(x, \tau)$ and $u_{xx} + u_{zz} - c^{-2}u_{tt} = u_{xx} + 4u_{\sigma\tau} = u_{xx}$, whence $u_{xx} = 0$.

(ii) The complex-valued function $v = e^{ik\sigma}u(x, \tau)$, $\sigma = z + ct$, $\tau = z - ct$, is a solution of the wave equation if and only if u satisfies the *Schrödinger equation* in $\mathbb{R}$,

$$u_{xx} + 4iku_\tau = 0 \qquad (i = \sqrt{-1}).$$

One such localized solution, important in applications to optics, is the "Gaussian beam,"

$$u = \frac{\exp[-kx^2/(a + i\tau)]}{\sqrt{a + i\tau}}, \qquad a > 0.$$

(iii) Similarly, the complex-valued function $v = e^{ik\sigma}u(x, y, \tau)$, $\sigma = z + ct$, $\tau = z - ct$, is a C^2 solution of the wave equation in three space variables,

$$u_{xx} + u_{yy} + u_{zz} - \frac{1}{c^2}u_{tt} = 0, \qquad \mathbf{x} = (x, y, z) \in \mathbb{R}^3, \ t \in \mathbb{R},$$

if and only if u satisfies the *Schrödinger equation* in $\mathbb{R}^2$,

$$u_{xx} + u_{yy} + 4iku_\tau = 0.$$

A particular solution is the Gaussian beam in $\mathbb{R}^2$,

$$u = \frac{\exp[-k\rho^2/(a + i\tau)]}{\sqrt{a + i\tau}},$$

where $\rho = (x^2 + y^2)^{1/2}$.

If $k = 0$, u is *harmonic* in (x, y), and if we add the localization properties, u is *zero* (Chapter 4). *Hint*: $u_{xx} + u_{yy} + u_{zz} - c^{-2}u_{tt} = u_{xx} + u_{yy} + 4u_{\sigma\tau} = 0$, and $u_{\sigma\tau} = iku_\tau$.

5.15. From Theorems 1.2, 5.3, and 5.4, derive formally the "Stokes rule": If u', u'' are the solutions of (CP) with $u_0 = 0$, $u_1 = 0$, respectively, then $u'' = u'_t$. Thus, all that is needed is to solve the "standard Cauchy problem" for u'.

5.16. Show that the wave equation $u_{tt} = \Delta u$ in $\mathbb{R}^n$ admits spherical wave solutions of the form $u = a(r)f[t - b(r)]$ for *any* $f(t)$ and suitable $a(r)$, $b(r)$ if and only if $n = 1$ [then $u = f(t \pm r)$] or $n = 3$ [then $u = f(t \pm r)/r$].

References

1. JOHN, F., *Partial Differential Equations*, Springer-Verlag, Berlin, Germany, 1982.
2. TIHONOV, A. N., and SAMARSKII, A. A., *Equazioni della Fisica Matematica*, Mir, Moscow, Russia, 1981.
3. CHESTER, C. R., *Techniques of Partial Differential Equations*, McGraw–Hill, New York, New York, 1971.
4. FASANO, A., and MARMI, S., *Meccanica Analitica*, Bollati Boringhieri, Torino, Italy, 1994.
5. BASSANINI, P., *Interference in Thin Films*, SIAM Review, Vol. 28, pp. 381–384, 1986.
6. BASSANINI, P., *Wave Reflection from a System of Plane Layers*, Wave Motion, Vol. 8, pp. 311–319, 1986.
7. COURANT, R., and HILBERT, D., *Methods of Mathematical Physics*, Vol. 2, Interscience, New York, New York, 1965.
8. MIKHLIN, S. G., *Mathematical Physics, an Advanced Course*, North-Holland, Amsterdam, Netherlands, 1970.
9. HELLWIG, G., *Partial Differential Equations*, Teubner, Stuttgart, Germany, 1977.
10. CHEN, G., COLEMAN, M., and WEST, H., *Pointwise Stabilization in the Middle of the Span for Second Order Systems, Nonuniform and Uniform Exponential Decay of Solutions*, SIAM Journal of Applied Mathematics, Vol. 47 (4), pp. 751–780, 1987.

3

Heat Equation

Here we study the differential equation

$$u_t - a^2 \Delta u = 0 \tag{H}$$

that arises in heat conduction and diffusion problems. For definiteness we will think of t as time, u as the temperature at $(\mathbf{x}, t)$, and (H) as the heat equation. Then $a^2 = k/c$, k the (constant) thermal conductivity, c a specific heat (times the density). Whenever physics is not involved, we will normalize the t-variable so that $a^2 = 1$. Because here, contrary to what happens for the wave equation, there is no decisive dependence on the number, n, of space variables $\mathbf{x} = (x, y, z, \ldots) = (x_1, \ldots, x_n)$, we will often refer to the one-dimensional equation

$$u_t - a^2 u_{xx} = 0. \tag{H1}$$

We begin with a study of some special solutions of (H) and (H1), which are of interest in their own right and will also play an important role in the general theory.

1. Heat Kernel and Miscellaneous Solutions

Let us seek plane wave solutions of the heat equation of the form $u(\mathbf{x}, t) = e^{i\mathbf{x} \cdot \xi} T(t)$, for a fixed $\xi \in \mathbb{R}^n$. An easy computation yields $T'(t) = -a^2 |\xi|^2 T$, hence

$$u(\mathbf{x}, t) = e^{i\mathbf{x} \cdot \xi} e^{-a^2 |\xi|^2 t}$$

is a particular solution of (H) for every $\xi \in \mathbb{R}^n$ and every t. For $t > 0$, a further solution is obtained by superposing these plane waves after multiplying by a constant amplitude $(2\pi)^{-n}$,

$$u(\mathbf{x}, t) = (2\pi)^{-n} \int_{\mathbb{R}^n} e^{i\mathbf{x} \cdot \xi} e^{-a^2 |\xi|^2 t} \, d\xi,$$

in the form of the inverse Fourier transform (see Section 2 of Chapter 8) of a bell curve in ξ-space. An easy calculation yields a bell curve in $\mathbf{x}$-space $u(\mathbf{x}, t) = G(\mathbf{x}, t)$, where

$$G(\mathbf{x}, t) := \frac{1}{(4\pi a^2 t)^{n/2}} e^{-|\mathbf{x}|^2/4a^2 t}, \qquad t > 0 \tag{1}$$

(Exercise 1.1). This function is called the *heat kernel*, or *Gaussian kernel* in n dimensions. We may complete the definition by taking $G(\mathbf{x}, t) \equiv 0$ for $t < 0$. By construction, G is a solution of (H) for $t > 0$. A more general solution is obtained by superposition using a general amplitude function, say $(2\pi)^{-n}\hat{\varphi}(\xi)$. This yields the (formal) solution

$$u(\mathbf{x}, t) = (2\pi)^{-n} \int_{\mathbb{R}^n} e^{i\mathbf{x}\cdot\xi} e^{-a^2|\xi|^2 t} \hat{\varphi}(\xi)\, d\xi,$$

and by a formal application of the convolution theorem for the Fourier transform (Corollary 2.1 of Chapter 8) we find the formula

$$u(\mathbf{x}, t) = \int_{\mathbb{R}^n} G(\mathbf{x} - \mathbf{y}, t)\varphi(\mathbf{y})\, d\mathbf{y}, \tag{2}$$

where $\varphi(\mathbf{y})$ is the inverse Fourier transform of $\hat{\varphi}(\xi)$. We will justify this formula rigorously later on.

The function $G(\mathbf{x} - \mathbf{y}, t - \tau)$ represents the temperature at time t due to a concentrated unit heat source ("hot spot") at $\mathbf{y}$ at time τ. We observe the following properties of $G(\mathbf{x}, t)$:

i. G is positive and C^∞ for $t > 0$.
ii. For $t > 0$, $\int_{\mathbb{R}^n} G(\mathbf{y}, t)\, d\mathbf{y} = 1$. (3)
iii. For $\gamma > 0$, $\lim_{t\to 0^+} \int_{|\mathbf{y}|\geq\gamma} G(\mathbf{y}, t)\, d\mathbf{y} = 0$.

(Exercise 1.2.) They imply that $G(\mathbf{x} - \mathbf{y}, t)$ is a "delta-approximate" family: As $t \to 0^+$, $G \to 0$ a.e. (for all $\mathbf{x} \neq 0$), the L^1 norm of G is identically 1 [properties (i) and (ii)], and $G(\mathbf{x} - \mathbf{y}, t)$ tends to the "δ-function" $\delta(\mathbf{x} - \mathbf{y})$ in the sense of distributions (Chapter 8). Note that G is also integrable on $\mathbb{R}^n \times [0, T]$, for every $T > 0$.

Property (i) shows that, as the support of G is all of $\mathbb{R}^n$ for any $t > 0$, the speed of propagation of heat is infinite, a physical contradiction. However, one may argue (in fact, Maxwell did) that all temperatures below a threshold $|u| < \varepsilon$

(corresponding to the tails of G) are imperceptible if $\varepsilon > 0$ is sufficiently small. Consider the cutoff equation $G(\mathbf{x}, t) = \varepsilon$, or

$$|\mathbf{x}| = \sqrt{-4a^2 t \ln(\varepsilon(4\pi a^2 t)^{n/2})} \tag{4}$$

(see Fig. 1).

Then $|\mathbf{x}| \to 0$ as $t \to 0$ and for $t > 0 |\mathbf{x}|$ has a finite maximum $\rho_\varepsilon = \sqrt{n}/(\sqrt{2\pi e}\varepsilon^{1/n})$ at $t = t_\varepsilon = (4\pi a^2 e \varepsilon^{2/n})^{-1}$, so that $G(\mathbf{x}, t) \geq \varepsilon$ only if $|\mathbf{x}| \leq \rho_\varepsilon$. In a sense, the "practical" support of G is bounded and the "practical" speed of propagation predicted by the heat equation is finite.

The heat kernel in one dimension is

$$G(x, t) = \frac{1}{\sqrt{4\pi a^2 t}} e^{-x^2/4a^2 t}, \qquad t > 0 \tag{5}$$

($G \equiv 0$ for $t < 0$). As $G(x_1, \ldots, x_n, t) = \prod_{i=1}^{n} G(x_i, t)$, from (ii) it follows that

$$G(x_1, \ldots, x_{n-1}, t) = \int_{\mathbb{R}} G(x_1, \ldots, x_n, t)\, dx_n.$$

Suppose $G(x_1, x_2, x_3, t)$ describes the smoke density in the diffusion of a cloud due to a concentrated source of intensity Q at $\mathbf{x} = t = 0$ (e.g., a mine blast). Then the light intensity transmitted through the cloud can be written as

$$I = I_0 \exp\left(- \alpha_0 Q \int_{\mathbb{R}} G(x_1, \ldots, x_3, t)\, dx_3 \right) = I_0 \exp(-\alpha_0 Q G(x_1, x_2, t)),$$

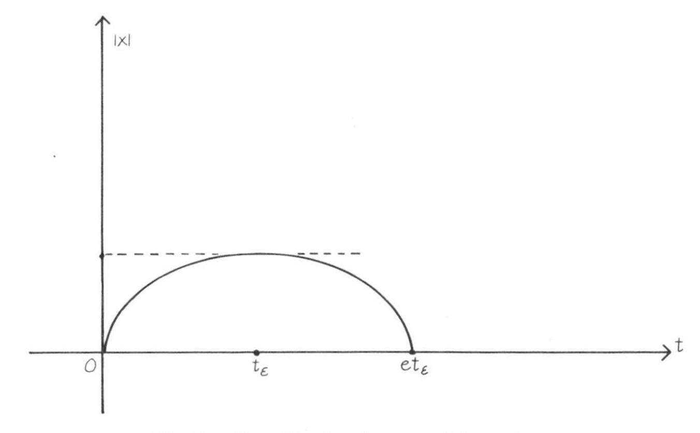

Fig. 1. The effective domain of dependence.

where x_3 is an axis in the direction of observation and α_0 is an absorption coefficient. The cloud will be visible only if I/I_0 is not too large, say $I \leq \delta I_0 < I_0$, so that $G(\mathbf{x}, t) \geq \varepsilon$, with $\delta = e^{-\alpha_0 Q\varepsilon}$. By applying (4) with $n = 2$ we see that the cloud cross section first expands up to a maximum radius ρ_ε at $t = t_\varepsilon$ and then shrinks and eventually disappears to $t = et_\varepsilon$. This predicted behavior agrees well with experience. From the value of (α_0, Q, I_0 and) t_ε, one can compute the value of the constant a^2 in G, which is interpreted here as a turbulent diffusion coefficient of smoke in air (Ref. 1).

Because the heat equation has constant coefficients, derivatives and antiderivatives of G are further solutions. Consider for simplicity the one-dimensional case. The error function

$$\Phi(z) = \operatorname{erf}(z) := \frac{2}{\sqrt{\pi}} \int_0^z e^{-\alpha^2}\, d\alpha, \qquad z = x/\sqrt{4a^2 t},$$

is given in terms of G by the formula $\Phi = 2\int_0^x G(y, t)\, dy$, and is a solution of (H1) for $t > 0$ with the properties:

$$\Phi \to \operatorname{sgn}(x)\ t \to 0^+\ (x \neq 0), \qquad \Phi \to 0\ x \to 0\ (t \neq 0).$$

Hence, $\Phi(x/\sqrt{4a^2 t})$ is a bounded solution of the Dirichlet problem (or initial–boundary value problem) for the quadrant

$$u_t - a^2 u_{xx} = 0\ x > 0, t > 0, \quad \lim_{t\to 0^+} u(x, t) = 1\ (x > 0), \quad \lim_{x\to 0^+} u(x, t) = 0\ (t > 0),$$

discontinuous at the origin. Notice that $\operatorname{erf}(2) \simeq 0.995 \simeq 1$ along the curve $x = 2\sqrt{4a^2 t}$, and the solution Φ is approximately constant ($\Phi \simeq 1$) outside the region (say) $x < 2\sqrt{4a^2 t}$, called *boundary layer*, where the influence of the boundary data $u = 0$ is significant. A problem of this kind governs the laminar Stokes–Rayleigh flow parallel to a flat wall, if u is the fluid velocity and $a^2 = \mu$ is the (kinematic) viscosity of the fluid (e.g., see Ref. 2). The boundary layer may be thought of as a region where the discontinuity in the data (here at $x = t = 0$) due to the initial slip velocity at the wall is smoothed out by the effect of viscosity (Fig. 2).

A similar problem with the initial and boundary data

$$\lim_{t\to 0^+} u(x, t) = T_+\ (x > 0), \qquad \lim_{x\to 0^+} u(x, t) = T_-\ (t > 0)$$

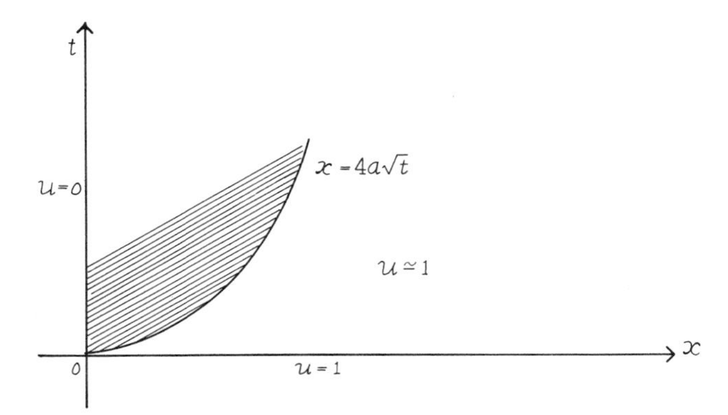

Fig. 2. The boundary layer in a discontinuous IBVP.

describes cooling of a substance (e.g., food), initially at constant temperature $T_+ > 0$, by a refrigerator enforcing the constant temperature $u = T_- < 0$ at the surface, $x = 0$. A (bounded) solution to this problem is given by

$$u = T_- + (T_+ - T_-)\,\mathrm{erf}(x/\sqrt{4a^2t}).$$

We see from this that $u = 0$ on the curve $x = \alpha\sqrt{t}$, where α is determined by the relation

$$\mathrm{erf}(\alpha/2a) = -T_-/(T_+ - T_-).$$

A more interesting version of this problem, taking into account the change of phase of the substance at the freezing point, is the Stefan problem. At the temperature $u = 0$ (degrees Celsius), the water contained in the food freezes and is replaced by ice. During this process, the temperature remains constant and a latent heat is generated at the rate $\lambda d\xi/dt$, if λ is the latent heat coefficient and $x = \xi(t)$ the separation curve between the two phases. The main goal is to find this separation curve, called *free boundary*, as it determines the freezing rate of the food. The thermal and heat conductivities k_1, a_1^2 for $0 < x < \xi(t)$ and k_2, a_2^2 for $\xi(t) < x < +\infty$ will in general be different. Then, the net heat flux at the free boundary must balance the released heat due to freezing:

$$k_1 \frac{\partial u_1}{\partial x}\bigg|_{x=\xi-} - k_2 \frac{\partial u_2}{\partial x}\bigg|_{x=\xi+} = \lambda d\xi/dt, \qquad t > 0, \tag{6}$$

while the temperature at the free boundary remains constant,

$$u_1(\xi(t), t) = u_2(\xi(t), t) = 0, \qquad t > 0. \tag{7}$$

The temperatures u_i in the two phases and the unknown function $x = \xi(t)$ must be determined from the equations

$$\begin{aligned} u_{1t} - a_1^2 u_{1xx} &= 0, \qquad 0 < x < \xi(t), \\ u_{2t} - a_2^2 u_{2xx} &= 0, \qquad \xi(t) < x < +\infty \end{aligned}$$

$(t > 0)$, from the initial and boundary data

$$\lim_{t \to 0^+} u_2(x, t) = T_+ \ (x > 0), \qquad \lim_{x \to 0^+} u_1(x, t) = T_- \ (t > 0),$$

and from the matching conditions (6), (7). Note that, despite the apparent linearity, the Stefan problem is inherently nonlinear, as the domain depends on the solution via the unknown location of the free boundary. Here, however, the constancy of the data enables us to find an explicit solution. Suppose for simplicity $T_+ = 0$. Then the cooling problem above suggests looking for a solution in the form

$$u_1 = T_- + B_1 \operatorname{erf}(x/2a_1\sqrt{t}), \qquad u_2 = 0,$$

which automatically satisfies the initial and boundary data (with $T_+ = 0$). Then from (7) we find $T_- + B_1 \operatorname{erf}(\xi(t)/2a_1\sqrt{t}) = 0$. As this condition holds for every $t > 0$, it follows that the free boundary must be of the form

$$\xi(t) = \alpha\sqrt{t},$$

where α is a constant, and then $B_1 = -T_-/\operatorname{erf}(\alpha/2a_1)$. (See Fig. 3.)

Here, however, the constant α is to be found by solving the transcendental equation

$$-2k_1 T_- e^{-\alpha^2/4a_1^2} = \lambda a_1 \sqrt{\pi}\alpha \operatorname{erf}(\alpha/2a_1)$$

following from (6). As $T_- < 0$, this equation determines a unique $\alpha > 0$, and finally

$$u_1 = T_-\left(1 - \frac{\operatorname{erf}(x/2a_1\sqrt{t})}{\operatorname{erf}(\alpha/2a_1)}\right).$$

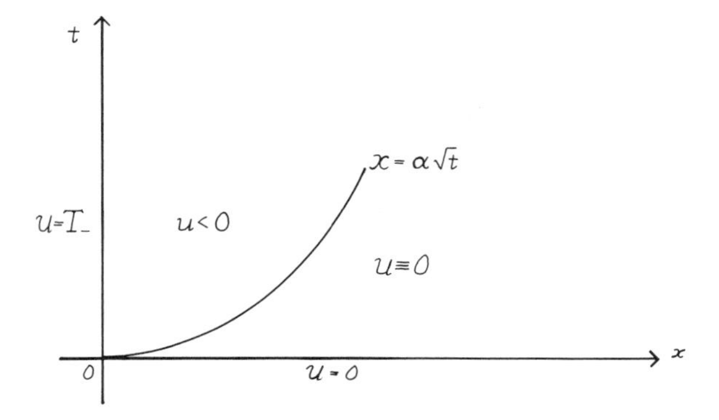

Fig. 3. Solution of a Stefan problem.

It is instructive to compare this with the solution of the one-phase problem (with $a = a_2$, $T_+ = 0$). In general, the temperature u_1 is higher in the two-phase problem due to the released latent heat.

The function $\Phi(z)$ also enables us to solve the Riemann problem for the one-dimensional heat equation (H1)

$$u_t - a^2 u_{xx} = 0 \qquad x > 0, t > 0,$$
$$\lim_{t \to 0^+} u(x, t) = \begin{cases} T_+, & x > 0, \\ T_-, & x < 0. \end{cases}$$

We easily find a bounded solution

$$u(x, t) = \tfrac{1}{2}(T_+ + T_-) + \tfrac{1}{2}(T_+ - T_-)\Phi(x/2a\sqrt{t})$$

and this solution is smooth for $t > 0$, the value of u at $x = 0$ being given by the average $u(0, t) = \frac{1}{2}(T_+ + T_-)$ for $t > 0$. Thus, the initial discontinuity is instantaneously smoothed out over a boundary layer region of finite thickness. Other aspects of this "smoothing property" of the heat operator will be encountered later on.

By taking the x-derivative of G we obtain the one-dimensional unit *heat dipole*

$$d(x, t) = -2a^2 G_x(x, t) = \frac{x}{2a\sqrt{\pi}t^{3/2}} e^{-x^2/4a^2 t}.$$

As $\Phi_x(x/2a\sqrt{t}) = 2G(x, t)$, this can also be written as $d(x, t) = -\Phi_t(x/2a\sqrt{t})$. The heat dipole satisfies $d(0, t) = 0$ for every $t > 0$, and

$$\lim_{t\to 0^+} d(x, t) = 0 \ (x \neq 0), \qquad \lim_{x\to 0} d(x, t) = 0 \ (t > 0),$$

while $d(2a\sqrt{t}, t) = (\sqrt{\pi e}t)^{-1} \to +\infty$ as $t \to 0$. Hence, $d(x, t)$ is an unbounded solution, discontinuous at $x = t = 0$, of the homogeneous Riemann problem and of the homogeneous Dirichlet problem for the quadrant.

Further properties of the heat dipole for $x > 0, t > 0$ are:

i. d is positive and C^∞ for $x > 0$.
ii. For $x > 0$, $\int_0^\infty d(x, t)\, dt = 1$.
iii. For $\gamma > 0$, $\lim_{x\to 0^+} \int_\gamma^\infty d(x, t)\, dt = 0$.

Thus, $d(x, t)$ is a "delta-approximate" family, $d(x, t - \tau) \to \delta(t - \tau)$ as $x \to 0^+$ in the sense of distributions. These properties suggest that the solution of the Dirichlet problem for the quadrant

$$u_t - a^2 u_{xx} = 0 \qquad (x > 0, t > 0),$$
$$\lim_{t\to 0^+} u(x, t) = 0 \ (x > 0), \qquad \lim_{x\to 0^+} u(x, t) = \mu(t) \ (t > 0)$$

should be represented as

$$u(x, t) = \int_0^t d(x, t - \tau)\mu(\tau)\, d\tau. \tag{8}$$

We will not verify this representation formula (e.g., see Cannon, Ref. 3, for details and proofs). If $\mu(t) = 1$, this solution reduces to $\operatorname{erfc}(x/2a\sqrt{t}) = 1 - \Phi(z)$. If, on the contrary, $\mu(t)$ is a positive pulse concentrated in the interval $0 < t < \varepsilon$ [i.e., $\mu(t) = 0$ for $t > \varepsilon$], then for $t \gg \varepsilon$,

$$u(x, t) = \int_0^\varepsilon d(x, t - \tau)\mu(\tau)\, d\tau \simeq d(x, t) \int_0^\varepsilon \mu(\tau)\, d\tau = \frac{Px}{2a\sqrt{\pi}t^{3/2}} e^{-x^2/4a^2 t}$$

does not depend on the shape of the pulse, but only on the "power" $P = \int_0^\varepsilon \mu(\tau)\, d\tau$. This special solution played an important role when the first transatlantic cables for telegraph transmissions were laid (Ref. 4).

We consider now a problem in which x is restricted to the half-space $x > 0$, t varies from $-\infty$ to $+\infty$, and u is prescribed at $x = 0$. Besides writing down the solution, we will now also prove a uniqueness theorem. The problem can be loosely described as that of determination of the temperature oscillations inside the Earth if the temperature oscillations at the surface are known as a function of time. (This is one of the original problems considered by Fourier in his study of heat conduction.) With appropriate idealizations this leads to the following mathematical formulation: Find $u(x, t)$ such that

$$\begin{aligned} u_t &= a^2 u_{xx}, & x > 0, t \in \mathbb{R}, \\ u(0, t) &= f(t), & t \in \mathbb{R}, \end{aligned} \tag{T}$$

where $f \in C^1(\mathbb{R})$. We will consider only the case in which f is periodic, with period p, and we seek solutions satisfying $u(x, t+p) = u(x, t)$. (In typical applications p might be 1 day or 365 days.) We see immediately that the problem is not well posed as $u(x, t) = Ax$ is a solution with $u(0, t) \equiv 0$. We add the condition that

$$u(x, t) = o(x) \qquad \text{as } x \to +\infty \tag{G}$$

uniformly in t. This growth condition enables us to prove uniqueness for the (pure boundary value) problem (T), (G).

Theorem 1.1. There is at most one periodic solution $u \in C^2((0, \infty) \times \mathbb{R}) \cap C([0, \infty) \times \mathbb{R})$ with prescribed period of (T), (G).

Proof. If u_1, u_2 are two solutions with period T, let $u = u_1 - u_2$, and define

$$E(x) = \frac{1}{2}\int_0^T u^2(x, t)\, dt.$$

Then $E \in C^2(0, \infty) \cap C[0, \infty)$, and for $x > 0$,

$$E'(x) = \int_0^T u u_x dt \leq \sqrt{2E(x)}\sqrt{\int_0^T u_x^2 dt}.$$

Also,

$$E''(x) = \int_0^T u_x^2 dt + \int_0^T u u_{xx}\, dt = \int_0^T u_x^2 dt + \frac{1}{a^2}\int_0^T u u_t dt = \int_0^T u_x^2 dt$$

as $u^2(x, 0) = u^2(x, T)$. Therefore, $(E')^2 \leq 2EE''$ for $x > 0$, and letting $E = U^2$, we have $U''(x) \geq 0$. As $U(0) = 0$, $U(x) \geq 0$, and $U(x) = o(x)$, we deduce that $U(x) \equiv 0$, and $u \equiv 0$. □

Suppose that $p = 2\pi/\omega$, and consider the special f given by

$$f(t) = Ae^{i\omega t}.$$

If we write $u(x, t) = X(x)e^{i\omega t}$, we obtain the problem

$$\begin{aligned} iX &= a^2 X'', \qquad x > 0, \\ X(0) &= A, \\ X(x) &= o(x) \qquad x \to \infty. \end{aligned}$$

The solution is then

$$u(x, t) = A \exp\left(-\sqrt{\frac{\omega}{2a^2}} x \right) \exp(i\omega(t - x/\sqrt{2\omega}a)). \tag{9}$$

Note that the solution is damped in x and the damping grows with ω. (This is known as the *skin effect*.) Also, the temperature at depth $x > 0$ has a *phase lag* equal to $x/\sqrt{2\omega}a$, which also depends on ω. All of these predictions are confirmed by measurements showing that annual oscillations in temperature are imperceptible at depths of the order of 10 m (this accounts for the permafrost, i.e., permanently frozen subsoil at high latitudes), and that daily variations are imperceptible at depths of the order of 50 cm.

We remark that taking $p = 365$ days, and using the appropriate value of a, the phase lag equals 6 months for x of the order of 3 to 4 m. This corresponds to the depth at which a good wine cellar should be built.

If we write, more generally,

$$f(t) = \sum_{-\infty}^{\infty} A_k e^{ik\omega t}, \qquad A_k = \frac{1}{p} \int_0^p f(t) e^{-ik\omega t} \, dt,$$

the formal solution is

$$u(x, t) = \sum_{-\infty}^{\infty} A_k e^{ik\omega t} \exp\left(-\sqrt{\frac{|k|\omega}{2a^2}}\, x - i \operatorname{sgn}(k) \sqrt{\frac{|k|\omega}{2a^2}} x \right). \tag{10}$$

The problem (T), (G) may be thought of as a *problem without initial conditions* in the same sense as the layer problem of Chapter 2. In contrast, the problem of the determination of the *age of the Earth* obviously requires some initial condition. An elementary quantitative model can be based on the one-dimensional heat equation with initial temperature (in degree Celsius) $u(x, 0) = T_0 \simeq 1200$ (the fusion temperature of rocks) for $x > 0$, and surface temperature $u(0, t) \simeq 0$ for $t > 0$. We have already met a problem of this kind. From the solution $u = T_0\Phi(x/2a\sqrt{t})$ we can compute the geothermic gradient $\gamma = u_x$ at $x = 0$ directly in terms of the heat kernel:

$$u_x(0, t) = T_0\Phi_x(x/2a\sqrt{t})\Big|_{x=0} = 2T_0G(0, t) = \frac{T_0}{\sqrt{\pi a^2 t}}.$$

This should be compared with the present measured value of the geothermic gradient at 2–3 km below the Earth's crust, $\gamma \simeq 3 \times 10^{-4}$ deg/cm. Then by using the value $a^2 \simeq 0.006$ cm^2/sec for the heat conductivity of granites and basalts we obtain the estimate $t \simeq 3 \times 10^7$ years for the age of the Earth. This value is by far too low. A more accurate estimate can be obtained by including the effect of radioactive decay on the Earth's crust (Refs. 1, 4).

All of these special solutions illustrate the irreversibility of diffusion phenomena with respect to time. The following example shows that the heat equation does in addition incorporate some irreversible features, peculiar to thermodynamics. Suppose that u is a *positive* solution of

$$u_t = a^2\Delta u + f(\mathbf{x}, t)$$

for $t > 0$, $\mathbf{x} \in \Omega$, Ω a bounded domain of $\mathbb{R}^3$ to which the divergence theorem can be applied. We may define the quantities

$$S = \int_\Omega \ln u \, d\mathbf{x}, \qquad \Phi = -a^2 \int_{\delta\Omega} \frac{1}{u}\frac{\partial u}{\partial n}\, dS, \qquad \Delta Q = \int_\Omega \frac{f}{u}\, d\mathbf{x}$$

as the entropy, the entropy flux, and the entropy rate of increase due to heat sources f in Ω, respectively. As u is positive, we can divide the equation by u and find

$$(u_t - a^2\Delta u - f)/u = u_t/u - a^2 \operatorname{div}(\operatorname{grad} u/u) - a^2|\operatorname{grad} u|^2/u^2 - f/u = 0,$$

so that $u_t/u - a^2 \operatorname{div}(\operatorname{grad} u/u) \geq f/u$. Integrating over Ω and applying the divergence theorem yields

$$\frac{dS}{dt} + \Phi \geq \Delta Q.$$

This inequality can be interpreted as the counterpart of the Clausius–Duhem inequality in thermodynamics (Ref. 5).

2. Maximum Principle

If D is a domain in the $(\mathbf{x}, t)$ space $\mathbb{R}^{n+1}$, we define for every constant T the open set

$$D_T = (\mathbf{x}, t) \in D : t < T$$

and the boundary sets

$$S_T = \{(x, t) \in \partial D : t < T\}, \qquad \Gamma_T = \{(x, t) \in D : t = T\}$$

(Fig. 4).

We can write ∂D_T as the disjoint union of $\bar{S}_T$ and Γ_T where $\bar{S}_t = (x, t) \in \partial D : t \leq T$ is closed. We consider solutions of the differential inequality $\Delta u - u_t \geq 0$ for the heat operator $\partial/\partial t - \Delta$ with $a^2 = 1$.

Theorem 2.1. Suppose that D_T is bounded and not empty. If $u \in C(\bar{D}_T)$, $u_{x_i x_i}, u_t \in C(D_T)$, and $\Delta u - u_t \geq 0$ in D_T, then u attains its maximum over $\bar{D}_T$ on $\bar{S}_T$.

Proof. Fix an arbitrary point $(\bar{\mathbf{x}}, \bar{t})$ in D_T, choose arbitrary constants T^*, ε with $\bar{t} < T^* < T$ and $\varepsilon > 0$, and consider the function $v(\mathbf{x}, t) = u(\mathbf{x}, t) = -\varepsilon t$ on $\bar{D}_{T^*}$. Suppose the maximum value on $\bar{D}_{T^*}$ of the (continuous) function v occurs at a point $(\mathbf{x}_0, t_0) \in D_{T^*} \cup \Gamma_{T^*}$. Then, from calculus, we have $\Delta v \leq 0$ and $v_t \leq 0$, so that $\Delta v - v_t \leq 0$, at $(\mathbf{x}_0, t_0)$. On the other hand,

$$\Delta v - v_t = \Delta u - u_t + \varepsilon \geq \varepsilon > 0$$

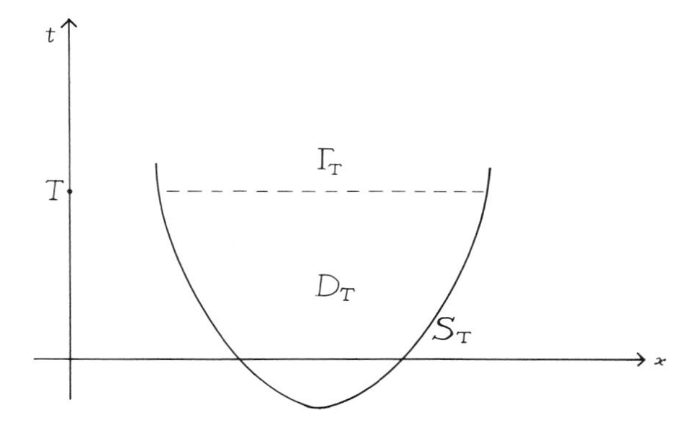

Fig. 4. The distinguished boundary in the maximum principle.

on all of $D_{T^*} \cup \Gamma_{T^*}$ and in particular at $(\mathbf{x}_0, t_0)$. Thus, we have a contradiction and it follows that the maximum of v can only occur on the boundary $\partial D_{T^*} \backslash \Gamma_{T^*} \subset S_T \subset \bar{S}_T$. At the point $(\bar{\mathbf{x}}, \bar{t})$ we then have

$$u(\bar{\mathbf{x}}, \bar{t}) = J + \varepsilon \bar{t} \leq \varepsilon T + \max{}_{\bar{S}_T} v \leq 2\varepsilon T + \max{}_{\bar{S}_T} u$$

and letting $\varepsilon \to 0$ we obtain the result $u(\bar{\mathbf{x}}, \bar{t}) \leq \max_{\bar{S}_T} u$. Finally, as the point $(\bar{\mathbf{x}}, \bar{t}) \in D_T$ was itself arbitrary, the result holds at every point of D_T and, by continuity, in $\bar{D}_T$. □

By applying the above result to $-u$ we can show that $\Delta u - u_t \leq 0$ implies the minimum of u is taken on $\bar{S}_T$.

Corollary 2.1. Under the conditions of Theorem 2.1, if $u(\mathbf{x}, t)$ satisfies the equation $\Delta u - u_t = 0$ in D_T, then

$$\min{}_{\bar{S}_T} u \leq u(\mathbf{x}, t) \leq \max{}_{\bar{S}_T} u \qquad \text{for every } (\mathbf{x}, t) \in \bar{D}_T.$$

In the next corollary we assume, without loss of generality, that $x_1 > 0$ for every $\mathbf{x} \in D_T$.

Corollary 2.2. Suppose again D_T is bounded and not empty, and let d denote the width of D_T in the x_1 direction. If $u(\mathbf{x}, t) \in C(\bar{D}_T)$ and $u_t, u_{x_i x_i} \in C(D_T)$, then we have the *a priori estimate*

$$|u(\mathbf{x}, t)| \leq \max{}_{\bar{S}_T} |u| + (e^d - 1) \sup{}_{D_T} |\Delta u - u_t|, \quad (\mathbf{x}, t) \in \bar{D}_T.$$

Proof. We may assume that $\sup_{D_T} |\Delta u - u_t| < +\infty$ for otherwise there is nothing to prove. Let $Lu := \Delta u - u_t$. We define the auxiliary functions

$$v_\pm(\mathbf{x}, t) = \pm u(\mathbf{x}, t) + \max{}_{\bar{S}_T} |u| + (e^d - e^{x_1}) \sup{}_{D_T} |Lu|,$$

where by definition $d \geq x_1$. Then $\Delta v_\pm - (v_\pm)_t = \pm Lu - e^{x_1} \sup_{D_T} |Lu| \leq 0$ in D_T, and

$$v_\pm(\mathbf{x}, t) = \pm u(\mathbf{x}, t) + \max{}_{\bar{S}_T} |u| + (e^d - e^{x_1}) \sup{}_{D_T} |Lu| \geq 0$$

on $\bar{S}_T$. Theorem 2.1 implies that $v_\pm(\mathbf{x}, t) \geq 0$ in all of $\bar{D}_T$. As $x_1 \geq 0$, the result follows. □

Remarks

2.1. Theorem 2.1 is a *weak maximum principle as it does not exclude the possibility that* u may also attain its extremal values at some point $(\mathbf{x}_1, t_1)$ of $D_T \cup \Gamma_T$. A *strong* maximum principle asserting that this is impossible unless u is constant (at least for $t \leq t_1$) is also true, but will not be needed in this book.

2.2. The fact that D_T is bounded is essential for Theorem 2.1. Without additional conditions the result is false if D_T is not bounded in the x_i-variables or not bounded below in t, as shown by the following examples:

i. The function $u = e^{-t}\cos x - \frac{1}{2}e^{-4t}\cos 2x$ is a solution of (H1) (with $a^2 = 1$) that vanishes on the curve $t = \frac{1}{3}\ln(\cos 2x/2\cos x)$. Hence, the maximum principle fails for the domain below the curve in the (x, t)-plane (this domain is bounded in x and unbounded below in t).
ii. The function $x^2 + 2a^2t$ is a solution of (H1) unbounded in every half-plane or horizontal strip and vanishes on the curve $x^2 = -2a^2t$.

2.3. The condition that $u \in C(\bar{D}_T)$ is also essential. Without additional conditions, Theorem 2.1 is false if u is not continuous on $\bar{D}_T$, as shown by the trivial example $u = 1$ in D, $u = 0$ on ∂D. The reader should also reflect on the example of the heat dipole, reported in the previous section.

2.4. The heat equation is t-irreversible: Theorem 2.1 is not valid if D_T is replaced by $(\mathbf{x}, t) \in D : t > T$ and S_T by $(\mathbf{x}, t) \in \partial D : t > T$ or, equivalently, if t is replaced by $-t$.

Corollary 2.2 implies the following theorem of uniqueness and continuous dependence for the Dirichlet problem for the heat equation.

Theorem 2.2. Let D_T be bounded and not empty for a fixed $T > 0$. Then the Dirichlet problem

$$\begin{aligned} u_t &= \Delta u + f(\mathbf{x}, t), \qquad (\mathbf{x}, t) \in D_T, \\ u(0, t) &= \varphi(\mathbf{x}, t), \qquad (\mathbf{x}, t) \in \bar{S}_T, \end{aligned} \tag{D}$$

has at most one solution $u \in C(\bar{D}_T)$, with $u_{x_ix_i}, u_t \in C(D_T)$, and the solution, if it exists, depends continuously on f and φ.

Proof. Suppose such a solution exists, then u satisfies the estimate

$$|u(\mathbf{x}, t)| \leq \max_{\bar{S}_T}|\varphi| + (e^d - 1)\sup_{D_T}|f|,$$

following from Corollary 2.2. This inequality implies uniqueness and continuous dependence. (Uniqueness also follows directly from the maximum principle.) □

Note that the Dirichlet data are assigned on a proper subset $\bar{S}_T$ of the boundary of D_T. A boundary value problem for the heat equation with Dirichlet data prescribed *over an entire closed boundary* is not well posed (Exercise 2.1). In the simplest situation, D is a cylinder domain $\Omega \times (0, +\infty)$, where $\Omega \subset \mathbb{R}^n$ is an open bounded set, $D_T = \Omega \times (0, T)$, $T > 0$, and the Dirichlet problem for D_T can be interpreted as an *initial–boundary value* (IBV) problem for the domain Ω. Theorem 2.2 then shows that prescribing initial and boundary values for u guarantees uniqueness. From the physical point of view, the maximum principle implies that (in the absence of heat sources) the material in D cannot get hotter or colder than a temperature either occurring initially or applied to the walls. We see then that $\bar{S}_T$ is a natural set on which to prescribe Dirichlet data for a solution.

In several applications it is useful to ascertain whether the normal derivative of a solution u of the heat equation can vanish at a boundary point where the maximum is attained. The following result, given without proof, answers this question.

Theorem 2.3 (Boundary point lemma). Suppose u is a solution of (H) in $D \subset \mathbb{R}^{n+1}$. Suppose that u is continuously differentiable at a boundary point $(\mathbf{x}_0, t_0) \in \partial D$, where $u(\mathbf{x}_0, t_0) = M$, that $u(\mathbf{x}, t) < M$ for $(\mathbf{x}, t) \in D_{t_0}$, that $(\mathbf{x}_0, t_0)$ lies on the boundary of a sphere S tangent to ∂D and such that the part of its interior where $t \leq t_0$ lies in D_{t_0}. Also, suppose that the radial direction from the center of the sphere to $(\mathbf{x}_0, t_0)$ is not parallel to the t-axis. Then

$$\frac{\partial u}{\partial n} > 0 \qquad \text{at } (\mathbf{x}_0, t_0),$$

where $\mathbf{n}$ denotes the outward normal from D_{t_0} at $(\mathbf{x}_0, t_0)$.

For the proof, we refer the reader to Protter and Weinberger (Ref. 6) or Friedman (Ref. 7).

The maximum principle can be extended to domains that are unbounded in the space variables by imposing suitable additional growth conditions on the functions at the boundaries at infinity. Such results are generally termed *Phragmèn–Lindelöf theorems* (Refs. 7, 8). The case when $D = \mathbb{R}^n \times (0, +\infty)$ is of particular interest for the initial value problem.

Theorem 2.4. Suppose $D_T = \mathbb{R}^n \times (0, T)$. If $u \in C(\bar{D}_T)$, $u_{x_i x_i}, u_t \in C(D_T)$, $\Delta u - u_t \leq 0$ in D_T, $u(\mathbf{x}, 0) \geq 0$ for all $\mathbf{x} \in \mathbb{R}^n$ [i.e., $u(\mathbf{x}, t) \geq 0$ on S_T] and if, in addition,

$$\liminf_{|\mathbf{x}| \to \infty} u(\mathbf{x}, t) \geq 0 \qquad \text{(PL)}$$

uniformly for all $t \in [0, T]$ $(\mathbf{x} \in D_T)$, then $u(\mathbf{x}, t) \geq 0$ in $\bar{D}_T$.

Proof. Fix an arbitrary point $(\mathbf{x}_0, t_0)$ in $\bar{D}_T = \mathbb{R}^n \times [0, T]$ and let an arbitrary $\varepsilon > 0$ be prescribed. By hypothesis, there exists a radius $R_\varepsilon > |\mathbf{x}_0|$ such that $u(\mathbf{x}, t) \geq -\varepsilon$ for all $(\mathbf{x}, t)[(\mathbf{x}, t) \in \bar{D}_T]$ with $|\mathbf{x}| \geq R_\varepsilon$, $0 \leq t \leq T$. Now consider the function $w(\mathbf{x}, t) = u(\mathbf{x}, t) + \varepsilon$ on the truncated domain $\bar{B}$ where $B := |\mathbf{x}| < R_\varepsilon, 0 < t < T$. We have

$$\begin{aligned} &\Delta w - w_t = \Delta u - u_t \leq 0 && \text{in } B, \\ &w(\mathbf{x}, t) \geq 0 && \text{for every } (\mathbf{x}, t) \in \partial B \text{ with } |\mathbf{x}| = R_\varepsilon, \\ &w(\mathbf{x}, 0) \geq \varepsilon > 0 && \text{for all } \mathbf{x} \in \mathbb{R}^n. \end{aligned}$$

It follows from Theorem 2.1 that $w \geq 0$ in $\bar{B}$ and in particular that $w(\mathbf{x}_0, t_0) = u(\mathbf{x}_0, t_0) + \varepsilon \geq 0$. Letting ε approach zero we obtain $u(\mathbf{x}_0, t_0) \geq 0$, which concludes the proof. □

Corollary 2.3. Theorem 2.4 remains true if condition (PL) is replaced by the growth condition

$$u(\mathbf{x}, t) \geq -Ke^{A|\mathbf{x}|^2} \tag{G}$$

for some nonnegative constants A, K independent of $\mathbf{x} \in \mathbb{R}^n$ and t, $0 \leq t \leq T$.

Proof. Choose $A_1 > A$ and define the auxiliary function

$$v(\mathbf{x}, t) = \frac{1}{(1 - 4A_1 t)^{n/2}} e^{A_1 |\mathbf{x}|^2/(1-4A_1 t)}$$

in the closure $\bar{\mathscr{S}}$ of the slab $\mathscr{S} := \{\mathbf{x} \in \mathbb{R}^n : 0 < t < 1/8A_1\}$. We may assume that A_1 is large enough so that $1/8A_1 \leq T$. Then $\Delta v - v_t = 0$, $v \geq e^{A_1|\mathbf{x}|^2}$ in $\bar{\mathscr{S}}$. Now let arbitrary $\varepsilon > 0$ be prescribed and consider the function $w = u + \varepsilon v$ in $\bar{\mathscr{S}}$. We have

$$\begin{aligned} &\Delta w - w_t = \Delta u - u_t \leq 0 && \text{in } \mathscr{S}, \\ &w(\mathbf{x}, 0) \geq 0 && \text{for every } \mathbf{x} \in \mathbb{R}^n. \end{aligned}$$

Moreover, it is easy to verify that $\liminf_{|\mathbf{x}| \to \infty} w(\mathbf{x}, t) \geq 0$ uniformly for all $t \in [0, 1/8A_1]$. From Theorem 2.4 we conclude that $w(\mathbf{x}, t) \geq 0$ in $\bar{\mathscr{S}}$, and letting $\varepsilon \to 0$ we obtain

$$u(\mathbf{x}, t) \geq 0 \qquad \text{in } \bar{\mathscr{S}}.$$

As the differential inequality $\Delta u - u_t \leq 0$ is invariant under a translation to the t-axis, we can repeat this argument in the slabs $\mathbb{R}^n \times [1/8A_1, 2/8A_1]$, $\mathbb{R}^n \times [2/8A_1, 3/8A_1] \ldots$ until the entire set $\mathbb{R}^n \times [0, T]$ has been covered. $\square$

The growth condition (G) can be rephrased in the form

$$u^-(\mathbf{x}, t) \leq Ke^{A|\mathbf{x}|^2},$$

where $u^-(\mathbf{x}, t) := \max\{0, -u(\mathbf{x}, t)\}$.

From Corollary 2.3 it follows that if $\Delta u - u_t = 0$ in $\mathbb{R}^n \times (0, T)$, $u(\mathbf{x}, 0) = 0$, $|u(\mathbf{x}, t)| \leq Ke^{A|\mathbf{x}|^2}$ for $\mathbf{x} \in \mathbb{R}^n$, $0 \leq t \leq T$, then $u(\mathbf{x}, t) \equiv 0$ in $\mathbb{R}^n \times [0, T]$.

Corollary 2.4. If $u \in C(\mathbb{R}^n \times [0, T])$, $u_{x_i x_i}$, $u_t \in C(\mathbb{R}^n \times (0, T))$, and

$$|u(\mathbf{x}, t)| \leq Ke^{A|\mathbf{x}|^2} \tag{GC}$$

for $\mathbf{x} \in \mathbb{R}^n$, $0 \leq t \leq T$ with some nonnegative constants A, K independent of $\mathbf{x}$ and t, then we have the a priori estimate

$$|u(\mathbf{x}, t)| \leq \sup_{\mathbb{R}^n} |u(\mathbf{x}, 0)| + t \sup_{\mathbb{R}^n \times (0,T)} |\Delta u - u_t|$$

for all $(\mathbf{x}, t) \in \mathbb{R}^n \times [0, T]$.

Proof. Assume that the right-hand side is finite (for otherwise there is nothing to prove), and consider the functions

$$w(\mathbf{x}, t) = \sup_{\mathbb{R}^n} |u(\mathbf{x}, 0)| + t \sup_{\mathbb{R}^n \times (0,T)} |Lu| \pm u(\mathbf{x}, t),$$

where $Lu := \Delta u - u_t$. We have

$$\begin{aligned} \Delta w - w_t &= -\sup_{\mathbb{R}^n \times (0,T)} |Lu| \pm Lu \leq 0 \qquad && \text{on } \mathbb{R}^n \times (0, T), \\ w(\mathbf{x}, 0) &= \sup_{\mathbb{R}^n} |u(\mathbf{x}, 0)| \pm u(\mathbf{x}, 0) \geq 0 && \text{for } \mathbf{x} \in \mathbb{R}^n. \end{aligned}$$

Furthermore,

$$w(\mathbf{x}, t) \geq \sup_{\mathbb{R}^n} |u(\mathbf{x}, 0)| + t \sup_{\mathbb{R}^n \times (0,T)} |Lu| - |u(\mathbf{x}, t)| \geq -Ke^{A|\mathbf{x}|^2}$$

From Corollary 2.3 we conclude that $w(\mathbf{x}, t) \geq 0$ in $\mathbb{R}^n \times [0, T]$, and the estimate follows. $\square$

Tikhonov's growth condition (GC) is certainly satisfied (with $A = 0$) if u is *bounded*, $|u(\mathbf{x}, t)| \leq K$.

Remark 2.5. Theorem 2.4 can be easily extended to domains $D \subset \mathbb{R}^n \times (0, +\infty)$ such that Γ_{t^*} is unbounded for at least one value of t^* in the interval $0 \leq t^* < T$ (see Exercise 2.2). One such domain is the *quadrant* $x > 0, t > 0$ considered for $n = 1$ in Section 1. It follows that the IBV (or Dirichlet) problem,

$$\begin{aligned} &u_t - u_{xx} = 0 \qquad x > 0, t > 0, \\ &u(x, 0) = \varphi(x)\ (x > 0), \qquad u(0, t) = \psi(t)\ (t > 0), \end{aligned}$$

has at most one solution continuous and *bounded* on the closed quadrant $[0, \infty) \times [0, \infty)$. For the existence of such a solution it will obviously be required that φ, ψ be bounded continuous functions with $\varphi(0) = \psi(0)$. We have seen in Section 1 examples where this compatibility condition is *not* fulfilled, say $\varphi(0) - \psi(0) = \delta \neq 0$. Set

$$u = v + w + \psi(0),$$

where w satisfies

$$\begin{aligned} &w_t - w_{xx} = 0 \qquad x > 0, t > 0, \\ &w(x, 0) = 0 (x \geq 0), \qquad w(0, t) = \mu(t) := \psi(t) - \psi(0)\ (t \geq 0), \end{aligned} \tag{11}$$

and v is a solution of

$$\begin{aligned} &v_t - v_{xx} = 0 \qquad x > 0, t > 0, \\ &v(x, 0) = \varphi(x) - \psi(0)\ (x > 0), \qquad v(0, t) = 0\ (t > 0). \end{aligned} \tag{12}$$

Now the IBV problem (11) has bounded and continuous data and hence has a unique solution [given by equation (8) in Section 1], continuous and bounded on the closed quadrant $[0, \infty) \times [0-, \infty)$. On the other hand, the boundary data in the problem (12) are discontinuous at the origin, as $\varphi(0) - \psi(0) = \delta$. In the next section we will show that this problem has a unique bounded solution, discontinuous at the origin.

3. Initial Value Problem

Here we consider the initial value problem for (H) on an infinite strip $D_T = \mathbb{R}^n \times (0, T)$,

$$\begin{aligned} u_t - \Delta u &= f(\mathbf{x}, t), \qquad (\mathbf{x}, t) \in \mathbb{R}^n \times (0, T), \\ u(\mathbf{x}, 0) &= \varphi(\mathbf{x}), \qquad \mathbf{x} \in \mathbb{R}^n, \end{aligned} \tag{IVP}$$

where $u \in C(\mathbb{R}^n \times [0, T])\, u_t, u_{x_i x_i} \in C(\mathbb{R}^n \times (0, T))$. We will assume in this section that $f(\mathbf{x}, t) \equiv 0$.

We present an example showing that solutions of (IVP) are not unique without further conditions on u (and φ), by constructing a nonzero solution for $\varphi \equiv 0$.

Example 3.1. Suppose $n = 1$ and $T = \infty$, and consider the function

$$u(x, t) := \begin{cases} \sum_{k=0}^{\infty} \dfrac{1}{(2k)!} f^{(k)}(t) x^{2k} & \text{on } \mathbb{R} \times (0, \infty), \\ 0 & \text{on } \mathbb{R} \times 0, \end{cases} \tag{13}$$

where $f(z)$ is the function

$$f(z) := \begin{cases} 0 & \text{if } z = 0, \\ e^{-1/z^2} & \text{if } z \in \mathbb{C}\backslash\{0\}, \end{cases}$$

analytic in the complex plane $\mathbb{C}\backslash\{0\}$. Proceeding formally we have

$$\lim_{t \to 0} u(x, t) = \lim_{t \to 0} \sum_{k=0}^{\infty} \frac{1}{(2k)!} f^{(k)}(t) x^{2k} = \sum_{k=0}^{\infty} \frac{1}{(2k)!} f^{(k)}(0) x^{2k} = 0,$$

while in $\mathbb{R} \times (0, \infty)$,

$$u_{xx} = \sum_{k=1}^{\infty} \frac{1}{(2k-2)!} f^{(k)}(t) x^{2k-2} = \sum_{k=0}^{\infty} \frac{1}{(2k)!} f^{(k+1)}(t) x^{2k} = u_t.$$

To justify these operations and thus show that (13) is indeed a solution vanishing for $t = 0$, we need only verify that the series involved converge uniformly in a neighborhood of each point of $R \times [0, \infty)$. This will require estimates of the derivatives $f^{(k)}(t)$ for $t > 0$. Because for any fixed $t > 0 f(z)$ is analytic on the disk centered at $(t, 0)$ with radius $t/2$, by Cauchy's theorem we find

$$f^{(k)}(t) = \frac{k!}{2\pi i} \oint_C \frac{f(z)\, dz}{(z-t)^{k+1}} \qquad k = 0, 1, 2, \ldots,$$

where $C: z - t = (1/t)e^{i\theta}$, $0 \leq \theta \leq 2\pi$. An easy calculation yields $1/z = 2(2 + e^{i\phi})/3t$ for some angle $\phi, 0 \leq \phi \leq 2\pi$. For $z \in C$ we have

$$\operatorname{Re}[1/z^2] = \frac{16}{9t^2}\left(1 + \cos\phi + \frac{1}{4}\cos 2\phi\right) \geq \frac{4}{9t^2}$$

and from the Cauchy formula we can estimate

$$|f^{(k)}(t)| \leq \frac{k!}{2\pi} \oint_C \frac{|e^{-1/z^2}|\,|dz|}{|z-t|^{k+1}} \leq \frac{k!}{2\pi} \int_0^{2\pi} \frac{e^{-4/9t^2}}{2|t/2|^{k+1}}\, t\, d\theta = k!(2/t)^k e^{-4/9t^2}.$$

As $k!2^k/(2k)! \leq 1/k!$, the series in (13) is dominated in any finite interval $|x| \leq L$

by the series $e^{-4/9t^2} \sum_{k=0}^{\infty} (k!)^{-1}(L^2/t)^k$, which is seen to converge uniformly to the continuous function

$$\begin{cases} e^{-4/9t^2} e^{L^2/t}, & t \neq 0, \\ 0, & t = 0 \end{cases} \tag{14}$$

(Exercise 3.1). Similar considerations show the uniform convergence of the derived series and justify the previous formal manipulations.

Remarks

3.1. The Tikhonov example (13) is a valid solution in all of $\mathbb{R}^2$ and hence demonstrates nonuniqueness for the initial value problem "backwards" in time as well as "forward."

3.2. Many other examples have been constructed to demonstrate nonuniqueness, e.g., the example (proposed by Rosenbloom and Widder)

$$u(x, t) = \int_0^{\infty} a(y) g(xy, ty^2)\, dy,$$

where $g(x, t) = e^x \cos(x + 2t) + e^{-x} \cos(x - 2t)$, and $a(y) = e^{-y^{4/3}} y \cos(\sqrt{3}y^{4/3})$.

Of course, uniqueness can be restored in (IVP) by imposing additional restrictions that will be sufficient to eliminate the above examples. We have seen in Corollary 2.2 that the Tikhonov growth condition restores not only *uniqueness* but also *continuous dependence* on the data (φ and f). We will now show that it is also sufficient to guarantee *existence* of the solution of the initial value problem

$$\begin{aligned} u_t - \Delta u &= 0 \qquad (\mathbf{x}, t) \in \mathbb{R}^n \times (0, T), \\ u(\mathbf{x}, 0) &= \varphi(\mathbf{x}) \qquad \mathbf{x} \in \mathbb{R}^n. \end{aligned} \tag{15}$$

[The inhomogeneous case, $f(\mathbf{x}, t) \neq 0$, will be considered in Section 4.] We first prove existence of a *bounded* solution for $t \in (0, +\infty)$ $(T = \infty)$.

Theorem 3.1. Suppose $\varphi \in C(\mathbb{R}^n)$ and is bounded, $|\varphi(\mathbf{x})| \leq K$. Then the solution of (15) is given by the Gauss (–Poisson) representation formula

$$u(\mathbf{x}, t) = \int_{\mathbb{R}^n} G(\mathbf{x} - \mathbf{y}, t)\varphi(\mathbf{y})\, d\mathbf{y} := G * \varphi \tag{16}$$

for all $t > 0$, where $G(\mathbf{x}, t) := [1/(4\pi t)^{n/2}]e^{-|\mathbf{x}|^2/4t}$ is the heat kernel [see equation (2) in Section 1]. u is continuous and bounded on $\mathbb{R}^n \times [0, \infty)$ and satisfies

$$|u(\mathbf{x}, t)| \leq \sup_{\mathbb{R}^n} |\varphi(\mathbf{y})| = K. \tag{17}$$

Moreover, $u \in C^\infty(\mathbb{R}^n \times (0, \infty))$.

Proof. Using properties (i)–(iii) of the heat kernel (see Section 1), we immediately see that $G(\mathbf{x} - \mathbf{y}, t)\varphi(\mathbf{y}) \in L^1(\mathbb{R}^n)$ for every $(\mathbf{x}, t)$ in $\mathbb{R}^n \times (0, \infty)$, and that $|u| \leq K$. If we now restrict $|\mathbf{x}| \leq R$, $t_1 \geq t \geq t_0 > 0$, we have the estimate

$$|G(\mathbf{x} - \mathbf{y}, t)\varphi(\mathbf{y})| \leq \frac{Ke^{-R^2/4t_o}}{(4\pi t_o)^{n/2}} e^{R|\mathbf{y}|/2t_0} e^{-|\mathbf{y}|^2/4t_1} := F_{\mathscr{K}}(\mathbf{y}), \tag{18}$$

showing that the integrand in (17) is dominated by an integrable function $F_{\mathscr{K}}(\mathbf{y}) \in L^1(\mathbb{R}^n)$ on any compact subset $\mathscr{K}$ of $\mathbb{R}^n \times (0, \infty)$. By Lebesgue's theorem we conclude that $u(\mathbf{x}, t)$ is a continuous function on $\mathbb{R}^n \times (0, \infty)$. Similar bounds can be found for any differentiated kernel $|D^m G(\mathbf{x} - \mathbf{y}, t)\varphi(\mathbf{y})|$

(with respect to x_i, t). It follows that every such differentiation can be done under the integral, so that $u \in C^\infty(\mathbb{R}^n \times (0, \infty))$. But then

$$(\partial/\partial t - \Delta)u = \int_{\mathbb{R}^n} \varphi(\mathbf{y})(\partial/\partial t - \Delta)G(\mathbf{x} - \mathbf{y}, t)\, d\mathbf{y} = 0$$

and u is a solution of the heat equation for $t > 0$.

We now verify that $\lim_{(\mathbf{x},t)\to(\mathbf{x}_0,0)} u(\mathbf{x}, t) = \varphi(\mathbf{x}_0)$ at any point $\mathbf{x}_0 \in \mathbb{R}^n$. Let $\varepsilon > 0$ be prescribed. As $\varphi(\mathbf{y})$ is continuous at $\mathbf{x}_0$, we can find $\delta > 0$ such that $|\varphi(\mathbf{y}) - \varphi(\mathbf{x}_0)| \leq \varepsilon$ for $|\mathbf{y} - \mathbf{x}_0| \leq \delta$. Then, using the properties of the heat kernel,

$$\begin{aligned}
|u(\mathbf{x}, t) - \varphi(\mathbf{x}_0)| &= \left| \int_{\mathbb{R}^n} G(\mathbf{x} - \mathbf{y}, t)\varphi(\mathbf{y})\, d\mathbf{y} - \varphi(\mathbf{x}_0) \right| \\
&= \left| \int_{\mathbb{R}^n} G(\mathbf{x} - \mathbf{y}, t)[\varphi(\mathbf{y}) - \varphi(\mathbf{x}_0)] d\mathbf{y} \right| \\
&\leq \int_{\mathbb{R}^n} G(\mathbf{x} - \mathbf{y}, t)|\varphi(\mathbf{y}) - \varphi(\mathbf{x}_0)| d\mathbf{y} \\
&= \int_{|\mathbf{y}-\mathbf{x}_0|<\delta} \cdots + \int_{|\mathbf{y}-\mathbf{x}_0|\geq\delta} \cdots .
\end{aligned}$$

As $|\varphi(\mathbf{y}) - \varphi(\mathbf{x}_0)| \leq \varepsilon$ in the first integral, $|\varphi(\mathbf{y}) - \varphi(\mathbf{x}_0)| \leq 2K$ in the second integral,

$$|u(\mathbf{x}, t) - \varphi(\mathbf{x}_0)| \leq \varepsilon + 2K \int_{|\mathbf{y}-\mathbf{x}_0|\geq\delta} G(\mathbf{x} - \mathbf{y}, t)\, d\mathbf{y}. \tag{19}$$

If we choose $|\mathbf{x} - \mathbf{x}_0| \leq \delta/2$, then $|\mathbf{y} - \mathbf{x}_0| \geq \delta$ implies $|\mathbf{y} - \mathbf{x}| \geq \delta/2$. Setting $\gamma = \delta/2$ in property (iii) of the heat kernel and letting $t \to 0$, we obtain the result. □

We said that uniqueness and continuous dependence follow from the results of the previous section. In fact, the a priori estimate in Corollary 2.2 shows that if $\varphi(\mathbf{x}) \equiv 0$, then $u(\mathbf{x}, t) \equiv 0$. Once uniqueness is established, continuous dependence also follows directly from the Gauss formula (16) [see (17)]. Thus, the initial value problem (15) is well posed in the class of bounded *continuous* solutions.

More generally, we will show that the initial value problem is also well posed in the class of *bounded* solutions, *discontinuous* at some points of the boundary, $t=0$.

Corollary 3.1. Suppose φ is bounded, $|\varphi(\mathbf{x})| \leq K$, and continuous on $\mathbb{R}^n$ with the exception of some points $\{\mathbf{x}_1, \ldots, \mathbf{x}_m\}$. Then the Gaussian representation formula (16) yields a bounded solution of the initial value problem (15) with the initial values taken on in the "a.e." sense,

$$\lim_{t\to 0} u(\mathbf{x}_0, t) = \varphi(\mathbf{x}_0) \qquad \text{at all points } \mathbf{x}_0 \text{ where } \varphi \text{ is continuous} \tag{20}$$

(i.e., for $\mathbf{x} \neq \mathbf{x}_i$). This "generalized" solution $u(\mathbf{x}, t)$ satisfies (17) and is unique in the class $C^\infty(\mathbb{R}^n \times (0, \infty)) \cap L^\infty(\mathbb{R}^n \times [0, \infty))$.

Proof. The fact that such a solution exists and is given by the Gauss formula (16) can be proven exactly as for Theorem 3.1. To demonstrate uniqueness, we need only show that *every* bounded solution is necessarily represented by (16). As u is continuous for $t > 0$, from Theorem 3.1 it follows that u can be represented as

$$u(\mathbf{x}, t) = \int_{\mathbb{R}^n} G(\mathbf{x} - \mathbf{y}, t - \varepsilon) u(\mathbf{y}, \varepsilon)\, d\mathbf{y}$$

for every $\varepsilon, 0 < \varepsilon \leq t_1 < t$. Let $\varepsilon \to 0$. As $G(\mathbf{x} - \mathbf{y}, t - \varepsilon)u(\mathbf{y}, \varepsilon) \to G(\mathbf{x} - \mathbf{y}, t)\varphi(\mathbf{y})$ almost everywhere and $|G(\mathbf{x} - \mathbf{y}, t - \varepsilon)u(\mathbf{y}, \varepsilon)| \leq e^{-|\mathbf{x}-\mathbf{y}|^2/4t}/(4\pi(t - t_1))^{n/2}$, Lebesgue's dominated convergence theorem implies

$$u(\mathbf{x}, t) = \int_{\mathbb{R}^n} \lim_{\varepsilon\to 0}[G(\mathbf{x} - \mathbf{y}, t - \varepsilon)u(\mathbf{y}, \varepsilon)]d\mathbf{y} = \int_{\mathbb{R}^n} G(\mathbf{x} - \mathbf{y}, t)\varphi(\mathbf{y})\, d\mathbf{y},$$

as asserted. Therefore, if $\varphi(\mathbf{x}) \equiv 0$, then $u(\mathbf{x}, t) \equiv 0$, and uniqueness follows. □

Observe that the solution is still smooth, even if the initial data are discontinuous. As a consequence of this corollary, the bounded solution of the Riemann problem for $u_t = a^2 u_{xx}$ written in Section 1 is unique. The example of the heat dipole $d(x, t)$ shows that the boundedness requirement (or some other condition at infinity, like Tikhonov's growth condition) is essential.

We can now prove a similar uniqueness result for the IBV problem for the quadrant

$$
\begin{aligned}
&v_t - v_{xx} = 0 \qquad x > 0, t > 0,\\
&\lim_{t\to 0} v(x, t) = \varphi(x)\ (x > 0), \qquad \lim_{x\to 0} v(x, t) = 0;\ (t > 0),
\end{aligned}
$$

where $\varphi(x) \in C([0, \infty))$, $|\varphi(x)| \leq K$, $\varphi(0) = \delta \neq 0$, see (12). We convert this problem to a pure initial value problem by extending $\varphi(x)$ to negative x as an odd function. As $\varphi(0) \neq 0$, the extended data will be discontinuous at the origin. The unique bounded solution is then given by the Gauss formula as

$$u(x, t) = \int_0^\infty (G(x - y, t) - G(x + y, t))\varphi(y)\, dy \tag{21}$$

(Exercise 3.3).

In some applications, the solution u of the initial value problem is not bounded, but satisfies the Tikhonov growth condition. Then the following variant of Theorem 3.1 applies.

Theorem 3.2. Suppose $\varphi \in C(\mathbb{R}^n)$ satisfies the growth condition

$$|\varphi(\mathbf{x})| \leq Ke^{A|\mathbf{x}|^2}$$

for some nonnegative constants, K, A. let T be fixed, with $0 < T < 1/4A$. Then (15) is well posed in the class of functions $u \in C(\mathbb{R}^n \times [0, T])$ satisfying a growth condition

$$|u(\mathbf{x}, t)| \leq K_1 e^{A_1|\mathbf{x}|^2} \tag{22}$$

with constants A_1, K_1 independent of t (but possibly depending on u). The solution is represented by the Gaussian formula (16), and is of class $C^\infty(\mathbb{R}^n \times (0, T])$.

Proof. The proof is very similar to that of Theorem 3.1 and we only summarize the main modifications. The details are left as an exercise.

a. The estimate

$$|G(\mathbf{x}-\mathbf{y},t)\varphi(\mathbf{y})| \le \frac{Ke^{-|\mathbf{x}-\mathbf{y}|^2/4t+A|\mathbf{y}|^2}}{(4\pi t)^{n/2}} \le \frac{Ke^{-(|\mathbf{x}|-|\mathbf{y}|)^2/4t+A|\mathbf{y}|^2}}{(4\pi t)^{n/2}}$$
$$\le \frac{Ke^{-|\mathbf{x}|^2/4t}}{(4\pi t)^{n/2}} e^{|\mathbf{x}||\mathbf{y}|/2t} e^{-(1/4T-A)|\mathbf{y}|^2}$$
$$\le \frac{Ke^{-|\mathbf{x}|^2/4t}}{(4\pi t)^{n/2}} e^{|\mathbf{x}||\mathbf{y}|/2t} e^{-(1/4T-A)|\mathbf{y}|^2}$$

shows that $G(\mathbf{x}-\mathbf{y},t)\varphi(\mathbf{y}) \in L^1(\mathbb{R}^n)$ for $(\mathbf{x},t) \in \mathbb{R}^n \times (0,T]$, with $4AT < 1$.

b. If we now restrict $|\mathbf{x}| \le R$, $T \ge t \ge t_0 > 0$, the estimate (18) becomes

$$|G(\mathbf{x}-\mathbf{y},t)\varphi(\mathbf{y})| \le \frac{Ke^{-R^2/4T}}{(4\pi t_0)^{n/2}} e^{R|\mathbf{y}|/2t_0} e^{-(1/4T-A)|\mathbf{y}|^2} := F_{\mathscr{K}}(\mathbf{y}) \in L^1(\mathbb{R}^n),$$

and we conclude as before that $u \in C^\infty(\mathbb{R}^n \times (0,T])$ and is a solution of the heat equation for $0 < t \le T$.

c. To verify that $\lim_{(\mathbf{x},t)\to(\mathbf{x}_0,0)} u(\mathbf{x},t) = \varphi(\mathbf{x}_0)$ at any point $\mathbf{x}_0 \in \mathbb{R}^n$, we proceed as before to obtain the inequality

$$|u(\mathbf{x},t)-\varphi(\mathbf{x}_0)| \le \varepsilon + \int_{|\mathbf{y}-\mathbf{x}_0|\ge\delta} G(\mathbf{x}-\mathbf{y},t)|\varphi(\mathbf{y})-\varphi(\mathbf{x}_0)|d\mathbf{y},$$

which replaces the inequality (19). Choosing $|\mathbf{x}-\mathbf{x}_0| \le \delta/2$, from the fact that $G\varphi \in L^1(\mathbb{R}^n)$ [point (a)] and the absolute continuity of the Lebesgue integral, we have

$$\int_{|\mathbf{y}-\mathbf{x}_0|\ge\delta} G(\mathbf{x}-\mathbf{y},t)|\varphi(\mathbf{y})-\varphi(\mathbf{x}_0)|d\mathbf{y}$$
$$\le \varepsilon + \int_{|\mathbf{y}-\mathbf{x}|\ge\delta} G(\mathbf{x}-\mathbf{y},t)|\varphi(\mathbf{y})-\varphi(\mathbf{x}_0)|d\mathbf{y}$$

if δ is small enough. The growth condition for φ implies

$$|\varphi(\mathbf{y})-\varphi(\mathbf{x}_0)| \le Ke^{A(|\mathbf{x}|+|\mathbf{y}-\mathbf{x}|)^2} + Ke^{A|\mathbf{x}_0|^2}.$$

If we now define the new variable $\rho = |\mathbf{x} - \mathbf{y}|/2\sqrt{t}$ and use the fact that $t \leq T$ and $|\mathbf{x}| \leq |\mathbf{x} - \mathbf{x}_0| + |\mathbf{x}_0| \leq (\delta/2 + |\mathbf{x}_0|) := C_1$, we find

$$G(\mathbf{x} - \mathbf{y}, t)|\varphi(\mathbf{y}) - \varphi(\mathbf{x}_0)| \leq \frac{Ke^{-\rho^2}}{(4\pi t)^{n/2}}(C_2 + e^{4AT\rho^2 + AC_1^2 + 4A\sqrt{T}C_1\rho})$$
$$\leq \frac{K(C_2+1)}{(4\pi t)^{n/2}} \exp(-\rho^2 + 4AT\rho^2 + AC_1^2 + 4A\sqrt{T}C_1\rho),$$

where $C_2 = e^{A|\mathbf{x}_0|^2}$. The estimate

$$\int_{|\mathbf{y}-\mathbf{x}|\geq\delta} G(\mathbf{x} - \mathbf{y}, t)|\varphi(\mathbf{y}) - \varphi(\mathbf{x}_0)|d\mathbf{y}$$
$$\leq C_3 \int_{\delta/2\sqrt{t}}^{+\infty} \exp(-\rho^2(1 - 4AT) + AC_1^2 + 4A\sqrt{T}C_1\rho)\rho^{n-1}\, d\rho$$

follows, with $C_3 = 2^n\omega_n K(C_2 + 1)/(4\pi)^{n/2}$. As $4AT < 1$, the integral on the right-hand side tends to zero as $t \to 0$.

d. It remains to prove that u satisfies (22). Setting $\mathbf{y} - \mathbf{x} = 2\sqrt{t}\mathbf{z}$, $K' = 1 - 4AT$, $K'' = 2A(T/(K' - \varepsilon))^{1/2}$, we find the estimate

$$|u(\mathbf{x}, t)| \leq \frac{K}{(\pi)^{n/2}} \int_{\mathbb{R}^n} e^{A|\mathbf{x}+2\sqrt{t}\mathbf{z}|^2 - |\mathbf{z}|^2}\, d\mathbf{z} \leq \frac{Ke^{A|\mathbf{x}|^2}}{(\pi)^{n/2}} \int_{\mathbb{R}^n} e^{4A\sqrt{T}|\mathbf{x}||\mathbf{z}| - K'|\mathbf{z}|^2}\, d\mathbf{z}$$
$$\leq \frac{Ke^{A_1|\mathbf{x}|^2}}{(\pi)^{n/2}} \int_{\mathbb{R}^n} e^{-\varepsilon|\mathbf{z}|^2} e^{-(\sqrt{K'-\varepsilon}|\mathbf{z}| - k''|\mathbf{x}|)^2}\, d\mathbf{z} \leq K_1 e^{A_1|\mathbf{x}|^2},$$

where

$$A_1 = A(1 - \varepsilon)/(1 - 4AT - \varepsilon), \qquad K_1 = \frac{K}{(\pi)^{n/2}} \int_{\mathbb{R}^n} e^{-\varepsilon|\mathbf{z}|^2}\, d\mathbf{z},$$

and ε is any fixed number satisfying $0 \leq \varepsilon < 1 - 4AT$. □

Remarks.

3.3. The a priori estimate in Corollary 2.2 still implies the inequality

$$|u(\mathbf{x}, t)| \leq \sup_{\mathbb{R}^n} |\varphi(\mathbf{y})| \equiv \|\varphi\|_{L^\infty(\mathbb{R}^n)}.$$

Hence, if $\|\varphi\|_{L^\infty(\mathbb{R}^n)} < \infty$ u is bounded, if $\|\varphi\|_{L^\infty}$ is small $|u|$ is small (continuous dependence), and if φ is zero u is zero (uniqueness).

3.4. The Gaussian representation (16) shows that the solution $u(\mathbf{x}, t)$ depends on all of the values of the data $\varphi(\mathbf{y})$ for $t > 0$, hence the *domain of dependence* $\mathscr{D}(\mathbf{x}, t)$ is all of $\mathbb{R}^n$ and the *domain of influence* $\mathscr{I}(K)$ of any subset K of $\mathbb{R}^n$ is all of $\mathbb{R}^n \times (0, T]$. The support of $u(\mathbf{x}, t)$ with respect to the $\mathbf{x}$-variable "expands in time with infinite speed" (see, however, the remarks in Section 1).

3.5. The solution u is bounded for every $0 \le t \le T$ and $\mathbf{x}$ in compact subsets of $\mathbb{R}^n$; for example,

$$|u(\mathbf{x}, t)| \le K_1 e^{A_1 R^2} \qquad \text{for } |\mathbf{x}| \le R,\ 0 \le t \le T,$$

for each $R > 0$.

3.6. The initial value problem that runs *backwards* in time is not well posed. If a solution exists in some strip $\mathbb{R}^n \times [-T, 0]$, then, from Theorem 3.2, we must have $u(\mathbf{x}, 0) \in C^\infty(\mathbb{R}^n)$. Even in this case, however, continuous dependence on the data can be lost as shown by the example

$$u(x, t) = \varepsilon e^{-t/\varepsilon^2} \sin(x/\varepsilon).$$

For $\varepsilon > 0$, $u(x, t)$ is an analytic solution of $u_t = u_{xx}$ with initial data $\varphi(x) = \varepsilon \sin(x/\varepsilon)$. As $\varepsilon \to 0$, we have $\sup_{\mathbb{R}} |\varphi(x)| \to 0$, but $\lim\sup_{\varepsilon\to 0} u(x, t) \to \infty$ for every $t < 0$ and every $x \neq 0$.

Note that Corollary 3.1 can be extended to other classes of "generalized solutions" allowing even less initial regularity for u. We may consider, for example, solutions $u(\mathbf{x}, t) \in L^2(\mathbb{R}^n)$ for every $t \ge 0$ such that the initial values are taken on in the sense of the L^2 norm

$$\lim_{t\to 0} \int_{\mathbb{R}^n} |u(\mathbf{y}, t) - \varphi(\mathbf{y})|^2 d\mathbf{y} = 0. \tag{24}$$

These L^2 solutions still are given by the Gaussian representation formula (16), are of the class $C^\infty(\mathbb{R}^n \times (0, \infty))$, are unique and depend continuously on φ in the L^2 norm. In fact, suppose $\varphi \in L^2(\mathbb{R}^n)$. Taking the Fourier transform $\hat{u}(\xi, t)$, from the heat equation we obtain $\hat{u}_t(\xi, t) = -|\xi|^2 \hat{u}(\xi, t)$, hence $\hat{u}(\xi, t) = \hat{\varphi}(\xi) e^{-|\xi|^2 t}$. As

$$\hat{G}(\xi, t) = e^{-|\xi|^2 t}$$

(see Section 1), the inverse Fourier transform

$$u(\mathbf{x}, t) = (2\pi)^{-n} \int_{\mathbb{R}^n} e^{i\mathbf{x}\cdot\xi} e^{-|\xi|^2 t} \hat{\varphi}(\xi)\, d\xi$$

together with the convolution theorem retrieves the Gaussian representation (16). Moreover,

$$\|u(t)\|_s^2 = \int_{\mathbb{R}^n} (1 + |\xi|^2)^s |\hat{u}(\xi, t)|^2 d\xi = \int_{\mathbb{R}^n} (1 + |\xi|^2)^s |\hat{\varphi}(\xi)|^2 e^{-2|\xi|^2 t_1}\, d\xi \quad (25)$$

is finite for any $s \geq 0$ and $t > 0$. In particular, the function u obtained by Fourier inversion is a C^∞ function of $\mathbf{x}$ for each fixed $t > 0$. As

$$\hat{u}(\xi, t_2) - \hat{u}(\xi, t_1) = \hat{\varphi}(\xi)(e^{-|\xi|^2 t_2} - e^{-|\xi|^2 t_1}),$$

Plancherel's theorem and Lebesgue's dominated convergence theorem imply that $u \in C([0, T], L^2)$ for any $T > 0$ and $u \in C((b, T), H^s)$ for any $T > b > 0$, $s \geq 0$, [see Chapter 8]. In particular, $\|u(\mathbf{x}, t) - \varphi(\mathbf{x})\|_0^2 \to 0$ as $t \to 0$, so that the initial values are taken on in the L^2 sense (24). We can rewrite (25) to get more useful information,

$$\|u(t)\|_s^2 = \int_{\mathbb{R}} (1 + |\xi|^2)^{s-r} |\hat{\varphi}(\xi)|^2 g(\xi, t)\, d\xi,$$

where $s \geq r \geq 0$ and $g(\xi, t) = (1 + |\xi|^2)^r e^{-2|\xi|^2 t}$. A first easy result is $\|u(t)\|_s \leq \|\varphi\|_s$ obtained by taking $r = s$. Considering the function $g(\xi, t)$, we find $g(0, t) = 1$, $g(\infty, t) = 0$, and for $r < 2t$, $g(\xi, t)$ is decreasing for $|\xi| \geq 0$, whereas for $r > 2t$ it has a maximum at $|\xi| = \bar{\xi}$, $1 + \bar{\xi}^2 = r/2t$, which is less than or equal to $(r/2t)^r > 1$. Hence, we have that

$$\|u(t)\|_s^2 \leq \max\{1, (r/2t)^r\} \|\varphi\|_{s-r}^2.$$

If we write $s - r = \sigma$, this becomes

$$\|u(t)\|_{\sigma+r}^2 \leq \max\{1, (r/2t)^r\} \|\varphi\|_\sigma^2$$

for all $\sigma \geq 0$, $r \geq 0$, $t > 0$. This shows another aspect of the smoothing property of (H).

A further and more drastic generalization of the concept of solution will be presented in Chapter 6.

We have seen that u given by (16) is C^∞ for $t > 0$. We can also give results about analyticity of these solutions. For simplicity we restrict our attention to one space dimension.

Lemma 3.1. Suppose D is a domain in the complex plane and $F(z, y)$ is continuous for $(z, y) \in D \times I$, I an interval on the real axis, and F is analytic as a function of z for each fixed y. Suppose that

$$\int_I |F(z, y)| dy \tag{Hp}$$

converges uniformly for z varying over any compact subset of D. Then

$$G(z) = \int_I F(z, y)\, dy$$

is analytic in D.

Proof. For $J \subset I$ compact and $K \subset D$ compact, $\int_J |F(z, y)| dy$ is analytic on K as it is the uniform limit of Riemann sums. The hypothesis (Hp) implies that G is the uniform limit of analytic functions on K, hence analytic. $\square$

Suppose now that we extend x in (13) to complex values $z = x + i\xi$. We write

$$F(z, y) = \exp(\xi^2/4t) G(x - y, t)\varphi(y) \exp(-2i(x - y)\xi/4t)$$

and observe that $|F(z, y)| \leq \exp(\xi^2/4t) G(x - y, t)|\varphi(y)|$. Lemma 3.1 then immediately implies the following result.

Theorem 3.3. Suppose that

$$|\varphi(y)| \leq K \exp(A|y|^{1+\alpha}), \qquad 0 \leq \alpha < 1.$$

Then, for each $t > 0$,

$$u(z, t) = \int_{-\infty}^{\infty} G(z - y, t)\varphi(y)\, dy$$

is an analytic function of z for all z.

We also observe that if φ is bounded, i.e., $A = 0$, $|u(z, t)| \leq k \exp(\xi^2/4t)$.

We can also give a result on analyticity of u in t. Let $\eta = t + i\tau$ and, for $\mu > 0$,

$$D_\mu = \{\eta : |\tau| < \mu t\}.$$

Choose $\eta^{1/2}$ so that it reduces to $\sqrt{t}$ when $\tau = 0$. Then, as $|\eta| \leq (1+\mu^2)^{1/2}|t|$, and $\exp(-(x-y)^2/4\eta) \leq \exp[-(x-y)^2/4(1+\mu^2)t]$ for $\eta \in D_\mu$, we have

$$|G(x-y, \eta)| \leq G(x-y, (1+\mu^2)t)(1+\mu^2)^{1/2}.$$

The integral defined $u(x, \eta)$ is therefore majorized by

$$\int_{-\infty}^{\infty} G(x-y, (1+\mu^2)t)|\varphi(y)|\,dy.$$

Theorem 3.4. If φ satisfies the hypothesis of Theorem 3.3, then

$$u(z, \eta) = \int_{-\infty}^{\infty} G(z-y, \eta)\varphi(y)\,dy$$

is an analytic function of η for $\eta \in D_\mu$ for each fixed z.

A solution of (H1) is not generally an analytic function of (x, t), however. In particular, Tikhonov's solution (13) vanishes for $t = 0$ and is not analytic in a neighborhood of any point on $t = 0$.

4. Inhomogeneous Initial Value Problem

We will consider now the initial value problem (IVP) for the inhomogeneous equation with zero initial data $\varphi(\mathbf{x})$. In order to deal with this problem we need to get some estimates about the behavior of the Gauss integral and its derivatives as t approaches zero (and t approaches $+\infty$). These estimates are of interest in their own right and are crucial for many applications.

Lemma 4.1. Suppose that $u \in C(\mathbb{R}^n \times [0, \infty)) \cap C^\infty(\mathbb{R}^n \times (0, \infty))$ is the unique bounded solution of

$$\begin{aligned} &u_t - \Delta u = 0, && (\mathbf{x}, t) \in \mathbb{R}^n \times (0, T), \\ &u(\mathbf{x}, 0) = \varphi(\mathbf{x}), && \mathbf{x} \in \mathbb{R}^n, \end{aligned}$$

given by the Gauss convolution integral (16),

$$u(\mathbf{x}, t) = G * \varphi \equiv \int_{\mathbb{R}^n} G(\mathbf{x}-\mathbf{y}, t)\varphi(\mathbf{y})\,d\mathbf{y}, \tag{26}$$

under the conditions of Theorem 3.1, in particular $|\varphi(\mathbf{x})| \leq K$. Then there is a positive constant M_1 such that

i. $|u_{x_i}(\mathbf{x}, t)| \leq M_1 t^{-1/2}$ for $\mathbf{x} \in \mathbb{R}^n$, $t > 0$.

If, in addition, φ satisfies a uniform Hölder condition

$$|\varphi(\mathbf{x}) - \varphi(\mathbf{y})| \leq \alpha |\mathbf{x} - \mathbf{y}|^{\mu}, \qquad 0 \leq \mu \leq 1,$$

on compact subsets of $\mathbb{R}^n$, then there are constants M_2, M_3, M_4, M_5, M_6 such that

ii. $|u_{x_i x_i}(\mathbf{x}, t)| \leq M_2 t^{-1+\mu/2} + M_4$,
iii. $|u_t(\mathbf{x}, t)| \leq M_3 t^{-1+\mu/2} + M_5$, and, if $\mu = 1$, also
iv. $|u_{x_i}(\mathbf{x}, t)| \leq M_6$

for every $t > 0$ and for every $\mathbf{x}$ in compact subsets of $\mathbb{R}^n$.

Proof. We have already shown in Theorem 3.1 that $|u(\mathbf{x}, t)| \leq K$ for $t \geq 0$. We denote below by M a generic positive constant that depends only on numerical factors and may differ in different formulas. As differentiation under the integral sign in the Gaussian representation formula is permissible, we have

$$|u_{x_i}| = \left| \int_{\mathbb{R}^n} G_{x_i}(\mathbf{x} - \mathbf{y}, t)\varphi(\mathbf{y})\, d\mathbf{y} \right| = \left| \frac{1}{(4\pi t)^{n/2}} \int_{\mathbb{R}^n} \frac{y_i - x_i}{2t} e^{-|\mathbf{x}-\mathbf{y}|^2/4t} \varphi(\mathbf{y})\, d\mathbf{y} \right|$$

for $t > 0$, and setting $|\mathbf{x} - \mathbf{y}| = 2\sqrt{t}\rho$ we obtain

$$|u_{x_i}| \leq MKt^{-n/2-1} \int_{\mathbb{R}^n} |\mathbf{x} - \mathbf{y}| e^{-|\mathbf{x}-\mathbf{y}|^2/4t}\, d\mathbf{y} \leq \frac{MK}{\sqrt{t}} \int_0^{\infty} \rho^n e^{-\rho^2}\, d\rho = \frac{M_1}{\sqrt{t}}.$$

This proves (i). As $\int_{\mathbb{R}^n} G(\mathbf{x} - \mathbf{y}, t)\, d\mathbf{y} = 1$, we have

$$\int_{\mathbb{R}^n} G_{x_i}(\mathbf{x} - \mathbf{y}, t)\, d\mathbf{y} = \frac{\partial}{\partial x_i} \int_{\mathbb{R}^n} G(\mathbf{x} - \mathbf{y}, t)\, d\mathbf{y} = 0 \tag{27}$$

and

$$\int_{\mathbb{R}^n} G_{x_i x_i}(\mathbf{x} - \mathbf{y}, t)\, d\mathbf{y} = \frac{\partial^2}{\partial x_i^2} \int_{\mathbb{R}^n} G(\mathbf{x} - \mathbf{y}, t)\, d\mathbf{y} = 0. \tag{28}$$

We can then write, using equation (27),

$$|u_{x_i}| = \left| \int_{\mathbb{R}^n} G_{x_i}(\mathbf{x}-\mathbf{y}, t)(\varphi(\mathbf{y}) - \varphi(\mathbf{x}))\, d\mathbf{y} \right|,$$

and if φ is Lipschitz continuous, we get the estimate

$$|u_{x_i}| \leq M\alpha t^{-n/2-1} \int_{|\mathbf{x}-\mathbf{y}|\leq R} |\mathbf{x}-\mathbf{y}|^2 e^{-|\mathbf{x}-\mathbf{y}|^2/4t}\, d\mathbf{y}$$
$$+ MKt^{-n/2-1} \int_{|\mathbf{x}-\mathbf{y}|>R} |\mathbf{x}-\mathbf{y}| e^{-|\mathbf{x}-\mathbf{y}|^2/4t}\, d\mathbf{y}$$

for every $|\mathbf{x}| \leq R$, where $R > 0$ is any constant and $\alpha = \alpha(R)$. Performing the change of variable $|\mathbf{x}-\mathbf{y}| = 2\sqrt{t}\rho$ we obtain

$$|u_{x_i}| \leq M\alpha \int_0^\infty \rho^{n+1} e^{-\rho^2}\, d\rho + \frac{MK}{\sqrt{t}} \int_{R/2\sqrt{t}}^\infty \rho^n e^{-\rho^2}\, d\rho.$$

As $t^{-1/2} \leq 2\rho/R$ in the second integral, we have

$$|u_{x_i}| \leq M\alpha \int_0^\infty \rho^{n+1} e^{-\rho^2}\, d\rho + \frac{MK}{R} \int_0^\infty \rho^{n+1} e^{-\rho^2}\, d\rho \leq M_6,$$

which proves (iv) for every $|\mathbf{x}| \leq R,\ t > 0$, with $M_6 = M_6(R)$.

Suppose $|\varphi(\mathbf{x}) - \varphi(\mathbf{y})| \leq \alpha|\mathbf{x}-\mathbf{y}|^\mu,\ 0 \leq \mu \leq 1$, on compact subsets of $\mathbb{R}^n$. Note first that for $t > 0$ we have

$$|G_{x_i x_i}| = \left| \left(\frac{(y_u - x_i)^2}{4t^2} - \frac{1}{2t} \right) G \right| \leq \frac{1}{t(4\pi t)^{n/2}} \left(\frac{|\mathbf{x}-\mathbf{y}|^2}{4t} + \frac{1}{2} \right) e^{-|\mathbf{x}-\mathbf{y}|^2/4t}$$

$(i = 1, \ldots, n)$. For any $\mathbf{x} \in \mathbb{R}^n,\ t > 0$, using equation (28) we can then estimate

$$|u_{x_i x_i}(\mathbf{x}, t)| = \left| \int_{\mathbb{R}^n} G_{x_i x_i}(\mathbf{x}-\mathbf{y}, t)(\varphi(\mathbf{y}) - \varphi(\mathbf{x}))\, dy \right|$$
$$\leq \frac{1}{t(4\pi t)^{n/2}} \int_{\mathbb{R}^n} \left(\frac{|\mathbf{x}-\mathbf{y}|^2}{4t} + \frac{1}{2} \right) e^{-|\mathbf{x}-\mathbf{y}|^2/4t} |\varphi(\mathbf{y}) - \varphi(\mathbf{x})| d\mathbf{y}.$$

We write the integral as a sum of integrals over $|\mathbf{x}-\mathbf{y}| \leq R$ and $|\mathbf{x}-\mathbf{y}| > R$, the corresponding terms on the right being denoted as I_1 and I_2. The Hölder condition assumed for φ implies

$$I_1 \leq \frac{\alpha(R)}{t(4\pi t)^{n/2}} \int_{|\mathbf{x}-\mathbf{y}|\leq R} \left(\frac{|\mathbf{x}-\mathbf{y}|^2}{4t} + \frac{1}{2} \right) e^{-|\mathbf{x}-\mathbf{y}|^2/4t} |\mathbf{x}-\mathbf{y}|^\mu\, dy$$

for $|\mathbf{x}| \leq R, t > 0$. Setting $\rho = |\mathbf{x} - \mathbf{y}|/2\sqrt{t}$, we get

$$I_1 \leq M\alpha(R)t^{-1+\mu/2} \int_0^\infty \left(\rho^2 + \frac{1}{2}\right) e^{-\rho^2} \rho^{\mu+n-1}\, d\rho := M_2 t^{-1+\mu/2}.$$

In order to bound I_2 we observe that $|\varphi(\mathbf{y}) - \varphi(\mathbf{x})| \leq 2K$, and with ρ as above we obtain

$$I_2 \leq \frac{MK}{t} \int_{R/2\sqrt{t}}^\infty \left(\rho^2 + \frac{1}{2}\right) e^{-\rho^2} \rho^{n-1}\, d\rho \leq \frac{MK}{R^2} \int_0^\infty \left(\rho^2 + \frac{1}{2}\right) e^{-\rho^2} \rho^{n+1}\, d\rho := M_4$$

as $t^{-1} \leq 4\rho^2/R^2$ in this integral. Combining the inequalities for I_1 and I_2 we obtain (ii) for every $|\mathbf{x}| \leq R, t > 0$, with $M_2 = M_2(\mu, R)$, $M_4 = M_4(R)$.

The inequality (iii) for u_t follows from the differential equation. □

Remarks

4.1. If φ is bounded and Lipschitz continuous, then $|u_{x_i}(\mathbf{x}, t)|$are bounded (locally in $\mathbf{x}$) for $t > 0$ and tend to zero (uniformly in $\mathbf{x}$) as $t \to +\infty$.

4.2. If, in addition, the partial derivatives φ_{x_i} are bounded and Lipschitz continuous, then u is C^1 up to the boundary, $u \in C^1(\mathbb{R}^n \times [0, \infty))$ (Exercise 4.5.).

We can prove a variant of this lemma with the boundedness condition replaced by the Tikhonov growth condition. This is useful in certain applications.

Lemma 4.2. Suppose that $u \in C(\mathbb{R}^n \times [0, T]) \cup C^\infty(\mathbb{R}^n \times (0, T])$ is the unique solution of

$$\begin{aligned} u_t - \Delta u &= 0, & (\mathbf{x}, t) &\in \mathbb{R}^n \times (0, T), \\ u(\mathbf{x}, 0) &= \varphi(\mathbf{x}), & \mathbf{x} &\in \mathbb{R}^n, \end{aligned}$$

for $T < 1/4A$, given by the Gauss formula (26) under the conditions of Theorem 3.2, in particular $|\varphi(\mathbf{x})| \leq Ke^{A|\mathbf{x}|^2}$. Then for each $R > 0$, there is a constant $M_1 = M_1(A, T, R)$ such that (i) holds for $|\mathbf{x}| \leq R,\ 0 < t \leq T$. If, in addition, for each $R > 0$ φ satisfies the Hölder condition $|\varphi(\mathbf{x}) - \varphi(\mathbf{y})| \leq K|\mathbf{x} - \mathbf{y}|^\mu$, $0 \leq \mu \leq 1$, for $|\mathbf{x}|, |\mathbf{y}| \leq R$ and $K = K(R)$, then there are constants $M_i = M_i(A, T, R, \mu)$, $i = 2, \ldots, 5$, such that (ii) and (iii) hold for $|\mathbf{x}| \leq R,\ 0 < t \leq T$.

Proof. The proof is very similar to that of Lemma 4.1 and we only summarize the main modifications. The details are left as an exercise.

i. We have already shown, in equation (23), that $|u(\mathbf{x}, t)| \leq M_0 := k_1 e^{A_1 R^2}$ for $|\mathbf{x}| \leq R,\ 0 \leq t \leq T$. For $|\mathbf{x}| \leq R,\ 0 < t \leq T$, we have the inequality for $|u_{x_i}|$

$$|u_{x_i}(\mathbf{x}, t)| \leq M t^{-n/2+1} \int_{\mathbb{R}^n} |\mathbf{x} - \mathbf{y}| \exp\left(-\frac{|\mathbf{x} - \mathbf{y}|^2}{4t} + A|\mathbf{y}|^2 \right) d\mathbf{y}.$$

By majorizing $e^{A|\mathbf{y}|^2} \leq e^{A(|\mathbf{y}-\mathbf{x}|^2 + R^2 + 2R|\mathbf{y}-\mathbf{x}|}$ and setting $|\mathbf{x} - \mathbf{y}| = 2\sqrt{t}\rho$, we obtain

$$|u_{x_i}| \leq \frac{M}{\sqrt{t}} \int_0^\infty \rho^n \exp(-(1 - 4AT)\rho^2 + 4A\sqrt{T}R\rho + AR^2)\, d\rho = \frac{M_1(A, T, R)}{\sqrt{t}},$$

where the constant $M_1 = M_1(A, T, R)$ is finite owing to the condition $T < 1/4A$.

ii. Suppose $|\varphi(\mathbf{x}) - \varphi(\mathbf{y})| \leq \alpha |\mathbf{x} - \mathbf{y}|^\mu,\ 0 \leq \mu \leq 1$, on compact subsets of $\mathbb{R}^n$. The estimate

$$\begin{aligned}
|u_{x_i x_i}(\mathbf{x}, t)| &\leq \frac{1}{t(4\pi t)^{n/2}} \int_{|\mathbf{x}-\mathbf{y}| \leq R} \left(\frac{|\mathbf{x} - \mathbf{y}|^2}{4t} + \frac{1}{2} \right) e^{-|\mathbf{x}-\mathbf{y}|^2/4t} |\varphi(\mathbf{y}) - \varphi(\mathbf{x})| d\mathbf{y} \\
&\quad + \frac{1}{t(4\pi t)^{n/2}} \int_{|\mathbf{x}-\mathbf{y}| > R} \left(\frac{|\mathbf{x} - \mathbf{y}|^2}{4t} + \frac{1}{2} \right) e^{-|\mathbf{x}-\mathbf{y}|^2/4t} |\varphi(\mathbf{y}) - \varphi(\mathbf{x})| d\mathbf{y} \\
&:= I_1 + I_2
\end{aligned}$$

holds for $|\mathbf{x}| \leq R,\ 0 < t \leq T$, with $I_1 \leq M_2(\mu, R) t^{-1+\mu/2}$ as before. In order to bound I_2 we observe that, for $|\mathbf{x}| \leq R$,

$$|\varphi(\mathbf{y}) - \varphi(\mathbf{x})| \leq |\varphi(\mathbf{y})| + |\varphi(\mathbf{x})| \leq K e^{AR^2} + K e^{A|\mathbf{y}|^2} \leq K' e^{A|\mathbf{y}|^2},$$

where $K' = K'(R)$. Then, with $\rho = |\mathbf{x} - \mathbf{y}|/2\sqrt{t}$,

$$I_2 \leq \frac{MK'(R)}{t} \int_{R/2\sqrt{t}}^\infty \left(\rho^2 + \frac{1}{2} \right) e^{-\rho^2 + A(4t\rho^2 + R^2 + 4R\sqrt{t}\rho)} \rho^{\mu+n-1}\, d\rho,$$

where $|\mathbf{y}|^2 \leq 2R|\mathbf{x} - \mathbf{y}| + R^2 + 4t\rho^2 = 4R\sqrt{t}\rho + R^2 + 4t\rho^2$ has been used. As $t^{-1} \leq 4\rho^2/R^2$ in this integral,

$$I_2 \leq \frac{MK'(R)}{R^2} e^{AR^2} \int_0^\infty \left(\rho^2 + \frac{1}{2} \right) e^{-(1-4AT)\rho^2 + 4AR\sqrt{T}\rho} \rho^{\mu+n-1}\, d\rho := M_4(A, T, R, \mu).$$

Combining the inequalities obtained for I_1 and I_2, we have (ii). The inequality (iii) for u_t follows again from the differential equation. □

We consider the inhomogeneous initial value problem

$$\begin{aligned} &u_t - \Delta u = f(\mathbf{x}, t), \qquad (\mathbf{x}, t) \in \mathbb{R}^n \times (0, T), \\ &u(\mathbf{x}, 0) = 0, \qquad \mathbf{x} \in \mathbb{R}^n, \end{aligned} \tag{29}$$

and we suppose that, for each $R > 0$, $f(\mathbf{x}, t)$ satisfies the hypotheses

$$\begin{aligned} &f(\mathbf{x}, t) \in C(\mathbb{R} \times [0, T]), \\ &|f(\mathbf{x}, t) - f(\mathbf{y}, t)| \leq \alpha|\mathbf{x} - \mathbf{y}|^{\mu}, \ 0 < \mu \leq 1, \text{ for } |\mathbf{x}|, |\mathbf{y}| \leq R; \text{ and all } 0 \leq t \leq T, \\ &|f(\mathbf{x}, t)| \leq K_0 e^{A|\mathbf{x}|^2} \text{ for every } \mathbf{x} \in \mathbb{R}^n, \ 0 \leq t \leq T, \end{aligned}$$

where $\alpha = \alpha(R)$, $K_0 \geq 0$, and $T < 1/4A$.

Theorem 4.1. If f satisfies (F), then

$$u(\mathbf{x}, t) = \int_0^t d\tau \int_{\mathbb{R}^n} G(\mathbf{x} - \mathbf{y}, t - \tau) f(\mathbf{y}, \tau) \, d\mathbf{y} \tag{D}$$

is a solution of (29) for $0 < t \leq T \leq 14A$ and is unique in the class of functions $u \in C(\mathbb{R}^n \times [0, T])$ satisfying the growth condition $|u(\mathbf{x}, t)| \leq K_1 e^{A_1|\mathbf{x}|^2}$ [Section 2, equation (10)] and such that u_t, $u_{x_i x_i} \in C(\mathbb{R}^n \times (0, T])$.

Proof. Uniqueness has already been proved in Section 3. In order to show that (D) is the required solution, consider, for fixed τ, $0 \leq \tau \leq T$, the problem

$$\begin{aligned} &v_t - \Delta v = 0, \qquad (\mathbf{x}, t) \in \mathbb{R}^n \times [0, T], \\ &v(\mathbf{x}, 0; \tau) = f(\mathbf{x}, \tau) \end{aligned}$$

for $v(\mathbf{x}, t; \tau)$. From our previous results,

$$v(\mathbf{x}, t; \tau) = \int_{\mathbb{R}^n} G(\mathbf{x} - \mathbf{y}) f(\mathbf{y}, \tau) \, d\mathbf{y}$$

and we have to prove that the integral

$$u(\mathbf{x}, t) = \int_0^t v(\mathbf{x}, t - \tau; \tau) \, d\tau \tag{30}$$

gives the required solution (Duhamel principle). As v satisfies the growth condition $|v| \leq K_2 e^{A_1|\mathbf{x}|^2}$ for some constant $K_2 > 0$, we see that the Duhamel integral is finite and $|u(\mathbf{x}, t)| \leq tK_2 e^{A_1|\mathbf{x}|^2}$. Hence, $u(\mathbf{x}, 0) = 0$ and $|u(\mathbf{x}, t)| \leq K_1 e^{A_1|\mathbf{x}|^2}$ with $K_1 = TK_2$. From Lemma 4.2, we can assert, for each compact set $K \subset \mathbb{R}^n$, the existence of constants A, B, and C, depending only on K, A, T, and μ, such that

$$|v_{x_i}(\mathbf{x}, t; \tau)| \leq \frac{A}{\sqrt{t}}; \qquad |v_{x_i x_i}(\mathbf{x}, t; \tau)|, |v_t(\mathbf{x}, t; \tau)| \leq Bt^{-1+\mu/2} + C$$

for every $(\mathbf{x}, t) \in K \times (0, T]$. This implies that in each of the integrals

$$\int_0^t v_{x_i}(\mathbf{x}, t-\tau; \tau)\, d\tau, \qquad \int_0^t v_{x_i x_i}(\mathbf{x}, t-\tau; \tau)\, d\tau, \qquad \int_0^t v_t(\mathbf{x}, t-\tau; \tau)\, d\tau,$$

the integrand function is dominated by a summable function of $\eta := t - \tau$ for all $\mathbf{x}, t$ in each compact subset of $\mathbb{R}^n \times (0, T]$. By the Lebesgue dominated convergence theorem, the Duhamel integral (30) can then be differentiated under the integral sign for $t > 0$ and $\mathbf{x} \in \mathbb{R}^n$. Then

$$\begin{aligned} u_t(\mathbf{x}, t) &= \int_0^t v_t(\mathbf{x}, t-\tau; \tau)\, d\tau + v(\mathbf{x}, 0; t) = \int_0^t \Delta v(\mathbf{x}, t-\tau; \tau)\, d\tau + f(\mathbf{x}, t) \\ &= \Delta u(\mathbf{x}, t) + f(\mathbf{x}, t) \end{aligned}$$

and the theorem is proven. □

The expression (D) is the Duhamel integral for this problem.

5. Initial–Boundary Value Problems

We consider in this section solutions $u(\mathbf{x}, t)$ of the heat equation in a cylindrical domain $D = \Omega \times (0, +\infty)$ satisfying an initial condition for $\mathbf{x} \in \Omega$, $t = 0$ and a homogeneous boundary condition for $\mathbf{x} \in \partial\Omega$, $t > 0$, where $\Omega \subset \mathbb{R}^n$ is an open bounded set. The boundary data for $(\mathbf{x}, t) \in \partial\Omega \times (0, +\infty)$will be of Dirichlet, Neumann, or Robin type:

$$\begin{aligned} &\text{i. } u = 0, \\ &\text{ii. } \partial u/\partial n = 0, \\ &\text{iii. } \partial u/\partial n = -\alpha u \ (\alpha \geq 0), \end{aligned} \tag{31}$$

where $\mathbf{n}$ is the outer normal to $\partial\Omega$ in $\mathbb{R}^n$. The physical interpretation of these boundary conditions in the context of heat conduction phenomena is explained in Chapter 1. In the case of Dirichlet boundary conditions (i), the IBV problem

$$\begin{aligned} u_t - \Delta u = 0 \qquad & (\mathbf{x}, t) \in D_T = \Omega \times (0, T), \\ u(\mathbf{x}, 0) = \varphi(\mathbf{x}) \qquad & \mathbf{x} \in \Omega, \end{aligned}$$

with $u(\mathbf{x}, t) = 0$ for $(\mathbf{x}, t) \in \partial\Omega \times (0, T)$, has at most one solution $u(\mathbf{x}, t) \in C(\bar{D}_T)$ with $u_t, u_{x_i x_i} \in C(D_T)$ for each $T > 0$, by virtue of the a priori estimate in Corollary 2.2. [For the existence of such a solution it will obviously be required that $\varphi \in C_0(\Omega)$, i.e., $\varphi \in C(\Omega)$ and $\varphi = 0$ on $\partial\Omega$.]

An alternative way of proving uniqueness, which also applies to Neumann and Robin boundary conditions and to "backwards" domains $\Omega \times (-T, 0)$, is the energy method. We define the *energy* integral

$$E(t) := \int_\Omega (u(\mathbf{x}, t))^2 d\mathbf{x} \tag{E}$$

as the $L^2(\Omega)$-norm of u, and we suppose that $\partial\Omega$ is sufficiently smooth so that the Gauss lemma can be applied.

Theorem 5.1. (a) Let $u \in C(\bar{\Omega} \times [t_1, t_2]) \cap C^1(\bar{\Omega} \times (t_1, t_2))$, with $u_{x_i x_i} \in C(\Omega \times (t_1, t_2))$ satisfy the heat equation

$$u_t = \Delta u \qquad \text{for } (\mathbf{x}, t) \in \Omega \times (t_1, t_2) \qquad (-\infty < t_1 < t_2 < +\infty)$$

and one of the boundary conditions (i)–(iii) for $(\mathbf{x}, t) \in \partial\Omega \times (t_1, t_2)$. The $E(t)$ is C^1 on (t_1, t_2), continuous, and nonincreasing on $[t_1, t_2]$.

(b) If in addition $u \in C^2(\bar{\Omega} \times (t_1, t_2)) \cap C^3(\Omega \times (t_1, t_2))$, then E is C^2 on (t_1, t_2) and satisfies the inequality

$$E(t) \le (E(t_1))^{(t_2 - t)/(t_2 - t_1)} (E(t_2))^{(t - t_1)/(t_2 - t_1)} \qquad \text{for } t \in [t_1, t_2]. \tag{LC}$$

Proof. The regularity properties of $E(t)$ are obvious. For $t \in (t_1, t_2)$ we have, by the divergence theorem,

$$\frac{1}{2} E'(t) = \int_\Omega u u_t \, d\mathbf{x} = \int_\Omega u \Delta u \, d\mathbf{x} = \int_{\delta\Omega} u \frac{\partial u}{\partial n} \, dS - \int_\Omega |\text{grad } u|^2 \, d\mathbf{x}.$$

Thus,

$$E'(t) = -2\int_\Omega |\text{grad } u|^2\, d\mathbf{x} \qquad \text{for the boundary conditions (i), (ii),}$$

$$E'(t) \le -2\int_\Omega |\text{grad } u|^2\, d\mathbf{x} \qquad \text{for the boundary condition (iii),}$$

which shows that $E(t)$ is nonincreasing on (t_1, t_2) and, by continuity, on $[t_1, t_2]$. [In fact, Poincaré's inequality implies that $E(t)$ decays exponentially, see below.]

Again for $t \in (t_1, t_2)$ we have, under the additional assumptions on u,

$$\begin{aligned} E''(t) &= 2\int_\Omega (uu_t)_t d\mathbf{x} = 4\int_\Omega u_t^2 d\mathbf{x} + 2\int_\Omega (uu_{tt} - u_t^2)\, d\mathbf{x} \\ &= 4\int_\Omega u_t^2 d\mathbf{x} + 2\int_\Omega (u\Delta u_t - u_t \Delta u)\, d\mathbf{x} \\ &= 4\int_\Omega u_t^2 d\mathbf{x} + \int_{\delta\Omega}\left(u\frac{\partial u_t}{\partial n} - u_t\frac{\partial u}{\partial n}\right) dS = 4\int_\Omega u_t^2\, d\mathbf{x}. \end{aligned}$$

Therefore,

$$(E'(t))^2 = \left(\int_\Omega 2uu_t\, d\mathbf{x}\right)^2 \le E(t)E''(t).$$

Note that $E(t) \ge 0$ and is nonincreasing. If $E(t) \equiv 0$ on $[t_1, t_2]$, then clearly (LC) holds. Otherwise $E(t)$ will be positive for $t_1 \le t < t^*$, where $t^* := \sup[t : E(t) > 0] \le t_2$, and either $t^* = t_2$ with $E(t_2) > 0$, or else $E(t) \equiv 0$ on $[t^*, t_2]$. In either case the function $y(t) = \ln E(t)$ is well defined on the interval $[t_1, \bar{t}]$ for every $\bar{t}$, $t_1 < \bar{t} < t^*$, and satisfies

$$y'' = \frac{E''E - E'^2}{E^2} \ge 0 \qquad \text{on } (t_1, \bar{t}),$$

i.e., $\ln E(t)$ is a convex function. Thus, $E(t)$ is *logarithmically convex*,

$$\ln E(t) \le \frac{\bar{t} - t}{\bar{t} - t_1}\ln E(t_1) + \frac{t - t_1}{\bar{t} - t_1}\ln E(\bar{t}) \qquad \text{on } [t_1, \bar{t}],$$

and hence

$$0 < E(t) \le [E(t_1)]^{(\bar{t}-t)/(\bar{t}-t_1)}[E(\bar{t})]^{(t-t_1)/(\bar{t}-t_1)} \qquad \text{on } [t_1, \bar{t}].$$

Fixing t and letting $\bar{t} \to t^*$ we see that $t^* < t_2$ leads to a contradiction, $E(t_2) > 0$, and the result (LC) holds on the entire interval $[t_1, t_2]$. To summarize, either E is identically zero, or it never vanishes on $[t_1, t_2]$. □

The first result (a) of Theorem 5.1 (with $t_1 = 0,\ t_2 = T$) is sufficient to establish *uniqueness* for the IBV problem *forward* in time, while the second (b) (with $t_1 = -T,\ t_2 = 0$) is needed to get uniqueness *backward* in time.

Corollary 5.1. Let $u \in C(\bar{\Omega} \times [0, T]) \cap C^1(\bar{\Omega} \times (0, T))$, with $u_{x_i x_i} \in C(\Omega \times (0, T))$ satisfy the IBV problem (31), (32) with $\varphi \equiv 0$. The $u \equiv 0$ in $\bar{\Omega} \times [0, T]$.

Proof. As $E(t)$ is continuous and nonincreasing, with $E(0) = 0$, we have $E(t) \equiv 0$ in $[0, T]$ and the result follows. □

Remark 5.1. This uniqueness result extends to generalized solutions of the type considered for example in Corollary 3.1. If $u(x, t)$ is a bounded solution of $u_t = u_{xx}$ such that $u(x, t) \to \varphi(x)$ a.e. as $t \to 0$, then $E(t)$ is nonincreasing, $E(t) \to E(0)$ as $t \to 0$ by the Lebesgue dominated convergence theorem, and this suffices to prove uniqueness for the IBV problem forward in time.

Corollary 5.2. Let

$$u \in C(\bar{\Omega} \times [-T, 0]),\ C^1(\bar{\Omega} \times (-T, 0)),\ C^2(\bar{\Omega} \times (-T, 0)),\ C^3(\Omega \times (-T, 0))$$

satisfy the IBV problem for $-T < t < 0$, with $\varphi \equiv 0$. Then $u \equiv 0$ in $\bar{\Omega} \times [-T, 0]$.

Proof. As $E(t_2) = E(0) = 0$, we have, from (LC), that $E(t) \equiv 0$ in $[-T, 0]$. □

Remark 5.2. The fact that the domain is cylindrical $\Omega \times (-T, 0)$ is essential in Corollary 5.2, as shown by the Example $u = e^{-t}\cos x - \frac{1}{2}e^{-4t}\cos 2x$ in Section 2. Besides, the IBV problem backward in time may not be well posed, in the sense that continuous dependence on the data can be lost, as shown by the one-dimensional example

$$u(x, t) = \frac{1}{n} e^{-n^2 t}\sin(nx), \qquad 0 < x < \pi,\ -\infty < t < 0$$

by letting the integer n approach ∞.

The solution of the IBV problem can be constructed by means of separation of variables and eigenfunction expansions. We illustrate the method here in the case of one space variable. Suppose we seek a solution of the IBV problem

$$\begin{aligned} &u_t - a^2 u_{xx} = 0, \qquad (x, t) \in (0, l) \times (0, \infty), \\ &u(x, 0) = \varphi(x), \qquad x \in (0, l); \qquad u(t, 0) = u(t, l) = 0, \qquad t > 0 \end{aligned} \tag{33}$$

of the form $X(x)\mathscr{T}(t)$. We find $X''(x)/X(x) = \mathscr{T}'(t)/a^2\mathscr{T}(t) = \text{const} \equiv -\lambda$, so that the equality for X leads to the same eigenvalue problem

$$X'' + \lambda X = 0 \qquad (0 < x < l), \qquad X(0) = X(l) = 0$$

for the operator $-d^2/dx^2$ with zero boundary data at $x = 0, l$, considered in Section 2 of Chapter 2. We have seen that $X_n = \sin(n\pi x/l)$ are the eigenfunctions corresponding to the eigenvalues $\lambda = \lambda_n = (n\pi/l)^2$ for $n = 1, 2, \ldots$. The equation for T yields then $\mathscr{T} = \mathscr{T}_n = a_n e^{-a^2\lambda_n t}$. Adding multiples of these we obtain the general solution

$$u(x, t) = \sum_{n=1}^{\infty} a_n e^{-a^2\lambda_n t} X_n \equiv \sum_{n=1}^{\infty} a_n e^{-a^2 n^2(\pi/l)^2 t} \sin(n\pi x/l), \tag{34}$$

where $X_n = \sin(n\pi x/l)$. Formal satisfaction of the initial conditions requires

$$\varphi(x) = \sum_{n=1}^{\infty} a_n X_n.$$

This is a Fourier sine series and for $\varphi \in L^2(0, l)$ the series is convergent in $L^2(0, l)$ with

$$a_n = \left(\varphi, \frac{X_n}{\|X_n\|}\right)_{L^2} \equiv \frac{2}{l}\int_0^l \varphi(y) \sin(n\pi y/l)\, dy \tag{35}$$

the Fourier sine coefficients of φ. If $\varphi \in L^2(0, l)$, it is easy to see that, due to the presence of the factor $\exp(-a^2n^2\pi^2t/l^2)$, $u(x, t)$ is a smooth function of (x, t), and hence satisfies the heat equation, for $t > 0$. Thus, the "smoothing" property of (H) that we have observed before holds for solutions of the initial–boundary value problem as well. As

$$\|u(x, t) - \varphi(x)\|_{L^2}^2 = \frac{l}{2}\sum_{n=1}^{\infty} a_n^2(1 - e^{-a^2\lambda_n t})^2, \qquad \frac{l}{2}\sum_{n=1}^{\infty} a_n^2 = \|\varphi\|_{L^2}^2 < \infty,$$

$u(x, t)$ converges to $\varphi(x)$ in L^2 as $t \to 0$, and $E(t) \to E(0)$, so that this "L^2 solution" is unique. Sufficient conditions for uniform convergence can also be established (see the exercises). This method can be extended to the IBV problem in n space dimensions by taking λ_n, X_n as the eigenvalues and eigenfunctions of the Laplace–Dirichlet operator in Ω (see Chapter 4). We can proceed similarly in the case of Neumann boundary conditions (ii) (Exercise 5.3).

We can use the solution (34) to illustrate the ill-posedness of the *inverse problem*. Suppose we want to determine the initial temperature $\varphi(x) = u(x, 0)$ from the knowledge of the final temperature $f(x) = u(x, T)$, for $T > 0$. Taking $l = \pi$ and $a = 1$ for simplicity, we can write

$$u(x, t) = \sum_{n=1}^{\infty} a_n e^{n^2(T-t)} \sin nx,$$

where $a_n = (2/\pi) \int_0^\pi u(y, T) \sin ny \, dy$ are the Fourier coefficients of the given final temperature. Then

$$\varphi(x) = \sum_{n=1}^{\infty} a_n e^{n^2 T} \sin nx$$

and this series converges only if $a_n = o(e^{-n^2 T})$ as $n \to +\infty$ (see Exercise 5.5). In particular, there is no positive constant C such that $\|\varphi\|_{L^2} \leq C\|f\|_{L^2}$. This ill-posedness of the inverse problem reflects the time irreversibility of the heat equation.

If we substitute the expression (35) for a_n into the series (34) for $u(x, t) = f(x)$ (with $l = \pi$ and $a = 1$), we obtain the integral formula

$$u(x, t) = \int_0^\pi \mathcal{G}(x, y, t)\varphi(y) \, dy, \qquad 0 \leq x \leq \pi, \ t > 0,$$

with kernel $\mathcal{G}(x, y, t) = (2/\pi) \sum_{n=1}^{\infty} e^{-n^2 t} \sin nx \sin ny$ (Green's function). In the context of the inverse problem, this may be viewed as an integral equation of the first kind in φ for given f at $t = T$,

$$\int_0^\pi K(x, y)\varphi(y) \, dy = f(x), \qquad 0 \leq x \leq \pi,$$

with (smooth) kernel $K(x, y) = \mathcal{G}(x, y, T)$. From this particular example we may infer the fact that integral equations of this kind are ill posed (see Kress, Ref. 9).

We can also apply the eigenfunction expansion method to the simple one-dimensional model of a *chain reaction*

$$v_t - a^2 v_{xx} = \gamma v,$$

where v is the neutron density, a^2 the neutron diffusion coefficient of the body, and $\gamma > 0$ the rate of neutron generation by nuclear scission. We suppose that u satisfies initial and boundary conditions as in (33). Setting $v(x, t) = e^{\gamma t} u(x, t)$ we find that u is the solution of the IBV problem (33), hence

$$v(x, t) = \sum_{n=1}^{\infty} a_n e^{(\gamma - a^2\lambda_n)t} \sin(n\pi x/l).$$

Thus, if $\gamma < a^2\lambda_1$, the neutron density decays exponentially, whereas if $\gamma > a^2\lambda_1$, the density v in general ($a_1 \neq 0$) blows up exponentially as $t \to +\infty$ (chain reaction). At the threshold value $\gamma = a^2\lambda_1$ the neutron density remains bounded (if $a_1 \neq 0$) and we say that the domain has critical dimensions (the body has *critical mass*). As $\lambda_1 = (\pi/l)^2$, the critical length for this one-dimensional chain reaction is

$$l_{\text{cr}} = a\pi/\sqrt{\gamma}.$$

The following general decay estimates imply that the temperature of a heat-conducting body approaches equilibrium as $t \to +\infty$ if the boundary is isothermal or adiabatic or, more generally, if the temperature at the boundary is independent of time. We denote below by $\|u\|$ the $L^2(\Omega)$-norm of a function u.

Theorem 5.2. Let $u \in C^1(\bar{\Omega} \times [0, \infty)) \cap C^2(\bar{\Omega} \times (0, \infty)) \cap C^3(\Omega \times (0, \infty))$ be a solution of

$$u_t = \Delta u \text{ for } (\mathbf{x}, t) \in \Omega \times (0, \infty), \qquad u = 0 \text{ for } (\mathbf{x}, t) \in \delta\Omega \times (0, \infty) \tag{36}$$

and let $u_0(\mathbf{x}) = u(\mathbf{x}, 0)$. Then either $u \equiv 0$ in $\bar{\Omega} \times [0, \infty)$ or $\|u_0\| > 0$ and the estimate

$$\|u_0\| e^{-t/\beta^2} \leq \|u\| \leq \|u_0\| e^{-t/\sigma^2} \tag{DE}$$

holds for $t \geq 0$, with constants $\sigma = \sigma(\Omega) \geq \beta := \|u_0\| / \|\text{grad } u_0\|$.

Proof. All of the conditions of Theorem 5.1 are satisfied with $t_1 = 0$, $t_2 = +\infty$, hence either the energy $E = \|u\|^2$ is identically zero or never vanishes. In the second case, $E(t) > 0$ for $t \geq 0$ and the function $y(t) = \ln E(t)$ satisfies

$y'' = (E''E - E'^2)/E^2 \geq 0$ for $t > 0$, so that $y'(t) \geq y'(0)$, and $y(t) \geq y(0) + ty'(0)$ for $t \geq 0$. This implies

$$E(t) \geq E(0)e^{E'(0)t/E(0)},$$

and as $E(0) = \|u_0\|^2$, $E'(0) = -2\|\text{grad } u_0\|^2$ (see proof of Theorem 5.1), the left side of inequality (DE) follows. As by assumption $u \in C_0^1(\Omega)$ for each $t \geq 0$, Poincaré's inequality (Lemma 3.1 of Chapter 8) yields

$$\|u\| \leq \sigma\|\text{grad } u\|, \tag{P}$$

so that

$$E'(t) \leq -2\int_\Omega |\text{grad } u|^2 d\mathbf{x} \leq -2\sigma^{-2}E(t)$$

and the right-hand side of (DE) follows. □

Remarks

5.3. If $u = 0$ on $\partial\Omega$, the energy decays exponentially and is bounded above and below by two exponentials, so that $u \to 0$ (the equilibrium solution) in $L^2(\Omega)$ as $t \to +\infty$. The slowest possible decay is obtained by choosing u_0 such that $\beta = \sigma$ (see the exercises).

5.4. If $\partial u/\partial n = 0$ on $\partial\Omega$, then the right-hand side of (DE) becomes

$$\|u\| \leq (|\Omega|\bar{u}^2 + \|u_0\|^2 e^{-2t/\alpha^2})^{1/2},$$

where $\bar{u} = \int_\Omega u_0(\mathbf{x})\, d\mathbf{x}/|\Omega|$ is the average of u_0, and $\alpha = \alpha(\Omega) > 0$ is a suitable constant. Moreover, $u \to \bar{u}$ (equilibrium solution) in $L^2(\Omega)$ as $t \to +\infty$.

5.5. More refined decay estimates can be found in Ref. 10.

Corollary 5.3. Under the assumptions of Theorem 5.2, a solution $u(\mathbf{x}, t)$ of

$$u_t = \Delta u \text{ for } (\mathbf{x}, t) \in \Omega \times (0, \infty), \qquad u = \phi(\mathbf{x}) \text{ for } (\mathbf{x}, t) \in \partial\Omega \times (0, \infty)$$

relaxes as $t \to +\infty$ to a solution $v(\mathbf{x})$ of

$$\Delta v = 0 \text{ for } \mathbf{x} \in \Omega, \qquad v = \phi(\mathbf{x}) \text{ for } \mathbf{x} \in \partial\Omega. \tag{37}$$

Proof. Set $u = v + w$ where w satisfies (36), then $w \to 0$ as $t \to +\infty$ by Theorem 5.2. □

Exercises

1.1. Perform the integration leading to (1). In general, show that the Fourier transform of a bell curve is a bell curve with "inverted parameters.".

1.2. Prove relations (3). *Hint*: Let ω_n denote the total solid angle in $\mathbb{R}^n$. Then the change of variables $z = |\mathbf{y}|^2/4a^2t$ yields

$$\int_{|\mathbf{y}|\geq\gamma} \frac{1}{(4\pi a^2 t)^{n/2}} e^{-|\mathbf{y}|^2/4a^2t}\, d\mathbf{y} = \frac{\omega_n}{2\pi^{n/2}} \int_{\frac{\gamma^2}{4a^2t}}^{+\infty} e^{-z} z^{n/2-1}\, dz = \frac{\omega_n \Gamma(n/2)}{2\pi^{n/2}} = 1$$

if $\gamma = 0$, while the limit as $t \to 0^+$ is zero if $\gamma > 0$.

1.3. Work out the details in the Stefan problem for generic data T_+, T_-.

1.4. In one dimension the formula $u = e^{i\mathbf{x}\cdot\xi} e^{-a^2|\xi|^2 t}$ becomes $u = e^{ix\xi} e^{-a^2\xi^2 t}$. This solution is periodic in x and damped for $t \to +\infty$. The solution (9) is obtained from this by taking a complex $\xi = \sqrt{i\omega}/a$ and choosing the root $\sqrt{i} = (1+i)/\sqrt{2}$. Thus, $\xi^2 = i\omega$, u is periodic in t and is damped for $x \to +\infty$.

1.5. Construct solutions of the wave equation $u_{tt} = c^2(u_{xx} + u_{yy})$ that are damped as $x \to +\infty$ ("evanescent waves"). *Hint*: $u = e^{-\lambda x} e^{i\omega t} \exp(i(\lambda^2 + \omega^2/c^2) y)$.

1.6. Verify (9) and (10). Show that (10) holds for $f \in H^1(0,p)$, and then $u_{xx}(0,t) \in L^2(0,p)$.

1.7. If we assume only that $f \in L^1(0,p)$, then $u(x,t)$ given by (10) is in C^∞ for $x > 0$. If

$$\sum_{-\infty}^{\infty} |k||A_k|^2 < \infty$$

[i.e., $f \in H^{1/2}(0,p)$], then

$$u_x(0,t) = -\sum_{-\infty}^{\infty} \sqrt{\frac{\omega}{2a^2}}(1+i)\sqrt{|k|}A_k e^{ik\omega t}$$

exists in $L^2(0,p)$.

1.8. Show that daily temperature oscillations are damped as fast as annual temperature oscillations at 19 times the depth.

1.9. The condition (G) can be replaced by $u(L,t) = 0$ at some finite depth L. Formulate and solve this (pure boundary value) problem.

1.10. Prove the identity

$$\operatorname{erf}(x/\sqrt{4a^2t}) \equiv \int_0^{+\infty} (G(x-y,t) - G(x+y,t))\,dy.$$

1.11. Show that all solutions of the heat equation $u_t = a^2u_{xx}$ of the form $u(x/2a\sqrt{t})$ are given by $a\Phi(z) + b(a, b$ constants).

2.1. The Dirichlet problem for the one-dimensional heat equation with continuous boundary data

$$u_t = u_{xx}, \qquad (x,t) \in R = (0,\pi) \times (0,\pi),$$

$$u = \begin{cases} \sin x, & 0 < x < \pi,\ t = 1, \\ 0, & \text{on the remaining part of } \partial R, \end{cases}$$

has no solution. *Hint*: Use the maximum principle.

2.2. Show that the solution of the Dirichlet problem

$$u_t = u_{xx}, \qquad (\mathbf{x},t) \in D_T = (0,l) \times (0,T),$$
$$u(x,0) = u(l,t) = 0, \qquad u(0,t) = \begin{cases} 0, & 0 \le t \le t^*, \\ t - t^*, & t^* \le t \le T, \end{cases}$$

is identically zero for $(x,t) \in (0,l) \times (0,t^*]$, but not for $t > t^*$.

2.3. Extend Theorem 2.4 as indicated in the text. *Hint*: Condition (PL) holds for $(\mathbf{x},t) \in D_T$, and $u(\mathbf{x},t) \ge -\varepsilon$ for all $(\mathbf{x},t) \in \bar{D}_t$ with $|\mathbf{x}| \ge R_\varepsilon$. B is replaced by $B \cap D_T$, and $w(\mathbf{x},t) \ge \varepsilon > 0$ in S_T.

2.4. Show by an example that the boundary point lemma may fail if the normal to ∂D is parallel to the t-axis at a maximum point. *Hint*: Consider a solution of $u_t = u_{xx}$ in $D = (0,1) \times (0,\infty)$ constant for $t \le t_0$.

3.1. Prove the limit (2). *Hint*: By Taylor's theorem

$$\left| \sum_{k=0}^{N} (k!)^{-1}(L^2/t)^k - e^{L^2/t} \right| = e^s(L^2/t)^{n+1}/(n+1)! \text{ for some } s,\ 0 < s < L^2/t,$$

and $e^s(L^2/t)^{n+1} < e^{2L^2/t} \le ce^{4/9t^2}$, c some constant depending only on L.

3.2. (Corollary 3.1). Prove the *group relation*

$$G(\mathbf{x}-\mathbf{y},t) = \int_{\mathbb{R}^n} G(\mathbf{x}-\mathbf{z},t-\tau)G(\mathbf{z}-\mathbf{y},\tau)\,d\mathbf{z}$$

for every τ, $0 < \tau < t$. *Hint*; Use the property $e^{-a^2|\xi|^2 t} = e^{-a^2|\xi|^2(t-\tau)}e^{-a^2|\xi|^2\tau}$ and the convolution theorem for the Fourier transform, Corollary 2.1 of Chapter 8.

3.3. Verify equation (21) and show that the function $u(x, t)$ satisfies $u(0, t) = 0$ for every $t > 0$.

3.4. Let $u = u_\mu(x, t)$ be the solution of the initial value problem

$$\begin{aligned} u_t &= \mu u_{xx}, \qquad (x, t) \in \mathbb{R} \times (0, T), \\ u(x, 0) &= \varphi(x), \qquad x \in \mathbb{R}, \end{aligned}$$

where $\mu > 0$. Show that $\lim_{\mu \to 0}(u_\mu(x, t) = \varphi(x)$, the solution of the IVP for $\mu = 0$.

3.5. Work out the details in the proof of Theorem 3.2.

3.6. Show that $G(\mathbf{x}, t) \in L^1(\mathbb{R}^n) \cap L^2(\mathbb{R}^n)$ for every $t > 0$. *Hint*: $\int_{\mathbb{R}^n} G^2 d\mathbf{x} = O(t^{-n/2})$. Hence, show directly from equation (16) that $\varphi \in L^2$ implies $u \in L^2$.

3.7. Using the Gaussian formula, prove that the solution of the Cauchy problem for the heat equation in $\mathbb{R}^n$

$$\begin{aligned} u_t &= \Delta u, \qquad \mathbf{x} \in \mathbb{R}^n, t > 0, \\ u(\mathbf{x}, 0) &= \varphi(\mathbf{x}), \qquad \mathbf{x} \in \mathbb{R}^n, \end{aligned}$$

with initial data $\varphi(\mathbf{x}) \in C^2(\mathbb{R}^n)$ satisfying $\Delta\varphi = 0$, is given by $u(\mathbf{x}, t) = \varphi(\mathbf{x})$. *Hint*: From $u = G * \varphi$ we find $\Delta u = \Delta(G * \varphi) = G * (\Delta\varphi) = 0$, hence $u_t = 0$, $u = \varphi$.

3.8. Using the Gaussian formula and the properties of the heat kernel, prove that if the initial data of the Cauchy problem

$$\begin{aligned} u_t &= \Delta u, \qquad \mathbf{x} = (x_1, \ldots, x_n) \in \mathbb{R}^n, t > 0, \\ u(\mathbf{x}, 0) &= \varphi(\mathbf{x}), \qquad \mathbf{x} \in \mathbb{R}^n, \end{aligned}$$

do not depend on a variable x_j, the same is true of u.

4.1. Work out the details in the proof of Lemma 4.2.

4.2. If f is bounded ($A = 0$), the Duhamel integral yields the unique solution bounded on each finite interval $0 < t < T$.

4.3. Show that if $\varphi \in L^p(\mathbb{R}^n)(p = 1, 2)$, then (26) gives $u = O(t^{-n/2p})$ and $u \to 0$ uniformly in $\mathbf{x}$ as $t \to +\infty$.

4.4. Under the hypotheses of Lemma 4.2 and Theorem 4.1 the problem

$$\begin{aligned} u_t - u_{xx} &= f(x, t), \qquad (x, t) \in \mathbb{R} \times (0, T], \qquad 0 < T < 1/4A, \\ u(x, 0) &= \varphi(x, 0), \end{aligned}$$

has a unique solution given by

$$u(x, t) = \int_{\mathbb{R}} G(x - y, t)\varphi(y)\, dy + \int_0^t d\tau \int_{\mathbb{R}} G(x - y, t - \tau) f(y, \tau)\, dy.$$

Deduce the estimate

$$|u(x, t)| \leq \|\varphi\|_{L^\infty} + t \sup_{\mathbb{R}\times[O,T]} |f(x, t)|$$

and compare this with the a priori estimate in Section 2.

4.5. Show that if, in addition to the assumptions of Lemma 4.1, the partial derivatives φ_{x_i} are bounded and Lipschitz continuous, then $u \in C^1(\mathbb{R}^n \times [0, \infty))$. *Hint*: Proceed as in the proof of Lemma 4.1 replacing u by u_{x_i}, u_t.

5.1. Show that a sufficient condition for the uniform convergence of $u(x, t)$ to $\varphi(x)$ as $t \to 0$ in the Fourier series (34) is $\varphi \in H_0^1(0, l)$.

5.2. Prove that if $\varphi \in L^2(0, l)$, the energy integral corresponding to the series (34) for u satisfies

$$E(t) = \int_0^l (u(x, t))^2 dx = \frac{l}{2} \sum_{n=1}^{\infty} a_n^2 e^{-2a^2\lambda_n t} \leq E(0)e^{-2a^2\pi^2 t/l^2}$$

so that $E(t) \to E(0) = \int_0^l (\varphi(x))^2 dx$ as $t \to 0$, and $E(t)$ decays exponentially as $t \to +\infty$.

5.3. If the boundary conditions at the ends of $I = (0, l)$ are $u_x = 0$ (an insulated rod):

$$\begin{aligned} &u_t - u_{xx} = 0, && (x, t) \in (0, l) \times (0, \infty), \\ &u_x(0, t) = u_x(l, t) = 0, && t > 0, \\ &u(x, 0) = \varphi(x), && 0 < x < l, \end{aligned}$$

the solution is given by the Fourier cosine series

$$u(x, t) = \frac{a_0}{2} + \sum_{n=1}^{\infty} a_n e^{-a^2 n^2 \pi^2 t/l^2} \cos(n\pi x/l),$$

$$a_n = \frac{2}{l} \int_0^l \varphi(x) \cos(n\pi x/l)\, dx$$

and $u(x, t) \to \varphi(x)$ as $t \to 0$ uniformly on $(0, l)$ if $\varphi \in H^1(0, l)$.

5.4. Suppose $l = 1$ in Exercise 5.3. If the initial temperature φ is positive and has an absolute maximum at $x_0 \in I = (0, 1)$, does the temperature $u(x, t)$ have a maximum at or near x_0 for $t > 0$? Intuitively, the temperature should

be decreasing away from the "hot spot" at x_0 so the answer should be yes. The (formal) solution

$$\frac{1}{2}a_0 + \sum_{n=0}^{\infty} a_n X_n(x) \exp(-\lambda_n t)$$

with $\lambda_n = n\pi$, $X_n = \cos(n\pi x)$ obtained in Exercise 5.3 can be written as

$$\tfrac{1}{2}a_0 + a_1 \cos(\pi x)e^{-\pi^2 t} + O(e^{-4\pi^2 t})).$$

This indicates that for $x_0 < \frac{1}{2}$ the maximum of u moves to 0, and for $x_0 > \frac{1}{2}$ it moves to 1. Make all of this rigorous.

5.5. Show that the solution of the initial value problem

$$u_t - u_{xx} = 0, \qquad -\infty < x < +\infty, t > 0,$$
$$u(x, 0) = \varphi(x), \qquad -\infty < x < +\infty,$$

with initial data

$$\varphi(x) = \sum_{n=1}^{\infty} a_n \sin nx,$$

where $|a_n| = O(e^{-Tn^2})$ as $n \to \infty (T > 0)$, can be extended for negative t in the strip $-T < t \leq 0$. *Hint*: The solution is given by the Fourier series (34) with $l = \pi$, $a = 1$.

5.6. The equation

$$v_t - v_{xx} + bv = f(t, x), \qquad b \in \mathbb{R},$$

describes heat conduction in a rod taking into account radiative losses (for $b > 0$). Verify that the equation reduces to $u_t - u_{xx} = e^{bt}f$ by the transformation $v = e^{-bt}u$.

5.7. In one space dimension, Poincaré's inequality (P) reads

$$\int_0^l u^2 dx < \lambda_1^{-1} \int_0^l u_x^2 dx = \frac{l^2}{\pi^2} \int_0^l u_x^2 dx \tag{F1}$$

for $u \in C_0^1(0, l)[u \in H_0^1(0, l)]$, and this implies the exponential decay of $E(t)$ in Exercise 5.2.

5.8. Show that the inequality (F1) becomes an equality for $u = \sin(\pi x/l)$.

5.9. For u a solution of (33), equation (34) implies that $\int_0^l u(x, t)\, dx \to 0$ as $t \to +\infty$. Prove it using the decay estimates for $E(t)$. *Hint*: $\int_0^l u(x, t)\, dx \leq$

$\sqrt{E(t)}]$. What is the interpretation of this in the context of population diffusion?

5.10. The constant in Poincaré's inequality (P) satisfies $\sigma = 1/\sqrt{\lambda_1}$, where λ_1 is the first eigenvalue of the Laplace operator with Dirichlet homogeneous boundary conditions (Chapter 4). Hence, the decay estimate (DE) agrees with the estimate of Exercise 5.2 in the one-dimensional case. Besides, the slowest possible decay occurs for $u_0 = X_1$, the first Laplace–Dirichlet eigenfunction, as then $\beta = \sigma = 1/\sqrt{\lambda_1}$.

5.11. Work out the details in Remark 2. *Hint*: prove a variant of Poincaré's inequality for C^1 functions.

References

1. TIHONOV, A. N., and SAMARSKII, A. A., *Equazioni della Fisica Matematica*, Mir, Moscow, Russia, 1981.
2. CHORIN, A. J., and MARSDEN, J. E., *A Mathematical Introduction to Fluid Mechanics*, Springer-Verlag, Berlin, Germany, 1990.
3. CANNON, J. R., *The One-Dimensional Heat Equation*, Addison–Wesley, London, England, 1984.
4. KÖRNER, T. W., *Fourier Analysis*, Cambridge University Press, London, England, 1993.
5. DAY, W. A., *Entropy and Elliptic Equations*, Quarterly of Applied Mathematics, Vol. 51, 1, pp. 191–200, 1993.
6. PROTTER, M. H., and WEINBERGER, H. F., *Maximum principles in Differential Equations*, Springer-Verlag, Berlin, Germany, 1984.
7. FRIEDMAN, A., *Partial Differential Equations of Parabolic Type*, Prentice–Hall, Englewood Cliffs, New Jersey, 1964.
8. ODDSON, J. K., *Differential Equations of Mathematical Physics*, unpublished lecture notes.
9. KRESS, R., *Linear Integral Equations*, Springer-Verlag, Berlin, Germany, 1989.
10. ODDSON, J. K., *On the Rate of Decay of Solutions of Parabolic Differential Equations*, Pacific Journal of Mathematics, Vol. 29, pp. 389–396, 1969.

4

Laplace Equation

For a function $u(\mathbf{x}) = u(x_1, \ldots, x_n)$ the differential equation

$$\sum_{i=1}^{n} u_{x_i x_i} := \Delta u = 0 \tag{L}$$

is called the Laplace equation (in $\mathbb{R}^n$). Its solutions are called *harmonic functions*.

Equation (L) arises in diverse physical applications. In the population diffusion example given in Chapter 1, a time-independent equilibrium density satisfies (L) with $n = 2$. In ideal fluid flow the velocity vector is the gradient of a solution of (L) with $n = 2$ or 3. In electrostatics the force in a charge-free region of space is the gradient of a solution of (L) with $n = 3$. For definiteness we will focus on this last situation for motivation.

It is useful to observe that solutions depending only on $r = |\mathbf{x} - \mathbf{x}_0|(\mathbf{x}_0$ a fixed point) are, apart from an additive constant, given by multiples of the function

$$E(\mathbf{x}, \mathbf{x}_0) = \begin{cases} |\mathbf{x} - \mathbf{x}_0|^{2-n}/(n-2)\omega_n, & n \geq 3, \\ (2\pi)^{-1}\ln(1/r), & n = 2, \end{cases}$$

where ω_n is the solid angle in $\mathbb{R}^n$, taking the value 4π for $n = 3$ and 2π for $n = 2$ (see Ref. 1 and Section 3 of Chapter 8). This function $E = E(r)$ is called the *fundamental solution* of Laplace's equation. For $n = 3$, $E(r)$ represents the potential at the point $\mathbf{x}$ due to a point charge concentrated at $\mathbf{x}_0$ (or vice versa). We note explicitly that

$$\text{grad } E(\mathbf{x}, \mathbf{x}_0) = -\frac{\mathbf{x} - \mathbf{x}_0}{\omega_n |\mathbf{x} - \mathbf{x}_0|^n}, \qquad n \geq 2,$$

so that $E(\mathbf{x}, \mathbf{x}_0)$ and its first partial derivatives are locally integrable in $\mathbb{R}^n$.

Some results of this chapter are formulated for $n = 3$ and can be extended easily to $n \geq 3$ dimensions. The case $n = 2$ is special and the main differences are dealt with in the text and the exercises.

1. Potential Theory: Basic Notions

We assume, as known, Coulomb's law, which says (in appropriate units, used hereupon) that the force $\mathbf{E}$ exerted on a positive unit charge at a point $\mathbf{x}$ in $\mathbb{R}^3$ by a charge of magnitude ρ (positive or negative) at the point $\mathbf{y}$ is given by $\mathbf{E} = -\text{grad}\, u$, where $u = \rho/r$ is the Coulomb (or Newtonian) potential, and $r = |\mathbf{x} - \mathbf{y}|$ is the distance from $\mathbf{x}$ to $\mathbf{y}$. The starting point of our discussion is the determination of the force exerted by a distribution of charges through a region G given by a density $\rho(\mathbf{y})$. If we assume that ρ is nearly constant in a small element of volume $d\mathbf{y}$, then the force exerted at $\mathbf{x}$ by that charge in $d\mathbf{y}$ is minus the gradient of $r^{-1}\rho(\mathbf{y})d\mathbf{y}$ and the total force exerted at $\mathbf{x}$ should be given by $E = -\text{grad}\, u(\mathbf{x})$, where u is the *volume potential*

$$u(\mathbf{x}) = \int_G \frac{\rho(\mathbf{y})d\mathbf{y}}{r}.$$

That the force exerted by the charge distribution ρ is, in fact, obtain in this way may be taken as the statement of Coulomb's law in this amplified context. The function ρ must, of course, be such that the integral exists in some sense. Suppose, for example, that G and ρ are bounded. Then (see the Appendix to Section 1) the integral exists for all $\mathbf{x}$ and differentiation under the integral for $\mathbf{x} \notin \bar{G}$ implies, because $\Delta(1/r) = 0$ (differentiation with respect to $\mathbf{x}$), that $\Delta u = 0$ (u is harmonic) outside of G.

Analogously to the above, we may define potentials of line and surface densities by

$$u = \int_C \frac{\lambda}{r}\, ds, \qquad u = \int_S \frac{\mu}{r}\, dS,$$

where λ and μ are the densities of charge on the curve C and surface S, respectively. As before, elementary theorems show that these potentials are harmonic functions for $\mathbf{x}$ not on C, S, respectively. The determination of these potentials for given densities and geometric configurations now becomes a problem in integration. Although a lengthy study of examples that can be done explicitly is off of our path, it is useful to give the results in a few examples. For details the reader is referred to Kellogg (Ref. 2). In each of these the density μ, λ, or ρ will be constant. It is convenient also to give the results in terms of the components of $\mathbf{E} = -\text{grad}\, u$, and we write the components of the electrostatic field $\mathbf{E}$ as X, Y, and Z.

Consider first a line charge with density λ along the segment $0 \leq z \leq l$ of the z-axis. The potential is given, for $\mathbf{x} = (x, y, z)$ not in the segment, by

$$u = -\lambda \ln\left|\left(-z + \sqrt{x^2 + y^2 + z^2}\right)\Big/\left(l - z + \sqrt{(z-l)^2 + y^2 + x^2}\right)\right|$$

and $\mathbf{E}$ can be easily calculated. Suppose we consider, however, a point that lies on the z-axis, $z > l$, $x = y = 0$. It then follows easily that

$$Z = \frac{\lambda l}{z(z-l)} = \frac{Q}{c^2}, \qquad X = Y = 0,$$

where Q is the total charge and $c = \sqrt{z(z-l)}$ is the geometric mean of the distance to the ends of the segment. We point out explicitly that the potential and force are infinite on the segment. Moreover, Z tends to $-\lambda/z$ as $l \to \infty$, while the potential u diverges in this limit. If we suppose now that the line charge of constant density λ is distributed on the segment $-l \leq z \leq l$, then letting $l \to \infty$ we find that $\mathbf{E}(x, y, 0)$ tends to $2\lambda\mathbf{r}/\sqrt{x^2+y^2}$ where $\mathbf{r}$ is the unit vector from the origin to the point (x, y) of the z-plane. Hence, in this limit $\mathbf{E} = -\text{grad } u$, where u is the logarithmic potential

$$u = -\lambda \ln(x^2 + y^2)$$

in the (x, y)-plane.

Consider next a conducting sphere with radius a centered at the origin with constant surface charge density σ. By symmetry there is no loss in generality in assuming $\mathbf{x} = (0, 0, z)$, and we take $z \neq a$. After introducing spherical coordinates and exercising some care in the integration, one finds $X = Y = 0$, and

$$Z = \begin{cases} 4\pi a^2\sigma/z^2 = Q/z^2, & z > a, \\ 0, & 0 < z < a \end{cases}.$$

The electrostatic field outside the sphere is equal to that due to a point charge at the center (with charge equal to that on the sphere), and is zero inside. (What happens when $z = a$?)

In addition to the concept of positive and negative charges exerting forces in space, we need the concept of *dipoles* or doublets. Suppose charges with magnitude $-m$ and m are placed at $\xi = (\xi, \eta, \zeta)$ and $\xi_h = (\xi + h, \eta, \zeta)$. The potential of the field induced by these at the point $\mathbf{x}$ is

$$u_h(x) = \frac{-m}{|\mathbf{x} - \xi|} + \frac{m}{|\mathbf{x} - \xi_h|}.$$

Suppose that we allow h to approach zero while holding $\nu = mh$ constant. Because

$$u_h(x) = \nu\left(\frac{1}{|\mathbf{x} - \xi_h|} - \frac{1}{|\mathbf{x} - \xi|}\right)\frac{1}{h},$$

we obtain

$$\nu \frac{\partial}{\partial \xi}\left(\frac{1}{|\mathbf{x}-\xi|}\right) = \nu \frac{\partial}{\partial \xi}\left(\frac{1}{r}\right)$$

in the limit. We think of a "dipole" located at ξ that exerts a force that is the gradient of this limiting potential. The x-axis is the axis of the dipole and ν is its moment; more generally, a dipole with axis along α (α is a unit vector) and moment ν has the potential

$$\nu \frac{\partial}{\partial \alpha}\left(\frac{1}{r}\right) = \nu\alpha \cdot \operatorname{grad}_{\xi}\left(\frac{1}{|\mathbf{x}-\xi|}\right).$$

We remark that this potential may also be written as $-\nu \cos(\alpha, \mathbf{r})/r^2$, where $\cos(\alpha, \mathbf{r})$ is the cosine of the angle between α and the unit vector pointing toward ξ from $\mathbf{x}$. in particular, if a unit dipole is located at a point $\mathbf{y}$ of a surface with axis along the normal $\mathbf{n}(\mathbf{y})$, its potential is

$$\mathbf{n}(\mathbf{y}) \cdot \operatorname{grad}_y\left(\frac{1}{r}\right) = \frac{\mathbf{n}(\mathbf{y}) \cdot (\mathbf{x}-\mathbf{y})}{|\mathbf{x}-\mathbf{y}|^3} = -\frac{\cos\varphi}{r^2},$$

with φ the angle between $\mathbf{n}$ and the vector $\mathbf{y}-\mathbf{x}$. Typical configurations are depicted in Fig. 1.

In analogy to the potential of a distribution of charges with density μ over a surface S, the *single layer potential*

$$\int_S \frac{\mu}{r}\, dS$$

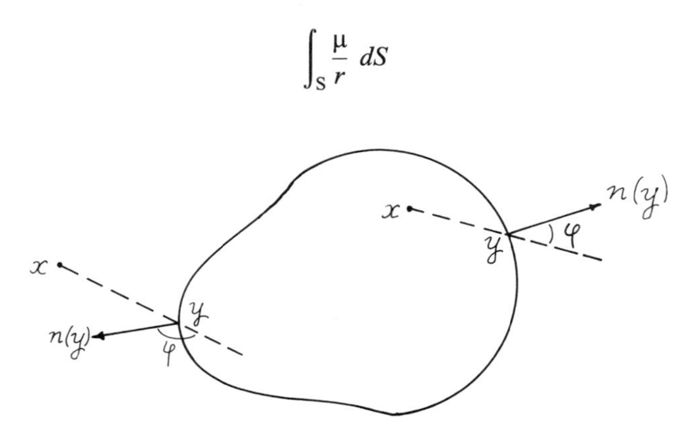

Fig. 1. Unit dipoles on a surface.

($r = |\mathbf{x} - \mathbf{y}|$), we may now define the potential of a surface distribution of dipoles with density ν, called the *double layer potential*

$$\int_S \nu \frac{\partial}{\partial n}\left(\frac{1}{r}\right) dS,$$

where $\mathbf{n}$ is the unit normal to the (orientable) surface S in the direction of the dipole axis. (If S is a closed surface, we always take $\mathbf{n}$ to be the exterior normal and take the sign of the dipole moment accordingly.) It is easily seen that this potential satisfies Laplace's equation off of S. In two dimensions, the integrals

$$\int_C \mu \ln\left(\frac{1}{r}\right) ds, \qquad \int_C \nu \frac{\partial}{\partial n} \ln\left(\frac{1}{r}\right) ds$$

define single and double layer potentials over a curve C, and

$$\int_G \rho \ln\left(\frac{1}{r}\right) d\mathbf{y}$$

is the "volume" potential over a two-dimensional region, G. Physically, we may interpret these as arising from newtonian potentials of infinite cylinders with generators perpendicular to the x–y-plane ("logarithmic charges" and "logarithmic dipoles"). Because $\ln(1/r)$ is harmonic in the plane, these are all harmonic away from the respective distributions of "charge." in general, we will take the integrals

$$V_\mu(\mathbf{x}) = \int_S \mu(\mathbf{y}) E(\mathbf{x}, \mathbf{y}) dS_y,$$

$$W_\nu(\mathbf{x}) = \int_S \nu(\mathbf{y}) \frac{\partial E(\mathbf{x}, \mathbf{y})}{\partial n(\mathbf{y})} dS_y,$$

$$\mathscr{V} \varrho(\mathbf{x}) = \int_G \varrho(\mathbf{y}) E(\mathbf{x}, \mathbf{y}) d\mathbf{y}$$

as definitions of single layer, double layer, and volume potentials, respectively, for $\mathbf{x} \in \mathbb{R}^n$. [With these definitions, a factor ω_n is incorporated in the densities: for example, $\varrho(\mathbf{y}) = 4\pi\rho(\mathbf{y})$ for $n = 3$.]

In dealing with physical problems we are often asked to solve boundary value problems for partial differential equations. In particular, in electrostatics the electric field, $\mathbf{E}$, is given by $\mathbf{E} = -\text{grad}\, u$, where u is the potential, and $-\Delta u = \text{div}\, \mathbf{E} = 4\pi\rho$, where $\rho(\mathbf{x})$ is the "volume" charge density. In a region of space in which there are no charges ($\text{div}\, \mathbf{E} = 0$), $\Delta u = 0$. Suppose that we have a closed conducting surface S and that there are no other conductors present. It is a physical fact that the charges are concentrated on the surface and that $\mathbf{E} \cdot \mathbf{n} = 4\pi\sigma$ on the surface exterior, where σ is the surface charge density. We may assume that the potential is zero at infinity. For the charges to be in equilibrium, grad u must be orthogonal to the surface, so that $\mathbf{E} \cdot \mathbf{n} \equiv |\mathbf{E}|$, and u must have a constant value V there. If V is prescribed, we obtain u by solving an *exterior Dirichlet problem* for u:

$$\Delta u = 0 \text{ outside } S, \qquad u = f \text{ on } S, \qquad u = 0 \text{ at infinity},$$

with constant boundary data $f = V$. The charge density $\sigma = -(4\pi)^{-1}\partial u/\partial n$ ($\mathbf{n}$ being the outer normal to S) can be obtained after the fact. It is easy to see that u is linear in V; u/V (the solution for $V = 1$) is called the *conductor* or *capacitary potential* of S. If instead the total charge $Q = \int_{\partial\Omega} \sigma \, dS$ is prescribed, as frequently occurs in actual physical problems, the constant value V can be determined from the prescription of Q. In any case, for a given conductor Q is proportional to V and the capacitance of the conductor is defined by Q/V (see Exercise 2.8). Suppose that we denote the conductor by Ω, so that $S = \partial\Omega$. Then we define

$$c(S) = \int_{\partial\Omega} \sigma \, dS = -\frac{1}{4\pi}\int_{\partial\Omega} \frac{\partial u}{\partial n} \, dS,$$

where u is the capacitary potential of S, and we call this the functional *capacity* of S (or of Ω).

The potential inside the conductor Ω follows by solving the *interior Dirichlet problem* for u:

$$\Delta u = 0 \text{ in } \Omega, \qquad u = f \text{ on } S = \partial\Omega,$$

with constant value $f = V$. Because on physical grounds the electric field $\mathbf{E} = -\text{grad } u$ must vanish inside the conductor, we expect the solution to be a constant, $u = V$.

Boundary value problems of this kind also arise from heat conduction problems. A solution of $\Delta u = 0$ can be interpreted as an equilibrium temperature, $u = u(\mathbf{x})$. Then the interior Dirichlet problem

$$\Delta u = 0 \text{ in } \Omega, \qquad u = f \text{ on } \partial\Omega,$$

where Ω is a bounded domain, asks for the equilibrium temperature in a heat-conducting body Ω for given surface temperature, f (in general, a function of $\mathbf{x}$). In the (interior) *Neumann problem*

$$\Delta u = 0 \text{ in } \Omega, \qquad \partial u/\partial n = g \text{ on } \partial\Omega,$$

the heat flux is prescribed at the boundary. More generally, a boundary relation $\partial u/\partial n + \alpha u = g$ can be prescribed as in Newton's law of cooling (see Chapter

1). In all cases, the heat flux vector (i.e., grad u) does not necessarily vanish, but is solenoidal: A little physical intuition suggests that the temperature cannot have extrema inside the body.

2. General Properties of Harmonic Functions

2.1. Green's Identities. The ideas that we pursue here are built around the divergence theorem, which says

$$\int_{\Omega} \operatorname{div} \mathbf{w}\, d\mathbf{V} = \int_{\partial\Omega} \mathbf{w} \cdot \mathbf{n}\, dS \tag{D}$$

for a vector field $\mathbf{w}$ that belongs, say, to $C^1(\bar{\Omega})$ on a bounded domain Ω in $\mathbb{R}^n$, with outward normal $\mathbf{n}$ on $\partial\Omega$ (see Chapter 1). Some hypothesis on $\partial\Omega$ is needed for this to hold. Certainly $\partial\Omega \in C^1$ suffices and we will simply assume that any region under discussion is such that (D) holds. Immediate consequences are the so-called Green's identities

$$\int_{\Omega} v\Delta u\, dV + \int_{\Omega} \operatorname{grad} u \cdot \operatorname{grad} v\, dV = \int_{\partial\Omega} v \frac{\partial u}{\partial n}\, dS \tag{G1}$$

and

$$\int_{\Omega} (v\Delta u - u\Delta v) dV = \int_{\partial\Omega} \left(v\frac{\partial u}{\partial n} - u\frac{\partial v}{\partial n} \right) dS, \tag{G2}$$

which hold for all $u, v \in C^2(\bar{\Omega})$. If $\mathbf{w} = \operatorname{grad} u,\ u \in C^2(\bar{\Omega})$, in (D) we obtain

$$\int_{\Omega} \Delta u\, dV = \int_{\partial\Omega} \frac{\partial u}{\partial n}\, dS. \tag{CC}$$

If $u, v \in C^2(\Omega) \cap C^1(\bar{\Omega})$, the integrals on Ω can be defined as improper integrals. Green's identities (G1) and (G2) have several important consequences. First of all, direct differentiation of the fundamental solution $E(r) = E(|\mathbf{x} - \mathbf{y}|)$ shows that

$$E'(r) = -r^{1-n}/\omega_n, \qquad |DE(\mathbf{x} - \mathbf{y})| \leq |\mathbf{x} - \mathbf{y}|^{1-n}/\omega_n,$$

where D denotes any $\mathbf{x}$ derivative. Suppose that $\mathbf{x} \in \Omega$. If we apply (G2) to $u \in C^2(\bar{\Omega})$ and $E(\mathbf{x}-\mathbf{y})$ on $\Omega \backslash B(\mathbf{x}, \rho)$, where ρ is small enough that $\overline{B(\mathbf{x}, \rho)} \subset \Omega$, we obtain

$$\int_{\Omega \backslash B} E\Delta u \, d\mathbf{y} = \int_{\Sigma} \left(E\frac{\partial u}{\partial n} - u\frac{\partial E}{\partial n}\right) dS_y - \int_{\partial B} \left(E\frac{\partial u}{\partial \rho} - u\frac{\partial E}{\partial \rho}\right) dS_y \tag{1}$$

[$\Sigma = \partial\Omega$, $B = B(\mathbf{x}, \rho)$, and the integration variable is $\mathbf{y}$]. We wish to take the limit as ρ goes to zero. We have that

$$\left|\int_{\partial B} E\frac{\partial u}{\partial n}\, dS\right| = \left|E(\rho)\int_{\partial B} \frac{\partial u}{\partial n}\, dS\right|$$

is majorized by $\omega_n \rho^{n-1} E(\rho) \sup_B |Du|$,

$$\int_{\partial B} u\frac{\partial E}{\partial \rho}\, dS = E'(\rho)\int_{\partial B} u \, dS = -\frac{1}{\omega_n \rho^{n-1}}\int_{\partial B} u\, dS \to -u(\mathbf{x})$$

as $\rho \to 0$, and $E\Delta u$ is integrable on Ω, so that in the limit we find the "Green representation theorem," or Green's third identity,

$$u(\mathbf{x}) = \int_{\partial\Omega} \left(E\frac{\partial u}{\partial n} - u\frac{\partial E}{\partial n}\right) dS_y - \int_{\Omega} E\Delta u \, d\mathbf{y} \qquad \text{for } \mathbf{x} \in \Omega. \tag{R}$$

An arbitrary smooth function on a bounded domain can be represented in terms of single and double layer potentials on the boundary and a volume potential. In particular, if $\Delta u = 0$, we have a representation solely in terms of layer potentials on the boundary.

If the point $\mathbf{x} = \mathbf{x}_0$ is on the boundary, we have to take in (1) $\Sigma = \{\mathbf{y} \in \partial\Omega \colon |\mathbf{y} - \mathbf{x}_0| \geq \rho\}$ and B the intersection of $B(\mathbf{x}, \rho)$ with $\bar{\Omega}$. If $\partial\Omega$ is smooth, ∂B tends to a hemisphere as $\rho \to 0$, and

$$\int_{\partial B} u\frac{\partial E(\mathbf{x}_0, \mathbf{y})}{\partial \rho}\, dS_y = E'(\rho)\int_{\partial B} u\, dS = -\frac{1}{\omega_n \rho^{n-1}}\int_{\partial B} u\, dS \to -\frac{1}{2}u(\mathbf{x}_0) \tag{2}$$

as $\rho \to 0$. Because $E\Delta u$ is integrable on Ω, $\int_{\Omega\backslash B} E\Delta u\, dV \to \int_\Omega E\Delta u \, dV$ so that the integral $\int_{\partial\Omega}[E(\partial u/\partial n) - u(\partial E/\partial n)dS$ exists at least in the Cauchy principal value sense (actually, it exists as an ordinary Lebesgue integral if $\partial\Omega$ is smooth, see Section 4). As a result we obtain

$$\frac{1}{2}u(\mathbf{x}) = \int_{\partial\Omega}\left(E\frac{\partial u}{\partial n} - u\frac{\partial E}{\partial n}\right)dS_y - \int_{\Omega} E\Delta u\, d\mathbf{y} \qquad \text{for } \mathbf{x} \in \partial\Omega. \tag{R$'$}$$

Finally, because $E(\mathbf{x}, \mathbf{y})$ is regular harmonic for $\mathbf{x} \in \Omega^c = \mathbb{R}^3\backslash\bar{\Omega}$ and $\mathbf{y} \in \partial\Omega$, we find

$$0 = \int_{\partial\Omega}\left(E\frac{\partial u}{\partial n} - u\frac{\partial E}{\partial n}\right)dS_y - \int_{\Omega} E\Delta u\, d\mathbf{y} \qquad \text{for } \mathbf{x} \in \Omega^c. \tag{R$''$}$$

A further important consequence is the following result; The (interior) *Neumann problem*

$$\Delta u = F \text{ in } \Omega, \qquad \partial u/\partial n = g \text{ on } \partial\Omega$$

has at most one solution $u \in C^2(\Omega) \cap C^1(\bar{\Omega})$ determined up to an arbitrary additive constant. (This is as good as we can do, because the problem only involves derivatives of u.) The proof follows immediately from (G1) applied with $v = u$. As the problem is linear, it suffices to consider $g \equiv F \equiv 0$. Then

$$\int_{\Omega} |\text{grad } u|^2\, d\mathbf{x} = \int_{\partial\Omega} u\frac{\partial u}{\partial n}\, dS \equiv 0$$

so that grad $u=0$ on Ω, and u is constant. Frequently an additional condition such as $\int_{\Omega} u\, d\mathbf{x} = 0$ or $\int_{\partial\Omega} u\, dS = 0$ or $u(\mathbf{x}_0) = 0$, $\mathbf{x}_0$ a point of $\bar{\Omega}$, is used to single out a solution.

Concerning existence, with $g \in C(\partial\Omega)$ and (say) $F \in C(\bar{\Omega})$, we observe that (CC) implies that

$$\int_{\partial\Omega} g\, dS = \int_{\Omega} F\, d\mathbf{x},$$

i.e., a compatibility condition between g and F must be satisfied. If $F \equiv 0$, this takes the form

$$\int_{\partial\Omega} g\, dS = 0. \tag{3}$$

Physically, this condition states that the net (heat) flux is zero at equilibrium if there are no (heat) sources.

2.2. Maximum Principle. Our first result is an immediate consequence of Green's identities.

Theorem 2.1. Suppose $u \in C^2(\Omega)$ and $\Delta u = 0$ in Ω. Then, if $\mathrm{B}(\mathbf{x}, \mathrm{R}) \subset \Omega$, $r = |\mathbf{x} - \mathbf{y}|$,

$$u(\mathbf{x}) = \frac{1}{\omega_n \rho^{n-1}} \int_{r=\rho} u(\mathbf{y}) dS_y = \frac{n}{\omega_n \rho^n} \int_{r \leq \rho} u\, dV \qquad \text{(MV)}$$

for $0 < \rho \leq R$.

Proof. Apply the identity (CC) to u on $\mathrm{B}_\rho = \mathrm{B}(\mathbf{x}, \rho)$, $\rho \leq \mathrm{R}$; introduce $\omega = (\mathbf{y} - \mathbf{x})/\rho$, $u(\mathbf{y}) = u(\mathbf{x} + \rho\omega)$ to obtain

$$0 = \int_{r=\rho} \frac{\partial u}{\partial \rho}\, dS = \rho^{n-1} \int_{|\omega|=1} \frac{\partial u}{\partial \rho}(\mathbf{x} + \rho\omega) d\omega = \rho^{n-1} \frac{\partial}{\partial \rho} \int_{|\omega|=1} u(\mathbf{x} + \rho\omega) d\omega$$
$$= \rho^{n-1} \frac{\partial}{\partial \rho}\left(\rho^{1-n} \int_{r=\rho} u(\mathbf{y}) dS_y\right)$$

so that

$$\rho^{1-n} \int_{r=\rho} u\, dS = \mathrm{R}^{1-n} \int_{r=R} u\, dS, \qquad 0 < \rho \leq \mathrm{R}.$$

By continuity the left-hand side approaches $\omega_n u(\mathbf{x})$ as ρ goes to zero. The second equality follows by integrating the first with respect to ρ. □

If instead $\Delta u \geq 0$, then we obtain the corresponding inequality for u:

$$u(\mathbf{x}) \leq \frac{1}{\omega_n \rho^{n-1}} \int_{r=\rho} u(\mathbf{y}) dS_y.$$

We call functions in $C^2(\Omega)$ that have this property *subharmonic*. In the opposite case $\Delta u < 0$, u is *superharmonic*.

Theorem 2.2. Suppose $u \in C^2(\Omega)$ is subharmonic and $u(\mathbf{y}) = \sup_\Omega u$ for some $\mathbf{y} \in \Omega$. Then u is constant in Ω.

Proof. Let $N = \sup_\Omega u$, and $\Omega_M = \{\mathbf{x} \in \Omega : u = M\}$. By hypothesis Ω_M is nonempty, and continuity of u implies Ω_M is closed in Ω. For $\mathbf{z} \in \Omega_M$, $B = B(\mathbf{z}, R) \subset \Omega$,

$$0 \leq \int_{\partial B} u\, ds - \omega_n R^{n-1} u(\mathbf{z}) = \int_{\partial B} (u - u(\mathbf{z}))dS.$$

As $u - u(\mathbf{z}) = u - M \leq 0$ and u is continuous, this implies $u \equiv M$ on $B(\mathbf{z}, R)$. Hence, Ω_M is both open and closed and it must be all of Ω. □

Thus, either u is constant or $u(\mathbf{y}) < \sup_\Omega u$ for all $\mathbf{y} \in \Omega$. This theorem is called the *strong maximum principle*. As an immediate consequence we have

Corollary 2.1. if $u \in C^2(\Omega) \cap C(\bar{\Omega})$, $\Delta u \geq 0$ in Ω, then either u is constant or $u(\mathbf{y}) < \max_{\partial\Omega} u$ for all $\mathbf{y} \in \Omega$.

The weaker result that $u \in C^2(\Omega) \cap C(\bar{\Omega})$, $\Delta u \geq 0$ in Ω implies $\max_{\bar{\Omega}} u = \max_{\partial\Omega} u$, is called the *weak maximum principle* (and can be proved independently). The analogous statements with inf in place of sup hold if u is superharmonic. If u is harmonic, then u is both subharmonic and superharmonic. Hence, if $u \in C(\bar{\Omega})$ and $\Delta u = 0$ in Ω,

$$\max_{\bar{\Omega}} |u| = \max_{\partial\Omega} |u|.$$

This implies a uniqueness theorem.

Corollary 2.2. Suppose $u, v \in C^2(\Omega) \cap C(\bar{\Omega})$, $u = v$ on $\partial\Omega$, and $\Delta u = \Delta v$ in Ω. Then $u = v$ in Ω.

A similar simple, but important, consequence is that u harmonic, v subharmonic in Ω, $u = v$ on $\partial\Omega$ implies $u \geq v$ on Ω.

A further consequence of the maximum principle is the following a priori estimate.

Theorem 2.3. Suppose $u \in C^2(\Omega) \cap C(\bar{\Omega})$, and let $l := \operatorname{diam}(\Omega)$. Then

$$|u(\mathbf{x})| \leq \max_{\partial\Omega} |u| + \frac{l^2}{2n} \sup_\Omega |\Delta u| \qquad \text{for every } \mathbf{x} \in \bar{\Omega}. \tag{AP}$$

Proof. Let $w_\pm(\mathbf{x}) = \pm u(\mathbf{x}) + \max_{\partial\Omega}|u| + (1/2n)(l^2 - |\mathbf{x} - \mathbf{x}_0|^2) \sup_\Omega|\Delta u|$, where $\mathbf{x}_0$ is an arbitrary point of $\partial\Omega$. Then $\Delta w_\pm \leq 0$ in Ω, $w_\pm \geq 0$ on $\partial\Omega$, the maximum principle implies $w_\pm \geq 0$ in $\bar{\Omega}$, that is, $\pm u(\mathbf{x}) \leq \max_{\partial\Omega} |u| + (1/2n)l^2 \sup_\Omega |\Delta u|$. □

Corollary 2.3. The interior Dirichlet problem

$$\Delta u = F \text{ in } \Omega, \qquad u = f \text{ on } \partial\Omega$$

with $f \in C(\partial\Omega)$ and (say) $F \in C(\bar{\Omega})$, has at most one solution $u \in C^2(\Omega) \cap C(\bar{\Omega})$ that depends continuously on f and F.

Proof. From (AP) we have $|u(\mathbf{x})| \leq \max_{\partial\Omega} |f| + (l^2/2n) \max_{\Omega} |F|$. □

Another important consequence of (MV) is the following.

Theorem 2.4 (Harnack's inequality, general form). if u is a nonnegative harmonic function in Ω and K is a compact subdomain, there is a constant $C = C(K, \Omega)$ such that

$$\sup_K u \leq C \inf_K u.$$

Proof. First consider a $\mathbf{y} \in \Omega$ such that $\mathrm{B}(\mathbf{y}, 4\mathrm{R}) \subset \Omega$. Then for $\mathbf{x}_1, \mathbf{x}_2 \in \mathrm{B}(\mathbf{y}, \mathrm{R})$,

$$u(\mathbf{x}_1) = \frac{1}{|B_R|} \int_{\mathrm{B}(\mathbf{x}_1, \mathrm{R})} u \, dV \leq \frac{1}{|B_R|} \int_{\mathrm{B}(\mathbf{y}, 2\mathrm{R})} u \, dV$$

and

$$u(\mathbf{x}_2) = \frac{1}{|B_{3R}|} \int_{\mathrm{B}(\mathbf{x}_2, 3\mathrm{R})} u \, dV \geq \frac{1}{|B_{3R}|} \int_{\mathrm{B}(\mathbf{y}, 2\mathrm{R})} u \, dV$$

so that $u(\mathbf{x}_1) \leq 3^n u(\mathbf{x}_2)$ and $\sup_{\mathrm{B}} u \leq 3^n \inf_{\mathrm{B}} u$, $B = \mathrm{B}(\mathbf{y}, \mathrm{R})$. Now if $\mathbf{x}_1, \mathbf{x}_2 \in K$ with $\sup_K u = u(\mathbf{x}_1)$ and $\inf_k u = u(\mathbf{x}_2)$, suppose Γ is an arc connecting $\mathbf{x}_1$ and $\mathbf{x}_2$ in K. Suppose $R < \frac{1}{4} \operatorname{dist}(K, \partial\Omega)$. The balls $\mathrm{B}(\mathbf{x}, \mathrm{R})$, $\mathbf{x} \in K$ cover K and there is a finite subcover by the Heine–Borel theorem. It follows that there is a finite chain of intersecting balls connecting $\mathbf{x}_1$ and $\mathbf{x}_2$ along Γ, and the number of balls required, say N, depends only on K and Ω. It follows that $u(\mathbf{x}_1) \leq 3^{nN} u(\mathbf{x}_2)$ and the theorem holds with $\mathrm{C} = 3^{nN}$. □

If $\partial\Omega$ is somewhat smooth at a boundary point, the maximum principle can be refined.

Definition 2.1. $\partial\Omega$ is said to satisfy the interior sphere condition at Q if there is a ball $B \subset \Omega$ with $S = \partial B$ passing through Q. ■

Theorem 2.5 (Boundary point lemma). Suppose $\Delta u \geq 0$ in Ω, u is continuous at $\mathbf{x}_0 \in \partial\Omega$, and $u(\mathbf{x}) < u(\mathbf{x}_0)$ in Ω. Suppose that $\partial\Omega$ satisfies the interior sphere condition at $\mathbf{x}_0$. Then if $\partial u(\mathbf{x}_0)/\partial n$ exists, $\partial u(\mathbf{x}_0)/\partial n > 0$.

Proof. Let $\mathscr{V} := e^{-\alpha r^2} - e^{-\alpha R^2}$ where $B(\mathbf{y}, R)$ is given in the interior sphere condition and $r = |\mathbf{x} - \mathbf{y}|$ (Fig. 2).

Suppose $0 < \rho < R$ and choose α so that $\Delta\mathscr{V} \geq 0$ for $\rho \leq r \leq R$. Then choose $\varepsilon > 0$ such that $u(\mathbf{x}) - u(\mathbf{x}_0) + \varepsilon\mathscr{V} \leq 0$ for $r = \rho, R$. Because then $u(\mathbf{x}) - u(\mathbf{x}_0) + \varepsilon\mathscr{V} \leq 0$ for $\rho \leq r \leq R$ and vanishes at $\mathbf{x}_0$,

$$\frac{\partial u}{\partial n}(\mathbf{x}_0) + \varepsilon\frac{\partial \mathscr{V}}{\partial n}(\mathbf{x}_0) \geq 0,$$

that is, $\partial u(\mathbf{x}_0)/\partial n \geq -\varepsilon\mathscr{V}'(R) > 0$. □

2.3. Green's Function, Poisson's Integral, and Mean Value Theorem. If $h(\mathbf{y})$ is a harmonic function in $C^1(\bar{\Omega}) \cap C^2(\Omega)$, (G2) implies

$$\int_{\partial\Omega}\left(h\frac{\partial u}{\partial n} - u\frac{\partial h}{\partial n}\right)dS = \int_{\Omega} h\Delta u \, d\mathbf{y},$$

and if we write $G(\mathbf{x}, \mathbf{y}) = E(\mathbf{x}, \mathbf{y}) + h(\mathbf{y})$, we find from (R),

$$u(\mathbf{x}) = \int_{\partial\Omega}\left(G\frac{\partial u}{\partial n} - u\frac{\partial G}{\partial n}\right)dS_y - \int_{\Omega} G\Delta u \, d\mathbf{y}, \qquad \mathbf{x} \in \Omega.$$

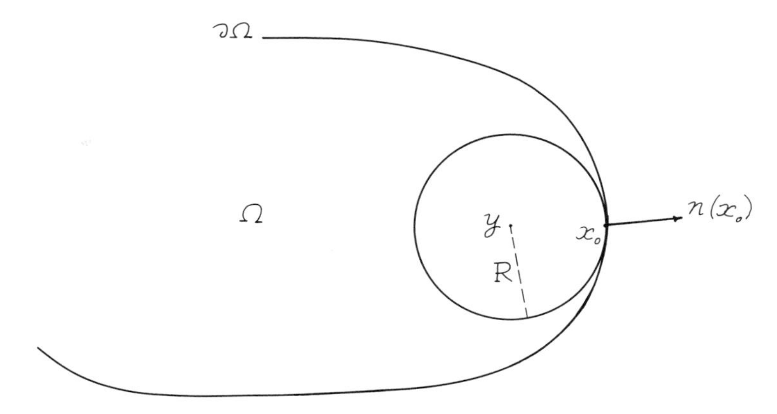

Fig. 2. The interior sphere condition in the boundary point lemma.

Of particular interest is the case in which there exists h such that $G(\mathbf{x}, \mathbf{y}) = 0$, or $h(\mathbf{y}) = -E(\mathbf{x}, \mathbf{y})$, for $y \in \partial\Omega$, $\mathbf{x} \in \Omega$. [This is true under mild smoothness assumptions on $\partial\Omega$; note that now $h = h(\mathbf{x}, \mathbf{y})$ depends also on $\mathbf{x}$.] We reserve the notation G for this situation and G is called the *Green's function* for Ω. We note that the maximum principle implies that G is unique and is positive for $\mathbf{x} \neq \mathbf{y}$. Another useful property is given in

Theorem 2.6. $G(\mathbf{x}, \mathbf{y}) = G(\mathbf{y}, \mathbf{x})$.

Proof. Apply (G2) to $u(\mathbf{y}) = G(\mathbf{x}_1, \mathbf{y})$ and $v(\mathbf{y}) = G(\mathbf{x}_2, \mathbf{y})$ on $\Omega \backslash B(\mathbf{x}_1, \rho) \cup B(\mathbf{x}_2, \rho)$ and invoke the ideas used in deriving (R). Because $E(\mathbf{x}, \mathbf{y}) = E(\mathbf{y}, \mathbf{x})$, what we actually prove is that $h(\mathbf{x}, \mathbf{y}) = h(\mathbf{y}, \mathbf{x})$. Thus, h is harmonic also in $\mathbf{x}$. □

We come to the problem of finding the harmonic function $h(\mathbf{x}, \mathbf{y})$ that determines the Green's function for a particular region Ω. In general, this is a difficult problem, but in some special cases the electrostatic interpretation (for $n = 3$) enables us to give a concrete construction. The Green's function may be thought of as the electrostatic potential at the point $\mathbf{y}$ of a unit charge located at $\mathbf{x}$ if the bounding surface of Ω is grounded, i.e., has a zero potential. This potential is the sum of the Coulomb potential $E(\mathbf{x}, \mathbf{y})$ of the unit charge plus the potential $h(\mathbf{x}, \mathbf{y})$ due to the "induced charges" on $\partial\Omega$. We will think of h as arising from imaginary charges located in Ω^c that are such that their potential on $\partial\Omega$ cancels that of the unit charge inside Ω. This leads to the *method of images*.

Example 2.1. $\Omega = \{\mathbf{x} : |\mathbf{x}| < a\}$. The appropriate image charge is located at $a^2\mathbf{x}/|\mathbf{x}|^2$ and has magnitude $-(a/|\mathbf{x}|)^{n-2}$. We have

$$\omega_n(n-2)G(\mathbf{x}, \mathbf{y}) = |\mathbf{y} - \mathbf{x}|^{2-n} - \left(\frac{a}{|\mathbf{x}|}\right)^{n-2} |\mathbf{y} - a^2\mathbf{x}|\mathbf{x}|^{-2}|^{2-n}.$$

A straightforward calculation shows that

$$-\frac{\partial G}{\partial n}(\mathbf{x}, \mathbf{y}) = \frac{1}{\omega_n a} \frac{a^2 - |\mathbf{x}|^2}{|\mathbf{y} - \mathbf{x}|^n} = K_a(\mathbf{x}, \mathbf{y})$$

on $|\mathbf{y}| = a$ (this formula holds also for $n = 2$). The expression on the right is called the *Poisson kernel*. It is an indispensable tool in deriving properties of harmonic functions, and more specifically in solving the Dirichlet problem

$$\Delta u = 0 \text{ in } \Omega, \qquad u \in C(\bar{\Omega}) \cap C^2(\Omega), \qquad u = f \text{ on } \partial\Omega \tag{DI}$$

for (the interior of) a sphere.

The Poisson kernel $K_a(\mathbf{x}, \mathbf{y})(\mathbf{y} \in \partial\Omega)$ has the following basic properties:

i. $K_a(\mathbf{x}, \mathbf{y}) > 0$, $K_a(\mathbf{x}, \mathbf{y}) \in C^\infty$ for $\mathbf{x} \in \Omega$,
ii. $\int_{|\mathbf{y}|=a} K_a(\mathbf{x}, \mathbf{y}) dS_y = 1$ for $\mathbf{x} \in \Omega$,
iii. $\lim_{|\mathbf{x}| \to a} \int_{S_T} K_a(\mathbf{x}, \mathbf{y}) dS_y = 0$ for each sufficiently small $\gamma > 0$,

where $S_\gamma = \{|\mathbf{y}| = a \colon |\mathbf{y} - \mathbf{x}| \geq \gamma\}$. Property (i) is obvious, (ii) follows from the fact that $u = 1$ is the unique solution with boundary values $f = 1$, and (iii) follows from the inequality

$$K_a(\mathbf{x}, \mathbf{y}) \leq \frac{1}{\omega_n a} \frac{a^2 - |\mathbf{x}|^2}{\gamma^n}$$

for $\mathbf{y} \in S_\gamma$. These properties imply that $K_a(\mathbf{x}, \mathbf{y})$ is a C^∞ "delta-approximate" family, i.e., $K_a(\mathbf{x}, \mathbf{y}) \to \delta(\mathbf{x} - \mathbf{y})$ in the sense of distributions as $|\mathbf{x}| \to a$ (Chapter 8).

Theorem 2.7. If f is continuous on the sphere $S_a : |\mathbf{y}| = a$, the unique solution of the Dirichlet problem on $|\mathbf{x}| < a$ is given by the Poisson integral

$$u(\mathbf{x}) = \int_{S_a} K_a(\mathbf{x}, \mathbf{y}) f(\mathbf{y}) dS_y. \tag{P}$$

Proof. The only thing to prove is that u given by (P) takes on the boundary values f. Consider $\mathbf{x}_0$ with $|\mathbf{x}_0| = a$. Take a ball B_0 about $\mathbf{x}_0$ of radius ρ and $\mathbf{x}$ such that $|\mathbf{x} - \mathbf{x}_0| \leq \rho/2$, $|\mathbf{x}| < a$. From the property (i) of the Poisson kernel we have

$$\begin{aligned} |u(\mathbf{x}) - f(\mathbf{x}_0)| &= \left| \int_{S_a} K_a(\mathbf{x}, \mathbf{y})(f(\mathbf{y}) - f(\mathbf{x}_0)) dS_y \right| \leq \int_{S_a} K_a(\mathbf{x}, \mathbf{y}) |f(\mathbf{y}) - f(\mathbf{x}_0)| dS_y \\ &\leq \int_{T_1} K_a(\mathbf{x}, \mathbf{y}) |f(\mathbf{y}) - f(\mathbf{x}_0)|\, dS_y + 2 \max_{S_a} |f(\mathbf{y})| \int_{T_2} K_a(\mathbf{x}, \mathbf{y}) dS_y, \end{aligned}$$

where $T_1 = S_a \cap B_0 : |\mathbf{y} - \mathbf{x}_0| \leq \rho$, $T_2 = S_a \backslash B_0 : |\mathbf{y} - \mathbf{x}_0| \geq \rho$, so that $|\mathbf{y} - \mathbf{x}| \geq \rho/2$ in the second integral. Because f is continuous, for arbitrary $\varepsilon > 0$ we can fix ρ so that $|f(\mathbf{y}) - f(\mathbf{x}_0)| \leq \varepsilon$ in the first integral. Then by applying property (ii) we have

$$|u(\mathbf{x}) - f(\mathbf{x}_0)| \leq \varepsilon + 2 \max_{S_a} |f(\mathbf{y})| \int_{T_2} K_a(\mathbf{x}, \mathbf{y}) dS_y,$$

and by property (iii) with $\gamma = \rho/2$ the last integral can also be made smaller than ε by letting $\mathbf{x}$ get close to $\mathbf{x}_0$, so that the result follows. □

Note that (P) is a special case of the general identity

$$u(\mathbf{x}) = -\int_{\partial\Omega} u(\mathbf{y}) \frac{\partial G}{\partial n_y}(\mathbf{x}, \mathbf{y}) dS_y \qquad (\mathbf{x} \in \Omega), \tag{G3}$$

which follows from (R) for any function u harmonic in Ω and continuously differentiable up to the boundary. Uniqueness is implied by this identity. However, the result in Theorem 2.7 is stronger because it shows that, given *arbitrary* continuous boundary values, the harmonic function exists and can be represented by the boundary integral.

Remark 2.1. Let $\mathbf{x}$ approach the boundary sphere along the normal direction, $\mathbf{x} = \lambda\mathbf{x}_0$, $\lambda \to 1$, for $|\mathbf{x}_0| = a$. Then $K_a(\mathbf{x}, \mathbf{y}) \to 0$ for every $u \neq \mathbf{x}_0$ and we have an example of a function, harmonic in a domain and approaching zero at every boundary point $\mathbf{y}$ except one, $\mathbf{x}_0$. For $\mathbf{y} - \mathbf{x}_0$ we have, for $r = |\mathbf{x}_0 - \mathbf{x}|$,

$$K_a(\mathbf{x}, \mathbf{x}_0) \sim 2/[\omega_n a^{n-1}(1-\lambda)^{n-1}], \qquad -dE(r)/dr \sim 1/[\omega_n a^{n-1}(1-\lambda)^{n-1}]$$

as $\lambda \to 1$. This shows that *the singularity of* $(\partial G/\partial n)(\mathbf{x}, \mathbf{x}_0)$ *is twice the singularity of* $(\partial E/\partial n)(\mathbf{x}, \mathbf{x}_0)$ *at* $\mathbf{x} = \mathbf{x}_0$. In fact, the right-hand side of (G3) tends to $u(\mathbf{x}_0)$ as $\mathbf{x}$ tends to the boundary point $\mathbf{x}_0$, whereas the integral

$$-\int_{\partial\Omega} u(\mathbf{y}) \frac{\partial E}{\partial n_y}(\mathbf{x}, \mathbf{y}) dS_y$$

tends to $\frac{1}{2}u(\mathbf{x}_0)$ [see equation (2)].

We pause here to establish an interesting and useful characterization of harmonic functions.

Definition 2.2. A function u has the *mean value property* (MVP) on a domain Ω if the relation (MV)

$$u(\mathbf{x}) = \frac{1}{\omega_n \rho^{n-1}} \int_{|\mathbf{y}-\mathbf{x}|=\rho} u(\mathbf{y}) dS_y$$

is satisfied for each $\mathbf{x} \in \Omega$ and for each $\rho > 0$ such that $|\mathbf{y} - \mathbf{x}| \leq \rho$ implies $\mathbf{y} \in \Omega$.

We have seen in Theorem 2.1 that a harmonic function has the mean value property. The converse is also true.

Theorem 2.8. A function u continuous in $\bar{\Omega}$ is harmonic in Ω if and only if u has the mean value property.

Proof. Suppose u has the mean value property on Ω, and consider a closed ball $\bar{B} \subset \Omega$. Let v be the harmonic function in B that has boundary values u on $S = \partial B$. Then $w = v - u$ is zero on S and has the mean value property in B. It suffices now to prove that a continuous function that has the mean value property does not take on its maximum or minimum at an interior point unless it is constant. We have actually already proven this, as the only property of harmonic functions required for the strong maximum principle is the mean value property. □

One interesting application of the mean value property is the Schwartz reflection principle stated here as a corollary to Theorem 2.8.

Corollary 2.4. Suppose u is harmonic in a domain Ω that has as part of its boundary a planar region L and that $u = 0$ on L. Then, if P* is the reflection of P across L, and Ω^* the reflected domain, $u(P^*) = -u(P)$ yields a harmonic function on $\Omega \cup \Omega^* \cup L$.

Proof. The extended u is continuous and has the mean value property on sufficiently small spheres. The result then follows from the fact that a function that has the mean value property on sufficiently small spheres must be harmonic. □

2.4. Other Consequences of the Poisson Formula and the Mean Value Property. The Poisson integral representation of a harmonic function can be translated to any sphere, and is a very useful tool.

Theorem 2.9 (Harnack's inequality). Suppose that u is harmonic and nonnegative in $B(\mathbf{x}_0, R)$. Then, if $\mathbf{x} \in B(\mathbf{x}_0, R)$,

$$u(\mathbf{x}_0)\frac{1 - |\mathbf{x} - \mathbf{x}_0|/R}{(1 + |\mathbf{x} - \mathbf{x}_0|/R)^{n-1}} \leq u(\mathbf{x}) \leq \frac{1 + |\mathbf{x} - \mathbf{x}_0|/R}{(1 - |\mathbf{x} - \mathbf{x}_0|/R)^{n-1}} u(\mathbf{x}_0).$$

Proof. Suppose, without loss of generality, that $\mathbf{x}_0 = 0$. Then, for $a < R$, $|\mathbf{x}| < a$, and $|\mathbf{y}| = a$, we have $a - |\mathbf{x}| \leq |\mathbf{y} - \mathbf{x}| \leq a + |\mathbf{x}|$ so that

$$\frac{1}{\omega_n a}\frac{a - |\mathbf{x}|}{(a + |\mathbf{x}|)^{n-1}} \leq K_1(\mathbf{x}, \mathbf{y}) \leq \frac{1}{\omega_n a}\frac{a + |\mathbf{x}|}{(a - |\mathbf{x}|)^{n-1}}.$$

Because u is nonnegative, the Poisson integral representation and the mean value property imply the result for the ball $B(\mathbf{0}, a)$. The conclusion of the theorem follows if we let a approach R. □

Corollary 2.5 (Liouville's theorem). A function v harmonic on $\mathbb{R}^n$ and bounded (above or below) is a constant.

Proof. For $v \geq a > -\infty$, set $u - v - a$. Then letting let $R \to \infty$ in Harnack's inequality for u, we find $v = a + u(\mathbf{x}_0)$. If v is bounded above, apply this argument to $-v$. □

One important use of Harnack's inequality is the establishment of the following two theorems, known as *Harnack's convergence theorems.*

Theorem 2.10. Suppose that $u_n \in C(\bar{\Omega})$, each u_n is harmonic in Ω, and $\underline{u}_n$ converges uniformly on $\partial\Omega$. Then, u_n converges uniformly to a function u on $\bar{\Omega}$, u is harmonic in Ω, and for any partial derivative D^k, $D^k u_n$ converges uniformly to $D^k u$ on compact subsets of Ω.

Proof. The maximum principle implies that u_n converges uniformly on Ω to a function u, and this function must satisfy the mean value property, hence it is harmonic. The conclusion about derivatives of u follows from the Poisson integral representation on a ball B with $\bar{B} \subset \Omega$, and the fact that the kernel and its derivatives are bounded on any ball B_0 with $B_0 \subset B$. □

Theorem 2.11. Suppose u_N is an increasing sequence of harmonic functions on Ω. Then, either $\lim u_N = \infty$ for each point of Ω, or $\lim u_N = u$ where u is harmonic in Ω and the convergence is uniform on compact subsets of Ω.

Proof. Suppose $u_N(\mathbf{x}_0)$. converges. Consider $\mathbf{x}$ with $|\mathbf{x} - \mathbf{x}_0| < \operatorname{dist}(\mathbf{x}_0, \partial\Omega) = R$. By Harnack's inequality, if $N > M$,

$$0 \leq u_N(\mathbf{x}) - u_M(\mathbf{x}) \leq \frac{1 + |\mathbf{x} - \mathbf{x}_0|/R}{(1 - |\mathbf{x} - \mathbf{x}_0|/R)^{n-1}} [u_N(\mathbf{x}_0) - u_M(\mathbf{x}_0)],$$

so u_N converges in $B(\mathbf{x}_0, R)$, and uniformly on any subball. Hence, the set of points of Ω where u_N converges is open. Suppose now that u_N diverges at $\mathbf{x}_1$, then the same argument shows that u_N diverges in a sufficiently small neighborhood of $\mathbf{x}_1$. Because the sets of points where u_N converges and diverges, respectively, are both open, one must be empty. □

It is appropriate to remark at this point that the Poisson integral representation implies that harmonic functions are infinitely differentiable at interior points of their domains. In fact, more is true.

Theorem 2.12. if u is harmonic in Ω, then u is (real) *analytic* in Ω.

Proof. We give the proof for $n = 3$ variables. Suppose that the ball B is such that $\bar{B} \subset \Omega$, and $\overline{B_0} \subset B$, B_0 a concentric ball (Fig. 3).

We assert that

$$|D^m u(\mathbf{x})| \leq 3^m e^{m-1} m!(\rho - \rho_0)^{-m} \max_S |u|$$

for $\mathbf{x} \in \overline{B_0}$, where D^m is any mth-order partial derivative, ρ_0, ρ are the radii of B_0, B, respectively, and $S = \partial B$. The proof is by induction. Suppose $m = 1$, $D = \partial/\partial x_i$. Then, as Du is harmonic and satisfies the MVP, we have

$$Du(\mathbf{x}) = \frac{3}{4\pi(\rho - \rho_0)^3} \int_{r \leq \rho - \rho_0} Du(\mathbf{y}) d\mathbf{y} = \frac{3}{4\pi(\rho - \rho_0)^3} \int_{r = \rho - \rho_0} n_i u(\mathbf{y}) dS_y$$

($r = |\mathbf{y} - \mathbf{x}|$), whence

$$|Du(\mathbf{x})| \leq \frac{3}{4\pi(\rho - \rho_0)^3} \int_{r = \rho - \rho_0} |u(\mathbf{y})| \, dS_y \leq \frac{3}{(\rho - \rho_0)} \max_S |u|.$$

Now, suppose the assertion is true for m. For $\theta \in (0, 1)$, let B_1 be the concentric ball of radius $\rho_1 = (1 - \theta)\rho + \theta\rho_0$. It follows that $\rho_0 < \rho_1 < \rho$. Suppose $\mathbf{x} \in B_0$, let Δ be the concentric ball of radius $\rho_1 - \rho_0 = (1 - \theta)(\rho - \rho_0)$, and apply

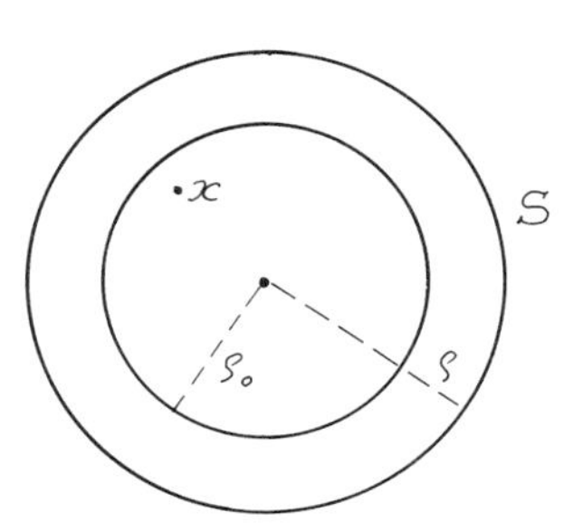

Fig. 3. Balls used for the proof of analyticity of harmonic functions.

the above to $D^m u$ on δ, to obtain

$$|D^{m+1}u(\mathbf{x})| \leq \frac{3}{(1-\theta)(\rho-\rho_0)} \max_{\partial\Delta} |D^m u|$$

for $\mathbf{x} \in \Delta$. The induction hypothesis implies that

$$|D^m u(\mathbf{x})| \leq \frac{3^m e^{m-1} m!}{(\rho-\rho_1)^m} \max_S |u|$$

for $\mathbf{x} \in B_1$. If these last two inequalities are combined, we obtain

$$|D^{m+1}u(\mathbf{x})| \leq \frac{3^{m+1} e^{m-1} m!}{\theta^m (1-\theta)(\rho-\rho_0)^{m+1}} \max_S |u|.$$

Suppose now that we choose $\theta = m/(m+1)$. Then, elementary considerations show that

$$\theta^{-m}(1=\theta)^{-1} < (m+1)e$$

and the assertion follows. The conclusion of the theorem can now be proven by showing that the remainder term R_m, defined by

$$u(\mathbf{x}+\mathbf{h}) = u(\mathbf{x}) + \sum_{k=1}^{m-1} \frac{1}{k!}\left(h_1 \frac{\partial}{\partial x_1} + h_2 \frac{\partial}{\partial x_2} + h_3 \frac{\partial}{\partial x_3}\right)^k u(\mathbf{x}) + R_m(\mathbf{x}, \mathbf{h}),$$

goes to zero as m goes to infinity. We use the expression

$$R_m(\mathbf{x}, \mathbf{h}) = \frac{1}{m!}\left(h_1 \frac{\partial}{\partial x_1} + h_2 \frac{\partial}{\partial x_2} + h_3 \frac{\partial}{\partial x_3}\right)^m u(\mathbf{x}+\theta\mathbf{h})$$

for this quantity. Suppose that $|\mathbf{h}| \leq d = \frac{1}{2}\, dist(\mathbf{x}, \partial\Omega)$. Then, if $B_0 = B(\mathbf{x}, d)$, we have

$$|D^m u(\mathbf{y})| \leq 3^m e^{m-1} m!\, d^{-m} \max_{B_1} |u|$$

for $\mathbf{y} \in \overline{B_0}$, where $B_1 = B(\mathbf{x}, 2d)$, $S_1 = \partial B_1$. This, together with the above expression for R_m, implies that

$$|R_m| \leq (9ed^{-1}|\mathbf{h}|)^m \max_{S_1} |u|.$$

If we choose σ so that $9\sigma e/d$ is less than 1, it follows that $R_m \to 0$ on $|\mathbf{h}| < \sigma$. □

The *strong unique continuation principle* holds as an immediate consequence of Theorem 2.12 and the identity theorem for real analytic functions (see the exercises).

Theorem 2.13. Suppose that u_1, u_2 are harmonic on Ω_1, Ω_2, respectively, and $u_1 = u_2$ on $\Omega_1 \cap \Omega_2$ (nonempty). Then $u_1 = u_2$ on $\Omega_1 \Omega_2$.

Another theorem is the following.

Theorem 2.14. Suppose that u and $\partial u/\partial n$ vanish on a smooth surface contained in the boundary of Ω and u is harmonic on Ω. Then u vanishes on Ω.

The proof requires a lemma.

Lemma 2.1. if Ω_1 and Ω_2 are two disjoint domains whose boundaries have a smooth surface T in common and u_1, u_2 are harmonic in Ω_1, Ω_2 with $u_1 = u_2$ and $\partial u_1/\partial n = \partial u_2/\partial n$ (same sense of normal), then u_1 and u_2 are harmonic continuations of each other.

The proof of the lemma is accomplished by showing that the representation

$$u(\mathbf{x}) = \int_S \left(E\frac{\partial u}{\partial n} - u\frac{\partial E}{\partial n} \right) dS_y$$

holds for $\mathbf{x}$ near T in either Ω_1 or Ω_2, where S is a sphere centered on T with $\mathbf{x}$ in its interior. The proof of the theorem then follows from unique continuation.

The following theorem shows that a harmonic function cannot have an isolated singularity weaker than that of the fundamental solution. In particular, an isolated singularity of a harmonic function is *removable* if the function is bounded.

Theorem 2.15. Suppose that u is harmonic in $B(\mathbf{x}_0, R)\backslash\{\mathbf{x}_0\}$ and $u = o(E(|\mathbf{x} - \mathbf{x}_0|))$ as $\mathbf{x} \to \mathbf{x}_0$. Then $\lim_{\mathbf{x}\to\mathbf{x}_0} u(\mathbf{x})$ exists and the extended function is harmonic.

Proof. We give the proof for $n = 3$ variables. Let v be the harmonic function in $B(\mathbf{x}_0, R/2)$ that coincides with u on $r = |\mathbf{x} - \mathbf{x}_0| = R/2$, and $w - v - u$. If $M_\varepsilon = \max|w|$ on $r = \varepsilon$, then the relation

$$|w(\mathbf{x})| \leq M_\varepsilon \varepsilon/(4\pi r)$$

holds throughout $\varepsilon \leq r \leq R/2$ by the maximum principle, as the function on the right-hand side is harmonic and the same relation holds on the boundary. By assumption the values of $\varepsilon|u|$ on $r = \varepsilon$ approach zero as $\varepsilon \to 0$, so that $\varepsilon M_\varepsilon \to 0$. Because we can hold r fixed and let $\varepsilon \to 0$, w vanishes for all $\mathbf{x} \neq \mathbf{x}_0$, u is bounded, and $\lim_{\mathbf{x}\to\mathbf{x}_0} u(\mathbf{x}) = v(\mathbf{x}_0)$. □

Suppose that $u(\mathbf{x})$ is defined and C^2 in a domain Ω that contains the ball $B(\mathbf{0}, R)$. Consider the spherical inversion transformation $I : \mathbf{x} \to \mathbf{x}' = R^2\mathbf{x}/|\mathbf{x}|^2$, $\mathbf{x} = (x_1, \ldots, x_n)$, so that $I^{-1} : \mathbf{x}' \to \mathbf{x} = R^2\mathbf{x}'/|\mathbf{x}'|^2$, $|\mathbf{x}||\mathbf{x}'| = R^2$. The function

$$v(\mathbf{x}') = \left(\frac{|\mathbf{x}|}{R}\right)^{n-2} u(\mathbf{x})|_{\mathbf{x}=R^2\mathbf{x}'/|\mathbf{x}'|^2} \equiv \left(\frac{R}{|\mathbf{x}'|}\right)^{n-2} u\left(\frac{R^2\mathbf{x}'}{|\mathbf{x}'|^2}\right),$$

defined on $I(\Omega)$, is called the *Kelvin transform* of u. If spherical coordinates are introduced, it is easily seen that

$$\Delta' v = \left(\frac{|\mathbf{x}|}{R}\right)^{n+2} \Delta u.$$

Thus, if u is harmonic in Ω, v is harmonic in $I(\Omega)$ and $v(\mathbf{x}') = O(|\mathbf{x}'|^{2-n})$ as $|\mathbf{x}'| \to \infty$. [It is clear that $I(\Omega)$ contains a neighborhood of infinity.] Note that inversion may be performed with respect to any sphere and the requirement that Ω contain a sphere about $\mathbf{0}$ is not necessary (Exercise 2.1).

Conversely, if $v(\mathbf{x}')$ is harmonic in $I(\Omega)$, then the (inverse) Kelvin transform

$$u(\mathbf{x}) = \left(\frac{|\mathbf{x}'|}{R}\right)^{n-2} v(\mathbf{x}')|_{\mathbf{x}'=R^2\mathbf{x}/|\mathbf{x}|^2} = \left(\frac{R}{|\mathbf{x}|}\right)^{n-2} v\left(\frac{R^2\mathbf{x}}{|\mathbf{x}|^2}\right)$$

is harmonic in $\Omega\backslash\{\mathbf{0}\}$. Theorem 2.15 for $n > 2$ shows that if in addition $v(\mathbf{x}') = o(|\mathbf{x}'|)$ as $|\mathbf{x}'| \to \infty$, $u(\mathbf{x})$ can be extended as a harmonic function in all of Ω. Then $u(\mathbf{x})$ is bounded near $\mathbf{x} = 0$, hence $v(\mathbf{x}') = O(|\mathbf{x}'|^{2-n})$ as $|\mathbf{x}'| \to \infty$. This motivates the following definition.

Definition 2.3. A function $u(\mathbf{x})$ harmonic in an exterior (unbounded) domain $\Omega^c = \mathbb{R}^n\backslash\bar{\Omega}$ is *regular* at infinity if

$$u(\mathbf{x}) = O(1/|\mathbf{x}|^{n-2}) \qquad \text{as } |\mathbf{x}| \to \infty.$$

Theorem 2.15 implies that u is regular at infinity for $n > 2$ if and only if

$$\lim_{|\mathbf{x}|\to\infty} u(\mathbf{x}) = 0$$

uniformly with respect to direction, and u is bounded for $n = 2$. If Ω contains a sphere, say centered at the origin, the Kelvin transform $v(\mathbf{x}')$ of $u(\mathbf{x})$ with respect to this sphere is a regular harmonic function near the origin. A straightforward calculation shows that

$$U_{x_h} = \left(\frac{R}{r}\right)^{n-1} R^{-1} \sum_{k=1}^{n} \left(\delta_{hk} - 2\frac{x_h x_k}{r^2}\right)(r' v_{x'_k} + (n-2)\frac{x'_k}{r'} v),$$

where $r := |\mathbf{x}|$, $r' := |\mathbf{x}'| = R^2/r$. Thus,

$$\operatorname{grad} u = O(1/|\mathbf{x}|^{n-1}) \text{ if } n > 2, \qquad \operatorname{grad} u = O(1/|\mathbf{x}|^2) \text{ if } n = 2$$

as $|\mathbf{x}| \to \infty$, if u is regular at infinity. [We note explicitly that the decay rate of the gradient at infinity is the same in two and three dimensions, $\operatorname{grad} u = O(1/|\mathbf{x}|^2)$.] The following remarks show the significance of this concept.

Remarks

2.2. if u, harmonic on Ω, is regular at infinity, then Green's identities (G1), (G2) hold in an unbounded external domain $\Omega^c = \mathbb{R}^n \backslash \bar{\Omega}$. (We need only apply these identities to a truncated domain bounded by $\partial\Omega$ and a large sphere Σ_R and observe that the integrals on the sphere vanish as R approaches infinity.) This implies that the *exterior Neumann problem*

$$\Delta u = 0 \text{ in } \Omega^c, \qquad \partial u/\partial n = g \text{ on } \partial\Omega, \qquad u \text{ regular at infinity} \qquad \text{(NE)}$$

has a unique solution $u \in C^2(\Omega^c) \cap C^1(\overline{\Omega^c})$, if $n > 2$.

The case $n = 2$ here is special. Because a regular harmonic function u for $n = 2$ does not necessarily vanish at infinity, the solution is determined up to an arbitrary additive constant, which can be taken as the limiting value $u_\infty := \lim_{|\mathbf{x}|\to\infty} u(\mathbf{x})$ of u at infinity. On the other hand, we have seen that $\operatorname{grad} u(\mathbf{x}) = O(|\mathbf{x}|^{-2})$ as $|\mathbf{x}| \to \infty$, so that (CC) holds also for an external domain in $\mathbb{R}^2$. This implies that a solution regular at infinity of (NE) for $n = 2$ exists only if the compatibility condition (3) is satisfied.

2.3. If $u(\mathbf{x}) \in C^2(\Omega^c) \cap C^1(\overline{\Omega^c})$ is harmonic in an exterior domain Ω^c and regular at infinity, Green's representation formulas (R)–(R″) hold for Ω^c,

$$\bar{\omega} u(\mathbf{x}) = -\int_{\partial\Omega} \left(E \frac{\partial u}{\partial n} - u \frac{\partial E}{\partial n} \right) dS_y, \tag{RE}$$

where $\mathbf{n}$ is the outer normal to Ω, and

$$\bar{\omega} = \bar{\omega}(\mathbf{x}) := \begin{cases} 1, & \mathbf{x} \in \Omega^c, \\ 1/2, & \mathbf{x} \in \partial\Omega, \\ 0, & \mathbf{x} \in \Omega. \end{cases}$$

The case $n = 2$ is, again, special because u given by (RE) must vanish at infinity (this implies that $\partial u/\partial n$ has zero average on $\partial\Omega$). Otherwise, the representation (RE) holds for $u - u_\infty$.

2.4. The maximum principle holds in an external domain Ω^c in $\mathbb{R}^n$, $n > 2$, if u is harmonic and regular at infinity (see Exercise 2.4). This implies uniqueness for the *exterior Dirichlet problem*

$$\Delta u = 0 \text{ in } \Omega^c, \qquad u = f \text{ on } \partial\Omega, \qquad u \text{ regular at infinity} \tag{DE}$$

in $\mathbb{R}^n$, $n > 2$. Uniqueness in the case $n = 2$ can be proven by inversion with respect to a circle.

2.5. Suppose we want to find a function harmonic in the exterior of the unit sphere in $\mathbb{R}^3$ with boundary values identically 1. For each a, $u = 1 - a + a/r$ is a solution of this problem, so, without a further restriction, this problem is not well posed. If the restriction that u be regular is added, then there is only one solution, $u = 1/r$. The proof of this fact is an application of the maximum principle after an inversion. The point is that constants are not regular, harmonic on $\mathbb{R}^3$. In contrast, the constants are regular in $\mathbb{R}^2$, and it suffices to require boundedness for the exterior Dirichlet problem there. A similar example in two dimensions is furnished by the function $u = 1 + a \ln r$.

2.5. Volume Potential and Poisson's Equation. Consider the convolution integral in $\mathbb{R}^n$ defining the volume potential

$$\mathscr{V}(\mathbf{x}) = \int_G E(\mathbf{x} - \mathbf{y}) \varrho(\mathbf{y}) d\mathbf{y} \equiv E * \varrho$$

with density $\varrho(\mathbf{y})$ having support in $\bar{G}$, G a bounded domain of $\mathbb{R}^n$ (see Section 1).

Lemma 2.2. If $\varrho(\mathbf{y})$ is bounded $\mathscr{V} \in C^1(\mathbb{R}^n)$, and the force $\mathbf{E} = -\operatorname{grad} \mathscr{V}$ can be obtained by differentiation under the integral.

The proof follows immediately from the expressions of $E(\mathbf{x}, \mathbf{x}_0)$ and grad $E(\mathbf{x}, \mathbf{x}_0)$ and from the results proven in the Appendix to Section 1.

It is useful for us to consider second derivatives of $\mathscr{V}$ because we want to show that $\mathscr{V}$ solves a differential equation inside G as well as outside, and this is more difficult as formal differentiation leads to an integral that is, in general, divergent. In fact, an additional hypothesis on $\varrho(\mathbf{y})$ is needed. Suppose, for example, that $\varrho(\mathbf{y})$ is *Hölder continuous* with exponent $0 < \alpha \leq 1$ at $\mathbf{x}_0 \in G$, i.e., that there is a ball $B(\mathbf{x}_0, r) \subset G$ and $A > 0$ such that

$$|\varrho(\mathbf{y}) - \varrho(\mathbf{x}_0)| \leq A|\mathbf{y} - \mathbf{x}_0|^\alpha, \qquad \mathbf{y} \in B(\mathbf{x}_0, r).$$

We then have

Theorem 2.16. Suppose $\varrho \in C(\bar{G})$ and ϱ is Hölder continuous at $\mathbf{x}_0$. The $\mathscr{V}$ has second derivatives at $\mathbf{x}_0$ and $\Delta\mathscr{V} = -\varrho$ there.

The proof will be given in an exercise with hints (Exercise 2.5). Here we give an easier proof using a stronger hypothesis.

Theorem 2.17. If $\varrho \in C^1(G)$, then $\mathscr{V} \in C^2(G)$ and $\Delta\mathscr{V} = -\varrho$ on G.

Proof. If $\mathbf{x}_0 \in G$ and $\overline{B(\mathbf{x}_0, r)} \subset G$, we can write

$$\begin{aligned}\mathscr{V}_{x_k} &= \int_G E_{x_k}(\mathbf{x} - \mathbf{y})\varrho(\mathbf{y})d\mathbf{y} = -\int_G E_{y_k}(\mathbf{x} - \mathbf{y})\varrho(\mathbf{y})d\mathbf{y} \\ &= -\int_B E_{y_k}(\mathbf{x} - \mathbf{y})\varrho(\mathbf{y})d\mathbf{y} - \int_{G_r} E_{y_k}(\mathbf{x} - \mathbf{y})\varrho(\mathbf{y})d\mathbf{y},\end{aligned}$$

where $B = B(\mathbf{x}_0, r)$ and $G_r = G\backslash\overline{B(\mathbf{x}_0, r)}$. The second integral is infinitely differentiable for any fixed $r > 0$. Using Gauss's lemma the first integral can be written as

$$-\int_{\partial B} E(\mathbf{x} - \mathbf{y})\varrho(\mathbf{y})n_k(\mathbf{y})dS_y + \int_B E(\mathbf{x} - \mathbf{y})\varrho_{y_k}(\mathbf{y})d\mathbf{y},$$

where $\partial B : |\mathbf{y} - \mathbf{x}_0| = r$. The first term is again infinitely differentiable, and the second is in $C^1(\mathbb{R}^n)$ as $\varrho_{y_k}(\mathbf{y})$ is bounded on B (see the Appendix to Section 1).

We see that $\mathscr{V}_{x_k}$ is continuously differentiable. Further,

$$\mathscr{V}_{x_k x_k}(\mathbf{x}_0) = \int_{G_r} E_{x_k x_k}(\mathbf{x}_0 - \mathbf{y})\varrho(\mathbf{y})d\mathbf{y} - \int_{\partial B} E_{x_k}(\mathbf{x}_0 - \mathbf{y})\varrho(\mathbf{y})n_k(\mathbf{y})dS_y$$
$$+ \int_B E_{x_k}(\mathbf{x}_0 - \mathbf{y})\varrho_{y_k}(\mathbf{y})d\mathbf{y}.$$

On ∂B we have $\mathbf{y} - \mathbf{x}_0 = r\mathbf{n}(\mathbf{y})$ and we can write

$$E_{x_k}(\mathbf{x}_0 - \mathbf{y})n_k(\mathbf{y}) = \frac{n_k^2(\mathbf{y})}{\omega_n r^{n-1}}.$$

The expression that we have derived for $\mathscr{V}_{x_k x_k}(\mathbf{x}_0)$ holds for any $r > 0$, so we can consider the limit as r tends to zero. Because $E_{x_k}(\mathbf{x}_0 - \mathbf{y})\varrho_{y_k}(\mathbf{y})$ is integrable over any compact subdomain of G, the last integral tends to zero and we have

$$\mathscr{V}_{x_k x_k}(\mathbf{x}_0) = \lim_{r \to 0}\left[\int_{G_r} E_{x_k x_k}(\mathbf{x}_0 - \mathbf{y})\varrho(\mathbf{y})d\mathbf{y} - \frac{\varrho(\mathbf{x}_0)}{\omega_n r^{n-1}}\int_{\partial B} n_k^2(\mathbf{y})dS_y\right]$$

(we have used here continuity of ϱ at $\mathbf{x}_0$ to obtain the second term in the brackets). Using $\sum_{k=1}^{n} n_k^2 = 1$ and symmetry we see that

$$\frac{1}{\omega_n r^{n-1}}\int_{|\mathbf{y}-\mathbf{x}_0|=r} n_k^2(\mathbf{y})dS_y = 1/n,$$

and because $\mathscr{V}_{x_k x_k}(\mathbf{x}_0)$ exists and does not depend on r, the limit of the first integral exists separately. We denote this limiting value by

$$PV\int_G E_{x_k x_k}(\mathbf{x}_0 - \mathbf{y})\varrho(\mathbf{y})d\mathbf{y},$$

the *principal value* of the integral. Then

$$\mathscr{V}_{x_k x_k}(\mathbf{x}_0) = PV\int_G E_{x_k x_k}(\mathbf{x}_0 - \mathbf{y})\varrho(\mathbf{y})d\mathbf{y} - \frac{1}{n}\varrho(\mathbf{x}_0). \tag{4}$$

Summing over k, the result follows. □

Suppose we consider now the Poisson equation

$$\Delta u = F$$

on a bounded domain G with, say, Dirichlet boundary values $u = f$ on ∂G. If $\varrho = -F$ satisfies the hypothesis of Theorem 2.16 or 2.17, we can write $U = u - \mathscr{V}$ and $\Delta U = 0$ in G, $U = f - \mathscr{V}$ on ∂G. If G is unbounded, the previous results can be extended provided $\varrho(\mathbf{y})$ vanishes sufficiently fast at infinity. In this way a boundary value problem for the Poisson equation can always be reduced to a problem for Laplace's equation.

Remarks.

2.6. We have shown that $-\Delta(E^*\varrho) = \varrho$, i.e., that E^* is the "inverse" of $-\Delta$ in $\mathbb{R}^n$, under a suitable assumption on ϱ. In the language of distributions,

$$-\Delta E(\mathbf{x}) = \delta(\mathbf{x}),$$

where $\delta(\mathbf{x})$ is the delta "function" (see Chapter 8).

2.7. The assumption $\varrho \in C(G)$ is not sufficient to ensure that the Poisson equation $\Delta u = -\varrho$ has a solution $u \in C^2(G)$. There exist functions $u \notin C^2(G)$ with $\Delta u \in C(G)$.

2.8. Equation (4) shows that second derivatives of $\mathscr{V}$ at a point $\mathbf{x}_0$ cannot be brought under the integral, unless the resulting integral is interpreted in the principal value sense *and* $\varrho(\mathbf{x}_0)$ vanishes.

3. Dirichlet Problem

3.1. Perron's Method. Suppose Ω is a bounded domain in $\mathbb{R}^n$ with boundary Γ. We will give a discussion of the question of solvability of the Dirichlet problem of Ω, i.e., given $f \in C(\Gamma)$, find $u \in C(\bar{\Omega}) \cap C^2(\Omega)$, u harmonic in Ω, and $u = f$ on Γ. The method described here is attributed to Perron. If B is a ball contained in $\bar{\Omega}$, and $v \in C(\bar{\Omega})$, we denote by $M_S[v]$ the function harmonic in B with boundary values on $S = \partial B$ equal to those of v, and equal to v elsewhere. in order to carry out our program we need a more general class of subharmonic functions.

Definition 3.1. v is *subharmonic* on Ω if (a) $v \in C(\bar{\Omega})$, (b) for every $B \subset \bar{\Omega}$, $M_S[v] \geq v$

A function is *superharmonic* if the inequality is reversed. Observe that the maximum principle implies that, if $v \in C^2(\Omega)$, this is a generalization of our earlier definition. The following easy results will be used repeatedly by us.

Lemma 3.1 (Properties of subharmonic functions).

a. $v \geq 0 \Rightarrow M_s[v] \geq 0$.
b. $v \geq w \Rightarrow M_S[v] \geq M_S[w]$.
c. v is subharmonic $\Rightarrow -v$ is superharmonic.

d. If v and w are subharmonic and α, β are nonnegative, then $\alpha v + \beta w$ is subharmonic.

The strong maximum principle also holds for subharmonic functions.

Theorem 3.1. If v is subharmonic in Ω and takes on its maximum in the interior of Ω, then v is constant.

Proof. Suppose $\max_\Omega v = v(P)$, $P \in \Omega$. Suppose B is any ball contained in Ω with center at P, and define $w = M_S[v]$. Then

$$\max_S w = \max_S v \equiv V \leq v(P) \leq w(P).$$

Because w is harmonic in B, the strong maximum principle (for harmonic functions) implies that w is constant in B, hence v is constant on S, and because this holds for all smaller concentric balls, v is constant in B. A repetition of our earlier argument then implies that v is constant in Ω. □

Theorem 3.2. if v_1, v_2 are subharmonic, $v = \max\{v_1, v_2\}$ is.

Proof. If $B \subset \bar{\Omega}$, $v_i \leq M_S[v_i] \leq M_S[v]$, so that $v \leq M_S[v]$. □

Theorem 3.3. If v is subharmonic on Ω, then $M_S[v]$ is.

Proof. Denote $M_S[v]$ by w, and consider an arbitrary ball $B' \subset \Omega$, $S' = \partial B'$. It suffices to prove that $w \leq M_{S'}[w]$. If $B' \cap B = \emptyset$ or $B' \subset B$, this is immediate. There remains the case $B' \cap B \neq \emptyset$ with $B' - B \neq \emptyset$. Suppose $P \in B' - B$. Then at P we have

$$w = v \leq M_{S'}[v] \leq M_{S'}[w].$$

Here we have used the subharmonicity of v and property (b) of the first lemma. Now consider $P \in B' \cap B$. Because w and $M_{S'}[w]$ are harmonic in $B' \cap B$, it suffices to examine $w - M_{S'}[w]$ on $\partial(B' \cap B) = S_1 \cup S_2$ (see Fig. 4).

The previous case implies, by continuity, that this function is nonpositive on S_2. Because $S_1 \subset S'$, $w = M_{S'}[w]$ there. The theorem follows. □

Now suppose $f \in C(\Gamma)$, and consider the corresponding Dirichlet problem.

Definition 3.2. $v \in C(\bar{\Omega})$ is a *subfunction* (for this Dirichlet problem) if v is subharmonic on Ω and $v \leq f$ on Γ.

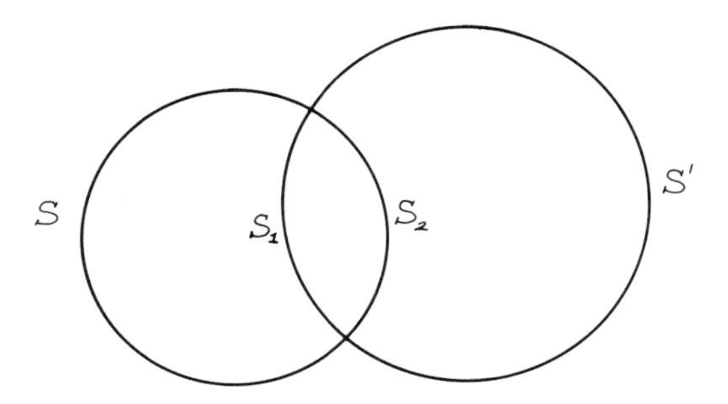

Fig. 4. Subharmonicity of $M_S(v)$.

Denote by $\mathscr{F}$ the class of subfunctions for f. It follows that $\mathscr{F}$ is a nonempty class of functions ($\min_\Gamma f$ is in $\mathscr{F}$) that are uniformly bounded above (by $\max_\Gamma f$). Also,

1. $v_1, v_2 \in \mathscr{F} \Rightarrow \max\{v_1, v_2\} \in \mathscr{F}$, and
2. $v \in \mathscr{F} \Rightarrow M_S[v] \in \mathscr{F}$ for any ball $B \subset \bar{\Omega}$, $S = \partial B$.

We define

$$u = \sup_{\mathscr{F}} v$$

and observe that if the Dirichlet problem with boundary data f has a solution, it must be u.

Theorem 3.4. u is harmonic inside Ω.

Proof. Suppose B is a ball, $\bar{B} \subset \Omega$, and $P \in B$. There is a sequence $\{u_n\}$ with $u_n \in \mathscr{F}$ and $\lim u_n(P) = u(P)$. If $U_n = \max\{u_1, \ldots, u_n\}$, then U_n is an increasing sequence in $\mathscr{F}$ with $\lim U_n(P) = u(P)$. Also, $M_S[U_n]$ is an increasing sequence in $\mathscr{F}$, each harmonic on B, with $\lim M_S[U_n](P) = u(P)$. Harnack's second convergence theorem implies that $M_S[U_n]$ converges uniformly to a harmonic function U on B. It remains to show that $U = u$ in B.

Suppose that $u(Q) > U(Q)$ for some Q in B. There exists $v_n \in \mathscr{F}$ with $\lim v_n(Q) = u(Q)$. Consider

$$V_n := \max\{u_1, \ldots, u_n, v_1, \ldots, v_n\}.$$

It follows that V_n is an increasing sequence with $\lim V_n(P) = u(P)$ and $\lim V_n(Q) = u(Q)$. Further, the sequence $M_S[V_n]$ is monotonically convergent to a harmonic limit V in B, with $V \geq U$ (because $V_n \geq U_n$), $V(P) = U(P) = u(P)$, and $V(Q) = u(Q) > U(Q)$. On the other hand, $V - U$ is harmonic in B, nonnegative, and vanishes at P. This implies that $V - U \equiv 0$ in B, and we have a contradiction. □

It suffices now to investigate the boundary behavior of u. To this end we introduce the following concept.

Definition 3.3. If $Q \in \Gamma$, a *barrier* at Q, w_Q, is a function with the following properties:

a. w_Q is superharmonic on Ω,
b. $w_Q > 0$ in $\Omega \cup \Gamma - \{Q\}$,
c. $w_Q(Q) = 0$.

A point $Q \in \Gamma$ is *regular* (for Ω) if there is a barrier at Q.

Remark 3.1. If $w_Q(\mathbf{x})$ is a barrier at Q, and $f \in C(\Gamma)$, then $f(Q) - \varepsilon - kw_Q(\mathbf{x}) < f(\mathbf{x})$, $f(Q) + \varepsilon + Kw_Q(\mathbf{x}) > f(\mathbf{x})$ for $\mathbf{x} \in \Gamma$, where $k = [M + f(Q)]/w_0$, $K = [M - f(Q)]/w_0$, $M = \max_\Gamma |f|$, $0 < w_0 = \min w_Q(\mathbf{x})$ on $\Gamma - I_\varepsilon$, where I_ε is an open neighborhood of Q on Γ where $|f(\mathbf{x}) - f(Q)| < \varepsilon$.

Theorem 3.5. If Q is regular (for Ω), then $\lim u(\mathbf{x}) = f(Q)$ as $\mathbf{x} \to Q$, $\mathbf{x} \in \Omega$.

Proof. Let $\varepsilon > 0$, and define $v(\mathbf{x}) = f(Q) - \varepsilon - kw_Q(\mathbf{x})$, where k is sufficiently large that $v \in \mathscr{F}$. Then, because $v \leq u$ (see Remark 3.1),

$$\underline{\lim}_{\mathbf{x}\to Q} u(\mathbf{x}) \geq \underline{\lim}_{\mathbf{x}\to Q} v(\mathbf{x}) = \lim_{\mathbf{x}\to Q} v(\mathbf{x}) = f(Q) - \varepsilon. \tag{5}$$

Similarly, let

$$V(\mathbf{x}) = f(Q) + \varepsilon + Kw_Q(\mathbf{x}),$$

and, if K is sufficiently large, $V \geq f$ on Γ and is superharmonic in Ω. The minimum principle then implies that V is an upper bound for any subfunction, hence $V \geq u$. It follows that

$$\overline{\lim}_{\mathbf{x}\to Q} u(\mathbf{x}) \leq \lim_{\mathbf{x}\to Q} V(\mathbf{x}) = f(Q) + \varepsilon, \tag{6}$$

and the inequalities (5) and (6) imply our result. □

Remark 3.2. The Dirichlet problem for a domain Ω is well posed (for all continuous data f) *if and only if* all points of the boundary Γ are regular. [The necessity of this condition follows from the fact that the solution of the problem with boundary values $f(\mathbf{x}) = |\mathbf{x} - Q|$, $Q \in \Gamma$, is a barrier at Q.]

Definition 3.4. Γ is said to satisfy the exterior sphere condition at Q if there is a ball $B \subset \Omega^c$ with $S = \partial B$ passing through Q.

Theorem 3.6. If Γ satisfies the exterior sphere condition at Q, Γ is regular at Q (for Ω).

Proof. Suppose R is the radius of the ball B, $\partial B = S$. We take the origin of coordinates at the center of S. Then $w_Q(\mathbf{x}) = E(R) - E(|\mathbf{x}|)$ is a (harmonic) barrier at Q. □

A necessary and sufficient condition for regularity of a boundary point (Wiener's test) will be given in Section 6.

Example 3.1. We will now study an example showing that the Dirichlet problem need not be well posed. Consider the line potential in $\mathbb{R}^3$ distributed along the segment from (0, 0, 0) to (1, 0, 0) given by

$$\phi(\mathbf{x}) = \int_0^1 \frac{\xi \, d\xi}{\sqrt{(x - \xi)^2 + \rho^2}},$$

where $\mathbf{x} = (x, y, z)$, $\rho^2 = y^2 + z^2$. This integral can be calculated explicitly, the result being

$$\phi(\mathbf{x}) = A(\mathbf{x}) - 2x \ln \rho,$$

where $A(\mathbf{x}) = A(x, \rho)$ is given by the expression

$$[(1 - x)^2 + \rho^2]^{1/2} - \sqrt{x^2 + \rho^2} + x \ln|1 - x + [(1 - x)^2 + \rho^2]^{1/2}(x + \sqrt{x^2 + \rho^2})|.$$

If $\mathbf{x}$ approaches $\mathbf{0}$ from $x > 0$, we find that $A(\mathbf{x})$ approaches 1. On the other hand,

$$\lim_{\mathbf{x} \to 0} [-2x \ln \rho].$$

does not exist, but has limiting values along certain paths. For example, along $\rho = |x|^\alpha$ $(\alpha > 0)$ the limiting value is zero, whereas along the surface $\rho = \exp(-c/2x)$, $c > 0$, which has an infinitely sharp peak at the origin, the limiting value is c. This means that all of the equipotential surfaces S_c, defined by

$$S_c = \{\mathbf{x} : \phi(\mathbf{x}) = 1 + c\},$$

meet at the origin. If the origin is approached by a sequence of values outside of S_c, the limit can exist and be any value between 1 and $1 + c$.

If we invert the region exterior to S_c with respect to $S : |\mathbf{x} - \mathbf{x}_0| = 1/2$, $\mathbf{x}_0 = (1/2, 0, 0)$, we obtain a bounded region Ω with an infinitely sharp interior peak at an irregular boundary point (Fig. 5), because the inversion of $\phi(\mathbf{x})$ is a function harmonic inside this region that is bounded and has boundary values $1 + c$ everywhere except at the exceptional point, where the limit can be any value between 1 and $1 + c$.

The significance of this example (due to Lebesgue) will be expanded on in what follows.

A simpler, related example that brings an important idea to the surface is the following. Suppose that Ω is the region in $\mathbb{R}^3$ obtained from the unit ball by deleting the segment $\mathfrak{G}$ from $(-1, 0, 0)$ to $(0, 0, 0)$. If f is a continuous function on the boundary of Ω, then consider the restriction of f to the unit sphere and the solution, u', of the corresponding Dirichlet problem. By comparison with the functions

$$u = u' + \varepsilon \int_{\mathfrak{G}} \frac{ds}{r},$$

it follows that the boundary values of a possible solution of the Dirichlet problem along $\mathfrak{G}$ are not arbitrary, but must coincide with u' along $\mathfrak{G}$ (Exercise 3.1).

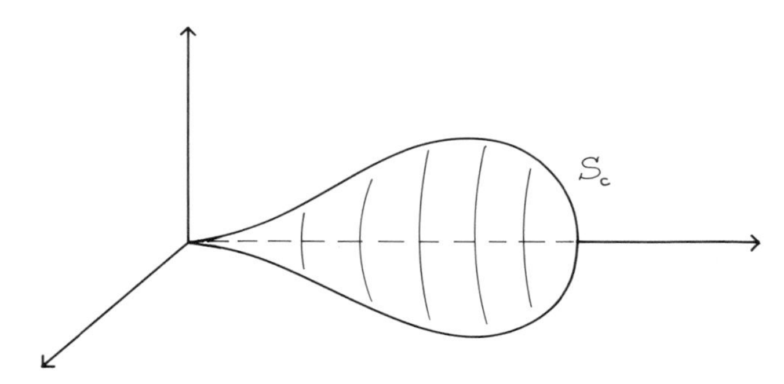

Fig. 5. The Lebesgue spike.

From these examples we see that a barrier may not exist if the boundary has "sharp" interior spines or components of lower dimension. We mention here that the exceptional nature of "sharp" points on the boundary of a surface has a manifestation in physical reality. The electrostatic field becomes very strong near such a point, and the equilibrium associated with solvability of Dirichlet's problem may not be possible.

The Perron solution of Dirichlet's problem may appear to be highly nonconstructive on the face of things as it arises from maximizing over an uncountable class of functions. There is, however, a related method that is constructive. The method of *balayage* ("sweeping out"), due to Poincaré, starts with a function u_1 that satisfies the given boundary values and through successive modifications moves all of the charge $\rho_1 = -\omega_n^{-1}\Delta u_1$ to the boundary, so that the final function u is harmonic (see Ref. 2). Suppose that the boundary function f is extended continuously to $\bar{\Omega}$ as a subharmonic function. Choose a sequence of open balls $\tilde{B}_n \subset \Omega$ such that $\cup\tilde{B}_n = \Omega$. Then form a new sequence $\{B_n\}$ in which each $\tilde{B}_n$ appears infinitely often. Let $u_1 = f$, $u_{n+1} = M_{S_n}[u_n]$ where $S_n = \partial B_n$. Then each u_n is continuous on $\bar{\Omega}$, $u_n = f$ on Γ and u_n is subharmonic on Ω and harmonic in B_n. Further, $\{u_n\}$ is an increasing, bounded sequence. If we consider the subsequence arising from this process on $\tilde{B}_n$, Theorem 2.11 implies that $u = \lim u_n$ is harmonic on $\tilde{B}_n$, and it follows that u is harmonic on Ω. The function u can be proven not to depend on the subharmonic extension to $\bar{\Omega}$ of f or on the choice of balls $\tilde{B}_n$. If Ω has a barrier at $Q \in \Gamma$, then u is continuous at Q and $u(Q) = f(Q)$.

A discrete version of the balayage method can be described as follows. If the derivatives in Laplace's equation (say, in two variables) are approximated by centered differences, e.g., if u_{xx} is replaced by $[u(x-h, y) - 2u(x, y) + u(x+h, y)]/h^2$ and similarly for u_{yy}, then a difference equation on a square grid is obtained:

$$u(x, y) = \tfrac{1}{4}[u(x, y-h) + u(x, y+h) + u(x-h, y, z) + u(x+h)] = 0, \quad (7)$$

which says the function value at a grid point $x = ih$, $y = jh$ is the average of the values at nearest neighbors. This is the discrete analogue of the mean value property, and it implies a discrete analogue of the weak maximum principle: The maximum and the minimum of $u_{ij} = u(ih, jh)$ is assumed at some boundary node $(\bar{i}, \bar{j})$. The set of all equations (7) for the relevant set of indices i, j yields a linear algebraic system in the unknowns u_{ij} at the interior nodes (i, j) for given $u_{ij} = f_{ij}$ at all boundary nodes, and the discrete maximum principle implies that the matrix of the system is nonsingular. Therefore, this discrete Dirichlet problem has a unique solution. An iterative method called *relaxation* consists of replacing values at grid points of an initial function u_1 by averages over nearest neighbors until the discrete mean value property is satisfied to required accuracy. The replacement starts at the node where the "residual," i.e., the value of the left-hand side of (7), is highest in absolute value ("worst first" rule). This discrete version of the balayage method is the ancestor of modern iterative methods for solving the linear algebraic systems that arise from discretization of partial differential equations.

Perron's method can also be extended to discontinuous boundary values. Suppose that f is a *bounded* function on Γ, say $|f| \leq M$. We define $\mathscr{F}$ to be the set of functions v that are subharmonic on Ω and satisfy $\overline{\lim_{P \to Q}} v(P) \leq f(Q)$, $Q \in \Gamma$. We define $u = \sup_{\mathscr{F}} v$ as before. ($\mathscr{F}$ is nonempty because constants $\leq -M$ are in $\mathscr{F}$.) Theorem 3.4 holds with the same proof. (One should prove first that $v \leq M$, for $v \in \mathscr{F}$, so that relevant sequences are known to be bounded.) At every regular boundary point Q at which f is continuous, $\lim_{P \to Q} u(P) = f(Q)$. The function provided by this procedure (Perron's method) may be thought of as a *generalized solution* of Dirichlet's problem. We will consider another kind of generalized solution in Section 5.

In order to guarantee uniqueness of this generalized solution, we need a growth condition near the discontinuity points on the boundary, in the context of so-called Phragmèn–Lindelöf theorems. We consider here a special case: Suppose that all points of Γ are regular and f is discontinuous at only one "exceptional" point $\mathbf{x}_0 \in \Gamma$. We will show that there is at most one *bounded* solution $u \in C^2(\Omega) \cap C(\bar{\Omega} \backslash \{\mathbf{x}_0\})$ of

$$\Delta u = 0 \text{ in } \Omega, \qquad \lim_{\mathbf{x} \to \mathbf{y}} u(\mathbf{x}) = f(\mathbf{y}) \text{ for all } \mathbf{y} \in \Gamma, \ \mathbf{y} \neq \mathbf{x}_0.$$

If u_1, u_2 are two solutions, then $v = u_1 - u_2$ is bounded, harmonic in Ω, and $\lim_{\mathbf{x} \to \mathbf{y}} u(\mathbf{x}) = 0$ for all $\mathbf{y} \in \Gamma$, $\mathbf{y} \neq \mathbf{x}_0$. Consider a ball $B_\delta : |\mathbf{x} - \mathbf{x}_0| < \delta$. Let $w := E(|\mathbf{x} - \mathbf{x}_0|)$ (for $n \geq 3$), and $U = \varepsilon w$, $\varepsilon > 0$, then U is harmonic and positive on $\Omega \backslash \overline{B_\delta}$. Choose δ so that $\varepsilon w > \max |v|$ on ∂B_δ. The maximum principle implies that $|v| < \varepsilon w$ on $\Omega \backslash \overline{B_\delta}$. In particular, for any fixed $\mathbf{x} \in \Omega$, $|v(\mathbf{x})| \leq \varepsilon w(\mathbf{x})$. Letting $\varepsilon \to 0$ we find $v(\mathbf{x}) \equiv 0$. If $n = 2$, $w = \ln(d/|\mathbf{x} - \mathbf{x}_0|)$ will do, with d the diameter of Ω.

The extension to a finite number of discontinuities for f is immediate. From the above proof we see that boundedness of u near $\mathbf{x}_0$ can be replaced by the more general *growth condition*

$$u(\mathbf{x}) = o(|E(|\mathbf{x} - \mathbf{x}_0|)|) \qquad \text{as } \mathbf{x} \to \mathbf{x}_0 \tag{GC}$$

and this condition is the best possible, in the sense that $u(\mathbf{x}) = O(|E(|\mathbf{x} - \mathbf{x}_0|)|)$ will not do (Exercises 3.7 and 3.8). Note that, with regard to uniqueness, the exceptional point $\mathbf{x}_0$ might well be an irregular point of Γ, irrespective of whether f is continuous at $\mathbf{x}_0$ or not. However, if $\mathbf{x}_0$ is isolated and $0 < |\mathbf{x} - \mathbf{x}_0| < \delta$ is contained in Ω for some $\delta > 0$, (GC) implies that $\lim_{\mathbf{x} \to \mathbf{x}_0} u(\mathbf{x})$ exists and the extended function is harmonic (see Theorem 2.15), so that no boundary data can be assigned at $\mathbf{x}_0$.

3.2. C^1 Regularity up to the Boundary. In many applications (e.g., see Sections 4, 6, and 7) it is important to know whether the solution of Dirichlet's problem is regular (C^1) up to the boundary. We need the following definition.

Definition 3.5. A closed (compact) surface S is of class $C^{1,\alpha}$, $S \in C^{1,\alpha}$, if for each point of S there is a neighborhood on which S is the graph of a function with first partial derivatives that are Hölder continuous with exponent $\alpha (0 < \alpha \leq 1)$.

Our goal here is to prove the following result, which will be used repeatedly in the sequel.

Theorem 3.7. Suppose $\Delta u = 0$ in Ω, Ω a bounded domain in $\mathbb{R}^3$, $\partial\Omega \in C^{1,\alpha}$, $u = f$ on $\partial\Omega$, $f \in C^{1,\alpha}(\partial\Omega)$. Then $u \in C^1(\bar{\Omega})$.

In fact, a refinement of the argument in the main theorem shows that $u \in C^{1,\alpha}(\bar{\Omega})$, and the result extends to $n \neq 3$. The proof, based on a sequence of auxiliary results, is somewhat technical and may be omitted in a first reading. A companion theorem (which will not be proven here) states that the solution of the Neumann problem with Hölder continuous boundary data is C^1 up to the boundary.

Lemma 3.2. Let $g(\mathbf{x}, \mathbf{y})$ be the Green's function of the half-space $x_3 > 0$. Then

$$g(\mathbf{x}, \mathbf{y}) \leq \begin{cases} K/|\mathbf{x} - \mathbf{y}|, \\ Ky_3/|\mathbf{x} - \mathbf{y}|^2, \\ Kx_3 y_3/|\mathbf{x} - \mathbf{y}|^3, \end{cases} \qquad g_{x_i}(\mathbf{x}, \mathbf{y}) \leq \begin{cases} K/|\mathbf{x} - \mathbf{y}|^2, \\ Ky_3/|\mathbf{x} - \mathbf{y}|^3, \end{cases}$$

and

$$g_{x_i y_j}(\mathbf{x}, \mathbf{y}) \leq \begin{cases} K/|\mathbf{x} - \mathbf{y}|^3, \\ Ky_3/|\mathbf{x} - \mathbf{y}|^4, \end{cases} \qquad g_{x_i x_j y_k}(\mathbf{x}, \mathbf{y}) \leq K/|\mathbf{x} - \mathbf{y}|^4.$$

This can be shown by direct calculation and is left as an exercise. Here and in what follows, K denotes positive pure constants whose exact values need not concern us.

We denote by $\mathbf{X}$ the planar vector (x_1, x_2) associated with $\mathbf{x} = (x_1, x_2, x_3)$. We need a special domain D defined by $|\mathbf{X}| < 1$, $-|\mathbf{X}|^{1+\alpha} < x_3 < 2$ (see Fig. 6).

Scaled versions sD of D will be used in making estimates in what follows. Because $\partial\Omega \in C^{1,\alpha}$, for every point on $\partial\Omega$ there is a neighborhood N in which $\partial\Omega$ is the graph of a function in $C^{1,\alpha}$. More precisely, if a coordinate system is introduced in which the point in question is the origin and the normal is along the x_3-axis, then $\partial\Omega \cap N$ is given by $x_3 = F(\mathbf{X})$ where $F(\mathbf{0}) = 0$, $F_{x_i}(\mathbf{0}) = 0$, and F_{x_i} are Hölder continuous with exponent α. A straightforward application of the mean value theorem shows that there is a constant C such that $|F(\mathbf{X})| < C|\mathbf{X}|^{1+\alpha}$ in N. Suppose that $D_s = sD$, $0 < s < 1$. Then if the special coordinate system used above is introduced, we can choose s so that $\partial\Omega \cap D$ lies "above" the curved part of ∂D and $\partial\Omega$ intersects this surface only at the origin (Fig. 7).

The compactness of $\partial\Omega$ implies that there is an s_0, $s_0 < 1$, such that D_s for all $0 < s \leq s_0$ can be used at all boundary points.

We need a general result about estimating the gradient of a function on the boundary of a domain.

Theorem 3.8. Suppose that $-\Delta u = \rho$ in $\mathscr{D}$, ρ bounded, $u = 0$ on $\partial\mathscr{D}$, and $\mathscr{D}$ satisfies the exterior sphere condition. Then there is a constant $C = C(\mathscr{D}, \rho)$ such that

$$|\text{grad } u| \leq C$$

on $\partial\mathscr{D}$. We are tacitly assuming that $u \in C^1(\bar{\mathscr{D}})$.

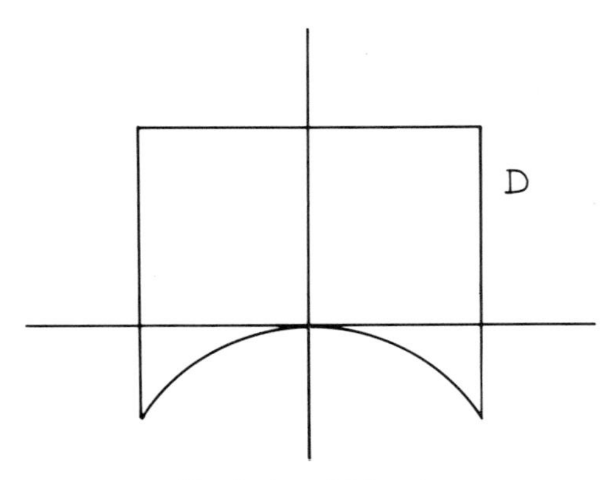

Fig. 6. A useful domain.

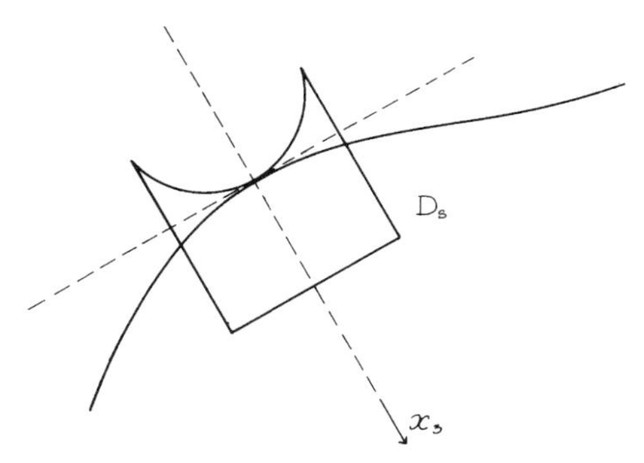

Fig. 7. A special boundary neighborhood.

Proof. Let $w(\mathbf{x}) = k[R^{-p} - |\mathbf{x} - \mathbf{x}^*|^{-p}]$ where $\mathbf{x}^*$ is the center of the ball of radius R touching $\partial\mathscr{D}$ at $\mathbf{x}_0 \in \partial\mathscr{D}$ and R is the radius in the external sphere condition. Then k and p can be chosen so that $-\Delta w \geq 1$ on $\mathscr{D}$. We bound $|\text{grad } u| = |\partial u/\partial n|$ at $\mathbf{x}_0$. Define $v = u - w \sup_{\mathscr{D}}|\rho|$ on $\bar{\mathscr{D}}$. Then $v \geq 0$ on $\partial\mathscr{D}$ and $-\Delta v = \rho + \Delta w \sup_{\mathscr{D}}|\rho| \leq 0$ in $\mathscr{D}$, and the maximum principle implies $v \leq 0$ in $\mathscr{D}$, that is, $u \leq w \sup|\rho|$. Because $u(\mathbf{x}_0) = w(\mathbf{x}_0) = 0$, we have

$$\frac{\partial u}{\partial n}(\mathbf{x}_0) \geq \frac{\partial w}{\partial n}(\mathbf{x}_0) \sup_{\mathscr{D}}|\rho| = -C.$$

The same argument with w replaced by $-w$ gives the upper bound $(\partial u/\partial n)(\mathbf{x}_0) \leq C$. □

Although the above statement is given a global form, the proof shows clearly that the result can be localized. Also, if we do not know in advance that $\partial u(\mathbf{x}_0)/\partial n$ exists, the same argument shows that

$$-C \leq \varliminf_{t\to 0} t^{-1}u(\mathbf{x}_0 + t\mathbf{n}) \leq \varlimsup_{t\to 0} t^{-1}u(\mathbf{x}_0 + t\mathbf{n}) \leq C.$$

We now define a harmonic function U on D by $U = 0$ on $x_3 = -|\mathbf{X}|^{1+\alpha}$ and 1 in the rest of ∂D. (Although U is discontinuous on ∂D, the Perron process guarantees the existence of such a function, and the Phragmen–Lindelöf-type result proved at the end of Section 3.1 guarantees uniqueness.) We will use U as a comparison function in deriving estimates for the Green's function of Ω. We point out explicitly that D does not satisfy the exterior sphere condition at the origin. This requires a limiting argument in the following lemma.

Lemma 3.3. $U(0, 0, x_3) \leq Kx_3, \ 0 \leq x_3 \leq 1.$

Proof. We form the domain D_r obtained by deleting a ball of radius r with center $(0, 0, -r)$ from D, $D_r = D \backslash B((0, 0, -r), r)$, for r sufficiently small (ultimately r will go to zero), and define U_r to be the harmonic function that is 1 on the lateral part of ∂D_r, 0 on the "curved" parts. The domain D_r satisfies an exterior sphere condition.

We represent U_r at $\mathbf{x} = (0, 0, x_3) \in D_r$ using the Green's function $g(\mathbf{x}, \mathbf{y})$ for the upper half-space $x_3 > 0$. Proceeding as in the deduction of Green's representation theorem we find the identity

$$U_r(\mathbf{x}) = \int_S \left(g(\mathbf{x}, \mathbf{y}) \frac{\partial U_r}{\partial n_y} - U_r(\mathbf{y}) \frac{\partial g}{\partial n_y} \right) dS_y + \int_C \frac{\partial g}{\partial y_3}(\mathbf{x}, \mathbf{Y}) U_r(\mathbf{Y}) dY,$$

where $S = \{\mathbf{y} \in \partial D_r : y_3 > 0\}$ and $C : |\mathbf{Y}| < 1$, $\mathbf{Y} = (y_1, y_2)$. We can give an estimate

$$0 \leq -\frac{\partial U_r}{\partial n} \leq K$$

on S using the boundary gradient estimate. We need only observe that U_r extends across flat parts of S to a harmonic function, so that existence of $\partial U_r / \partial n$ is not in question, and consider the equation satisfied by $U_r - 1$ multiplied by a cutoff function vanishing near $x_3 = 0$. If the estimate for g and g_{x_i} in Lemma 3.2 are applied, we find easily that

$$U_r(\mathbf{x}) \leq K x_3 + K x_3 \int_C |\mathbf{x} - \mathbf{Y}|^{-3} U_r(\mathbf{Y}) d\mathbf{Y}.$$

The values of $U_r(\mathbf{Y})$ can be estimated in terms of those of U_r on the x_3-axis using the maximum principle and a scaled copy of D_r. If we write the above integral as the sum of integrals over $|\mathbf{Y}| \leq \varepsilon$ and $\varepsilon \leq |Y| \leq 1$, we obtain

$$U_r(\mathbf{x}) \leq K x_3 + K x_3 \int_{|\mathbf{Y}| \leq \varepsilon} |\mathbf{x} - \mathbf{Y}|^{-3} U_r(\mathbf{Y}) d\mathbf{Y} + K x_3 / \varepsilon^3.$$

If a scaled and translated domain D_r' has $\mathbf{Y}$ along its axis, the maximum principle implies that the values of the transplanted functions U_r' are upper bounds for those of U_r if the lower boundary of D_r' lies below that of D_r. It can be shown that for ε sufficiently small this can be accomplished and

$$U_r(\mathbf{Y}) \leq U_r(0, 0, k|\mathbf{Y}|^{1+\alpha})$$

for k a large enough power of 2 (Exercise 3.11). Let

$$\mathscr{M}_4 = \sup_{0<x_3<1} [U_r(0, 0, x_3)/x_3].$$

Because D_r satisfies an exterior sphere condition, this quantity is finite for each $r > 0$. We can now write

$$\mathscr{M}_r \leq K + K \int_{|\mathbf{Y}|\leq \varepsilon} \frac{\mathscr{M}_r |\mathbf{Y}|^{1+\alpha}}{|\mathbf{x} - \mathbf{Y}|^3} \, d\mathbf{Y} + K/\varepsilon^3 \leq K + K\mathscr{M}_r \int_0^{\varepsilon} t^{\alpha-1} \, dt + K/\varepsilon^3.$$

Choose ε small enough and a bound for $\mathscr{M}_r$ that is independent of r results. The proof is completed by letting r go to zero. Because the functions $\{U_r\}$ have uniformly convergent subsequences on any compact subdomain of $D \cap \{x_3 > 0\}$, the result follows. □

Let $G(\mathbf{x}, \mathbf{y})$ denote the Green's function of Ω (G exists because all points of $\partial\Omega$ are regular). The above lemma is essential in proving that the following fundamental estimates hold. Here $\delta(\mathbf{x})$ is the distance of $\mathbf{x}$ to $\partial\Omega$.

Theorem 3.9. (i) $G(\mathbf{x}, \mathbf{y}) \leq K\delta(\mathbf{x})|\mathbf{x} - \mathbf{y}|^{-2}$, (ii) $|G_{x_i}(\mathbf{x}, \mathbf{y})| \leq K|\mathbf{x} - \mathbf{y}|^{-2}$, (iii) $|G_{x_i}(\mathbf{x}, \mathbf{y})| \leq K\delta(\mathbf{x})|\mathbf{x} - \mathbf{y}|^{-3}$, (iv) $|G_{x_i y_j}(\mathbf{x}, \mathbf{y})| \leq K|\mathbf{x} - \mathbf{y}|^{-3}$.

Proof. Let d be the diameter of Ω. Fix $\mathbf{y}$. If $\delta(\mathbf{x}) \geq s_0$ (see above discussion of $\partial\Omega \in C^{1+\alpha}$), then

$$\delta(\mathbf{x}) \geq s_0 \geq Kd \geq K|\mathbf{x} - \mathbf{y}|$$

and the elementary inequality $G(\mathbf{x}, \mathbf{y}) \leq K|\mathbf{x} - \mathbf{y}|^{-1}$ imply (i). If $\delta(\mathbf{x}) < s_0$ but $|\mathbf{x} - \mathbf{y}| < 2\delta(\mathbf{x})$, then $G(\mathbf{x}, \mathbf{y}) \leq K|\mathbf{x} - \mathbf{y}|^{-1} \leq 2K|\mathbf{x} - \mathbf{y}|^{-2}\delta(\mathbf{x})$ implies (i). Therefore, it suffices to consider $\delta(\mathbf{x}) < s_0$ and $2\delta(\mathbf{x}) \leq |\mathbf{x} - \mathbf{y}|$. let $\mathbf{x}^* \in \partial\Omega$, $|\mathbf{x}^* - \mathbf{x}| = \delta(\mathbf{x})$. We place a domain D' similar to D along the normal to $\partial\Omega$ at $\mathbf{x}^*$. The scale factor is taken to be $\min\{s_0, |\mathbf{x} - \mathbf{y}|/4\}$. If $\mathbf{z} \in \partial D'$, then $|\mathbf{x} - \mathbf{y}| \geq |\mathbf{x} - \mathbf{y}|/4$, so

$$G(\mathbf{z}, \mathbf{y}) \leq K|\mathbf{z} - \mathbf{y}|^{-1} \leq 4K|\mathbf{x} - \mathbf{y}|^{-1}.$$

Consider first the case $|\mathbf{x} - \mathbf{y}| < 4s_0$. Let u' be the function U of Lemma 3.3 transplanted to D'. The maximum principle implies that $G(\mathbf{x}, \mathbf{y})$ is bounded above by $\max_{\mathbf{z}} G(\mathbf{z}, \mathbf{y})u'(\mathbf{x})$, the max taken over $\partial D' \cap \Omega$. Then Lemma 3.3

implies $G(\mathbf{x}, \mathbf{y}) \le 4K|\mathbf{x}-\mathbf{y}|^{-1}\delta(\mathbf{x})/s$ where the scale factor s is, in this case, $|\mathbf{x}-\mathbf{y}|/4$ so that

$$G(\mathbf{x}, \mathbf{y}) \le K|\mathbf{x}-\mathbf{y}|^{-2}\delta(\mathbf{x}).$$

If $s = s_0$, the same comparison function shows that

$$G(\mathbf{x}, \mathbf{y}) \le 4K|\mathbf{x}-\mathbf{y}|^{-1}\delta(\mathbf{x})/s_0 \le K|\mathbf{x}-\mathbf{y}|^{-2}\delta(\mathbf{x}).$$

We have proven (i).

The proof of (ii) is accomplished by representing $G(\mathbf{x}, \mathbf{y})$ by a Poisson integral of its values over a sphere with center at $\mathbf{x}$ and differentiating with respect to $\mathbf{x}$. If $|\mathbf{x}-\mathbf{y}| < \delta(\mathbf{x})$, take the radius to be $|\mathbf{x}-\mathbf{y}|/2$, use $G(\mathbf{z}, \mathbf{y}) \le K|\mathbf{z}-\mathbf{y}|^{-1}$, and use elementary estimates. If $|\mathbf{x}-\mathbf{y}| \ge \delta(\mathbf{x})$, take the radius to be $\delta(\mathbf{x})/2$ and use (i) to deduce $G(\mathbf{z}, \mathbf{y}) \le |\mathbf{z}-\mathbf{y}|^{-2}\delta(\mathbf{z})$ and use elementary estimates.

In order to proceed further we need to show that, for fixed $\mathbf{x} \in \Omega$, $G_{x_i}(\mathbf{x}, \mathbf{y}) \to 0$ as $\mathbf{y}$ approaches $\partial\Omega$. First observe that if $|\mathbf{x}-\mathbf{y}| = \rho$, ρ small enough, then $\frac{1}{2}|\mathbf{x}-\mathbf{y}|^{-1} \le G(\mathbf{x}, \mathbf{y})$. Then, if $|\mathbf{h}| < \rho/2$, the mean value theorem and (ii) imply that

$$\left|\frac{G(\mathbf{x}+\mathbf{h}, \mathbf{y}) - G(\mathbf{x}, \mathbf{y})}{|\mathbf{h}|}\right| \le K|\mathbf{x}-\mathbf{y}|^{-2}$$

and, for $|\mathbf{x}-\mathbf{y}| = \rho$, the difference quotient is bounded by $K\rho^{-1}G(\mathbf{x}, \mathbf{y})$. Because, for fixed $\mathbf{x}$, $\mathbf{h}$, and ρ, $[G(\mathbf{x}+\mathbf{h}, \mathbf{y}) - G(\mathbf{x}, \mathbf{y})]/|\mathbf{h}|$ and $K\rho^{-1}G(\mathbf{x}, \mathbf{y})$ are harmonic functions on $\Omega\backslash\{|\mathbf{y}-\mathbf{x}| > \rho\}$ and both vanish on $\partial\Omega$, this inequality holds throughout Ω. Letting $|\mathbf{h}| \to 0$ we obtain

$$|G_{x_i}(\mathbf{x}, \mathbf{y})| \le K\rho^{-1}G(\mathbf{x}, \mathbf{y}).$$

(It should be pointed out here that ρ depends on $\mathbf{x}$, but that $\mathbf{x}$ is fixed in this discussion.) Because (i) implies

$$G(\mathbf{x}, \mathbf{y}) \le K|\mathbf{x}-\mathbf{y}|^{-1}\delta(\mathbf{y}),$$

the result follows. Now (iii) can be proven by applying the argument used in proving (i) to $G_{x_i}(\mathbf{x}, \mathbf{y})$ with the elementary inequality $G(\mathbf{x}, \mathbf{y}) \le K|\mathbf{x}-\mathbf{y}|^{-1}$ replaced by (ii).

Finally, (iv) can be proven, using (iii) in place of (i), as (ii) was. □

We are ready to prove our main theorem.

Proof of Theorem 3.7. The proof is accomplished by obtaining estimates for u and its derivatives near a fixed boundary point. By introducing the usual special coordinate system we may assume that the point in question is the origin, and that the boundary values satisfy

$$|u(\mathbf{x})| \leq K|\mathbf{x}|^{1+\alpha}$$

for $\mathbf{x} \in \partial\Omega$, $\mathbf{x}$ near the origin. (The global assumptions on $\partial\Omega$ are reflected in the uniformity of the constant K. The entire proof can, of course, be localized.)

Our first step is to show that $|\text{grad } u|$ is bounded. It suffices to consider $\mathbf{x}$ on the vertical axis; our choice of coordinate system guarantees that $\delta(\mathbf{x}) = |\mathbf{x}| = x_3$. Let B be a ball with radius $\rho > 2\delta(\mathbf{x})$ (but sufficiently small) centered at the origin. We want to represent u, and then u_{x_i}, using (G2) on $B \cap \Omega = \Sigma$. For $t > 0$ we write $u^t(\mathbf{x}) = u(\mathbf{X}, x_3 + t)$ so that, for t sufficiently small, $\partial u^t/\partial n$ exists on $\partial\Sigma$. We write, formally,

$$u^t(\mathbf{x}) = \int_{\partial\Sigma \cap \Omega} \left[\frac{\partial u^t(\mathbf{y})}{\partial n_y} G(\mathbf{x}, \mathbf{y}) - u^t(\mathbf{y}) \frac{\partial G(\mathbf{x}, \mathbf{y})}{\partial n_y}\right] dS_y - \int_{B \cap \partial\Omega} u^t(\mathbf{y}) \frac{\partial G(\mathbf{x}, \mathbf{y})}{\partial n_y}\, dS_y$$

and

$$\begin{aligned}\frac{\partial u^t(\mathbf{x})}{\partial x_i} = \int_{\partial\Sigma \cap \Omega} &\left[\frac{\partial u^t(\mathbf{y})}{\partial n_y} \frac{\partial G(\mathbf{x}, \mathbf{y})}{\partial x_i} - u^t(\mathbf{y}) \frac{\partial^2 G(\mathbf{x}, \mathbf{y})}{\partial x_i \partial n_y}\right] dS_y \\ &- \int_{B \cap \partial\Omega} u^t(\mathbf{y}) \frac{\partial^2 G(\mathbf{x}, \mathbf{y})}{\partial x_i \partial n_y}\, dS_y.\end{aligned}$$

This can be justified by first considering a translated domain Σ_τ and showing that the above makes sense with appropriate weak limits taking the role of the derivatives of G on $\partial\Omega$. (These functions satisfy the same inequalities that the derivatives of G do.) Then we let $t \to 0$, which causes no difficulty because $|\text{grad } u^t| \leq K/\delta(\mathbf{x})$, K independent of t, and (iii) of the previous theorem holds. We can now deduce that the integral $B \cap \partial\Omega$ is bounded by

$$K \int_{|\mathbf{Y}| \leq \rho} |\mathbf{Y}|^{1+\alpha} |\mathbf{Y}|^{-3}\, d\mathbf{Y} = K \int_0^\rho t^{-1+\alpha}\, dt < \infty.$$

Here we have used the fact that $|\mathbf{y}| \leq (1 + \max|\text{grad } \varphi|)|\mathbf{Y}|$ [if $\partial\Omega$ is given by $y_3 = \varphi(\mathbf{Y})$] and $|\mathbf{y} - \mathbf{x}| \geq |\mathbf{Y}|$ [recall that $\mathbf{x} = (0, 0, x_3)$]. The second integral is easily bounded using the estimates for G and $|\text{grad } u| \leq K/\delta(\mathbf{x})$.

The next step is to show that

$$|u_{x_ix_j}(\mathbf{x})| \leq K(\delta(\mathbf{x}))^{-1+\alpha}.$$

In order to do this we construct a domain E bounded by $x_3 = k|\mathbf{X}|^{1+\alpha}$ for $x_3 \leq s$ and a hemisphere of radius s placed on this surface as a "cap." With appropriate choices of k and s we can guarantee that a copy of E can be placed inside Ω touching $\partial\Omega$ only at a fixed boundary point and with symmetry axis along the normal at this point. As usual we take $\mathbf{x}$ along the normal and use our special coordinate system so that the boundary point is at the origin and $\mathbf{x} = (0, 0, x_3)$.

We need the estimate $|u(\mathbf{y})| \leq K|\mathbf{y}|^{1+\alpha}$ for $\mathbf{y} \in \partial E$. For this we need only consider $\mathbf{y}$ in a sufficiently small neighborhood of the origin. Because the angle of inclination of the normal to $\partial\Omega$ is continuous, we may assume $|\mathbf{y}^*| \leq K|\mathbf{y}|$ for $\mathbf{y} \in \partial E$ (in this neighborhood). Similarly, $|\mathbf{y}^* - \mathbf{y}|$ is less than the distance between the symmetric points $(\mathbf{Y}^*, \pm k|\mathbf{Y}^*|^{1+\alpha})$ so that $|\mathbf{y}^* - \mathbf{y}| \leq 2K|\mathbf{Y}^*|^{1+\alpha} \leq 2K|\mathbf{y}|^{1+\alpha}$. It follows that

$$|u(\mathbf{y})| \leq |u(\mathbf{y}^*) - u(\mathbf{y})| + |u(\mathbf{y}^*)| \leq K|\mathbf{y}^* - \mathbf{y}| + K|\mathbf{y}^*|^{1+\alpha} \leq K|\mathbf{y}|^{1+\alpha},$$

where we have used the boundedness of $|\text{grad}\, u|$ in the second inequality. We write $\partial E = S_1 \cup S_2 \cup S_3$ where $y_3 < s$ and $|\mathbf{Y}| < x_3$ on S_1, $y_3 < s$ and $|\mathbf{Y}| > x_3$ on S_2, and $y_3 > s$ on S_3. We can write, for any second derivative D^2 [recall $g(\mathbf{x}, \mathbf{y})$ is the Green's function for $y_3 > 0$],

$$D^2u(\mathbf{x}) = -\int_{\partial E} \left(u(\mathbf{y}) D^2 \frac{\partial g(\mathbf{x}, \mathbf{y})}{\partial n_y} - \frac{\partial u(\mathbf{y})}{\partial n_y} D^2 g(\mathbf{x}, \mathbf{y}) \right) dS_y.$$

We estimate the first of these integrals breaking it up into integrals on S_1, S_2, and S_3. Using the last estimate of Lemma 3.2 we have

$$\left| \int_{S_1} \cdots \right| \leq K \int_{S_1} \frac{|\mathbf{y}|^{1+\alpha}}{|\mathbf{x} - \mathbf{y}|^4} \, dS_y \leq K \int_{|\mathbf{Y}| \leq x_3} \frac{|\mathbf{Y}|^{1+\alpha}}{|\mathbf{x}|^4} \, d\mathbf{Y}$$

(as $|\mathbf{x} - \mathbf{y}| \geq K|\mathbf{x}|$ for $\mathbf{y} \in S_1$) and this is bounded by

$$Kx_3^{-3+\alpha} \int_{|\mathbf{Y}| \leq x_3} d\mathbf{Y} = Kx_3^{-1+\alpha}.$$

Then, denoting by d the diameter of E, and by $\mathscr{D}$ the set $x_3 \leq |\mathbf{Y}| \leq d$,

$$\left| \int_{S_2} \cdots \right| \leq K \int_{S_2} \frac{|\mathbf{y}|^{1+\alpha}}{|\mathbf{x}-\mathbf{y}|^4} \, dS_y \leq K \int_{\mathscr{D}} \frac{|\mathbf{Y}|^{1+\alpha}}{|\mathbf{Y}|^4} \, d\mathbf{Y} \leq K x_3^{-1+\alpha}.$$

As long as $x_3 < s$, which we may as well assume, the integral over S_3 is bounded in terms of $\max|u|$ and s.

In order to deal with the second term we observe that on S_1, $y_3 = |\mathbf{Y}|^{1+\alpha}$, so that

$$|D^2 g(\mathbf{x}, \mathbf{y})| \leq K y_3 |\mathbf{x}-\mathbf{y}|^{-4} = K|\mathbf{Y}|^{1+\alpha}|\mathbf{x}-\mathbf{y}|^{-4}.$$

We can then proceed as before. The other terms are dealt with similarly.

The (Hölder) continuity of grad u along the normal direction follows immediately from this estimate of the second derivatives. The theorem follows from continuity of the angle of inclination of the normal directions. ∎

4. Integral Equation Formulations of Dirichlet and Neumann Problems

In this section Ω is a bounded domain in $\mathbb{R}^n$ with compact boundary $\partial\Omega \in C^{1,\alpha}$, $0 < \alpha \leq 1$. We will refer for definiteness to the case $n = 3$. We recall that a closed (compact) surface S is of class $C^{1,\alpha}$, if for each point of S there is a neighborhood on which S is the graph of a function with first partial derivatives that are Hölder continuous with exponent α $(0 < \alpha \leq 1)$. A surface S of class $C^{1,\alpha}$ has the following important property.

Lemma 4.1. Suppose $\mathrm{S} \in C^{1,\alpha}$. Then

a. $|\mathbf{n}(\mathbf{x}) - \mathbf{n}(\mathbf{y})| \leq A|\mathbf{x}-\mathbf{y}|^\alpha$, for $\mathbf{x}, \mathbf{y} \in \mathrm{S}$ and some $A \geq 0$,
b. $\cos\varphi = O(|\mathbf{x}-\mathbf{y}|^\alpha)$, as $\mathbf{x}-\mathbf{y} \to 0$, $\mathbf{x}, \mathbf{y} \in \mathrm{S}$,

Where φ is the angle between $\mathbf{y}-\mathbf{x}$ and $\mathbf{n}(\mathbf{y})$, the unit normal to S at $\mathbf{y}$.

Proof. We can always choose a coordinate system in which the point $\mathbf{y}$ is the origin and the normal is along the positive z-axis. It is then easy to show that these statements hold at $\mathbf{y} = \mathbf{0}$ for points varying over S in this neighborhood. The invariance of these geometric quantities and the compactness of S then imply the results. ∎

We will suppose $\partial\Omega$ connected, so that $\Omega^c = \mathbb{R}^3\backslash\Omega$ is an unbounded (connected) domain. This assumption only simplifies the discussion and could be easily dispensed with. We recall that the normal $\mathbf{n}$ to a closed surface S is always assumed oriented toward the exterior of S.

4.1. Integral Operators with Weakly Singular Kernel. We begin by studying certain integral operators on the (compact) boundary S of our domain $\Omega \subset \mathbb{R}^n$ (Refs. 3, 4). The arguments work equally well if S is the closure of a bounded domain in $\mathbb{R}^m$, $m = n - 1$, and if n is any integer ≥ 2. Thus, we leave n unspecified here, although the case we have in mind is $n = 3$, $m = 2$.

Let A be a continuous function defined on $S \times S$. Because S is compact, A is bounded, $|A| \leq M$. Then the function

$$T(\mathbf{x}, \mathbf{y}) = A(\mathbf{x}, \mathbf{y})|\mathbf{x} - \mathbf{y}|^{-\beta}, \qquad \mathbf{x}, \mathbf{y} \in S,$$

is called a *weakly singular* (w.s.) *kernel* of order β if $0 \leq \beta < m = n - 1$, and the operator

$$Tf(\mathbf{x}) = \int_S T(\mathbf{x}, \mathbf{y}) f(\mathbf{y}) dS_y$$

is called a *weakly singular integral operator* of order β. The kernels $T(\mathbf{x}, \mathbf{y})$ are singular on the diagonal $\mathbf{x} = \mathbf{y}$ of $S \times S$, but the singularity is integrable (see the Appendix to Section 1).

Proposition 4.1. If T is a w.s. operator of order β, then T is defined and bounded on $L^2(S)$:

$$\|Tf\|_{L^2} \leq c\, d^{m-\beta} M \|f\|_{L^2}, \tag{8}$$

where $M = \|A\|_{L^\infty}$, $c = \omega_n/(m - \beta)$, $d = \operatorname{diam}(S)$, and ω_n the total solid angle (4π for $n = 3$, see Section 3 of Chapter 8).

Proof. First of all, we have

$$\int_S r^{-\beta}|A(\mathbf{x}, \mathbf{y})| dS_y \leq \omega_n d^{m-\beta} M/(m - \beta) = cM \qquad (r = |\mathbf{x} - \mathbf{y}|)$$

(see Lemma 1.1), and the variables $\mathbf{x}$, $\mathbf{y}$ can be interchanged in this inequality. Fubini's theorem and the Schwarz inequality then imply

$$\left| \int_S r^{-\beta} |A(\mathbf{x}, \mathbf{y})| |f(\mathbf{y})| dS_y \right|^2 \leq c\, d^{m-\beta} M \left| \int_S r^{-\beta} |A(\mathbf{x}, \mathbf{y})| |f(\mathbf{y})|^2 \, dS_y \right|,$$

so that

$$\|Tf\|_{L^2}^2 \leq c d^{m-\beta} M \|f\|_{L^2}^2 \left| \int_S r^{-\beta} |A(\mathbf{x}, \mathbf{y})| dS_y \right| \leq c^2 d^{2(m-\beta)} M^2 f_{L^2}^2. \qquad \square$$

Proposition 4.2. If T is a w.s. operator (of any order β), then T is compact on $L^2(S)$.

The proof requires a brief digression on Hilbert–Schmidt operators. A Hilbert–Schmidt operator is an integral operator with kernel $T(\mathbf{x}, \mathbf{y}) \in L^2(S \times S)$.

Lemma 4.2. If T is a Hilbert–Schmidt operator on $L^2(S)$, then T is compact and the norm of T is bounded by the $L^2(S \times S)$ norm of the kernel $T(\mathbf{x}, \mathbf{y})$.

Proof of Lemma 4.2. By the Schwarz inequality,

$$|Tf| \leq \left| \int_S |T(\mathbf{x}, \mathbf{y})| |f(\mathbf{y})| dS_y \right| \leq \|f\|_{L^2} \left| \int_S |T(\mathbf{x}, \mathbf{y})|^2 dS_y \right|^{1/2}.$$

This shows Tf is finite almost everywhere, and furthermore

$$\|Tf\|_{L^2} \leq \left| \int_S \int_S |T(\mathbf{x}, \mathbf{y})|^2 dS_x \, dS_y \right|^{1/2} \|f\|_{L^2},$$

hence T is bounded on $L^2(S)$ with norm bounded as stated. Let $\{u_i(\mathbf{x})\}$ be an orthonormal basis for $L^2(S)$. From Fubini's theorem it follows that $\{u_i(\mathbf{x})u_j(\mathbf{y})\}$ is an orthonormal basis for $L^2(S \times S)$, and we can expand $T(\mathbf{x}, \mathbf{y})$ in the norm convergent Fourier series

$$T(\mathbf{x}, \mathbf{y}) = \sum_{i,j=1}^{\infty} a_{ij} u_i(\mathbf{x}) u_j(\mathbf{y})$$

with $\sum_{i,j=1}^{\infty} |a_{ij}|^2 < \infty$. Then for $N = 1, 2, \ldots$, we have $Tf = T_n f + T'_N f$, where

$$T_N f = \sum_{i+j \leq N} a_{ij}(u_j, f)u_i(\mathbf{x}) \equiv \sum_{i+j \leq N} a_{ij}\left(\int_S u_j(\mathbf{y})f(\mathbf{y})dS_y\right)u_i(\mathbf{x})$$

is an operator of finite rank, while $\|T'_N\|_{L^2}^2 = \sum_{i+j>N}^{\infty} |a_{ij}|^2 \to 0$ as $N \to \infty$. By Lemma 1.2 of Chapter 8, T is compact. □

Proof of Proposition 4.2. Given $\varepsilon > 0$, set

$$T_\varepsilon(\mathbf{x}, \mathbf{y}) = \begin{cases} T(\mathbf{x}, \mathbf{y}), & \text{if } |\mathbf{x} - \mathbf{y}| > \varepsilon, \\ 0, & \text{otherwise,} \end{cases}$$

and set $T'_\varepsilon = T - T_\varepsilon$. Then $T_\varepsilon(\mathbf{x}, \mathbf{y})$ is bounded on $S \times S$, and the corresponding integral operator T_ε is Hilbert–Schmidt, hence compact on $L^2(S)$ by Lemma 4.2. On the other hand, T'_ε is a w.s. operator of order β with kernel supported in the ball $|\mathbf{x} - \mathbf{y}| \leq \varepsilon$, hence Proposition 4.1 holds with $d = \varepsilon$ and, because $\beta < m$, the operator norm of T'_ε tends to zero as $\varepsilon \to 0$. So again T is compact by the cited lemma in Chapter 8. □

Proposition 4.3. If T is a w.s. operator (of any order β), then T transforms bounded functions into continuous functions. In particular, $T : C(S) \to C(S)$.

Proof. We may assume $\beta > 0$, for a continuous kernel is also a weakly singular kernel of any order $\beta > 0$. Take an arbitrary point $\mathbf{x}_0 \in S$ and for $\delta > 0$ sufficiently small let $S_\delta = S \cap B(\mathbf{x}_0, \delta) = \{\mathbf{y} \in S : |\mathbf{y} - \mathbf{x}_0| < \delta\}$, $S'_\delta = S \backslash S_\delta$. Then for $\mathbf{x} \in S$ we have

$$\begin{aligned}
|Tf(\mathbf{x}) - Tf(\mathbf{x}_0)| &\leq \int_S |T(\mathbf{x}, \mathbf{y}) - T(\mathbf{x}_0, \mathbf{y})||f(\mathbf{y})|\, dS \\
&= \int_{S'_{2\delta}} |T(\mathbf{x}, \mathbf{y}) - T(\mathbf{x}_0, \mathbf{y})||f(\mathbf{y})|\, dS \\
&\quad + \int_{S_{2\delta}} |T(\mathbf{x}, \mathbf{y}) - T(\mathbf{x}_0, \mathbf{y})||f(\mathbf{y})|\, dS \\
&\leq \|f\|_{L^\infty} \int_{S'_{2\delta}} |T(\mathbf{x}, \mathbf{y}) - T(\mathbf{x}_0, \mathbf{y})|\, dS_y \\
&\quad + M\|f\|_{L^\infty} \int_{S_{2\delta}} (|\mathbf{x}_0 - \mathbf{y}|^{-\beta} + |\mathbf{x} - \mathbf{y}|^{-\beta})dS_y.
\end{aligned}$$

Suppose $|\mathbf{x} - \mathbf{x}_0| < \delta$. Because $\beta < m$, integrating in polar coordinates we see that the integral on $S_{2\delta}$ is $O(\delta^{m-\beta})$ and hence can be made small by suitably restricting δ. With $|\mathbf{y} - \mathbf{x}| > \delta$ on $S'_{2\delta}$, $|T(\mathbf{x}, \mathbf{y}) - T(\mathbf{x}_0, \mathbf{y})|$ can be made uniformly small over $S'_{2\delta}$ by taking $|\mathbf{x} - \mathbf{x}_0|$ sufficiently small. Thus, for any given $\varepsilon > 0$, we can fix δ so that for $|\mathbf{x} - \mathbf{x}_0| < \delta$, $|Tf(\mathbf{x}) - Tf(\mathbf{x}_0)| \leq \varepsilon$. □

By adapting the proof of Proposition 4.2, it cn be proved that a w.s. operator T is also compact on $C(S)$ (Ref. 5).

Proposition 4.4. Suppose T is a w.s. operator. If $u \in L^2(S)$ and $u + Tu \in C(S)$, then $u \in C(S)$.

Proof. Given $\varepsilon > 0$, choose $\phi \in C(S \times S)$ such that $0 \leq \phi \leq 1$, $\phi = 1$ on $S_{\varepsilon/2}$, $\phi = 0$ on S'_ε. Set $T_0(\mathbf{x}, \mathbf{y}) = \phi T(\mathbf{x}, \mathbf{y})$, $T_1(\mathbf{x}, \mathbf{y}) = (1 - \phi)T(\mathbf{x}, \mathbf{y})$. Then by the Schwarz inequality,

$$|T_1 u(\mathbf{x}) - T_1 u(\mathbf{x}_0)|^2 \leq \|u\|_{L^2}^2 \int_S |T_1(\mathbf{x}, \mathbf{y}) - T_1(\mathbf{x}_0, \mathbf{y})|^2 \, dS_y.$$

Because $T_1(\mathbf{x}, \mathbf{y})$ is continuous, the integral on the right tends to zero as $\mathbf{x} \to \mathbf{x}_0$; thus, $T_1 u(\mathbf{x})$ is continuous. Set $g = u + T_0 u \equiv u + Tu - T_1 u$. Then g is continuous, as both $u + Tu$ and $T_1 u$ are. Proposition 4.1 (with $d = \varepsilon$) tells us that the operator norm of T_0 on both L^2 and L^∞ will be less than 1 for ε sufficiently small. Then $I + T_0$ is invertible, and

$$u = (I + T_0)^{-1} g = \sum_{j=0}^{\infty} (-T_0)^j g,$$

the series being uniformly convergent. Because each term is continuous by Proposition 4.3, u is continuous on S. □

This crucial proposition ensures that solutions $u \in L^2$ of the Fredholm equation $\lambda u - Tu = f$ are continuous if f is continuous.

Remark 4.1. By Fubini's theorem, the integral operator T' with kernel $T(\mathbf{y}, \mathbf{x})$ is the *adjoint* T^* of T in $L^2(S)$ (see Section 1 of Chapter 8):

$$\begin{aligned} (Tf, g) &= \int_S g(\mathbf{x}) dS_x \int_S T(\mathbf{x}, \mathbf{y}) f(\mathbf{y}) dS_y \\ &= \int_S f(\mathbf{x}) dS_x \int_S T(\mathbf{y}, \mathbf{x}) g(\mathbf{y}) dS_y = (f, T'g). \end{aligned}$$

Thus, T' is a w.s. operator of order β if and only if T is.

4.2. Layer Potentials. We will discuss some properties of the single layer potential

$$V_\mu(\mathbf{x}) = \int_{\partial\Omega} E(\mathbf{x}, \mathbf{y})\mu(\mathbf{y})dS_y$$

and of the double layer (or dipole) potential

$$W_\nu(\mathbf{x}) = \int_{\partial\Omega} \frac{\partial E(\mathbf{x}, \mathbf{y})}{\partial n_y} \nu(\mathbf{y})dS_y$$

in three dimensions, so that the fundamental solution is $E(\mathbf{x}, \mathbf{y}) = 1/(4\pi|\mathbf{x} - \mathbf{y}|)$. We assume that the corresponding densities are continuous over $\partial\Omega$, μ, $\nu \in C(\partial\Omega)$. As mentioned in Section 1, both $V_\mu(\mathbf{x})$ and $W_\nu(\mathbf{x})$ are defined and harmonic (hence, analytic) in $\mathbb{R}^3\backslash\partial\Omega$. Moreover, they are regular at infinity. [Note the exception of $V_\mu(\mathbf{x})$ for $n = 2$; see the exercises.] What happens when $\mathbf{x}$ is near $\partial\Omega$? In general, one should distinguish two cases: (i) $\mathbf{x}$ goes along $\partial\Omega$ and (ii) $\mathbf{x}$ goes across $\partial\Omega$.

We begin with $W_\nu(\mathbf{x})$. A basic property of double layer potentials with unit density, $W_1(\mathbf{x})$, follows from Gauss's solid angle formula

$$\int_S \frac{\partial E(\mathbf{x}, \mathbf{y})}{\partial n_y} dS_y = \begin{cases} -1, & \mathbf{x} \in \Omega, \\ -1/2, & \mathbf{x} \in S, \\ 0, & \mathbf{x} \in \Omega^c, \end{cases} \tag{9}$$

where $\Omega^c = \mathbb{R}^3\backslash\bar{\Omega}$ and $S = \partial\Omega$. This formula (which holds in any number of dimensions) is an immediate consequence of Green's representation theorem (R), (R$'$), and (R$''$) in Section 2 for the harmonic function $u \equiv 1$ in Ω. This formula is named for the fundamental relation between a unit dipole distribution on a surface element dS centered at a point $\mathbf{y}$ on S and the solid angle (with sign) spanned by dS at a point $\mathbf{x}$ in $\mathbb{R}^3$,

$$-\frac{\partial}{\partial n_y}\left(\frac{1}{r}\right)dS = \frac{\cos\varphi}{r^2} dS = \pm d\omega,$$

where $\varphi = \varphi(\mathbf{x}, \mathbf{y})$ is the angle between $\mathbf{y} - \mathbf{x}$ and $\mathbf{n}(\mathbf{y})$, $r = |\mathbf{x} - \mathbf{y}|$, and $d\omega$ is the element of solid angle subtended at $\mathbf{x}$ (Fig. 8).

This relation implies that if the surface S is well behaved, e.g., if S can be subdivided into a finite number of portions on each of which $\cos\varphi$ retains the

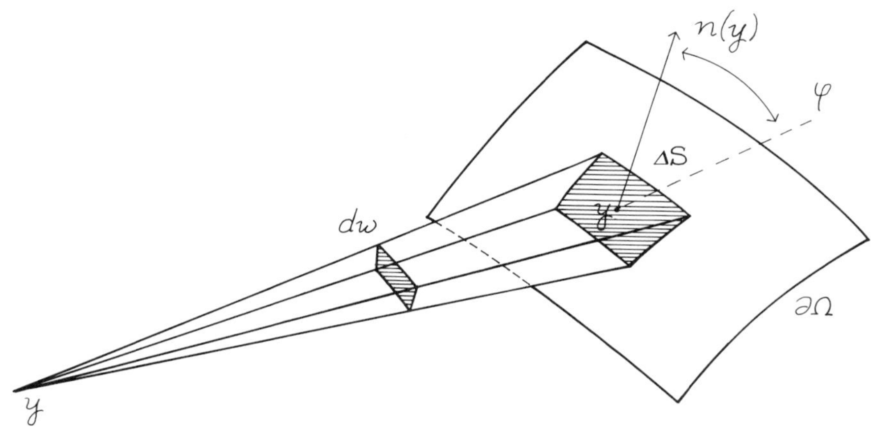

Fig. 8. The unit dipole on a surface and the subtended solid angle.

same sign (see, however, Remark 4.2), the integral $\int_S(\partial/\partial n_y)(1/r)dS_y$ yields the total "solid angle with sign" subtended at $\mathbf{x}$ by $\partial\Omega$. The Gauss formula then follows from elementary geometrical considerations.

Another consequence of the above relation is the following. Suppose $\mathbf{x} \in \partial\Omega$. We will use the notation $K\nu(\mathbf{x})$, $\mathbf{x} \in \partial\Omega$, for the integral operator on $\partial\Omega$ with kernel $T(\mathbf{x}, \mathbf{y}) = \cos\varphi(\mathbf{x}, \mathbf{y})/(4\pi|\mathbf{x} - \mathbf{y}|^2)$. Its adjoint in $L^2(\partial\Omega)$ is the integral operator K^* with kernel $T(\mathbf{y}, \mathbf{x})$.

Theorem 4.1. K, K^* are weakly singular operators of order $\beta = 2 - \alpha$ on $S = \partial\Omega$. Hence, $K, K^*: C(\partial\Omega) \to C(\partial\Omega)$ and $K, K^*: L^2(\partial\Omega) \to L^2(\partial\Omega)$ are compact.

Proof. If *both* $\mathbf{x}, \mathbf{y} \in \partial\Omega$, then $\cos\varphi(\mathbf{x}, \mathbf{y}) \to 0$ as $r = |\mathbf{x} - \mathbf{y}| \to 0$. In fact, as $\partial\Omega \in C^{1,\alpha}$ by assumption, Lemma 4.1 tells us that $\cos\varphi/r^2 = O(r^{-2+\alpha})$. From the results of Section 4.1 it follows that K, K^* are w.s. operators of order $2 - \alpha$, and propositions 4.1–4.4 hold. □

By Theorem 4.1, it is reasonable to extend the potential $W_\nu(\mathbf{x})$ to $\partial\Omega$ by setting

$$W_\nu(\mathbf{x}) = K\nu(\mathbf{x}), \qquad \mathbf{x} \in \partial\Omega,$$

and then the restriction of $W_\nu(\mathbf{x})$ to $\partial\Omega$ is continuous. However, $W_\nu(\mathbf{x})$ is not continuous on $\mathbb{R}^3$: There is a jump when a point $\mathbf{x} \in \mathbb{R}^3\backslash\partial\Omega$ approaches a point

$\mathbf{x}_0 \in \partial\Omega$. We have already seen this to happen when ν is constant, as W_ν is linear in ν and Gauss's formula says that

$$W_\nu(\mathbf{x}) = \begin{cases} -\nu, & \mathbf{x} \in \Omega, \\ -\nu/2, & \mathbf{x} \in \partial\Omega, \\ 0, & \mathbf{x} \in \Omega^c, \end{cases} \tag{9a}$$

when ν is constant. Thus, the limits

$$W_{\text{int}}(\mathbf{x}_0) = \lim_{\substack{\mathbf{x}\to\mathbf{x}_0 \\ \mathbf{x}\in\Omega}} W_\nu(\mathbf{x})$$

and

$$W_{\text{ext}}(\mathbf{x}_0) = \lim_{\substack{\mathbf{x}\to\mathbf{x}_0 \\ \mathbf{x}\in\Omega'}} W_\nu(\mathbf{x})$$

exist for constant ν. In order to deal with a general $\nu \in C(\partial\Omega)$, we need a further property of $C^{1,\alpha}$ surfaces.

(FV): *There is a constant N such that*

$$\int_{\partial\Omega} \left| \frac{\partial E(\mathbf{x}, \mathbf{y})}{\partial n_y} \right| dS_y \equiv \frac{1}{4\pi} \int_{\partial\Omega} \frac{|\cos\varphi|}{r^2}\, dS \le N$$

uniformly for $\mathbf{x} \in \mathbb{R}^3$.

For a fixed $\mathbf{x}$ and $\Sigma \subset \partial\Omega$ on which $\cos\varphi(\mathbf{x}, \mathbf{y})$ is positive (negative), $\int_\Sigma r^{-2} \cos\varphi\, dS$ is the solid angle (with sign) subtended at $\mathbf{x}$ by Σ. If there is an upper bound on the number of sign changes of $\cos\varphi$ on $\partial\Omega$ that is independent of $\mathbf{x}$ (e.g., if Ω is convex), (FV) certainly holds. For the proof concerning a general $C^{1,\alpha}$ surface, the reader is referred to Mikhlin (Ref. 6, p. 352).

We then have the following important result.

Theorem 4.2. if $\nu \in C(\partial\Omega)$ and $\partial\Omega \in C^{1,\alpha}$, the limits $W_{\text{int}}(\mathbf{x}_0)$, $W_{\text{ext}}(\mathbf{x}_0)$ exist and the jump relations

$$W_{\text{int}}(\mathbf{x}_0) = K\nu(\mathbf{x}_0) - \tfrac{1}{2}\nu(\mathbf{x}_0),$$
$$W_{\text{ext}}(\mathbf{x}_0) = K\nu(\mathbf{x}_0) + \tfrac{1}{2}\nu(\mathbf{x}_0)$$

hold as $\mathbf{x}$ tends to any point $\mathbf{x}_0 \in \partial\Omega$.

Proof. We define, for $\mathbf{x}_0 \in \partial\Omega$,

$$\Phi(\mathbf{x}) = \int_{\partial\Omega} \frac{\partial E(\mathbf{x}, \mathbf{y})}{\partial n_y}[\nu(\mathbf{y}) - \nu(\mathbf{x}_0)]dS_y.$$

Then for $\mathbf{x} \in \mathbb{R}^3$,

$$|\Phi(\mathbf{x}) - \Phi(\mathbf{x}_0)| \leq \int_{S_\delta} \left|\frac{\partial E(\mathbf{x}, \mathbf{y})}{\partial n_y} - \frac{\partial E(\mathbf{x}_0, \mathbf{y})}{\partial n_y}\right| |\nu(\mathbf{y}) - \nu(\mathbf{x}_0)| dS_y$$
$$+ \int_{T_\delta} \left|\frac{\partial E(\mathbf{x}, \mathbf{y})}{\partial n_y} - \frac{\partial E(\mathbf{x}_0, \mathbf{y})}{\partial n_y}\right| |\nu(\mathbf{y}) - \nu(\mathbf{x}_0)| dSy,$$

where $S_\delta = \partial\Omega \cap B(\mathbf{x}_0, \delta)$ and $T_\delta = \partial\Omega \backslash S_\delta$, so that $|\mathbf{y} - \mathbf{x}_0| > \delta$ on T_δ. Let $|\mathbf{x} - \mathbf{x}_0| < \delta/2$. If $M = \max_{\partial\Omega}|\nu|$, the second integral is bounded by

$$2M \int_{T_\delta} \left|\frac{\partial E(\mathbf{x}, \mathbf{y})}{\partial n_y} - \frac{\partial E(\mathbf{x}_0, \mathbf{y})}{\partial n_y}\right| dS_y,$$

where, because $|\mathbf{y} - \mathbf{x}| \geq |\mathbf{y} - \mathbf{x}_0| - |\mathbf{x} - \mathbf{x}_0| > \delta/2$, the integrand is continuous and vanishes at $\mathbf{x} = \mathbf{x}_0$. The first integral is bounded by $2N \max|\nu(\mathbf{y}) - \nu(\mathbf{x}_0)|$. For given $\varepsilon > 0$, we choose δ so that $|\mathbf{y} - \mathbf{x}_0| < \delta$ implies that $|\nu(\mathbf{y}) - \nu(\mathbf{x}_0)| < \varepsilon/4N$, and then choose $\mathbf{x}$ sufficiently close to $\mathbf{x}_0$ so that the bound for the second integral is less than $\varepsilon/2$. This shows that $\Phi(\mathbf{x})$ is continuous at $\mathbf{x}_0$. Then the assertion follows from formula (9a), with $\nu = \nu(\mathbf{x}_0)$. □

The jump relations can be stated in the form

$$K\nu(\mathbf{x}_0) = \tfrac{1}{2}(W_{\text{ext}}(\mathbf{x}_0) + W_{\text{int}}(\mathbf{x}_0)), \qquad \nu(\mathbf{x}_0) = W_{\text{ext}}(\mathbf{x}_0) - W_{\text{int}}(\mathbf{x}_0).$$

Remark 4.2. The definition of solid angle subtended at $\mathbf{x}$ can be generalized to Borel subsets of very general surfaces (finite area and zero volume), and this set function is a Borel measure whose variation is given by the integral in (FV). It has been shown by Maz'ya (Ref. 7) that, in this general context, (FV) is a necessary and sufficient condition for the jump relations to hold.

We now turn our attention to the single layer potential, $V_\mu(\mathbf{x})$.

Theorem 4.3. If $\mu \in C(\partial\Omega)$, $V_\mu \in C(\mathbb{R}^3)$.

Proof. The proof is very similar to that of Proposition 4.3. Suppose $\mathbf{x}_0 \in \partial\Omega$, $\mathbf{x} \in \mathbb{R}^3$. If $S_\delta = \partial\Omega \cap B(\mathbf{x}_0, \delta)$, S'_δ are as in the proof of proposition 4.3, and $m = \max_{\partial\Omega}|\mu|$, then

$$|V_\mu(\mathbf{x}) - V_{\mu(\mathbf{x}_0)}| \leq M \int_{T_{2\delta}} |E(\mathbf{x}, \mathbf{y}) - E(\mathbf{x}_0, \mathbf{y})| dS_y$$
$$+ M \int_{S_{2\delta}} |E(\mathbf{x}, \mathbf{y})| dS_y + M \int_{S_\delta} |E(\mathbf{x}_0, \mathbf{y})| dS_y.$$

Suppose $|\mathbf{x} - \mathbf{x}_0| \leq \delta$. Then the uniform absolute continuity of the integrals $\int |E(\mathbf{x}, \mathbf{y})| dS_y$ over subsets of $\partial\Omega$ implies that the last two integrals can be made small by suitably restricting δ, and, as $|\mathbf{y} - \mathbf{x}| > \delta$ on $T_{2\delta}$, $|E(\mathbf{x}, \mathbf{y}) - E(\mathbf{x}_0, \mathbf{y})|$ can be made uniformly small over $T_{2\delta}$ by taking δ sufficiently small. □

Thus, $V_\mu(\mathbf{x})$ is continuous everywhere, in particular the restriction of $V_\mu(\mathbf{x})$ to $\partial\Omega$ is continuous. If for this restriction we use the notation

$$V_\mu(\mathbf{x}) = V\mu(\mathbf{x}), \qquad \mathbf{x} \in \partial\Omega,$$

then V is an integral operator on $\partial\Omega$ with kernel $T(\mathbf{x}, \mathbf{y}) = E(\mathbf{x}, \mathbf{y})$, and, as $E(\mathbf{x}, \mathbf{y}) = E(\mathbf{y}, \mathbf{x})$, $V = V^*$ is self-adjoint in $L^2(\partial\Omega)$.

Theorem 4.4. V is a weakly singular operator of order 1 on $S = \partial\Omega$. Hence, propositions 4.1–4.4 hold, in particular $V: C(\partial\Omega) \to C(\partial\Omega)$ and $V: L^2(\partial\Omega) \to L^2(\partial\Omega)$ is compact.

We now consider normal derivatives on $\partial\Omega$ of $V_\mu(\mathbf{x})$. In order to do this with minimal hypotheses, we consider, for $\mathbf{x} \notin \partial\Omega$,

$$W'_\mu(\mathbf{x}) = \text{grad } V_\mu(\mathbf{x}) \cdot \mathbf{n}(\mathbf{x}_0) := \partial V_\mu(\mathbf{x})/\partial n(\mathbf{x}_0)$$

for $\mathbf{x}_0 \in \partial\Omega$. We can also define, for $\mathbf{x} \in \partial\Omega$,

$$K'\mu(\mathbf{x}) = \int_{\partial\Omega} \frac{\partial E(\mathbf{x}, \mathbf{y})}{\partial n_x} \mu(\mathbf{y}) dS_y.$$

As $\partial E(\mathbf{x}, \mathbf{y})/\partial n_x = \cos \varphi(\mathbf{y}, \mathbf{x})/(4\pi|\mathbf{x} - \mathbf{y}|^2)$ coincides with $\partial E(\mathbf{x}, \mathbf{y})/\partial n_y$ if $\mathbf{x}$ and $\mathbf{y}$ are exchanged, we see that K' coincides with the adjoint K^* of K and hence is a w.s. operator with all of the properties stated in Theorem 4.1 and in Propositions 4.1–4.4.

Suppose that $\mathbf{x}$ lies along the normal to $\partial\Omega$ at $\mathbf{x}_0$, $\mathbf{x} = \mathbf{x}_0 + \varepsilon\mathbf{n}(\mathbf{x}_0)$. We write

$$
\begin{aligned}
W'_\mu(\mathbf{x}) &= \mathbf{n}(\mathbf{x}_0)\cdot \operatorname{grad}_x V_\mu(\mathbf{x}) = -\int_{\partial\Omega} \mathbf{n}(\mathbf{x}_0)\cdot \operatorname{grad}_y E(\mathbf{x},\mathbf{y})\mu(\mathbf{y})dS_y \\
&= -\int_{\partial\Omega} [\mathbf{n}(\mathbf{x}_0) - \mathbf{n}(\mathbf{y})]\cdot \operatorname{grad}_y E(\mathbf{x},\mathbf{y})\mu(\mathbf{y})dS_y - W_\mu(\mathbf{x}) \\
&:= I_\varepsilon(\mathbf{x}_0) - W_\mu(\mathbf{x}), \qquad \mathbf{x} = \mathbf{x}_0 + \varepsilon\mathbf{n}(\mathbf{x}_0).
\end{aligned}
$$

As $\partial\Omega$ is a $C^{1,\alpha}$ surface and $\mathbf{x}$ is on the normal through $\mathbf{x}_0$, $|\mathbf{x}-\mathbf{y}| \geq C|\mathbf{x}_0 - \mathbf{y}|$ and the integrand in the first integral is $O(|\mathbf{y}-\mathbf{x}_0|^\alpha/|\mathbf{x}-\mathbf{y}|^2) \leq O(|\mathbf{y}-\mathbf{x}_0|^{\alpha-2})$ for $\mathbf{y}$ in a neighborhood of x_0. It follows that

$$
\lim_{\varepsilon\to 0} I_\varepsilon(\mathbf{x}_0) = -\int_{\partial\Omega} [\mathbf{n}(\mathbf{x}_0) - \mathbf{n}(\mathbf{y})]\cdot \operatorname{grad}_y E(\mathbf{x}_0,\mathbf{y})\mu(\mathbf{y})dS_y.
$$

On the other hand,

$$
W_\mu(\mathbf{x}) \to \pm\frac{1}{2}\mu(\mathbf{x}_0) + \int_{\partial\Omega} \mathbf{n}(\mathbf{y})\cdot \operatorname{grad}_y E(\mathbf{x}_0,\mathbf{y})\mu(\mathbf{y})dS_y
$$

as $\varepsilon \to 0\pm$. If $W'_{\text{ext}}(\mathbf{x}_0)$, $W'_{\text{int}}(\mathbf{x}_0)$ are the limits along the normal of $W'_\mu(\mathbf{x})$ as $\varepsilon \to 0\pm$, we have proven

Theorem 4.5. The limits $W'_{\text{ext}}(\mathbf{x}_0)$, $W'_{\text{int}}(\mathbf{x}_0)$ exist and the jump relations

$$
\begin{aligned}
W'_{\text{int}}(\mathbf{x}_0) &= K'\mu(\mathbf{x}_0) + \tfrac{1}{2}\mu(\mathbf{x}_0), \\
W'_{\text{ext}}(\mathbf{x}_0) &= K'\mu(\mathbf{x}_0) - \tfrac{1}{2}\mu(\mathbf{x}_0)
\end{aligned}
$$

hold at every $\mathbf{x}_0 \in \partial\Omega$ if $\mu \in C(\partial\Omega)$.

The functions W'_{int} and W'_{ext} are normal derivatives of V_μ on Ω, Ω^c if V_μ is in $C^1(\bar{\Omega})$, $C^1(\bar{\Omega}^c)$, respectively. This is true if it is known that $\mu \in C^{0,\alpha}(\partial\Omega)$ (see Refs. 8, 9). As a consequence, the solution of the Neumann problem with Hölder continuous boundary data is C^1 up to the boundary (see the analogous result for the Dirichlet problem in Section 3.2). The proof is somewhat involved and we omit it from our presentation. We assume only that μ is continuous and use normal derivatives of V_μ in the "principal value" sense that leads to W'_{int}, W'_{ext} above.

The previous considerations can be extended to two space dimensions, $n = 2$. In particular, V, K, K' are compact and the same jump relations for W_ν, W'_μ hold in this case (Exercise 4.1).

4.3. Layer Ansatz and Boundary Integral Equations. We will use the operators K, K' and the jump relations for W_ν, W'_μ to give integral equation formulations of Dirichlet and Neumann problems in Ω and Ω^c. As mentioned, we assume here that $\partial\Omega$ is connected; the general case requires some slight modifications (Ref. 8). Let

$$\text{(DI)}\begin{cases} \Delta u = 0 \text{ in } \Omega, \\ u = f \text{ on } \partial\Omega, \end{cases} \qquad \text{(NI)}\begin{cases} \Delta u = 0 \text{ in } \Omega, \\ \partial u/\partial n = g \text{ on } \partial\Omega, \end{cases}$$

$$\text{(DE)}\begin{cases} \Delta u = 0 \text{ in } \Omega^c, \\ u = f \text{ on } \partial\Omega, \\ u \text{ regular at } \infty, \end{cases} \qquad \text{(NE)}\begin{cases} \Delta u = 0 \text{ in } \Omega^c, \\ \partial u/\partial n = g \text{ on; } \partial\Omega, \\ u \text{ regular at } \infty. \end{cases}$$

We make the Ansatz $u = W_\nu$ for (DI) and (DE), and $u = V_\mu$ for (NI) and (NE). The jump relations then give rise, formally, to the equations

$$(K - \tfrac{1}{2}I)\nu = f, \qquad (K' - \tfrac{1}{2}I)\mu = g$$

for (DI) and (NE), respectively, and

$$(K + \tfrac{1}{2}I)\nu = f, \qquad (K' + \tfrac{1}{2}I)\mu = g$$

for (DE) and (NI), respectively. We envisage these as integral equations in the unknown densities μ, ν for given f, g in $C(\partial\Omega)$. As mentioned, K, K' are compact on $L^2(\partial\Omega)$ and $K' = K^*$ is the adjoint operator of K. We can therefore utilize the Fredholm alternative for compact operators on a Hilbert space. Further, if $f, g \in C(\partial\Omega)$, solutions of $(K \mp \tfrac{1}{2}I)\nu = f$, $(K' \pm \tfrac{1}{2}I)\mu = g$ in $L^2(\partial\Omega)$ belong to $C(\partial\Omega)$ (by proposition 4.4).

Consider first (DI). Solvability of $(K - \tfrac{1}{2}I)\nu = f$ for all $f \in C(\partial\Omega)$ is equivalent to $\tfrac{1}{2} \notin sp(K)$, which is in turn equivalent to $\tfrac{1}{2} \notin sp(K')$ by virtue of the Fredholm alternative. Suppose that $\mu \in C(\partial\Omega)$ is such that, for $\mathbf{x} \in \partial\Omega$,

$$\frac{1}{2}\mu(\mathbf{x}) = \int_{\partial\Omega} \frac{\partial E(\mathbf{x}, \mathbf{y})}{\partial n_x} \mu(\mathbf{y}) dS_y.$$

Let $u(\mathbf{x}) = V_\mu(\mathbf{x})$. Then the jump relations of W'_μ and the equation $K'\mu = \frac{1}{2}\mu$ imply that $W'_{\text{ext}} = 0$.

Lemma 4.3. On a $C^{1,\alpha}$ domain the boundary point lemma for Laplace's equation (Theorem 2.5) holds if normal derivatives are taken in the principal value sense.

In order to see that this is true we need only consider interior neighborhoods bounded by surfaces $x_3 = -|\mathbf{X}|^{1+\alpha}$ (notations of Section 3.2) in place of spheres in the usual proof. This lemma implies that a harmonic function, continuous on the closure of a domain, with zero normal derivative in the principal value sense must be constant. Because $u \in C(\mathbb{R}^3)$ and is harmonic on $\Omega \cup \Omega^c$, we can deduce that u is constant in Ω^c. Then $V_\mu(\mathbf{x}) = O(|\mathbf{x}|^{-2})$ implies that $u \equiv 0$ in Ω^c, and $u = 0$ on $\partial\Omega$ implies that $u \equiv 0$ in Ω also. Finally, $\mu(\mathbf{x}) = W'_{\text{int}}(\mathbf{x}) - W'_{\text{ext}}(\mathbf{x})$ on $\partial\Omega$ implies that $\mu \equiv 0$, and $\frac{1}{2} \notin sp(K')$.

We conclude that (DI) is uniquely solvable in a $C^{1+\alpha}$ domain, for continuous boundary values f, as a double layer potential with continuous density ν.

The exterior Neumann problem (NE) is solvable as a single layer potential with continuous density μ, for every continuous boundary value g, if $(K' - \frac{1}{2}I)\mu = 0$ has only the trivial solution. We have seen that this is true for any $C^{1+\alpha}$ domain. We observe that the solution μ of $(K' - \frac{1}{2}I)\mu = g$ satisfies the relation

$$\int_{\partial\Omega} \mu \, dS = -\int_{\partial\Omega} g \, dS$$

(Exercise 4.3).

Consider now (DE) and (NI). Suppose that μ_0 is a solution of $K'\mu = -\frac{1}{2}\mu$, and let $u = V_{\mu_0}$. Then $W'_{\text{int}} \equiv 0$ and we can deduce that u is constant in Ω. Suppose that μ_1 is another solution and let $\mu_2 = \mu_1 - c\mu_0$. If $u_2 = V_{\mu_2}$, u_2 is also constant in Ω and we can choose c so that $u_2 \equiv 0$ in Ω. Then $u_2 = O(|\mathbf{x}|^{-1})$ implies that $u_2 \equiv 0$ in $\mathbb{R}^3$. As before, $\mu_2 = 0$ on $\partial\Omega$ and we have $\mu_1 = c\mu_0$. On the other hand, we know from the Gauss solid angle formula (9a) that constants automatically satisfy $K\nu = -\frac{1}{2}\nu$. It follows that $\dim \ker(K' + \frac{1}{2}I) = 1$ and all eigensolutions of K are constant. Suppose that we choose a nonzero element μ_0 in $\ker(K' + \frac{1}{2}I)$. Then the Fredholm alternative implies that $(K + \frac{1}{2}I)\nu = f$ is solvable if and only if

$$\int_{\partial\Omega} f\mu_0 \, dS = 0 \tag{10}$$

and $(K' + \frac{1}{2}I)\mu = g$ is solvable if and only if

$$\int_{\partial\Omega} g\, dS = 0. \tag{11}$$

The (nonunique) solutions can be written in the form $\mu = \bar{\mu} + a\mu_0$, $\nu = \bar{\nu} + a$ for an arbitrary constant a. The corresponding solutions of (DE) and (NI) are

$$u = W_{\bar{\nu}}(\mathbf{x}) + aW_1(\mathbf{x}) \qquad (\mathbf{x} \in \Omega^c), \qquad u = V_{\bar{\mu}}(\mathbf{x}) + aV_{\mu_0}(\mathbf{x}) \qquad (\mathbf{x} \in \Omega).$$

Because $W_1(\mathbf{x})$ vanishes on Ω^c, $u = W_{\bar{\nu}}(\mathbf{x})$ is the *unique* solution of (DE). Uniqueness up to constants for the interior Neumann problem implies that $V_{\mu_0}(\mathbf{x})$ is constant on Ω and, as the constant can be shown to be nonzero (see the corollary below), we may normalize $V_{\mu_0}(\mathbf{x})$ so that

$$V_{\mu_0}(\mathbf{x}) = 1, \qquad \mathbf{x} \in \Omega. \tag{12}$$

It follows that V_{μ_0} is nothing else but the *capacitary potential* of Ω. The function μ_0 is called the *Robin density* of Ω.

Theorem 4.6. $\int_{\partial\Omega} \mu_0\, dS \neq 0$.

Proof. Because $V_{\mu_0} = u_0$ is constant on $\bar{\Omega}$, the jump relations for W'_μ imply that $\mu_0 = -W'_{\text{ext}}$ on $\partial\Omega$. Theorem 3.7 implies that $u_0 \in C^1(\overline{\Omega^c})$ so that W'_{ext} is an exterior normal derivative in the ordinary sense and, because u_0 is regular at infinity, Green's identities hold in Ω^c. It follows that

$$\int_{\Omega^c} |\text{grad}\, u_0|^2 dV = -\int_{\partial\Omega} u_0 W'_{\text{ext}}\, dS = c_0 \int_{\partial\Omega} \mu_0\, dS, \tag{13}$$

where $c_0 = u_0|_{\bar{\Omega}}$, so that, in case $\int_{\partial\Omega} \mu_0\, dS = 0$, u_0 is constant on Ω^c. Because $u_0 = O(|\mathbf{x}|^{-1})$ as $|\mathbf{x}| \to \infty$, this constant is necessarily zero and by the jump relations for W'_{μ_0} we find $\mu_0 \equiv 0$, a contradiction. ☐

Corollary 4.1. $V_{\mu_0}(\mathbf{x}) \equiv c_0 \neq 0\ \forall \mathbf{x} \in \bar{\Omega}$, and $\int_{\partial\Omega} \mu_0\, dS$ has the sign of c_0.

Proof. Follows immediately from (13). ☐

To summarize, the interior Neumann problem is solvable as a single layer potential for any continuous boundary data g satisfying the compatibility condition (11), as expected on the basis of Green's identities [see equation (3) in Section 2].

On the other hand, because the exterior Dirichlet problem can be solved for any continuous f, as may be seen, for example, by applying Perron's method, the question of the meaning of the necessary condition (10) must be answered. We must remember that this arose from seeking a solution represented as a double layer potential, $u = W_\nu(\mathbf{x})$, so that necessarily $u = O(|\mathbf{x}|^{-2})$, whereas a regular harmonic function need only be $O(|\mathbf{x}|^{-1})$. For instance, if $f = c \neq 0$ on $\partial\Omega$, the solution of (DE) cannot be represented as a double layer potential, because otherwise $\int_{\partial\Omega} f\mu_0 = c\int_{\partial\Omega}\mu_0 = 0$. In fact, the unique solution is $u = cV_{\mu_0}(\mathbf{x})$. For a general f we can retrieve the solution by subtracting $c = \int_{\partial\Omega} f\mu_0 / \int_{\partial\Omega}\mu_0$, representing the solution with boundary values $f - c$ as a double layer potential and adding $cV_{\mu_0}(\mathbf{x})$.

There are certain modifications to the above theory that need to be made when the *space dimension is* 2, and $E(\mathbf{x}, \mathbf{y}) = (-1/2\pi)\ln|\mathbf{x} - \mathbf{y}|$. We outline the main ones, leaving the proofs as an exercise.

i. u harmonic and regular at infinity has a (uniform) limit u_∞ as $|\mathbf{x}| \to \infty$.

ii. $u = V_\mu$ is regular at infinity if and only if $\int_{\partial\Omega}\mu = 0$. In general,

$$V_\mu(\mathbf{x}) = -A\ln|\mathbf{x}| + O(|\mathbf{x}|^{-1}),$$

where $A = (1/2\pi)\int_{\partial\Omega}\mu\, dS$. Hence, the Robin potential V_{μ_0} is not regular at infinity in two dimensions. *Hint*: $\ln|\mathbf{x} - \mathbf{y}| = \frac{1}{2}\ln|\mathbf{x} - \mathbf{y}|^2 = \frac{1}{2}\ln[|\mathbf{x}|^2(1 + (|\mathbf{y}|^2 - 2\mathbf{x}\cdot\mathbf{y})/|\mathbf{x}|^2)]$.

iii. In (NE) a necessary condition for existence of a (regular) solution is $\int_{\partial\Omega} g = 0$. The solution $u = V_\mu + u_\infty = u_\infty + O(|\mathbf{x}|^{-1})$ is determined up to the constant u_∞.

iv. For (DE) the unique solution for $f = c$ is $u = c$, and this function cannot be represented as a double layer potential.

v. For (DE), if $\int_{\partial\Omega} f\mu_0 = c|\partial\Omega|$, we have $\int_{\partial\Omega}\tilde{f}\mu_0 = 0$, where $\tilde{f} = f - c$, and the solution of (DE) is given by $u = W_{\tilde{\nu}} - cV_{\mu_0}$, where $\tilde{\nu}$ is a solution of $K\tilde{\nu} + \frac{1}{2}\tilde{\nu} = \tilde{f}$ and μ_0 is the Robin density of Ω.

vi. If $\partial\Omega$ is the circle of radius R around the origin, then $K(\mathbf{x}, \mathbf{y}) = K'(\mathbf{x}, \mathbf{y}) = -1/4\pi R$ (hence, $K = K'$ is self-adjoint). The solutions of $(k - \frac{1}{2}I)\nu = f$ and $(K - \frac{1}{2}I)\mu = g$ are given by $\nu = -2f + \int_{\partial\Omega} f/2\pi R$, $\mu = -2g + \int_{\partial\Omega} g/2\pi R$. The solution $u = W_\nu(\mathbf{x})$ reduces to the Poisson integral of f.

vii. In the case of a circle, (DE) and (NI) have solutions only if $\int_{\partial\Omega} f = \int_{\partial\Omega} g = 0$, and the Robin density μ_0 is constant.

viii. $c_0 = 0$ for Ω a domain bounded by a "gamma contour" (see exercises for Section 1).

ix. If the condition of regularity at infinity is given up, then we have $u = u_\infty - A \ln|\mathbf{x}| + O(|\mathbf{x}|^{-1})$. The solution of (NE) is still determined up to u_∞, while the solution of (DE) is determined up to a term $(1 - V_{\mu_0})2\pi A / \int_{\partial\Omega} \mu_0$, $A \in \mathbb{R}$ (for gamma contours, $1 - V_{\mu_0}$ is replaced by V_{μ_0}). More generally, we can add the (multiple valued) potential $\kappa \arctan(x_2/x_1)$. If $\mathbf{v} = \operatorname{grad} u$ is thought of as a velocity field, then A is the flux and κ is the circulation at infinity.

4.4. Direct Method. A different but related way of solving the Dirichlet and Neumann problems by reduction to integral equations on the boundary is based on Green's identities. In contrast to the layer Ansatz, this "direct" method makes no special assumption on the form of the solutions. The resulting boundary integral equations can be either of the first or of the second kind and their unknowns are directly related to u or the normal derivative of u. Before giving a brief account of this method, we present a useful variant of the Green representation theorem.

If $u \in C^1(\bar{\Omega})$ is harmonic in Ω, Green's identities combined with the jump relations for the double layer potential (Theorem 4.1) imply the form of Green's representation theorem

$$\bar{\omega} u(\mathbf{x}) = V_{\partial u/\partial n|_i}(\mathbf{x}) - W_{u_i}(\mathbf{x}) \tag{IR}$$

valid on $\mathbb{R}^3$, where the index i denotes interior boundary values, and $\bar{\omega}$ is the "solid angle factor"

$$\bar{\omega} = \bar{\omega}(\mathbf{x}) := \begin{cases} 1, & \mathbf{x} \in \Omega, \\ 1/2, & \mathbf{x} \in \partial\Omega, \\ 0, & \mathbf{x} \in \Omega^c. \end{cases}$$

If $u \in C^1(\bar{\Omega}^c)$ is harmonic and regular at infinity, we obtain, similarly, that

$$(1 - \bar{\omega}) u(\mathbf{x}) = -V_{\partial u/\partial n|_e}(\mathbf{x}) + W_{u_e(\mathbf{x}),} \tag{ER}$$

where the index e denotes exterior boundary values.

Suppose that u is harmonic on $\Omega \cup \Omega^c$, regular at infinity, and has continuous derivatives up to $\partial\Omega$ from Ω and Ω^c. Then, adding, we obtain the formula

$$u(\mathbf{x}) = W_{[u]}(\mathbf{x}) - V_{[\partial u/\partial n]}(\mathbf{x}), \qquad \mathbf{x} \in \Omega \cup \Omega^c,$$

and for $\mathbf{x} \in \partial\Omega$,

$$\tfrac{1}{2}(u_i(\mathbf{x}) + u_e(\mathbf{x})) = W_{[u]}(\mathbf{x}) - V_{[\partial u/\partial n]}(\mathbf{x}),$$

where $[u] = u_e - u_i$ denotes the jump across $\partial\Omega$. If either $[u] = 0$ or $[\partial u/\partial n] = 0$, we retrieve the layer Ansatz for (NI)–(NE) (with $\mu = -[\partial u/\partial n]$) or (DI)–(DE) (with $\nu = [u]$), respectively. (In the two-dimensional case, the formula holds provided u vanishes uniformly at infinity.)

This representation formula is very useful in applications. For instance, flow past a wing profile in aerodynamics with velocity potential u is often matched to a given fictitious potential flow inside the wing.

Suppose now we want to solve the interior Neumann problem, (NI). By "collocating" (IR) on the boundary we obtain the integral equation

$$(K + \tfrac{1}{2}I)u_i = V\partial u/\partial n|_i \tag{14}$$

in the unknown u_i, with given data $\partial u/\partial n|_i = g$. Note that this is the adjoint of the equation previously obtained for (NI) via the layer Ansatz. Hence, by the Fredholm alternative theorem Vg must be orthogonal to μ_0. Using (12) we find that the solvability condition

$$0 = \int_{\partial\Omega} \mu_0(Vg)dS = \int_{\partial\Omega} g(V\mu_0)dS = \int_{\partial\Omega} g\, dS$$

coincides with (11). If this condition is satisfied, all solutions of the integral equation are given by $u_i = \bar{u}_i + c$ [$\bar{u}_i$ a particular solution, c the eigensolution of $(K + \frac{1}{2}I)u_i = 0$]. Inserting this in (IR) and using the Gauss solid angle formula, one finds that the general solution of (NI) in Ω,

$$u(\mathbf{x}) = V_{\partial u/\partial n|_i}(\mathbf{x}) - W_{\bar{u}_i}(\mathbf{x}) + c, \qquad \mathbf{x} \in \Omega,$$

is determined up to an arbitrary constant c, as expected.

Similarly, for the exterior Neumann problem (NE) we find from (ER) the boundary integral equation

$$(K - \tfrac{1}{2}I)u_e = V\partial u/\partial n|_e \tag{15}$$

in the unknown u_e for given data $g = \partial u/\partial n|_e$. This is the adjoint of the equation previously obtained for (NE) via the layer Ansatz and the Fredholm theory shows that the solution of the integral equation u_e exists and is unique for every g. Inserting this boundary function u_e in (ER) yields the solution u of the boundary value problem (NE) in Ω^c.

Equations (14), (15) may be envisaged as integral equations of the first kind for solving (DI) or (DE), respectively. Taking $v = \partial u/\partial n$ as unknown and $(K \pm \frac{1}{2}I)u = f$ as given, we can rewrite both of them in the common form

$$Vv = f. \tag{16}$$

The theory of this kind of integral equation is outside the aims and limits of this book. It is worthwhile, however, to discuss a simple example. Consider equation (16) on the whole of $\mathbb{R}^2$ and suppose that the two-dimensional Fourier transform of both sides exists (possibly in the distributional sense). Proceeding formally, and using results from Sections 2 and 3 of Chapter 8, we find

$$\hat{v}(\xi) = c|\xi|\hat{f}(\xi), \qquad \xi \in \mathbb{R}^2,$$

where $c > 0$ is a fixed constant. This suggests that the solution v exists for every f, and is unique (as $\hat{f} = 0$ implies $\hat{v} = 0$). However, the solution is less regular than the data: Suppose $f \in H^{s+1}(\mathbb{R}^2)$, then $v \in H^s(\mathbb{R}^2)$, and the inequality

$$\frac{1}{4\pi^2}\int_{\mathbb{R}^2}(1 + |\xi|^2)^s|\hat{v}(\xi)|^2 d\xi \leq \frac{c^2}{4\pi^2}\int_{\mathbb{R}^2}(1 + |\xi|^2)^{s+1}|\hat{f}(\xi)|^2 d\xi$$

shows that $\|v\|_s \leq c\|f\|_{s+1}$. In order to control the H^s norm of v, one needs to bound the H^{s+1} norm of f, and the problem is *ill posed*. The operator V is compact (Theorem 4.4), hence V^{-1} is unbounded.

Remarks

4.3. In the direct method no "spurious" compatibility conditions such as (10) arise and the boundary integral equations are solvable exactly as and when the corresponding boundary value problem is solvable.

4.4. The fact that equation (16) is ill posed often goes unnoticed in numerical computations (e.g., see Ref. 10).

4.5. Poincaré's Identity and Harmonic Vector Fields. Many physical problems are formulated in terms of a vector field $\mathbf{v}$, a function of the space variables $\mathbf{x}$ and possibly of time, t. This is the case, for example, of the Euler, Navier–Stokes, or Maxwell equations. Here we present an interesting representation theorem for vector fields, which is the counterpart of the Green representation theorem for scalar functions.

Suppose that $\mathbf{v}(\mathbf{x})$ is a vector-valued function defined in the closure of a bounded domain Ω in $\mathbb{R}^3$ with a smooth boundary, $\partial\Omega$, and suppose $\mathbf{v} \in C^1(\bar{\Omega})$. Then from the properties of the volume potential we have, for $\mathbf{x} \in \Omega$,

$$\mathbf{v}(\mathbf{x}) = -\Delta \int_\Omega E(\mathbf{x}, \mathbf{y})\mathbf{v}(\mathbf{y})d\mathbf{y}.$$

Using the vector identities $\Delta \equiv$ grad div $-$ curl curl and

$$\text{curl}_x[E(\mathbf{x}, \mathbf{y})\mathbf{v}(\mathbf{y})] \equiv \text{grad}_x\, E \wedge \mathbf{v}(\mathbf{y}), \qquad \text{div}_x[E(\mathbf{x}, \mathbf{y})\mathbf{v}(\mathbf{y})] \equiv \text{grad}_x\, E \cdot \mathbf{v}(\mathbf{y}),$$

we obtain

$$\mathbf{v}(\mathbf{x}) = \text{curl}_x \int_\Omega \text{grad}_x\, E(\mathbf{x}, \mathbf{y}) \wedge \mathbf{v}(\mathbf{y})d\mathbf{y} - \text{grad}_x \int_\Omega \text{grad}_x\, E(\mathbf{x}, \mathbf{y}) \cdot \mathbf{v}(\mathbf{y})d\mathbf{y}.$$

Then

$$\begin{aligned} \text{grad}_x\, E(\mathbf{x}, \mathbf{y}) \wedge \mathbf{v}(\mathbf{y}) &= -\text{grad}_y\, E(\mathbf{x}, \mathbf{y}) \wedge \mathbf{v}(\mathbf{y}) = -\text{curl}_y[E(\mathbf{x}, \mathbf{y})\mathbf{v}(\mathbf{y})] \\ &\quad + E(\mathbf{x}, \mathbf{y})\, \text{curl}_y \mathbf{v}(\mathbf{y}) \\ \int_\Omega \text{grad}_x\, E(\mathbf{x}, \mathbf{y}) \wedge \mathbf{v}(\mathbf{y})d\mathbf{y} &= -\int_{\partial\Omega} E(\mathbf{x}, \mathbf{y})\mathbf{n}(\mathbf{y}) \wedge \mathbf{v}(\mathbf{y})dS_y \\ &\quad + \int_\Omega E(\mathbf{x}, \mathbf{y})\, \text{curl}_y\, \mathbf{v}(\mathbf{y})d\mathbf{y} \end{aligned} \tag{17}$$

[this is true, by the Gauss lemma, even though $E(\mathbf{x}, \mathbf{y})$ is singular when $\mathbf{x} = \mathbf{y}$], and

$$\begin{aligned} \text{grad}_x\, E \cdot \mathbf{v}(\mathbf{y}) &= -\text{grad}_y\, E \cdot \mathbf{v}(\mathbf{y}) = -\text{div}_y[E(\mathbf{x}, \mathbf{y})\mathbf{v}(\mathbf{y})] \\ &\quad + E(\mathbf{x}, \mathbf{y})\, \text{div}_y \mathbf{v}(\mathbf{y}) \\ \int_\Omega \text{grad}_x\, E(\mathbf{x}, \mathbf{y}) \cdot \mathbf{v}(\mathbf{y})d\mathbf{y} &= -\int_{\partial\Omega} E(\mathbf{x}, \mathbf{y})\mathbf{n}(\mathbf{y}) \cdot \mathbf{v}(\mathbf{y})dS_y \\ &\quad + \int_\Omega E(\mathbf{x}, \mathbf{y})\, \text{div}_y \mathbf{v}(\mathbf{y})d\mathbf{y} \end{aligned} \tag{17a}$$

imply the identity (Ref. 11) in $\mathbb{R}^3$, abbreviated as (P3),

$$\mathbf{v}(\mathbf{x}) = \text{grad}_x\left[\int_{\partial\Omega} E(\mathbf{x},\mathbf{y})\mathbf{n}(\mathbf{y})\cdot\mathbf{v}(\mathbf{y})dS_y - \int_{\Omega} E(\mathbf{x},\mathbf{y})\,\text{div}_y\mathbf{v}(\mathbf{y})d\mathbf{y}\right]$$
$$+ \text{curl}_x\left[-\int_{\partial\Omega} E(\mathbf{x},\mathbf{y})\mathbf{n}(\mathbf{y})\wedge\mathbf{v}(\mathbf{y})dS_y + \int_{\Omega} E(\mathbf{x},\mathbf{y})\,\text{curl}_y\mathbf{v}(\mathbf{y})d\mathbf{y}\right]. \quad \text{(P3)}$$

This form of the Clebsch–Helmholtz decomposition is variously known as *Poincaré's identity* or the *fundamental theorem of vector analysis*. It also holds in an exterior domain Ω^c (with $\mathbf{n}$ replaced by $-\mathbf{n}$, so that $\mathbf{n}$ is always the outer normal to Ω) if $\mathbf{v}(\mathbf{x}) = O(|\mathbf{x}|^{-2})$ and div $\mathbf{v}$, curl $\mathbf{v} = O(|\mathbf{x}|^{-3})$ as $|\mathbf{x}| \to \infty$. When the vector field admits a potential u, $\mathbf{v} = -\text{grad}\, u$, Poincaré's formula reduces to Green's theorem for u, because

$$\text{grad}\int_{\partial\Omega} E\frac{\partial u}{\partial n}\,dS - \text{curl}\int_{\partial\Omega} E\mathbf{n}\wedge\text{grad}\,u\,dS \equiv \text{grad}\int_{\partial\Omega}\left(E\frac{\partial u}{\partial n} - u\frac{\partial E}{\partial n}\right)dS$$

(Exercise 4.8).

A number of consequences can be derived from this formula. By letting $\mathbf{x}$ tend to the boundary $\partial\Omega$ and using the jump relations above, we find two alternative boundary integral equations for either $\mathbf{n}\cdot\mathbf{v}$ or $\mathbf{n}\wedge\mathbf{v}$ that serve to determine $\mathbf{v}$ when div $\mathbf{v}$, curl $\mathbf{v}$ are given in Ω and the boundary data $\mathbf{n}\wedge\mathbf{v}$ or $\mathbf{n}\cdot\mathbf{v}$ are assigned on $\partial\Omega$ (see Ref. 12).

Example 4.1. We can retrieve the fact that the conductor potential of a domain Ω is given by a single layer potential from this identity applied to an exterior domain. Suppose that $\mathbf{v}$ is the electrostatic field in Ω^c due to some distribution of charge on $\partial\Omega$. The div $\mathbf{v} = $ curl $\mathbf{v} = 0$ in Ω^c, and $\mathbf{n}\wedge\mathbf{v} = 0$ on $\partial\Omega$ imply that

$$\mathbf{v}(\mathbf{x}) = -\text{grad}_x\int_{\partial\Omega} E(\mathbf{x},\mathbf{y})\mu(\mathbf{y})dS_y, \qquad \mathbf{x}\in\Omega^c, \tag{18}$$

where $\mu = \mathbf{n}\cdot\mathbf{v}$ (with $\mathbf{n}$ oriented outwards from Ω). Because the potential $\int_{\partial\Omega} E\mu\, dS_y$ is constant on $\partial\Omega$, μ is a Robin density μ_0 (see Corollary 4.1). According to Exercise 2.8, $\mathbf{v}(\mathbf{x})$ is uniquely determined if the total surface charge $(1/4\pi)\int_{\partial\Omega}\mathbf{n}\cdot\mathbf{v}\,dS$ is assigned. Alternatively, we may determine $\mathbf{v}$ by fixing μ. By definition of the conductor potential, $\mu_0(\mathbf{y})$ must be the solution of the integral equation of the first kind $V\mu_0 = 1$, and the *capacitance* of $\partial\Omega$

$$\frac{1}{4\pi}\int_{\partial\Omega}\mu_0(\mathbf{y})dS_y$$

is positive by Corollary 4.1.

In the two-dimensional case, $\mathbf{x} = (x_1, x_2)$, $\mathbf{v} = (v_1, v_2)$, the Poincaré identity reads

$$\mathbf{v}(\mathbf{x}) = \text{grad}_x \left[\int_{\partial\Omega} E(\mathbf{x}, \mathbf{y})\mathbf{n}(\mathbf{y}) \cdot \mathbf{v}(\mathbf{y}) ds_y - \int_{\Omega} E(\mathbf{x}, \mathbf{y}) \, \text{div}_y \mathbf{v}(\mathbf{y}) d\mathbf{y} \right]$$
$$+ J \, \text{grad}_x \left[- \int_{\partial\Omega} E(\mathbf{x}, \mathbf{y})\mathbf{t}(\mathbf{y}) \cdot \mathbf{v}(\mathbf{y}) ds_y + \int_{\Omega} E(\mathbf{x}, \mathbf{y}) \omega(\mathbf{y}) d\mathbf{y} \right], \quad \text{(P2)}$$

where $E(\mathbf{x}, \mathbf{y}) = (-1/2\pi) \ln|\mathbf{x} - \mathbf{y}|$, $\omega(\mathbf{y}) = \partial v_2/\partial x_1 - \partial v_1/\partial x_2$ is the scalar vorticity, $\mathbf{n}$ is the outer normal to the curve $\partial\Omega$ and $\mathbf{t}$ the unit tangent vector, and J grad $= (\partial/\partial x_2, -\partial/\partial x_1)$. This identity also holds for an exterior domain Ω^c (with $\mathbf{n}$ replaced by $-\mathbf{n}$, so that $\mathbf{n}$ is always the outer normal to Ω) provided $\mathbf{v}(\mathbf{x}) = O(|\mathbf{x}|^{-1})$ and div $\mathbf{v}, \omega = O(|\mathbf{x}|^{-2})$ at infinity. Letting $\mathbf{x}$ tend to the boundary and taking the scalar product $\mathbf{n}(\mathbf{x}) \cdot \mathbf{v}(\mathbf{x})$ or $\mathbf{t}(\mathbf{x}) \cdot \mathbf{v}(\mathbf{x})$ yields two boundary integral equations that are uniquely solvable for either $\mathbf{n}(\mathbf{x}) \cdot \mathbf{v}(\mathbf{x})$ or $\mathbf{t}(\mathbf{x}) \cdot \mathbf{v}(\mathbf{x})$ (see below and Ref. 13).

If $\mathbf{v}$ is irrotational and solenoidal, $\mathbf{v} = -\text{grad}\, u$ and div $\mathbf{v} = 0$, then u is harmonic and $\mathbf{v}$ is called a *harmonic vector field*. An example (with $\mathbf{n} \wedge \mathbf{v} = 0$) is given by (18). The two equivalent boundary integral equations obtained from (P2) for harmonic vector fields are

$$\left(K' \pm \frac{1}{2} I\right) \mathbf{t} \cdot \mathbf{v} = -\frac{\partial}{\partial s} V \mathbf{n} \cdot \mathbf{v}, \qquad \left(K' \pm \frac{1}{2} I\right) \mathbf{n} \cdot \mathbf{v} = \frac{\partial}{\partial s} V \mathbf{t} \cdot \mathbf{v} \qquad (19)$$

(s is the arc-length along $\partial\Omega$). Here the $-$ sign holds for Ω and the $+$ sign for Ω^c. It is instructive to compare these equations with those obtained for u from the layer Ansatz and from the direct method, equations (14) and (15) (see the exercises).

5. Variational Theory

We have seen that Dirichlet's problem can be approached in more than one way. The Perron method permits very general domains and arbitrary continuous boundary values, and it can be generalized to include discontinuity points at the boundary. The method of integral equations is very useful in applications, but the domain must be $C^{1,\alpha}$. (Generalizations to more general domains exist, but the technical demands raise the level of difficulty by an order of magnitude; see Refs. 5, 12.) Here we present another method that is in some ways more general, and that has far-reaching ramifications. In this method the Dirichlet problem is reformulated as a simple geometrical problem of orthogonal projections in a

Hilbert space, that of finding the normal to a hyperplane, and the assumption of boundary values is interpreted in a suitably generalized sense.

5.1. Variational Solutions of Dirichlet's Problem. If $u(\mathbf{x}) \in C^2(\Omega)$ is harmonic in a bounded domain Ω of $\mathbb{R}^n$, then (G1) implies that

$$\int_{\Omega} u\Delta v \, d\mathbf{x} = 0 \tag{W}$$

for all $v(\mathbf{x}) \in C^2(\Omega)$ with compact support in Ω. If $u \in L^1_{\text{loc}}(\Omega)$ satisfies (W), we say that u is a *weak solution* of $\Delta u = 0$. [This implies that u is a distribution solution of $\Delta u = 0$. Conversely, if $u \in L^1_{\text{loc}}(\Omega)$ is a distribution solution, an approximation of v by functions in $C_0^\infty(\Omega)$ shows that u is a weak solution.] The following theorem shows that weak solutions of Laplace's equation are in fact much more.

Theorem 5.1. If $u \in L^1_{\text{loc}}(\Omega)$ satisfies (W) for all $v \in C_0^2(\Omega)$, then $u(\mathbf{x})$ (after a possible modification on a set of zero Lebesgue measure) is harmonic in Ω.

Proof. It suffices to prove that u is harmonic in the neighborhood of every point $\mathbf{y} \in \Omega$. We give the proof here for $n = 3$. Assume first $u(\mathbf{x}) \in C(\Omega)$. Let $B_\varepsilon(\mathbf{y}) = \{|\mathbf{x} - \mathbf{y}| \leq \varepsilon\} \subset \Omega$, and choose

$$v(\mathbf{x}) = \begin{cases} (|\mathbf{x} - \mathbf{y}|^2 - \varepsilon^2)^3, & |\mathbf{x} - \mathbf{y}| \leq \varepsilon, \\ 0, & |\mathbf{x} - \mathbf{y}| > \varepsilon. \end{cases}$$

Then

$$\Delta v(\mathbf{x}) = \begin{cases} 6(7|\mathbf{x} - \mathbf{y}|^4 - 10\varepsilon^2|\mathbf{x} - \mathbf{y}|^2 + 3\varepsilon^4, & |\mathbf{x} - \mathbf{y}| \leq \varepsilon, \\ 0, & |\mathbf{x} - \mathbf{y}| > \varepsilon, \end{cases}$$

and (W) yields

$$\int_0^\varepsilon dr \int_{|\mathbf{x}-\mathbf{y}|=r} u(\mathbf{x})(7|\mathbf{x} - \mathbf{y}|^4 - 10\varepsilon^2|\mathbf{x} - \mathbf{y}|^2 + 3\varepsilon^4) dS_x = 0.$$

Differentiating with respect to ε and dividing by 4ε we find

$$\int_0^\varepsilon dr \int_{|\mathbf{x}-\mathbf{y}|=r} u(\mathbf{x})(-5|\mathbf{x} - \mathbf{y}|^2 + 3\varepsilon^2) dS_x = 0.$$

One further differentiation yields

$$(4\pi\varepsilon^2)^{-1}\int_{|\mathbf{x}-\mathbf{y}|=\varepsilon} u(\mathbf{x})dS_x = \left(\frac{4}{3}\pi\varepsilon^3\right)^{-1}\int_{|\mathbf{x}-\mathbf{y}|\le\varepsilon} u(\mathbf{x})d\mathbf{x},$$

and finally

$$u(\mathbf{y}) = (4\pi\varepsilon^2)^{-1}\int_{|\mathbf{x}-\mathbf{y}|=\varepsilon} u(\mathbf{x})dS_x,$$

i.e., u has (locally) the mean value property and is continuous, hence is harmonic in Ω (see Section 2). Now suppose that $u \in L^1_{\text{loc}}(\Omega)$. From Proposition 3.3 of Chapter 8, if η_δ is an even mollifier, and $u_\delta = \eta_\delta * u$,

$$\int_\Omega u_\delta \Delta v\, d\mathbf{x} = \int_\Omega u(\Delta v)_\delta\, d\mathbf{x} = \int_\Omega u\Delta v_\delta\, d\mathbf{x},$$

and v_δ has compact support in Ω for δ sufficiently small. It follows that u_δ is harmonic in Ω, and satisfies the mean value property

$$u_\delta(\mathbf{y}) = \frac{3}{4\pi\varepsilon^3}\int_{|\mathbf{x}-\mathbf{y}|\le\varepsilon} u_\delta(\mathbf{x})d\mathbf{x} \tag{20}$$

for ε sufficiently small. Because $u_\delta \to u$ in $L^1_{\text{loc}}(\Omega)$ as $\delta \to 0$, we have

$$\lim_{\delta\to 0}\int_{|\mathbf{x}-\mathbf{y}|\le\varepsilon} u_\delta(\mathbf{x})d\mathbf{x} = \int_{|\mathbf{x}-\mathbf{y}|\le\varepsilon} u(\mathbf{x})d\mathbf{x}$$

for any $\mathbf{y} \in \Omega$ and $\varepsilon > 0$ such that $\overline{B(\mathbf{y}, \varepsilon)} \subset \Omega$. Then

$$\lim_{\delta\to 0}\frac{3}{4\pi\varepsilon^3}\int_{|\mathbf{x}-\mathbf{y}|\le\varepsilon} u_\delta(\mathbf{x})d\mathbf{x} = \frac{3}{4\pi\varepsilon^3}\int_{|\mathbf{x}-\mathbf{y}|\le\varepsilon} u(\mathbf{x})d\mathbf{x} := \hat{u}(\mathbf{y}).$$

(The right-hand side is independent of ε because the left-hand side is.) As $u \in L^1_{\text{loc}}(\Omega)$, the function $\hat{u}(\mathbf{y})$ is continuous. (This is essentially the uniform absolute continuity of the Lebesgue integral.) On the other hand, we can choose a subsequence u_{δ_j} such that $u_{\delta_j} \to u$ a.e. on Ω. It follows from (20) that $u = \hat{u}$ a.e. □

This amazing result is known as *Weyl's lemma* (Ref. 14).

Suppose now that u is a weak solution of $\Delta u = 0$ in a bounded domain Ω and that $u \in H^1(\Omega)$. Using Proposition 3.17 of Chapter 8, we see that (W) can be integrated by parts, and

$$\int_\Omega \text{grad } u \cdot \text{grad } v \, d\mathbf{x} = 0 \tag{V}$$

for all $v \in C_0^2(\Omega)$. Because such v are dense in $H_0^1(\Omega)$, this identity holds for all $v \in H_0^1(\Omega)$. This leads to our new formulation of the Dirichlet problem:

To find a function $u \in H^1(\Omega)$ such that (V) *holds for all $v \in H_0^1(\Omega)$ and $(u - \Phi) \in H_0^1(\Omega)$, where $\Phi \in H^1(\Omega)$ is given.*

We call this u a *variational solution* of the Dirichlet problem. We will show below that u is determined by the "trace" f of Φ on $\partial\Omega$ (as defined in Proposition 3.17 of Chapter 8).

It is useful to observe that Poincaré's inequality (Section 3 of Chapter 8) implies that, for $u \in H_0^1(\Omega)$,

$$\int_\Omega u^2 \, d\mathbf{x} \leq \sigma^2 \int_\Omega |\text{grad } u|^2 \, d\mathbf{x}, \tag{PI}$$

where $\sigma = \sigma(\Omega) > 0$. It follows that we can give $H_0^1(\Omega)$ the equivalent norm $\|\cdot\|$, where

$$\|u\|^2 = \int_\Omega |\text{grad } u|^2 \, d\mathbf{x} := \mathbb{D}(u).$$

The functional $\mathbb{D}(u)$ is called *Dirichlet's integral* and this norm derives from the inner product

$$[u, v] := \int_\omega \text{grad } u \cdot \text{grad } v \, d\mathbf{x}.$$

The following result shows that the variational Dirichlet problem,

$$[u, v] = 0 \qquad \text{for all } v \in H_0^1(\Omega)$$

and

$$(u - \Phi) \in H_0^1(\Omega),$$

has a unique solution in $H^1(\Omega)$.

Theorem 5.2. The variational solution u exists and is unique, and satisfies

$$\mathbb{D}(u) = L := \inf \mathbb{D}(w) : w \in H^1(\Omega),\ w - \Phi \in H_0^1(\Omega)\} \tag{DP}$$

(Dirichlet's principle).

Proof. Suppose $U = u_1 - u_2$ is the difference of two solutions, then $U \in H_0^1(\Omega)$ and $[U, v] = 0$ for every $v \in H_0^1(\Omega)$. Choosing $v = U$ yields $\mathbb{D}(U) = 0$, whence $U = \text{const}$ and, as the only constant in H_0^1 is zero, $u_1 = u_2$. This proves uniqueness. We will prove that the extremum L, defined in (DP), is attained for a certain element $w = u$ and that this minimizing element is the solution of the variational Dirichlet problem. Let u_k be a minimizing sequence:

$$u_k \in H^1(\Omega), \qquad u_k - \Phi \in H_0^1(\Omega), \qquad \mathbb{D}(u_k) \downarrow L \qquad (k \to \infty).$$

Because $\mathbb{D}$ is quadratic, we have

$$\mathbb{D}\left(\frac{u_k - u_j}{2}\right) = \frac{1}{2}[\mathbb{D}(u_k) + \mathbb{D}(u_j)] - \mathbb{D}\left(\frac{u_k + u_j}{2}\right)$$

and $\frac{1}{2}(u_k + u_j) - \Phi \in H_0^1(\Omega)$, so that, by convexity,

$$\mathbb{D}\left(\frac{u_k + u_j}{2}\right) \geq L.$$

It follows that

$$\mathbb{D}(u_k - u_j) \leq 2[\mathbb{D}(u_k) + \mathbb{D}(u_j)] - 4L \to 0 \qquad k, j \to \infty,$$

where $u_k - u_j \in H_0^1(\Omega)$. By (PI), $\{u_k\}$ is a Cauchy sequence in $H^1(\Omega)$ and therefore converges in the H^1-norm to $u \in H^1(\Omega)$, and $\mathbb{D}(u_k) \to \mathbb{D}(u) = L$; u also verifies the boundary condition because H_0^1 is closed in this norm. Let $\psi \in C_0^\infty(\Omega)$, $\alpha \in \mathbb{R}$. The $u + \alpha\psi - \Phi \in H_0^1(\Omega)$, and the function $\mathscr{F}(\alpha)$

$$\mathscr{F}(\alpha) := \mathbb{D}(u + \alpha\psi) - \mathbb{D}(u) = 2\alpha[u, \psi] + \alpha^2 \mathbb{D}(\psi) \geq 0$$

has a minimum at $\alpha = 0$. Therefore, $\mathscr{F}'(0) = 0$ and we get the variational equation $[u, \psi] = 0$. Because $C_0^\infty(\Omega)$ is dense in $H_0^1(\Omega)$, the result follows. □

Remarks

5.1. By Weyl's lemma, $u \in C^\infty(\Omega)$.

5.2. If u' is the variational solution corresponding to Φ', with $\Phi - \Phi' \in H_0^1(\Omega)$, then $u = u'$. This follows from uniqueness and $u - u' = (u - \Phi) - (u' - \Phi') \in H_0^1(\Omega)$. Hence, the variational solution is independent of the values taken by the particular "lifting" Φ in Ω and is determined solely by the trace f of Φ on $\partial\Omega$.

5.3. In general, $\mathbb{D}(w)$ is not defined for $w \in C(\bar{\Omega}) \cap C^2(\Omega)$ and has no minimum in $C^1(\bar{\Omega}) \cap C^2(\Omega)$.

5.4. From the physical point of view $\mathbb{D}(u)$ represents a potential energy (see Chapter 1), and the variational solutions are solutions with "finite energy."

The problem of the existence of a lifting Φ in $H^1(\Omega)$ for given boundary data f requires further discussion.

Example 5.1. Let $u(r, \theta) := \sum_{k=1}^{\infty} r^k(a_k \cos k\theta + b_k \sin k\theta)$ in the unit circle $\Omega(0 \leq r < 1, -\pi \leq \theta \leq \pi)$. Then, as

$$\mathbb{D}(u) \equiv \int_\Omega |\text{grad } u|^2 dS = \pi \sum_{k=1}^{\infty} k(a_k^2 + b_k^2),$$

$u \in H^1(\Omega)$ if and only if the trace $Tu = \sum_{k=1}^{\infty}(a_k \cos k\theta + b_k \sin k\theta)$ satisfies $\sum_{k=1}^{\infty} k(a_k^2 + b_k^2) < \infty$. Functions f with this property are said to be of class $H^{1/2}(\partial\Omega)$, and they are not necessarily continuous on the circumference $\partial\Omega$ (Example 5.2). Example 5.3 shows that, conversely, a continuous and 2π-periodic function $f(\theta)$ may not belong to this class.

Example 5.2. The function

$$f(\theta) = \begin{cases} \ln|\ln|\theta||, & |\theta| < 1/e, \\ 0, & 1/e \leq |\theta| \leq \pi, \end{cases}$$

has Fourier coefficients $b_k = 0, |a_k| = O(1/k \ln k)$ as $k \to \infty$. Hence, $\sum_{k=1}^{\infty} k(a_k^2 + b_k^2)$ converges while $\sum_{k=1}^{\infty}(|a_k| + |b_k|)$ diverges, $\mathbb{D}(u)$ is finite but f is discontinuous at $\theta = 0$.

Example 5.3. We can construct a continuous function f on the boundary of the unit circle whose harmonic extension u inside the circle has infinite Dirichlet integral. Choose the coefficients

$$a_k = 0, \qquad b_k = \begin{cases} n^{-2} & \text{if } k = 2^n \\ 0, & \text{if } k \neq 2^n \end{cases} \qquad (n = 1, 2, \ldots)0$$

in the above series for $u(r, \theta)$ ("lacunary" Fourier series). Then $\sum_{k=1}^{\infty}(|a_k| + |b_k|) = \sum_{n=1}^{\infty} n^{-2}$ converges, while the series for $\mathbb{D}[u]$

$$\pi \sum_{k=1}^{\infty} k(a_k^2 + b_k^2) = \pi \sum_{n=1}^{\infty} 2^n n^{-4}$$

diverges.

These examples show that the solution of the Dirichlet problem cannot be determined from Dirichlet's principle for all continuous boundary values. The variational method places some restrictions on admissible boundary values.

5.2. Variational Theory for Poisson's Equation. We now briefly consider the Dirichlet problem for the Poisson equation $\Delta u = F$ in a bounded domain Ω of $\mathbb{R}^n$. We will restrict our attention to zero boundary data. Then, by integrating by parts the equation over Ω after multiplying by a test function v, we see that the appropriate weak formulation is to find, for a given F in $L^2(\Omega)$, $u \in H_0^1(\Omega)$ such that

$$[u, v] = -(F, v) \qquad \text{for every } v \in H_0^1(\Omega), \tag{VP}$$

where $(F, v) := \int_\Omega Fv d\mathbf{x}$.

Theorem 5.3. The solution u of (VP) exists and is unique in $H_0^1(\Omega)$.

Proof. Because

$$|(F, v)| \leq \|F\|_{L^2}\|v\|_{L^2} \leq \sigma\|F\|_{L^2}\|v\|,$$

the linear functional (F, v) is continuous in v. Riesz's theorem then implies that there exists $g = g_F$ in $H_0^1(\Omega)$ such that

$$(F, v) = [-g_F, v] \equiv [GF, v],$$

where $g_F = -GF$ defines a linear operator G. Substituting in (VP) we find $[u - g_F, v] = 0$ for every $v \in H_0^1(\Omega)$. Taking $v = u - g_F$ we get

$$u = g_F \equiv -GF,$$

so $G : L^2 \to H_0^1$ is the "inverse" of $-\Delta$. It is called the *Green's operator*. This solution is unique. (The proof can be carried out as in Theorem 5.2 and is left as an exercise.) If GF' is the solution corresponding to another source term F', we have

$$\|GF - GF'\|^2 = (F - F', GF - GF') \leq \sigma \|F - F'\|_{L^2} \|GF - GF'\|,$$

which implies continuous dependence of $u = GF$ on F,

$$\|GF - GF'\| \leq \sigma \|F - F'\|_{L^2}$$

in the appropriate norms. □

From the inequalities in the proof of Theorem 5.3, we see that the Green's operator G is bounded with norm $\leq \sigma$ and strictly monotone (positive), in agreement with the Green's function being positive (see Section 2).

Remark 5.5. Because $v \in H_0^1(\Omega)$, the scalar product (F, v) can be extended by duality to $F \in H^{-1}(\Omega)$. Proceeding by a similar argument as in Theorem 5.3, and replacing the inequality $|(F, v)| \leq \|F\|_{L^2} \|v\|_{L^2}$ by

$$|(F, v)| \leq c \|F\|_{-1} \|v\|,$$

we see that for $F \in H^{-1}(\Omega)$ the weak solution u still exists in $H_0^1(\Omega)$.

As for Laplace's equation, we will show that the weak solution is the solution of a variational problem. Define the quadratic functional

$$\mathbb{Q}(v) := \mathbb{D}(v) + 2(F, v) \equiv \|v\|^2 + 2(F, v)$$

for $v \in H_0^1(\Omega)$. Then the weak solution u minimizes $\mathbb{Q}(v)$ over $H_0^1(\Omega)$. In fact, if $u = g_F = -GF$ is the weak solution, then $(F, v) = -[g_F, v]$ and we find

$$\mathbb{Q}(v) := [v, v] - 2[g_F, v] = [v - g_F, v - g_F] - [g_F, g_F] = \|v - g_f\|^2 - \|g_F\|^2,$$

which shows that $\mathbb{Q}(v)$ has a minimum at $v - u$. Conversely, if u is a minimum of $\mathbb{Q}(v)$, then u is a weak solution. (The proof can be carried out as for the Laplace equation and is left as an exercise.)

5.3. Laplace–Dirichlet Eigenvalue Problem. The eigenvalue problem for the Laplace operator in a bounded domain Ω of $\mathbb{R}^n$ with homogeneous Dirichlet boundary data:

$$-\Delta u = \lambda u \text{ in } \Omega, \qquad u = 0 \text{ on } \partial\Omega$$

has the following weak (variational) formulation [obtained formally from (VP) by replacing F by $-\lambda u$]: to find $u \in H_0^1(\Omega)$ such that

$$[u, v] = \lambda(u, v) \qquad \text{for every } v \in H_0^1(\Omega). \tag{21}$$

We are interested in finding the values of λ (eigenvalues) such that nonzero u (eigenfunctions) exist. It follows immediately from (21) that the eigenvalues, if they exist, are *positive*; thus, we may write $\lambda = \mu^2 > 0$.

We start by determining the smallest eigenvalue and a corresponding eigenfunction. We may assume that u is normalized in L^2, i.e., that u belongs to the unit sphere S in L^2. Consider the set $S_0 := \{u \in H_0^1 : \|u\|_{L^2} = 1\} \subset S$, and define $\mu_1 = \inf_{S_0} \|u\|$.

Theorem 5.4. (i) There exists $z_1 \in S_0$ such that $\|z_1\| = \mu_1 > 0$, and (ii) z_1 satisfies (21) with $\lambda = \lambda_1 \equiv \mu_1^2$.

Proof. (i) Let $\{u_n\}$ be a minimizing sequence, satisfying

$$\mu_1 \leq \|u_n\| \leq \mu_1 + n^{-1} \leq \mu_1 + 1. \tag{22}$$

This sequence is bounded (by $M = \mu_1 + 1$) in H_0^1, hence weakly compact in H_0^1. There exists a subsequence and an element $z_1 \in H_0^1$ such that $z_1 = w - \lim u_n$. On the other hand, Rellich's lemma (Proposition 3.13 of Chapter 8) implies that $u_n \to z_1$ in the strong topology of L^2 and this implies $1 = \|u_n\|_{L^2} \to \|z_1\|_{L^2}$. It follows that $z_1 \in S_0$ and

$$\|z_1\| \geq \mu_1.$$

From the fact that $z_1 = w - \lim u_n$ and $\|u_{n+p}\| \leq \mu_1 + p^{-1}$ we have

$$\|z_1\| \leq \mu_1 + p^{-1} \tag{23}$$

(Exercise 5.4), and we conclude that

$$\|z_1\| = \mu_1 = \lim\|u_n\|. \tag{24}$$

It remains to prove that z_1 is also the strong limit of u_n in H_0^1. This follows from weak convergence and the convergence of the norms (24) (see Exercise 5.5). Finally, we remark that $\mu_1 > 0$ for otherwise (PI) would imply $\|z_1\|_{L^2} = 0$, which contradicts $z_1 \in S_0$.

ii. We will now show that z_1 is an eigenfunction corresponding to the eigenvalue $\lambda_1 = \mu_1^2$. Define

$$\Lambda(t) = \|Z\|^2, \qquad Z := \frac{z_1 + tv}{\|z_1 + tv\|_{L^2}}$$

with an arbitrary $v \in H_0^1$. Because $Z \in S_0$, we have $\|Z\|^2 = \Lambda(t) \geq \|z_1\|^2 = \Lambda(0)$, and the function $\Lambda(t)$ has a minimum at $t = 0$; hence, $\Lambda'(0) = 0$, and by an easy calculation we find

$$\Lambda'(0) = 2[[z_1, v] - \lambda_1(z_1, v)] = 0,$$

which proves (ii). □

From Theorem 5.4 it follows that the "imbedding constant" $\sigma(\Omega)$ is related to λ_1 by $\sigma(\Omega) = 1/\sqrt{\lambda_1}$.

Similar constructions allow us to find all other eigenfunctions and the corresponding eigenvalues. Let us consider the space $K_1 = \{z \in H_0^1 : [z, z_1] = 0\}$, i.e., the orthogonal complement in H_0^1 of the linear manifold spanned by z_1. K_1 is a Hilbert space and every $v \in H_0^1$ can be decomposed as

$$v = z + \gamma z_1, \tag{25}$$

where $z \in K_1$ and γ are uniquely determined. We remark that if $u \in K_1$, then u is orthogonal to z_1 also in L^2: This follows from the relation $0 = [z_1, u] = \lambda_1(z_1, u)$. Let $S_1 = \{u \in K_1 : \|u\|_{L^2} = 1\} \subset S_0$ and $\mu_2 = \inf_{S_1} \|u\|$.

Theorem 5.5. There exists $z_2 \in S_1$ such that (i) $\mu_2 = \|z_2\|$ and (ii) z_2 is an eigenfunction belonging to $\lambda_2 = \mu_2^2$.

Proof. Before sketching the proof, we note that as $S_1 \subset S_0$, we will certainly have $\lambda_2 \geq \lambda_1$. The proof of (i) is accomplished as in the previous theorem, and as observed above we have $(z_1, z_2) = 0$. For (ii), we arrive at the relation

$$[z_2, v] - \lambda_2(z_2, v) = 0$$

for every $v = z \in K_1$, and all we have to prove is that this relation holds also for every $v \in H_0^1$. This follows easily from the decomposition (25):

$$[z_2, v] = [z_2, z + \gamma z_1] = [z_2, z] = \lambda_2(z_2, z) = \lambda_2(z_2, v). \qquad \square$$

The construction then proceeds recursively, by replacing K_1 with K_2, the orthogonal complement in H_0^1 of the linear manifold spanned by z_1 and z_2, and so forth. In this way we arrive at a sequence of eigenfunctions

$$z_1, z_2, \ldots, z_n, \ldots$$

belonging to the eigenvalues

$$0 < \lambda_1 \leq \lambda_2 \leq \cdots \leq \lambda_n \leq \cdots$$

and satisfying the relations

$$[z_n, v] = \lambda_n(z_n, v)$$

for every $v \in H_0^1(\Omega)$. The sets $\{z_n\}$, $\{z_n/\sqrt{\lambda_n}\}$ are orthonormal in L^2 and H_0^1, respectively:

$$(z_i, z_j) = \left[z_i/\sqrt{\lambda_i}, z_j/\sqrt{\lambda_j}\right] = \delta_{ij}. \tag{26}$$

Theorem 5.6. The sequence of eigenvalues accumulates at infinity:

$$\lim_{n\to\infty} \mu_n = \lim_{n\to\infty} \sqrt{\lambda_n} = +\infty.$$

Proof. Suppose $\lim_{n\to\infty} \sqrt{\lambda_n} = M < \infty$. Then the set $\{z_n\}$ would be bounded in H_0^1, $\|z_n\|^2 \leq \lambda_n \leq M^2$. This would imply existence of a subsequence strongly convergent in the L^2 norm. The orthogonality relations (26) together with continuity of the inner product (,) then lead to a contradiction. □

Theorem 5.7. The sequence $\{z_n/\sqrt{\lambda_n}\}$ is complete in $H_0^1(\Omega)$.

Proof. Suppose the contrary, then there is an element $\bar{z} \neq 0$ in S_0 such that $[z_n, \bar{z}] = 0$ for every n. From Theorem 5.6 it follows that, for a sufficiently large integer p,

$$\|\bar{z}\| < \|z_{p+1}\| = (\lambda_{p+1})^{1/2} = \inf_{S_p} \|u\|,$$

hence $\bar{z} \notin S_p$. This implies that $\bar{z}$ cannot be orthogonal to all z_i, $i = 1, \ldots, p$, which is a contradiction. □

Thus, $\{z_n/\sqrt{\lambda_n}\}$ is a *complete orthonormal set in* $H_0^1(\Omega)$ and every function $u \in H_0^1$ can be expanded into the Fourier series

$$u = \sum_{n-1}^{\infty} a_n z_n/\sqrt{\lambda_n}, \qquad a_n := [u, z_n]/\sqrt{\lambda_n} \tag{27}$$

convergent in H_0^1. We can also consider the Fourier series of u in L^2,

$$u \sim \sum_{n=1}^{\infty} (u, z_n) z_n$$

convergent in L^2. The next theorem implies that it converges to u.

Theorem 5.8. $\{z_n\}$ is a complete orthonormal set in $L^2(\Omega)$.

Proof. We already know that $\|z_n\|_{L^2} = 1$ for every n. For $\varphi \in H_0^1(\Omega)$, $[\varphi, z_n] = \lambda_n(\varphi, z_n)$ and (27) reduces to $\varphi = \sum_{n=1}^{\infty} (\varphi, z_n) z_n$, convergent in H_0^1, and certainly in L^2. Suppose that there is a $\bar{z} \neq 0$ in L^2 orthogonal to all z_n. Then

$$(\bar{z}, \varphi) = \sum_{n=1}^{\infty} (\varphi, z_n)(\bar{z}, z_n) = 0$$

and this implies $\bar{z} = 0$, which is a contradiction. □

We now show that the above construction yields all possible eigenfunctions and eigenvalues. We do this by proving an independent orthogonality result.

Theorem 5.9. Eigenfunctions corresponding to different eigenvalues are orthogonal (in H_0^1 and in L^2).

Proof. Let $[z_n, v] = \lambda_n(z_n, v), [z_m, v] = \lambda_m(z_m, v)$ with $\lambda_n \neq \lambda_m$. Then taking $v = z_m$, $v = z_n$, respectively, and subtracting we find

$$(\lambda_n - \lambda_m)(z_n, z_m) = (\lambda_n - \lambda_m)\lambda_n^{-1}[z_n, z_m] = 0$$

and the result follows. □

We mention here that these techniques can be extended to eigenvalue problems with other boundary conditions. We illustrate this with a special case that we will use in the applications in Section 7. The details of the analysis will not be given.

Suppose that $\partial\Omega$ is the disjoint union of two sets $\mathfrak{D}$ and $\mathfrak{N}$. The eigenvalue problem to be considered is

$$-\Delta u = \lambda u \text{ in } \Omega, \qquad u = 0 \text{ on } \mathfrak{D}, \qquad \partial u/\partial n = 0 \text{ on } \mathfrak{N} \tag{M}$$

(*mixed* boundary conditions). We introduce the space $\tilde{H} = \tilde{H}(\Omega)$, which is the closure in H^1 of smooth functions vanishing near $\mathfrak{D}$. Under very general hypotheses on $\mathfrak{D}$, the inequality

$$\int_\Omega u^2 d\mathbf{x} \leq C \int_\Omega |\text{grad } u|^2 d\mathbf{x} \tag{PI'}$$

holds for $u \in \tilde{H}$. The variational theory of (M) proceeds as before with $H_0^1(\Omega)$ replaced by $H(\Omega)$. A formal calculation, assuming sufficient smoothness of u up to $\mathfrak{N}$, shows that the Neumann condition on $\mathfrak{N}$ is a *natural boundary condition*, i.e., the minimizer satisfies this boundary condition even though the functions in the competing set do not. [A rigorous proof of this would require assumptions about $\mathfrak{N}$ and theorems about smoothness up to the boundary for elliptic equations. We will simply assume the variational formulation of (M) without worrying about strict satisfaction of the boundary conditions by the eigenfunctions.]

6. Capacity

We have formally defined in Section 1 the capacity of a closed surface Γ (or of the interior domain Ω) by the formula

$$c(\Gamma) = -\frac{1}{4\pi}\int_{\Gamma}\frac{\partial u}{\partial n}\,dS$$

in terms of the capacitary potential of its exterior, the function u harmonic in the exterior of Γ, regular at infinity, and such that $u = 1$ on Γ. This requires that u and Γ be sufficiently regular that the integral makes sense. The results of Section 3.2 imply that the previous definition certainly makes sense if Γ is a $C^{1,\alpha}$ surface. On the other hand, we would like to define the conductor potential and capacity for more general conductors. In doing this an approximation by smooth domains proves useful.

Suppose that T is an open connected set containing a neighborhood of infinity. A sequence of open domains T_n is said to be a *nested approximation* of T if

a. $\bar{T}_n \subset T_{n+1}$ for all n, and
b. Every point of T is in T_n for n sufficiently large.

The following theorem is proven in the Appendix to Section 6.

Theorem 6.1. For any T, there exists a sequence of nested approximations of T having connected analytic boundaries without singularities.

Suppose now that we consider such a nested approximation T_n of T and let u_n be the conductor potential of ∂T_n. Define u_n to be 1 on T_n^c. Then u_n is a decreasing positive sequence of harmonic functions. Let u be the limit of this sequence. If $\mathbf{x} \in T$, then there is a ball B with center at $\mathbf{x}$ contained in T_n for sufficiently large n, and Harnack's second convergence theorem implies that u is harmonic on B. This function u, regular harmonic on T and taking values between zero and one, is called the conductor potential of T. Because a point Q of ∂T has points of ∂T_n in every neighborhood, we have $\overline{\lim_{P\to Q}}\, u(P) = 1$ and if u is continuous at Q it has the value 1 there. The conductor potential of T is independent of the nested sequence approximating T. We need only consider another nested approximation T_n' and the sequence $U_n = \max\{u_n, u_n'\}$ where u_n' is the conductor potential of $\partial T_n'$. We may assume without loss of generality that $T_0 = T_0'$. The functions u_n, u_n' are superharmonic on $\mathbb{R}^3$. Because both sequences are decreasing, $u_n \leq u_0$ and $u_n' \leq u_0$. As u_n', u are harmonic functions in T'_n with $u \leq u'_n$ on $\partial T'_n$, $u \leq u'_n$ on T'_n. Similarly, $u' \leq u_n$ on T_n. It follows that $u \leq u' = \lim u'_n$ and $u' \leq u = \lim u_n$ on T.

If A is a compact set, there is exactly one component of its complement that contains a neighborhood of infinity. Denote this set by $T(A)$.

Definition 6.1. Suppose that A is a *compact* set. The *capacity* of A is defined by

$$c(A) = -\frac{1}{4\pi}\int_S \frac{\partial u}{\partial n}\, dS, \tag{28}$$

where u is the conductor potential of $T(A)$ and S is any smooth closed surface surrounding A.

We see that if A is bounded by a smooth closed surface, then $c(A) = c(\partial A)$ is given by the previous expression. The capacity of a set is defined in terms of the conductor potential of its *exterior*.

We can define another set function $c'(A)$ for general sets A by

$$c'(A) = \inf \int_{\mathbb{R}^3} |\text{grad } v|^2 d\mathbf{x} \tag{29}$$

where the infimum is over nonnegative functions

$$v \in \mathscr{K} := \{v : A \subset \text{interior } \{v \geq 1\}, v \in L^6(\mathbb{R}^3), \text{grad } v \in L^2(\mathbb{R}^3)\}.$$

We will show below that c and c' coincide in compact sets. The definition (28) evolves naturally from the physical definition of capacitance whereas (29) is more convenient in establishing properties of capacity. (A bridge connecting these ideas is given only implicitly in sources known to the authors.) The Sobolev inequality (Section 3 of Chapter 8) implies that if $v \in W^{1,2}(\mathbb{R}^3)$, then $v \in L^6(\mathbb{R}^3)$. On the other hand, $\mathscr{K}$ includes functions that are not in $W^{1,2}(\mathbb{R}^3)$. In particular, if v is the conductor potential of a smooth surface, $v = O(|\mathbf{x}|^{-1})$, $\text{grad } v = O(|\mathbf{x}|^{-2})$ as $|\mathbf{x}| \to \infty$ so that $\text{grad } v \in L^2(\mathbb{R}^3)$ but, in general, $v \notin L^2(\mathbb{R}^3)$.

Lemma 6.1. Suppose that u is the conductor potential of an exterior domain T with smooth boundary. Then

$$\int_T |\text{grad } u|^2 d\mathbf{x} = \min \int_T |\text{grad } w|^2 d\mathbf{x},$$

where the minimum is over functions w such that grad $w \in L^2(T)$, $w = O(|\mathbf{x}|^{-1})$ as $|\mathbf{x}| \to \infty$, and $w = 1$ on ∂T (in the sense of traces).

Proof. If $h = w - u$,

$$\int_T |\text{grad } w|^2 d\mathbf{x} = \int_T |\text{grad } u|^2 d\mathbf{x} + \int_T |\text{grad } h|^2 d\mathbf{x} + 2\int_T \text{grad } u \cdot \text{grad } h \, d\mathbf{x}. \tag{30}$$

If B_R is a ball of radius R centered at the origin such that $\partial T \subset B_R$, and $T_R = T \cap B_R$, then, because $\Delta u = 0$, and the trace of h on ∂T is zero, proposition 3.17 of Chapter 8 implies

$$\int_{T_R} \text{grad } u \cdot \text{grad } h \, d\mathbf{x} = \int_{\partial D_R} \text{grad } u \cdot \mathbf{n} h \, dS.$$

Because grad $u = O(|\mathbf{x}|^{-2})$ and $h = O(|\mathbf{x}|^{-1})$, this integral vanishes as $R \to \infty$, and the last integral in (30) vanishes. □

Theorem 6.2. $c'(A) = c(A)$ if A is compact.

Proof. Suppose that T is the component of A^c containing a neighborhood of infinity, that T_n is a nested sequence of sets with analytic boundaries for T, and u_n is the capacitary potential of ∂T_n. If we extend u_n to be identically 1 on T_n^c, then $u_n \in L^6(\mathbb{R}^3)$, grad $u_n \in L^2(\mathbb{R}^3)$, $u_n \geq 0$, and $u_n \geq 1$ on a neighborhood of A, so

$$\int_{\mathbb{R}^3} |\text{grad } u_n|^2 dV = \int_{T_n} |\text{grad } u_n|^2 dV \geq c(A).$$

On the other hand, on any smooth surface surrounding A, grad u_n converges uniformly to grad u, and the left-hand side converges to $c(A)$. Therefore, $c'(A) \leq c(A)$. In order to show that $c'(A) \geq c(A)$, let $v \in \mathcal{K}$ and $w = \min\{v, 1\}$; then

$$\int_{\mathbb{R}^3} |\text{grad } v|^2 dV \geq \int_{R^3} |\text{grad } w|^2 dV \geq \int_{T_n} |\text{grad } w|^2 dV \geq \int_{T_n} |\text{grad } u_n|^2 dV$$

for n sufficiently large, where the last inequality follows from Lemma 6.1. □

The following two theorems are immediate consequences of the definition of c'.

Theorem 6.3. If $A \subset B$, then $c'(A) \leq c'(B)$.

Theorem 6.4 (Scaling). If $B = kA$, then $c'(B) = kc'(A)$.

Theorem 6.5. $c'(A \cup B) \leq c'(A) + c'(B)$.

Proof. First note that if $f, g \in L^6(\mathbb{R}^3)$, $\operatorname{grad} f$, $\operatorname{grad} g \in L^2(\mathbb{R}^e)$, and $h = \max\{f, g\} = f + (g - f)^+$, then $h \in L^6(\mathbb{R}^3)$ and

$$\operatorname{grad} h = \begin{cases} \operatorname{grad} f & \text{a.e. on } \{f \geq g\}, \\ \operatorname{grad} g & \text{a.e. on } \{f \leq g\}. \end{cases}$$

Choose f, g with A, B contained in the interiors of $\{f \geq 1\}$, $\{g \geq 1\}$, respectively, and

$$\int_{\mathbb{R}^3} |\operatorname{grad} f|^2 d\mathbf{x} \leq c'(A) + \varepsilon/2, \qquad \int_{\mathbb{R}^e} |\operatorname{grad} g|^2 d\mathbf{x} \leq c'(B) + \varepsilon/2.$$

Then

$$\begin{aligned} \int_{\mathbb{R}^3} |\operatorname{grad} h|^2 d\mathbf{x} &\leq \int_{\mathbb{R}^3} \max\{|\operatorname{grad} f|^2, |\operatorname{grad} g|^2\} d\mathbf{x} \\ &\leq \int_{\mathbb{R}^3} |\operatorname{grad} f|^2 d\mathbf{x} + \int_{\mathbb{R}^3} |\operatorname{grad} g|^2 d\mathbf{x} \leq c'(A) + c'(B) + \varepsilon. \quad \square \end{aligned}$$

A fundamental property of c' is

$$c'(A) = \inf\{c'(U) : U \text{ open}, \ U \supset A\}.$$

In order to see that this holds we need only consider the open sets $\operatorname{int}\{f \geq 1\}$ used in the definition of c'.

Theorem 6.6. If $A_n \supset A_{n+1}$, A_i compact, and $A = \bigcap_{i=1}^{\infty} A_i$, then $c'(A) = \lim_{n \to \infty} c'(A_n)$.

Proof. $c'(A) \leq \lim c'(A_n)$ follows from Theorem 6.3. For any open set U containing A, compactness of A_n implies that $A_n \subset U$ for n sufficiently large, so that $c'(A_n) \leq c'(U)$. It follows $c'(A) \geq \lim c'(A_n)$. $\square$

Theorem 6.7. Suppose u is the conductor potential of $T(A)$. Then

$$\frac{c'(A)}{r''} \leq u(\mathbf{x}) \leq \frac{c'(A)}{r'}, \qquad \mathbf{x} \notin A,$$

where $r' = \inf_A |\mathbf{x} - \mathbf{y}|$, $r'' = \sup_A |\mathbf{x} - \mathbf{y}|$.

Proof. In view of Theorem 6.6 it suffices to consider A with ∂A smooth. We can write

$$4\pi u(\mathbf{x}) = \int_{S_R} \left(\frac{1}{r}\frac{\partial u}{\partial r} - u\frac{\partial}{\partial r}\frac{1}{r} \right) dS_y + \int_{\partial A} \left(\frac{\partial}{\partial n}\frac{1}{r} - \frac{1}{r}\frac{\partial u}{\partial n} \right) dS_y,$$

where $r = |\mathbf{x} - \mathbf{y}|$, S_R is the surface of a ball containing A and $\mathbf{x}$, and $\mathbf{n}$ denotes the outward normal to ∂A. Because the integrands in the first two integrals behave like R^{-3} and the area of S_R like R^2, these integrals tend to zero as R becomes infinite. The third integral vanishes because $1/r$ is harmonic in A. Finally, $\partial u/\partial n \leq 0$ on ∂A (because $u = 1$ on ∂A, $u \leq 1$ outside) and the result follows. □

Before proceeding to our main theorem we will give several examples.

i. The *capacity of a sphere* is its radius, for, supposing the origin to be at its center, the conductor potential is $R/|\mathbf{x}|$where R is the radius of the sphere. Problems involving ellipsoids can also be solved in closed form (Ref. 2, Chapter 7) and can be used to deal with some special cases by taking limits. For example, the capacitance of a circular disk is $2/\pi$ times its radius.
ii. The *capacity of a Jordan arc* is zero. This may be seen by considering the line potential with constant density ε along this arc. The conductor potential of the arc is dominated by this potential for all ε because the potential approaches ∞ as the arc is approached. Therefore, the conductor potential of an arc is identically zero off the arc.
iii. We find the *capacity of two touching spheres* by using the method of images. Consider two touching spheres, each of radius R, with centers at the points $-R, R$ on the x-axis and touching at the origin. We will calculate the constant value of the potential on the spheres using an infinite series of image charges. First place two charges of magnitude $\mathcal{Q}_1$, one at the center of each sphere. The potential $u' = \mathcal{Q}_1(1/r_1 + 1/r_1')$ at a point P (say, of the first sphere) clearly depends on the location of P on the sphere. Adding an image charge $\mathcal{Q}_2 = -R\mathcal{Q}_1/2R = -\mathcal{Q}_1/2$ at the image point $x = -R/2$ (as $2R \cdot R/2 = R^2$) on the x-axis leaves the

potential $\mathcal{Q}_1/R$ on the first sphere, but modifies the potential on the second sphere. Similarly, adding the image charge $\mathcal{Q}_2$ at the point $x = R/2$ leaves the potential $\mathcal{Q}_1/R$ in the second sphere, but modifies the potential in the first. To restore the value $\mathcal{Q}_1/R$ for the potential, we add two further image charges $\mathcal{Q}_3 = -R\mathcal{Q}_2/(\frac{3}{2}R) = \mathcal{Q}_1/3$ at $x = \mp R/3$ (as $2R/3 \cdot 3R/2 = R^2$), and so on. To show that the process converges, we choose a convenient point P, namely, the point $x = -2R$ on the x-axis. Then

$$u_1'(P) = \frac{\mathcal{Q}_1}{2R - R} + \frac{\mathcal{Q}_1}{2R + R}$$

is the potential due to the first pair of charges (we have written the relative distance from P with respect to the point of contact). According to the recipe above, we add a sequence of pairs of image charges $(-1)^{n+1}\mathcal{Q}_1/n$ located at the symmetric points $x = \mp R/n$. The potential at P due to the nth pair is

$$u_n'(P) = \frac{(-1)^{n+1}\mathcal{Q}_1/n}{2R - R/n} + \frac{(-1)^{n+1}\mathcal{Q}_1/n}{2R + R/n}.$$

Thus,

$$u(P) = \sum_{n=1}^{\infty} u_n'(P) = -\frac{\mathcal{Q}_1}{R}\left[\sum_{n=0}^{\infty}\frac{(-1)^{n+1}}{2n+1} + \sum_{n=1}^{\infty}\frac{(-1)^n}{2n+1}\right] = \frac{\mathcal{Q}_1}{R},$$

as asserted. The total charge $\mathcal{Q} = (-1/4\pi)\int_\Gamma (\partial u/\partial n)dS$ on the surface is equal to the total inner charge

$$\mathcal{Q} = 2\mathcal{Q}_1 \sum_{n=1}^{\infty}(-1)^{n+1}n^{-1},$$

and, because $\sum_{n=1}^{\infty}(-1)^{n+1}n^{-1} = \ln 2$, $\mathcal{Q} = 2\mathcal{Q}_1 \ln 2$. Dividing by the constant value of the potential $u(P)$ yields for the capacity of the two touching spheres

$$c = 2R \ln 2.$$

The previous considerations can be applied to give a necessary and sufficient condition for existence of a barrier (Section 3). Consider a point P on the boundary of a domain Ω. Suppose that λ is a number between zero and one, and B_n is the ball about P of radius λ^n. We set

$$\gamma_n = c(\Omega^c \cap \bar{B}_n \cap B_{n+1}^c) \equiv c(\Sigma_n),$$

where the set $\Sigma_n = \Sigma_n(\lambda)$ is defined by this equation (see Fig. 9).

Theorem 6.8 (Wiener). *P is a regular boundary point of Ω if and only if*

$$\sum_{n=0}^{\infty} \frac{\gamma_n}{\lambda^n} = \infty \tag{31}$$

for $0 < \lambda < 1$.

The regularity of a boundary point P is seen to follow if the complement is not too "thin" near P.

We observe that if (31) holds for one value of λ, then it must hold for any other in (0,1). In fact, by using Theorems 6.4 and 6.5 one easily proves

Proposition 6.1. The series

$$\sum_{n=0}^{\infty} \frac{\gamma_n(\lambda)}{\lambda^n}, \qquad \sum_{n=0}^{\infty} \frac{\gamma_n(\lambda^2)}{\lambda^{2n}}$$

converge or diverge together.

Proof. Suppose

$$\sum_{n=0}^{\infty} \gamma_n(\lambda^2)\lambda^{-2n} < \infty.$$

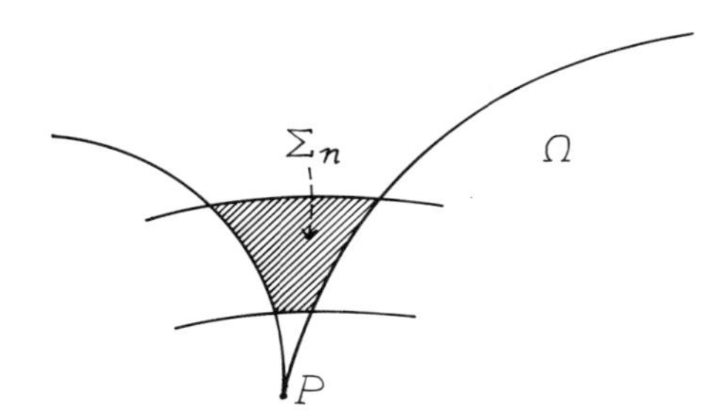

Fig. 9. A shell domain used to define Wiener's criterion.

If $\mu = \lambda^2$, $\Sigma_{2n}(\lambda) \subset \Sigma_n(\mu)$ so $\gamma_{2n}(\lambda) \leq \gamma_n(\mu)$, whence $\sum_{n=0}^{\infty} \gamma_{2n}(\lambda)\lambda^{-2n} < \infty$. Moreover, $\Sigma_{2n+1}(\lambda) \subset \Sigma_n(\mu)$ so that $\gamma_{2n+1}(\lambda) \leq \gamma_n(\mu)$ and

$$\lambda^{-1} \sum_{n=0}^{\infty} \gamma_{2n+1}(\lambda)\lambda^{-2n} < \infty.$$

To prove the converse, observe that $\Sigma_n(\mu) = \Sigma_{2n}(\lambda) \cup \Sigma_{2n+1}(\lambda)$, so $\gamma_n(\mu) \leq \gamma_{2n}(\lambda) + \gamma_{2n+1}(\mu)$. If $\sum_{n=0}^{\infty} \gamma_n(\lambda)\lambda^{-n} < \infty$, it follows that both

$$\sum_{n=0}^{\infty} \gamma_{2n}(\lambda)\lambda^{-2n} < \infty \quad \text{and} \quad \lambda \sum_{n=0}^{\infty} \gamma_{2n+1}(\lambda)\lambda^{-2n} < \infty$$

hold, and the result follows by addition. □

Lemma 6.2. Suppose $\Omega_a^c = B(P, a) \cap \Omega^c$ (Fig. 10). Then P is regular if and only if the conductor potential u_a of Ω_a^c has a limit at P, for any $a > 0$.

Proof. The condition is necessary because a barrier at P for Ω will serve as one at P for the complement of Ω_a^c. Suppose that a takes a sequence of values a_{n} approaching zero, and let

$$v(\mathbf{x}) = \sum_{n=1}^{\infty} 2^{-n} u_{a_{\mathrm{n}}}(\mathbf{x}).$$

As this series is uniformly convergent, it has limit 1 at P. At any point in Ω at least one term is less than 2^{-n} so that v is less than 1. The function $1 - v$ is a barrier for Ω at P. □

Proof of Theorem 6.8. Suppose that (31) diverges. We will show that u_a has a limit at P for any positive a. Let $0 < \varepsilon < \frac{1}{3}$. Then we will show that $u_a(\mathbf{x}) > 1 - \varepsilon$ if $\mathbf{x}$ is in a sufficiently small neighborhood of P. We choose $\lambda = 1 - \varepsilon/3$, and then choose k so that $\lambda^{k-1} < \varepsilon/3$. The series (31) can be

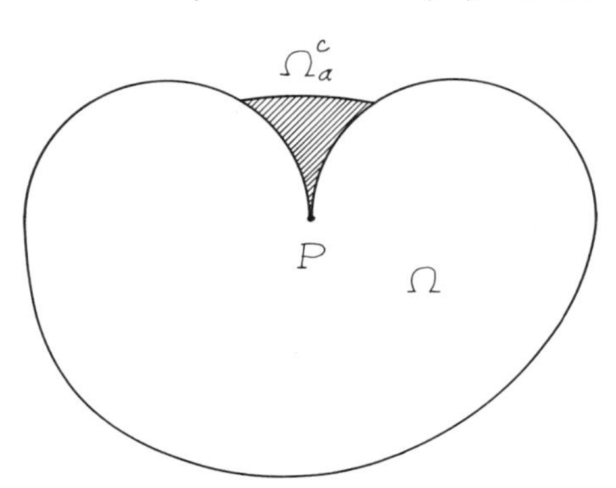

Fig. 10. A conductor related to boundary regularity.

broken up into k series by grouping terms mod k and at least one of these must diverge. Suppose

$$\sum_{i=0}^{\infty} \frac{\gamma_{ki+j}(\lambda)}{\lambda^{ki+j}} = \infty, \qquad 0 \le j < k.$$

Then if $\tilde{\lambda} = \lambda^j \lambda$, we have

$$\gamma_{ki+j} = c(\Sigma_{ki+j}(\lambda)) = c(\Sigma_{ki}(\{\tilde{\lambda})) = \lambda^j \gamma_{ki},$$

so that we may assume without loss of generality that

$$\sum_{i=0}^{\infty} \gamma_{ki} \lambda^{-ki} = \infty.$$

Suppose m is such that $\lambda^{km} < a$. We define $e_i = \Omega^c \cap \bar{B}_i \cap B_{i+1}^c$ and denote by v_i the conductor potential of e_i. If $n > m$, define

$$V_{m,n} = \sum_{i=m}^{n} v_{ki}$$

which is harmonic outside of

$$e_{m,n} = \bigcup_{i=m}^{n} e_{ki}.$$

Our choice of m implies that this set is contained in Ω_a^c. We wish to bound $V_{m,n}$ on $e_{m,n}$, as a bound on this set must also be a bound everywhere. Suppose $\mathbf{x} \in e_{m,n}$. The sets e_{ki} are disjoint so that $\mathbf{x}$ is in exactly one of these sets, say e_{kl}. On e_{kl} we have $v_{kl} \le 1$, whereas Theorem 6.7 implies that on e_{ki}, $i \ne l$, we have

$$v_{ki} \le \gamma_{ki}[\lambda^{ki+1} - \lambda^{k(i+1)}]^{-1} = \frac{1}{\lambda(1 - \lambda^{k-1})} \frac{\gamma_{ki}}{\lambda^{ki}}.$$

The situation when $i = l + 1$ is depicted in Fig. 11.

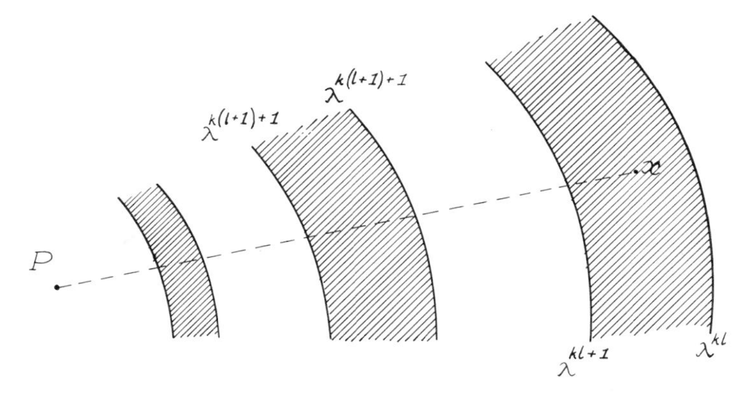

Fig. 11. The shells in Wiener's series (31).

Then we have

$$V_{m,n} \leq 1 + [\lambda(1 - \lambda^{k-1})]^{-1} \sum_{\substack{i=m \\ i \neq l}}^{n} \gamma_{ki} \lambda^{-ki}$$

$$< [\lambda(1 - \lambda^{k-1})]^{-1} \left[1 + \sum_{i=m}^{n} \gamma_{ki} \lambda^{-ki} \right].$$

We define

$$\tilde{V}_{m,n} = V_{m,n} \lambda(1 - \lambda^{k-1}) \left[1 + \sum_{i=m}^{n} \gamma_{ki} \lambda^{-ki} \right]^{-1}.$$

Then $\tilde{V}_{m,n}$ is always less than 1, and is harmonic on a domain containing the complement of Ω_a^c, so that $\tilde{V}_{m,n} \leq u_a$. If the other half of Theorem 6.7 is utilized, we find that

$$V_{m,n}(\mathbf{x}) \geq \sum_{i=m}^{n} \gamma_{ki} (|\mathbf{x}| + \lambda^{ki})^{-1},$$

and then

$$u_a(\mathbf{x}) \geq \lambda(1 - \lambda^{k-1}) \sum_{i=m}^{n} \gamma_{ki} (|\mathbf{x}| + \lambda^{ki}) d^{-1},$$

where $d = \sum_{i=m}^{n} \gamma_{ki}\lambda^{-ki} + 1$. If we impose our choice of λ, we find that

$$u_a(\mathbf{x}) > \left(1 - \frac{2\varepsilon}{3}\right) d^{-1} \sum_{i=m}^{n} \gamma_{ki}(|\mathbf{x}| + \lambda^{ki})^{-1}.$$

Note that the last term approaches $d - 1$ as $|\mathbf{x}| \to 0$, so that we may choose a neighborhood of P in which

$$u_a(\mathbf{x}) > \left(1 - \frac{2\varepsilon}{3}\right)\left(1 - \frac{2}{d}\right).$$

We need only make use of the fact that ε is arbitrary and the divergence of (31) to obtain $u_a(\mathbf{x}) > 1 - \varepsilon$.

Suppose now that (31) converges. We will show that for some value of a, u_a does not have a limit at P. In fact, suppose m is such that

$$\sum_{i=m}^{\infty} \gamma_i \lambda^{-1} < \frac{1}{4}\lambda.$$

We show that the assumption that $u_m = u_{\lambda^m}$ has a limit at P implies a contradiction. Suppose B is a ball about P such that $u_m > \frac{3}{4}$ in B. Then choose $n > m$ such that $u_n < \frac{1}{4}$ on the surface of B (Theorem 6.7), and denote by $u_{m,n}$ the conductor potential of $\Omega^c \cap \bar{B}_m \cap B_n^c$. By a limiting argument we find that $u_m \le u_n + u_{m,n}$, and this implies that on the surface of B we have

$$\tfrac{3}{4} < \tfrac{1}{4} + u_{m,n} \Rightarrow u_{m,n} > \tfrac{1}{2}.$$

The monotone sequence used to define $u_{m,n}$ has constant boundary values on surfaces approaching $\partial(\Omega^c \cap \bar{B}_m \cap B_n^c)$ so that the maximum principle implies they are greater than $\frac{1}{2}$ inside B, and therefore $u_{m,n}$ is. Now in contrast we have

$$u_{m,n}(P) \le \sum_{i=m}^{n} v_i(P) \le \sum_{i=m}^{n} \gamma_i \lambda^{-(i+1)} \le \lambda^{-1} \sum_{i=m}^{\infty} \gamma_i \lambda^{-1} < \frac{1}{4}.$$

If we now let n become large, we arrive at a contradiction. □

Finally, we give a theorem showing that sets of capacity zero are removable singularities for bounded harmonic functions.

Theorem 6.9. Suppose that Ω is a bounded domain such that all points of $\partial\Omega$ are regular and A is compact, $A \subset \Omega$, such that $\Omega \backslash A$ is connected. If $\mathscr{U}$ is a bounded harmonic function on $\Omega\backslash A$ that is continuous on $(\Omega\backslash A)\cup\partial\Omega$ and $c(A)=0$, then $\mathcal{U}$ extends to Ω as a harmonic function.

Proof. Let E_n be a nested sequence of domains with analytic boundaries converging to $E=A^c$ and u_n the corresponding capacitary potential. Because $c(A)=0$,

$$\int_{E_n} |\text{grad } u_n|^2 dV \to 0 \qquad n\to\infty.$$

If $\mathbf{x}\in\Omega\backslash A$, there is a ball B centered at $\mathbf{x}$ such that $B\subset E_n$ for n sufficiently large. It follows that

$$\int_B |\text{grad } u_n|^2 dV \to 0,$$

and $u=\lim u_n$ is constant on B. Unique continuation implies that u is constant on E, and the only possibility is $u=0$.

Now let v be the harmonic function on Ω with $v=\mathcal{U}$ on $\partial\Omega$. Let $w=\mathcal{U}-v$. Then the maximum principle implies that $|w|\leq Cu_n$ on $E_n\cap\Omega$ where C depends only on $\sup|\mathcal{U}|$. It follows that $w=0$, and the theorem is proven. □

7. Applications

7.1. A Problem Concerning Asymptotic Efficiency of Cooling (Crushed Ice). Here we study the asymptotic behavior of the solution of the heat equation in an insulated container in which a finite number of spherical "coolers" have been placed. Suppose Ω is an open, bounded region in $\mathbb{R}^3$, $B_1,\ldots,B_n$ are balls of radius r, and $\Omega_n=\Omega\backslash K_n$ where $K_n=\bigcup_{i=1}^{n} B_i$. Consider the solution of the problem

$$u_t = c\Delta u \qquad \mathbf{x}\in\Omega_n, \tag{H}$$

$$\begin{cases} \partial u/\partial n = 0, & \mathbf{x}\in\partial\Omega, \\ u=0, & \mathbf{x}\in\partial K_n. \end{cases} \tag{BC}$$

(We may assume, for simplicity, that $\bar{B}_n\subset\Omega$; overlap of the B_j's is allowed.) For given $u(\mathbf{x},0)$ the solution is given by an expansion $\sum a_j \exp(-c\lambda_j t)\varphi_j(\mathbf{x})$, where $\varphi_j(\mathbf{x})$ are the normalized eigenfunctions of $-\Delta$ in Ω_n subject to (BC). The first term dominates and the asymptotic behavior is given, to leading order, by knowledge of λ_1. We are concerned with the behavior of λ_1 as n increases [and $r = r(n)$ decreases]. The results are summarized in the following two theorems (due to J. Rauch).

Theorem 7.1. There is a constant $c > 0$, independent of n, such that

$$\lambda_1 \leq \frac{2nr}{|\Omega|}(c + O(nr))$$

as $nr \to 0$, provided $|\Omega_n| \geq \bar{c}|\Omega|$, $\bar{c}$ independent of n.

Theorem 7.2. There is a positive constant c', independent of n, such that

$$\lambda_1 \geq c'nr - 1$$

provided the spheres satisfy condition (E), defined below.

One sees that $\lambda_1 \to \infty$, i.e., the cooling efficiency becomes infinite, if $nr \to \infty$. In particular, if the total volume (proportional to nr^3) is held fixed as $r \to 0$, the total surface area (proportional to nr^2) goes to infinity and certainly $\lambda_1 \to \infty$. The natural conjecture that nr^2 is the appropriate measure of efficiency is seen to be incorrect, however. In order to give the proof of Theorem 7.1 we need to use the variational characterization of the eigenvalues of $-\Delta$. In particular,

$$\lambda_1(\Omega_n) = \inf\left(\int_{\Omega_n} |\text{grad}\, \psi|^2 \Big/ \int_{\Omega_n} \psi^2\right) \tag{VC}$$

for $\psi \in H^1(\Omega_n)$ with $\psi = 0$ on ∂K_n.

Proof of Theorem 7.1. The proof is obtained by using an appropriate test function in (VC). Let u_n be the capacitary potential of K_n. Recall that u_n is harmonic on $K_c^n = \mathbb{R}^3 \backslash K_n$, $u_n = 1$ on ∂K_n, and $u_n = O(|\mathbf{x}|^{-1})$ as $|\mathbf{x}| \to \infty$. Also, we can write

$$c(K_n) = \int_{K_n^c} |\text{grad}\, u_n|^2,$$

and, as $c(B_i) = r$ and c is subadditive, $c(\mathrm{K}_n) \leq \sum_i c(B_i) = nr$. It follows that

$$\int_{\Omega_n} |\text{grad } u_n|^2 \leq \int_{\mathrm{K}_n^c} |\text{grad } u_n|^2 \leq nr.$$

Consider any $\mathbf{x} \in \mathbb{R}^3 \backslash \bar{\mathrm{K}}_n$. The usual limiting argument implies that

$$u_n(\mathbf{x}) = \frac{1}{4\pi} \int_{\partial \mathrm{K}_n} |\mathbf{x} - \mathbf{y}|^{-1} \frac{\partial u_n}{\partial n} \, dS_y,$$

and

$$4\pi |u_n(\mathbf{x})| = 4\pi u_n(\mathbf{x}) \leq (\text{dist}(\mathbf{x}, \mathrm{K}_n))^{-1} \int_{\partial \mathrm{K}_n} \frac{\partial u_n}{\partial n} \, dS_y \leq nr(\text{dist}(\mathbf{x}, \mathrm{K}_n))^{-1}$$

follows. Suppose that $\bar{\Omega} \subset B_R$, where B_R is a ball of radius R. Let η be a smooth function with $0 \leq \eta \leq 1$, $\eta = 1$ on B_R and $\eta = 0$ on B_{2R}^c. Let σ_1 be the constant in Poincaré's inequality for B_{2R}, i.e.,

$$\int_{B_{2R}} u^2 \leq \sigma_1 \int_{B_{2R}} |\text{grad } u|^2$$

for all $u \in H_0^1(B_{2R})$. Then if we extend u_n to Ω by $u_n = \mathrm{I}$ on $\mathrm{K_n}$,

$$\begin{aligned}
\int_{\Omega_n} u_n^2 &\leq \int_{B_{2R}} |\eta u_n|^2 \leq \sigma_1 \int_{B_{2R}} |\text{grad } (\eta u_n)|^2 \\
&\leq 2\sigma_1 \left[\int_{B_{2R}} |\text{grad } u_n|^2 + \max_{B_{2R}} |\text{grad } \eta|^2 \int_{\mathscr{B}} u_n^2 \right] \\
&\leq C \left[\int_{B_{2R}} |\text{grad } u_n|^2 + \int_{\mathscr{B}} u_n^2 \right],
\end{aligned}$$

where $\mathscr{B} = B_{2R} \backslash B_R$, and C depends only on R. It follows that

$$\int_{\Omega_n} u_n^2 \leq C \left[nr + \frac{n^2 r^2}{(4\pi \, \text{dist}(\partial B_R, \mathrm{K}_n))^2} \right] \leq C[nr + (nr)^2].$$

Because we are concerned with nr small we can write $\int_{\Omega_n} u_n^2 \leq Cnr$, where, again, C is a constant depending only on R. The test function that we use in (VC) is $\psi = 1 - u_n$. We have

$$\int_{\Omega_n} |\text{grad } \pi|^2 = \int_{\Omega_n} |\text{grad } u_n|^2 \leq nr$$

and

$$\int_{\Omega_n} \psi^2 = |\Omega_n| - 2\int_{\Omega_n} u_n + \int_{\Omega_n} u_n^2.$$

Applying the Cauchy–Schwarz and arithmetic–geometric mean inequalities we obtain

$$\int_{\Omega_n} \psi^2 \geq |\Omega_n| - \left[\frac{1}{2}|\Omega_n| + 2\int_{\Omega_n} u_n^2\right] + \int_{\Omega_n} u_n^2 = \frac{1}{2}|\Omega_n| - \int_{\Omega_n} u_n^2$$
$$= \tfrac{1}{2}|\Omega_n| - O(nr)$$

for nr small. It follows that

$$\lambda_1 \leq \frac{Cnr}{\frac{1}{2}|\Omega_n| - O(nr)} = \frac{2nr}{|\Omega|} \frac{C}{\bar{c} - O(nr)/|\Omega|} = \frac{2nr}{|\Omega|}(C + O(nr)),$$

where, again, C denotes a constant depending only on R. The theorem is proven. $\square$

In order to establish the lower bound in Theorem 7.2 we need to impose an "evenly spaced" condition on B_i (Fig. 12).

(E) There is a number M independent of n and a number $R = \mathrm{R}(n) > \sqrt{3}r(n)$ such that

$$\Omega \subset \bigcup_{i=1}^{n} B_i, \qquad B_i = B(\mathbf{x}_i, R), \qquad \text{for all } n,$$

and each $x \in \Omega$ is in at most M of the B_i.

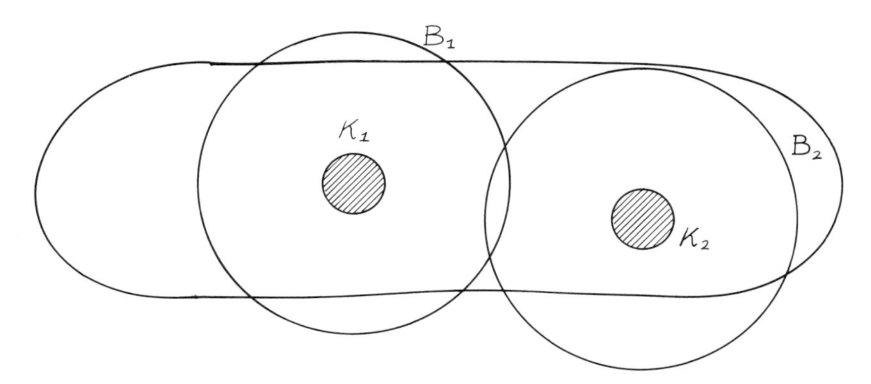

Fig. 12. The evenly spaced condition.

Lemma 7.1. If $f \in C^1[r, R]$, $f(r) = 0$, then

$$\int_r^R (f'(t))^2 t^2 \, dt \geq \frac{2r}{R^3} \int_r^R f^2(t) t^2 \, dt.$$

Proof. Apply the Cauchy–Schwarz inequality to $f^2(t) = (\int_r^t f'(t) t t^{-1} dt)^2$, and integrate □

Corollary 7.1. If A= $\{\mathbf{x} : r < |\mathbf{x}| < R\}$, then

$$\int_{\mathrm{A}} |\mathrm{grad}\ \psi|^2 \geq \frac{2r}{R^3} \int_{\mathrm{A}} \psi^2$$

for all $\psi \in H^1(\mathrm{A})$ with $\psi = 0$ on $|\mathbf{x}| = r$.

Proof. The infimum of the quotient $\int_A |\mathrm{grad}\ \psi|^2 / \int_A \psi^2$ is taken on for ψ an eigenfunction corresponding to the lowest eigenvalue of $-\Delta$ on Ω subject to $\psi = 0$ on $|\mathbf{x}| = r$, $\psi_r = 0$ on $|\mathbf{x}| = R$. Exercises 7.1 and 7.2 show that $\psi(\mathbf{x}) = f(|\mathbf{x}|)$. The result then reduces to the lemma.

We define $\Omega_e = \bigcup_{i=1}^n B_i$. We may assume that Ω_e is contained in a fixed compact subset of a fixed open set $\tilde{\Omega}$ for all n.

Proposition 7.1. There is an operator $\mathbb{E} \cdot H^1(\Omega) \to H^1(\mathbb{R}^3)$ such that $\mathbb{E}\psi = \psi$, $\operatorname{supp}(\mathbb{E}\psi) \subset \tilde{\Omega}$ for $\psi \in H^1(\Omega)$, and a constant K such that

$$\int_{\mathbb{R}^3} |\text{grad } \psi_e|^2 \leq \text{K}\left(\int_\Omega |\text{grad } \psi|^2 + \int_\Omega \psi^2\right),$$

where $\psi_e = \mathbb{E}\psi$.

The proof of Proposition 7.1 follows from Proposition 3.7 of Chapter 8.

Proof of Theorem 7.2. Suppose $\psi \in H^1(\Omega_n)$, $\psi = 0$ on ∂K_n. If we define $\psi = 0$ on K_n, the extended function is in $H^1(\Omega)$. Applying the proposition we can extend ψ to $\psi_e \in H^1(\mathbb{R}^3)$, with

$$\int_{\Omega_n} |\text{grad } \psi|^2 + \int_{\Omega_n} \psi^2 = \int_\Omega |\text{grad } \psi|^2 + \int_\Omega \psi^2 \geq \text{K}^{-1} \int_{\mathbb{R}_3} |\text{grad } \psi_e|^2.$$

If we consider ψ_e restricted to B_i, then, because $\psi = 0$ on K_i, Corollary 7.1 implies that

$$\int_{\text{B}_\text{i}} |\text{grad } \psi_e|^2 \geq \frac{2r}{R^3} \int_{B_i} \psi_e^2.$$

Condition (E) implies that

$$\int_{\mathbb{R}^3} |\text{grad } \psi_e|^2 \geq \text{M}^{-1} \sum_i \int_{\text{B}_\text{i}} |\text{grad } \psi_e|^2 \geq \frac{2r}{\text{MR}^3} \sum_i \int_{\text{B}_\text{i}} \psi_e^2 \geq \frac{2r}{\text{MR}^3} \int_{\Omega_e} \psi_e^2.$$

Then

$$\int_{\Omega_n} |\text{grad } \psi|^2 + \int_{\Omega_n} \psi^2 \geq \frac{2r}{\text{KMR}^3} \int_{\Omega_e} \psi_e^2 \geq \frac{2r}{\text{KMR}^3} \int_{\Omega_n} \psi^2.$$

We can write $2r/\text{KMR}^3 = (2/\text{KM})(nr/nR^3)$, and (E) implies

$$nR^3 \leq \text{M}|\Omega_\text{e}| \leq \text{M}|\tilde{\Omega}|.$$

The result follows with $c' = 2K\text{M}^3|\tilde{\Omega}|$. □

7.2. A Free Boundary Problem Modeling Separated Flow of an Incompressible Fluid. The velocity components in an incompressible two-dimensional fluid flow can be found in terms of a "stream function" $\psi = \psi(x, y) \equiv \psi(\mathbf{x})$ by

$$v_x = \psi_y, \qquad v_y = -\psi_x \tag{SF}$$

(Exercise 7.3). If the flow is irrotational, ψ is harmonic, and in regions where the "vorticity," i.e., the curl of the velocity $\mathbf{v}$, is nonzero, its magnitude is $-\Delta\psi$. The problem that we set can be thought of as that of determining a "separated" region where $-\Delta\psi$ is nonzero in a flow problem.

Suppose that Ω is a bounded, simply connected domain with $C^{1+\alpha}$ boundary. Suppose that $\partial\Omega = \Gamma_1 \cup \Gamma_2$, where Γ_1, Γ_2 are connected arcs, and that boundary values $\psi = f$, where $f \in C(\partial\Omega)$ is positive on Γ_1 and zero on Γ_2, are given. We seek an arc γ with endpoints on Γ_2 and a function $\psi \in C^1(\Omega)$ such that

$$\Delta\psi = \begin{cases} \omega & \text{"below } \gamma\text{,"} \\ 0 & \text{"above } \gamma\text{,"} \end{cases}$$

$$\psi = f \text{ on } \partial\Omega, \qquad \psi = 0 \text{ on } \gamma,$$

where ω is a given positive number (Fig. 13).

The arc γ divides Ω into two subdomains N ("below γ") and P ("above γ"). The maximum principle implies that $\psi < 0$ on N, $\psi > 0$ on P. A solution of this problem solves the nonlinear equation

$$\Delta\psi = \omega I_{\{\psi<0\}}.$$

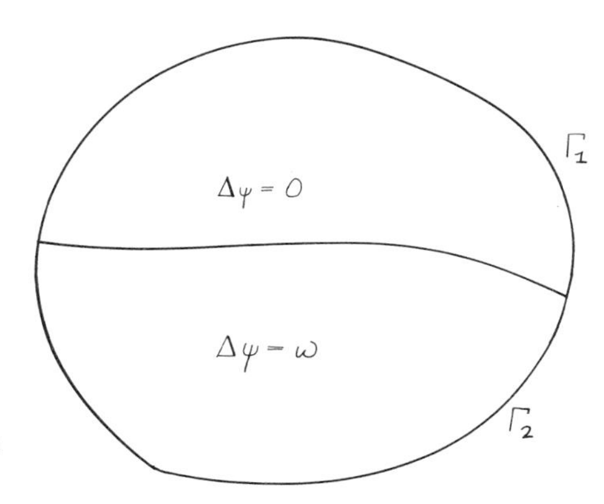

Fig. 13. The vortex subdomain bounded by γ.

If $g(\mathbf{x}, \mathbf{y})$ is the Green's function for Ω and ψ_0 is the harmonic function with $\psi_0 = f$ on $\partial\Omega$,

$$\psi(\mathbf{x}) = \psi_0 - \omega \int_N g(\mathbf{x}, \mathbf{y}) d\mathbf{y}.$$

We emphasize that this is a nonlinear equation for ψ as the right-hand side depends on ψ through the set N where ψ is negative. If M is a measurable subset of Ω, we define $\Phi[M] \equiv \Phi[M](\mathbf{x})$ by

$$\Phi[M] = \int_M g(\mathbf{x}, \mathbf{y}) d\mathbf{y}.$$

Our equation can then be written

$$\psi(\mathbf{x}) = \psi_0 - \omega\Phi[N] \tag{32}$$

and

$$\psi = 0 \text{ on } \gamma. \tag{33}$$

We will find a solution using successive approximations. In particular, suppose γ_0 is a smooth arc with endpoints a_0, b_0 on Γ_2, and let N_0 be the subdomain of Ω bounded by γ_0 and the connected arc σ_0 cut on Γ_2 (Fig. 14).

We define

$$\psi_1 = \psi_0 - \omega\Phi[N_0].$$

The maximum principle implies that ψ_0 and $\Phi[N_0]$ are positive on Ω. In order to avoid technical complications, we make the further assumption that $\partial\Omega$ satisfies

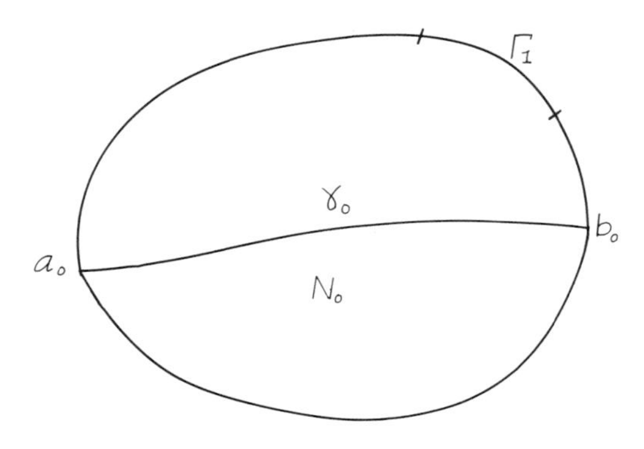

Fig. 14. An initial guess for the iteration.

the internal sphere condition, and that f extends into Ω as a function in $C^2(\bar{\Omega})$. This implies that $\psi_0 \in C^1(\bar{\Omega})$, $\Phi[N_0] \in C^1(\bar{\Omega})$ (Section 3.2), and the boundary point lemma can be used. On σ_0 both ψ_0 and $\Phi[N_0]$ are zero, but grad ψ_0 and grad $\Phi[N_0]$ are continuous there, and the boundary point lemma implies that they are positive. It follows that the ratio $\psi_0/\Phi[N_0]$ is well defined on $\bar{N}_0$.

We now choose

$$\omega > \sup_{\bar{N}_0} \psi_0/\Phi[N_0]. \tag{34}$$

It then follows immediately from the definition that $\psi_1 < 0$ on $N_0 \cup \gamma_0$. As $\psi_1 > 0$ near Γ_1, there must be a level curve $\psi_1 = 0$ for the harmonic function ψ_1 on $\Omega\backslash\bar{N}_0$. This curve must have endpoints on Γ_2 and cannot have branch points in Ω. (A subdomain of Ω on which ψ_1 is harmonic and has zero boundary values leads to a contradiction.) The only possibility is that this curve, denoted γ_1, is a simple arc with endpoints a_1, b_1 on Γ_2. Let N_1 be the domain bounded by γ_1 and the connected arc σ_1 of Γ_2 with endpoints a_1, b_1, and define

$$\psi_2 = \psi_0 - \omega\Phi[N_1].$$

As $N_1 \supset N_0$, say $N_1 = N_0 + D_1$, we can write $\psi_2 = \psi_1 - \omega\Phi[D_1] < \psi_1$ except on σ_1. We can continue recursively, letting γ_2 be the level curve of ψ_2, and so on, to obtain sets $N_0 \subset N_1 \subset N_2 \ldots$ and functions $\psi_1 \geq \psi_2 \geq \psi_3 \ldots$. The limits of these sequences clearly exist and it remains to show that they are solutions of our problem. Let $\psi = \lim \psi_n$, $N = \bigcup N_n$, and $R_n = N\backslash N_{n-1}$, $r_n = \omega\Phi[R_n]$. Then

$$\psi_n = \psi_0 - \omega\Phi[N] + r_n.$$

As $|R_n| \to 0$ implies $r_n \to 0$, we find that $\psi = \psi_0 - \omega\Phi[N]$, and it follows that $\psi \in C^1(\Omega) \cap C(\bar{\Omega})$. Dini's theorem now implies that ψ_n converges uniformly on $\bar{\Omega}$, and this can be used to show that $\psi < 0$ on N and $\psi = 0$ on $\gamma = \partial N \cap \Omega$. In fact, $\psi \leq 0$ on N is immediate. If $\mathbf{x} \in \gamma$, there are points, say $\mathbf{x}_n$, of γ_n in arbitrarily small neighborhoods of $\mathbf{x}$, and, given $\varepsilon > 0$, there is an N such that for $n \geq N$, $|\psi_n(\mathbf{x}) - \psi(\mathbf{x})| < \varepsilon/2$ (convergence of ψ_n), and $|\psi_n(\mathbf{x}_n) - \psi_n(\mathbf{x})| < \varepsilon/2(\{\psi_n(\mathbf{x})\}$ is equicontinuous as it is monotone and uniformly continuous). Hence,

$$|\psi(\mathbf{x})| = |\psi(\mathbf{x}) - \psi_n(\mathbf{x}_n)| < \varepsilon$$

and $\psi(\mathbf{x})$ vanishes on γ. The maximum principle now implies that $\psi < 0$ on N. The maximum principle implies also that $\psi > 0$ on $\Omega\backslash\bar{N}$. We may call ψ a solution to our problem.

It should be remarked here that, although we might expect γ to be a smooth arc, this has not been proven. This is an open problem and leaves a gap between what we have established and the problem as originally stated.

Suppose that we have found a solution ψ_1 corresponding to some value of ω, say ω_1, and we seek a solution for $\omega = \omega_2 > \omega_1$. If $N_1 = \{\psi_1 < 0\}$, then we can start the iteration used above with this as the initial set and a solution $\psi_2 < \psi_1$ is obtained, and letting $N_2 = \{\psi_2 < 0\}$, $N_2 \supset N_1$. As ω increases, the sets N increase.

We conclude by giving a particular configuration that is of some physical interest (Fig. 15). We take AB as Γ_2, BCDA as Γ_1, and $f = y$ on Γ_1, $f = 0$ on Γ_2.

This then is a model for flow past an obstacle in a channel and the region N is an eddy region caused by the disturbance to uniform flow by the obstacle.

8. Appendix to Section 1

Here we collect some useful results concerning continuity and differentiability of integrals. We begin with a well-known general theorem concerning differentiation of integrals.

Proposition 8.1. Suppose that for each fixed $\mathbf{y} \in G \subset \mathbb{R}^m$, $f(\mathbf{x}, \mathbf{y})$ has first $\mathbf{x}$ partial derivative $Df(\mathbf{x}, \mathbf{y})$ for almost all $\mathbf{x} \in \Omega \subset \mathbb{R}^n$, $\int_G |f(\mathbf{x}, \mathbf{y})| d\mathbf{y}$ is finite for $\mathbf{x} \in \Omega$, and $|Df(\mathbf{x}, \mathbf{y})| \le g(\mathbf{y}) \in L^1(G)$. Then if $F(\mathbf{x}) = \int_G f(\mathbf{x}, \mathbf{y}) d\mathbf{y}$,

$$DF(\mathbf{x}) = \int_G Df(\mathbf{x}, \mathbf{y}) d\mathbf{y}.$$

Stated briefly, the result of a formal calculation must be dominated by an integrable function.

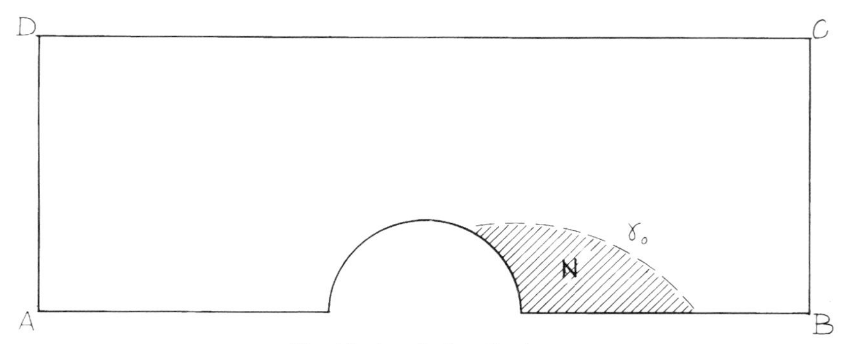

Fig. 15. A typical application.

Lemma 8.1. Let $\mathbf{x}_0 \in \mathbb{R}^n$, $r = |\mathbf{x} - \mathbf{x}_0|$. For any bounded region Ω (whose boundary has measure zero), $\int_\Omega r^{-\beta}\, d\mathbf{x}$ is finite if $\beta < n$ and among all regions that have the same volume the largest value is attained when Ω is a ball centered at $\mathbf{x}_0$ (a disk for $n = 2$).

Proof. Finiteness follows from polar coordinates and Lebesgue's dominated convergence theorem. If $B = B(\mathbf{x}_0, a)$ is a ball with the same volume, then the measures of $\Omega \backslash B$ and $B \backslash \Omega$ are the same by hypothesis,

$$\begin{aligned}\int_B r^{-\beta}\, d\mathbf{x} - \int_\Omega r^{-\beta}\, d\mathbf{x} &= \int_{B\backslash\Omega} r^{-\beta}\, d\mathbf{x} - \int_{\Omega\backslash B} r^{-\beta}\, d\mathbf{x} \\ &\geq a^{-\beta}\left(\int_{B\backslash\Omega} d\mathbf{x} - \int_{\Omega\backslash B} d\mathbf{x}\right) = 0,\end{aligned}$$

and the inequality is strict when $\Omega \backslash B$ has positive measure, so that the integral over B is larger. $\square$

Theorem 8.1. Suppose $f(\mathbf{y})$ is bounded on a bounded domain $\Omega \subset \mathbb{R}^n$, $k(\mathbf{x}, \mathbf{y}) \in C(\mathbb{R}^n \times \bar{\Omega})$, $|k(\mathbf{x}, \mathbf{y})| \leq M$, and $K(\mathbf{x}, \mathbf{y}) = k(\mathbf{x}, \mathbf{y})|\mathbf{x} - \mathbf{y}|^{-\beta}$. Then for any $\beta < n$ the integral

$$F(\mathbf{x}) = \int_\Omega K(\mathbf{x}, \mathbf{y}) f(\mathbf{y}) d\mathbf{y}$$

exists and defines a continuous function in $\mathbb{R}^n$.

Proof. Take an arbitrary point $\mathbf{x}_0 \in \mathbb{R}^n$ and for $\delta > 0$ sufficiently small let $\Omega_\delta = \Omega \cap B(\mathbf{x}_0, \delta) = \{\mathbf{y} \in \Omega : |\mathbf{y} - \mathbf{x}_0| < \delta\}$, $\Omega'_\delta = \Omega \backslash \Omega_\delta$ (Ω_δ may be empty if $\mathbf{x}_0 \notin \bar{\Omega}$.) Then for $\mathbf{x} \in \mathbb{R}^n$ we have

$$\begin{aligned}|F(\mathbf{x}) - F(\mathbf{x}_0)| &\leq \int_\Omega |K(\mathbf{x}, \mathbf{y}) - K(\mathbf{x}_0, \mathbf{y})||f(\mathbf{y})| d\mathbf{y} \\ &= \int_{\Omega'_{2\delta}} |K(\mathbf{x}, \mathbf{y}) - K(\mathbf{x}_0, \mathbf{y})||f(\mathbf{y})| d\mathbf{y} \\ &\quad + \int_{\Omega_{2\delta}} |K(\mathbf{x}, \mathbf{y}) - K(\mathbf{x}_0, \mathbf{y})||f(\mathbf{y})| d\mathbf{y} \\ &\leq \|f\|_{L^\infty} \int_{\Omega'_{2\delta}} |K(\mathbf{x}, \mathbf{y}) - K(\mathbf{x}_0, \mathbf{y})| d\mathbf{y} \\ &\quad + M \|f\|_{L^\infty} (|\mathbf{x}_0 - \mathbf{y}|^{-\beta} + |\mathbf{x} - \mathbf{y}|^{-\beta}) d\mathbf{y},\end{aligned}$$

and, using the lemma,

$$\int_{\Omega_{2\delta}} (|\mathbf{x}_0 - \mathbf{y}|^{-\beta} + |\mathbf{x} - \mathbf{y}|^{-\beta}) d\mathbf{y} \leq 2 \int_{\Omega_{2\delta}} |\mathbf{x}_0 - \mathbf{y}|^{-\beta}\, d\mathbf{y}.$$

Suppose $|\mathbf{x} - \mathbf{x}_0| < \delta$. Then because $\beta < n$, integrating in polar coordinates we see that the integral on $\Omega_{2\delta}$ is $O(\delta^{n-\beta})$ and hence can be made small by suitably restricting δ. Because $|\mathbf{y} - \mathbf{x}| > \delta$ on $\Omega'_{2\delta}$, $|K(\mathbf{x}, \mathbf{y}) - K(\mathbf{x}_0, \mathbf{y})|$ can be made uniformly small over $\Omega'_{2\delta}$ by taking $|\mathbf{x} - \mathbf{x}_0|$ sufficiently small. Thus, for any given $\varepsilon > 0$, we can fix δ so that for $|\mathbf{x} - \mathbf{x}_0| < \delta$, $|F(\mathbf{x}) - F(\mathbf{x}_0)| \leq \varepsilon$. □

Corollary 1.1. Suppose $f(\mathbf{y})$, Ω are as in Theorem 1.1 and $K(\mathbf{x}, \mathbf{y}) = |\mathbf{x} - \mathbf{y}|^{-\beta}$ with $\beta < n - 1$. Then $F(\mathbf{x}) \in C^1(\mathbb{R}^n)$ and

$$DF(\mathbf{x}) = \int_{\Omega} f(\mathbf{y}) D(|\mathbf{x} - \mathbf{y}|^{-\beta}) d\mathbf{y}.$$

9. Appendix to Section 6

Here we present a proof of Theorem 6.1.

First form the regions $R_1, R_2, R_3, \ldots$ as follows. Let $P_0 \in T$. Let C be a cube with P_0 as its center and s.t. $C \subset T$. Form the lattice of cubes whose sides have length a equal to one-third the length of the sides of C and such that the faces of C lie in the planes of the lattice. Now let R_1 be the cube of the lattice containing P_0 together with all other cubes C of the lattice such that:

a. C and all of its 26 adjacent cubes are in T, and
b. C is one of a succession of cubes each having a face in common with the next and the cube containing P_0 is one of the succession.

Now form a new lattice of cubes by adding the parallel planes that bisect the edges of the cubes of the first. Let R_2 be the cubes of the second lattice satisfying (a) and (b). Note that R_1 is entirely interior to R_2. If C is a cube of R_1, it is entirely surrounded by cubes of the first lattice in T, so it is entirely surrounded by cubes of the second lattice in T. Hence, the cubes of the second lattice surrounding C satisfy (a); clearly they also satisfy (b). Moreover, $C \subset R_2^0$, $R_1 \subset R_2^0$.

Similarly construct $R_3, R_4, R_5, \ldots$ in the same way: R_n is made up of cubes of the nth lattice whose sides have length $a/2^{n-1}$. As all R_n's are closed, $\bar{R}_n = R_n \subset R_{n+1}^0 \forall n$.

Now let $P \in T$. Because T is connected, there exists a polygonal line γ joining P and P_0. Let $3d$ denote the minimum distance from γ to ∂T. Now let n

be such that the diagonals of the nth lattice have length less than d. Now find a new curve γ' by redefining γ where necessary such that γ' does not meet an edge of the lattice except at P and possibly at P_0, while keeping the distance from γ to γ' less than d. The distance from γ' to ∂T is greater than $2d$. Thus, any cube containing points of γ' is surrounded by cubes of the lattice that stay within T, so cubes through which γ' passes satisfy (a). Also, because γ' passes from one cube to another through a face, all cubes containing points of γ' satisfy (b). Therefore, $P \in R_n$.

Now let $S_n = \partial R_n$. The function

$$F(P) = \int_{S_n} r^{-2}\, dS$$

is analytic off S_n and $F(P) \to \infty$ as $P \to S_n$. S_{n-1} is bounded away from S_n so $F(P)$ is bounded on S_{n-1} by some $M < \infty$. Let K_1, K_2 be such that $M < K_1 < K_2$.

Let E be the set of points in $\{P : K_1 \leq F(P) \leq K_2\}$ at which grad $F = 0$. E is closed, so E is contained in a finite number of neighborhoods (by the Heine–Borel theorem). By Sard's theorem, the set of values of K for which grad $F = 0$ has measure zero. Hence, there exists K, $K_1 \leq K \leq K_2$, such that the surface $F(P) = K$ is singularity free. The set $\{P : F(P) < K\}$ contains R_{n-1}. Let T_n be the component of this set that contains R_{n-1}. Then $\bar{R}_{n-1} \subset T_n \subset R_n$. It follows that for every K, $\bar{T}_{k-1} \subset T_k$, and the T_n's are nested. By construction of the R_n's, the T_n's approximate T. Because ∂R_{n-1} is a connected set, ∂T_n must be also. (Otherwise a polygonal arc connecting points of ∂R_{n-1} would have to pass through ∂T_n.) The theorem is proven.

It should also be stated the theorem assumes nothing about the topology of ∂T apart from the connectedness of the exterior domain T.

Exercises

1.1. The radial Laplace operator in $\mathbb{R}^n$ is given by $\Delta u(r) = u_{rr} + (n-1)u_r/r$. Show that all solutions of $\Delta u(r) = 0$ are given by $u(r) = c_1 E(r) + c_2$, with $E(r)$ the fundamental solution.

1.2. Show that the force due to a constant volume charge density ρ over a ball B of radius R is given by $\mathbf{F}(\mathbf{x}) = \frac{4}{3}\pi\rho R^3\mathbf{x}/r^3 = Q\mathbf{x}/r^3$ (i.e., by an equivalent charge concentrated at the origin) if $r = |\mathbf{x}| > R$, and by $\mathbf{F} = \frac{4}{3}\pi\rho\mathbf{x}$ if

$r = |\mathbf{x}| \leq R$. The spherical shell $|\mathbf{x}| \leq r \leq R$ exerts no force at $\mathbf{x}$. Hence, the potential is given by

$$u(\mathbf{x}) = \begin{cases} \frac{4}{3}\pi\rho R^3/r, & r = |\mathbf{x}| > R, \\ -\frac{2}{3}\pi\rho r^2 + 2\pi\rho R^2, & r = |\mathbf{x}| \leq R. \end{cases}$$

1.3. The double layer potential cannot be defined if S is a Möbius strip. Why?

1.4. If C is a smooth arc in the plane

$$\frac{\partial \ln|\mathbf{x} - \mathbf{x}_0|}{\partial n} = \frac{\partial\theta}{\partial s}, \qquad \frac{\partial \ln|\mathbf{x} - \mathbf{x}_0|}{\partial s} = -\frac{\partial\theta}{\partial n},$$

where θ is the angle of the vector $\mathbf{x} - \mathbf{x}_0$ with the x-axis ($\ln r$ and θ are "conjugate harmonic functions").

1.5. If C is a circle of radius a,

$$\int_C \ln|\mathbf{x} - \xi|\, ds_\xi = 2\pi a \ln(a) \qquad \text{for } |\mathbf{x}| \leq a.$$

In particular, a constant distribution of charge on the unit circle ($a = 1$) generates a logarithmic single layer potential that vanishes in the circle. (Closed curves with this property are called *gamma-contours*.)

1.6. If C is the interval $[0, a]$ of the x-axis,

$$\int_C \frac{\partial}{\partial n_\xi} \ln\left(\frac{1}{|\mathbf{x} - \xi|}\right) ds_\xi = \theta_a - \theta_0,$$

where θ_0, θ_a are the angles with the x-axis of the segments connecting $\mathbf{x}$ to the endpoints of the interval. Hence and from the fact that θ_0 and θ_a are multivalued, find jump relations across C.

1.7. The function $u = \theta_a$ (harmonic conjugate of the function $\ln|\mathbf{x} - \mathbf{A}|$) defines the potential of a unit hydrodynamic "vortex" located at the point $\mathbf{A}$: $x = a$, $y = 0$. Find the circulation of $\operatorname{grad} u$ around any contour surrounding $\mathbf{A}$. Prove that a logarithmic dipole distribution with constant density ν over a (smooth Jordan) arc C reduces to two opposite vortices at the endpoints of C. (What is the interpretation of this in the context of electromagnetism?)

1.8 (Gauss's law). We recall that in general the surface charge density on a closed surface $S = \partial\Omega$ is defined by $\sigma = (4\pi)^{-1}[\mathbf{E} \cdot \mathbf{n}] = -(4\pi)^{-1}[\partial u/\partial n]$, where $[\,\cdot\,]$ denotes the jump across S, $[\mathbf{E} \cdot \mathbf{n}] = (\mathbf{E} \cdot \mathbf{n})_e - (\mathbf{E} \cdot \mathbf{n})_i$. Suppose that u is regular inside S with the exception of a set of points $\mathbf{x}_i$ and that $u(\mathbf{x}) \sim \rho_i/r_i$ as $r_i = |\mathbf{x} - \mathbf{x}_i| \to 0$. Then

$$\frac{-1}{4\pi}\int_{|\mathbf{x}-\mathbf{x}_i|=\varepsilon} \mathbf{E}\cdot\mathbf{n}\,dS = \frac{-1}{4\pi}\int_{|\mathbf{x}-\mathbf{x}_i|=\varepsilon} \frac{\partial u}{\partial r_i}\,dS \to \rho_i \qquad \text{as } \varepsilon \to 0.$$

By definition, $-(1/4\pi)\Delta u = (1/4\pi)\operatorname{div}\mathbf{E} = \rho(\mathbf{x})$, the volume charge distribution in Ω. The divergence theorem applied to $\Omega\setminus\bigcup_i\{\mathbf{x}_i\}$ then implies

$$\frac{1}{4\pi}\int_S (\mathbf{E}\cdot\mathbf{n})_i\,dS = \frac{1}{4\pi}\int_\Omega \operatorname{div}\mathbf{E}\,d\mathbf{x} + \sum_i \rho_i = \int_\Omega \rho(\mathbf{x})d\mathbf{x} + \sum_i \rho_i.$$

Because, by the definition of σ,

$$\frac{1}{4\pi}\int_S (\mathbf{E}\cdot\mathbf{n})_e\,dS = \frac{1}{4\pi}\int_S (\mathbf{E}\cdot\mathbf{n})_i\,dS + \int_S \sigma\,dS,$$

we find

$$\frac{1}{4\pi}\int_S (\mathbf{E}\cdot\mathbf{n})_e dS \equiv \frac{-1}{4\pi}\int_S (\partial u/\partial n)_e dS = Q,$$

where $Q = \int_\Omega \rho(\mathbf{y})d\mathbf{y} + \sum_i \rho_i + \int_S \sigma\,dS$ is the *total charge* enclosed by the surface S (including the surface charge on S). This is Gauss's law in electrostatics.

1.9. The potential due to a charge q concentrated at a point $\mathbf{x}$ satisfies

$$-\frac{1}{4\pi}\int_\Gamma \frac{\partial u}{\partial n}\,dS = \frac{1}{4\pi}\int_{r=\varepsilon} \frac{\partial u}{\partial r}\,dS_y = q,$$

where $r = |\mathbf{y} - \mathbf{x}|$ and Γ is a surface surrounding q. (We express this by saying that Δu is a "delta function" centered at $\mathbf{x}$.)

2.1. If $\sigma(\mathbf{x}) = \lambda A\mathbf{x} + \mathbf{x}_0$ where $\lambda \in \mathbb{R}$, $\mathbf{x}_0 \in \mathbb{R}^n$, and A is an orthogonal matrix, then, if u is harmonic in Ω, $u(\mathbf{y}) = u(\sigma(\mathbf{x}))$ is harmonic in $\sigma(\Omega)$.

2.2. Give an electrostatic interpretation of Theorem 2.6 ("reciprocity theorem").

2.3. Find the Green's function of the half-space, $x_n > 0$. *Hint*: Take an image charge -1 at the reflected point $\mathbf{x}'$ of $\mathbf{x}$.

2.4. Prove that the maximum principle holds for a regular harmonic function in an exterior unbounded domain Ω^c in $n > 2$ variables. *Hint*: Apply the maximum principle to a truncated domain bounded by $\partial\Omega$ and a large sphere Σ_R and let R approach infinity.

2.5. (Proof of Theorem 2.16). We give the proof for $n = 3$ variables. First we observe that it suffices to consider a ball B centered at $\mathbf{x}_0$ as the potential of charges bounded away from $\mathbf{x}_0$ has continuous partial derivatives of all orders and satisfies Laplace's equation at $\mathbf{x}_0$. Then, because the potential of a constant density $\varrho(\mathbf{x}_0)$ over B is given by $u = c\varrho(\mathbf{x}_0)|\mathbf{x} - \mathbf{x}_0|^2$, where c is a fixed constant (see Exercise 1.2), we need only show that the potential of a Hölder continuous density over a ball that vanishes at the center of the ball has second derivatives that vanish at the center. It suffices to consider Z. Suppose for simplicity that $\mathbf{x}_0 = (0, 0, 0)$ and $\mathrm{B} = \mathrm{B}(\mathbf{0}, r_0)$ so that $|\varrho(\mathbf{x})| \leq A|\mathbf{x}|^\alpha$ on B. Consider

$$J = \int_B \varrho(\mathbf{x}) \frac{\partial^2}{\partial z^2}\left(\frac{1}{r}\right) d\mathbf{x} = \int_B \varrho(\mathbf{x}) \left[\frac{3z^2}{r^5} - \frac{1}{r^3}\right] d\mathbf{x},$$

where $\mathbf{x} = (x, y, z)$ and $r = (x^2 + y^2 + z^2)^{1/2}$. This is the result of formal differentiation of Z at (0, 0, 0). Because $|\varrho(\mathbf{x})| \leq \mathrm{A}r^\alpha$ this integral is finite (see the Appendix to Section 1). Consider

$$\frac{Z(\mathbf{h}) - Z(0)}{h} - J = I,$$

where $\mathbf{h} = (0, 0, h)$, $h \neq 0$. We need only show that I approaches 0 as h goes to zero. If we let $R = [x^2 + y^2 + (z - h)^2]^{1/2}$, then

$$I(h) = \int_B \varrho(\mathbf{x})[h^{-1}(R^{-3}(z - h) - r^{-3}z) - 3z^2 r^{-5} + r^{-3}] d\mathbf{x}.$$

With some algebra we find that the term involving h can be rewritten as

$$-\frac{1}{R^3} + \frac{z(2z - h)}{Rr(R + r)}\left(\frac{1}{R^2} + \frac{1}{Rr} + \frac{1}{r^2}\right)$$

and we see that the integrand in I vanishes when $h = 0$. It suffices then to show that I is continuous at 0. Consider a ball B_0 interior to B centered at $\mathbf{x}_0 = 0$ and containing $(0, 0, h)$. If we can show that the integral I_0 over B_0 can be made small, uniformly with respect to h, by choosing its radius small enough, the result follows. The bracketed part of the integrand can be broken into two parts, namely,

$$\left|-\frac{1}{R^3} + \frac{z(2z - h)}{Rr(R + r)R^2}\right|,$$

$$\left|\frac{z(2z - h)}{Rr(R + r)}\left(\frac{1}{Rr} + \frac{1}{R^2}\right) - \frac{3z^2}{r^5} + \frac{1}{r^3}\right| \leq \frac{1}{R^2 r} + \frac{1}{Rr^2} + \frac{4}{r^3}.$$

Thus, the entire integrand is bounded by $Ar^{\alpha}[3/R^2r + 1/Rr^2 + 4/r^3]$. On B_0, we can consider separately the points where $r \le R$ and $r > R$ and obtain the bounds $8Ar^{\alpha-3}$ and $8AR^{\alpha-3}$, respectively. By Lemma 1.1 we may deduce that I_0 is bounded above by

$$16A\int_{B_0} R^{\alpha-3}d\mathbf{x} = 64\pi A\varepsilon^{\alpha}/\alpha,$$

which goes to zero with the radius ε of B_0 uniformly with respect to h.

2.6. Theorem 2.13 requires the *identity theorem* for real analytic functions: *A real analytic function f on a domain Ω is determined by $D^kf(\mathbf{x})$, for all k, at any fixed $\mathbf{x}_0$ in Ω.* *Hint*: Let $\Omega_1 = \{\mathbf{x} \in \Omega : D^kf(\mathbf{x}) = 0 \text{ for all } k\}$, $\Omega_2 = \{\mathbf{x} \in \Omega : D^kf(\mathbf{x}) \neq 0 \text{ for some } k\}$. Both Ω_1 and Ω_2 are open. Then if $\Omega_1 \neq \emptyset, f \equiv 0$.

2.7. Show by an example that the identity theorem does not apply to C^{∞} functions that are not analytic. *Hint*: $\Omega = (-a, a)$, $f(x) = e^{-1/x^2}$, $x_0 = 0$. Here $\Omega_1 = \{0\}$ is not open.

2.8. Prove that there is at most one solution of the electrostatic charge problem

$$\Delta u = 0 \text{ in } \Omega^c, \quad u = V \text{ on } \partial\Omega, \quad \frac{-1}{4\pi}\int_{\partial\Omega}\frac{\partial u}{\partial n}\,dS = Q, \quad u \text{ regular at infinity}$$

($\Omega^c = \mathbb{R}^3\backslash\bar{\Omega}$), where the total surface charge Q is given and the constant V is unknown. Hence, V is proportional to Q, so that *the capacitance Q/V is constant.* *Hint*: By applying Green's identity (G1) to Ω^c, it follows that $u \equiv 0$ for $Q = 0$. Because the capacitary potential $v := u/V$ is uniquely determined by its boundary data $v = 1$ on $\partial\Omega$, the corresponding total charge $Q' = Q/V$ must be a fixed constant.

2.9. (Wave equation with harmonic initial data). Using the formula of spherical averages and the mean value property of harmonic functions, prove that the solution of the Cauchy problem for the wave equation in $\mathbb{R}^3$

$$u_{tt} = c^2\Delta u, \qquad \mathbf{x} \in \mathbb{R}^3,\ t > 0,$$
$$u(\mathbf{x}, 0) = u_0(\mathbf{x}), \qquad u_t(\mathbf{x}, 0) = u_1(\mathbf{x}), \qquad \mathbf{x} \in \mathbb{R}^3,$$

with initial data u_0, u_1 harmonic in $\mathbb{R}^3$, is given by $u(\mathbf{x}, t) = u_0(\mathbf{x}) + tu_1(\mathbf{x})$.

The result extends to $\mathbb{R}^n$. For $n = 1$ the initial data must be linear.

2.10. (Localized traveling wave solutions of the wave equation). Apply Liouville's theorem to show that solutions $u = u(x, y, z - ct)$ of the wave equation in three space variables

$$u_{xx} + u_{yy} + u_{zz} - \frac{1}{c^2}u_{tt} = 0, \qquad \mathbf{x} = (x, y, z) \in \mathbb{R}^3,\ t \in \mathbb{R}$$

with the properties

$$u(x, y, z - ct) \in C^2(\mathbb{R}^3), \quad u \text{ bounded}, \quad u(x, y, z - ct) \to 0 \ \rho \to \infty (\forall z, t),$$

where $\rho = (x^2 + y^2)^{1/2}$, are identically zero. *Hint*: Defining $\sigma = z + ct$, $\tau = z - ct$ we find, for $u = u(x, y, \tau)$, $u_{xx} + u_{yy} + u_{zz} - c^{-2}u_{tt} = u_{xx} + u_{yy} + 4u_{\sigma\tau} = u_{xx} + u_{yy} = 0$ for $(x, y) \in \mathbb{R}^2$. Hence, u is harmonic and bounded in $\mathbb{R}^2$ and vanishes at infinity. See Exercise 5.14 of Chapter 2.

2.11. Show that if u, $\Delta u \in H^s(\mathbb{R}^n)$, $s \geq 0$, then $u \in H^{s+2}(\mathbb{R}^n)$. *Hint*: Use the Fourier transform in Chapter 8.

3.1. Prove the statements in the text concerning $u = u' + \varepsilon \int_{\mathfrak{G}} ds/r$. *Hint*: $\int_{\mathfrak{G}} ds/r = \ln(1 + x + \sqrt{\rho^2 + (1+x)^2})/(x + \sqrt{x^2 + \rho^2})$ diverges on $\mathfrak{G}$, hence $u = u'$ in Ω.

3.2. if P is the vertex of a right circular cone contained in $\Omega^c = \mathbb{R}^3 \backslash \bar{\Omega}$, P is regular for Ω. Note that the algebraic "spike" is regular (as Example 3.1 provides a barrier for $\rho = |x|^\alpha$), so that this sufficient condition is not necessary. *Hint*: Seek a barrier in the form $w = r^\alpha f(\varphi)$, where φ is the azimuthal spherical coordinate.

In particular a C^1 boundary point is regular.

3.3. Suppose $P \in \partial\Omega$ satisfies a "flat cone" condition, i.e., there is a plane through P and a triangle in this plane with a vertex at P that is contained in $\Omega^c = \mathbb{R}^3 \backslash \bar{\Omega}$. Then P is regular for Ω. *Hint*: We may assume P is the origin and the plane is the (x, y)-plane. Choose $f(\theta)$, τ, and ε so that $w_0 = r^\tau f(\theta)$ is positive and $\Delta_2 w_0$ is negative at points of Ω in the (x, y)-plane for which $r < \varepsilon$. $w_0 + \mu z^2$ is a barrier at P if μ is sufficiently small.

This is a generalization of the cone condition of Exercise 3.2; it shows that a "lower-dimensional contact" suffices.

3.4. In $\mathbb{R}^2$ much weaker conditions suffice for regularity. In particular, if there is a line segment ending at Q all other points of which are in Ω^c, Q is regular for Ω. *Hint*: It suffices to consider a branch of $\sqrt{(z - Q)/(z - P)}$ where $P \in \Omega^c$. After a suitable rotation the imaginary part is a barrier.

3.5. Give the details of the justification for the balayage method.

3.6. Work out the details in extending Perron's method to discontinuous boundary values.

3.7. The function $A(1 - 1/|\mathbf{x}|^{n-2})$ is harmonic inside the punctured unit sphere $0 < |\mathbf{x}| < 1$ in $\mathbb{R}^n$ and vanishes for $|\mathbf{x}| = 1$, for all real A. Similarly for the function $A \ln|\mathbf{x}|$ in the punctured disk $0 < |\mathbf{x}| < 1$ in $\mathbb{R}^2$. For boundary data $u = 0$ for $|\mathbf{x}| = 1$, $u = 1$ for $\mathbf{x} = 0$, the Dirichlet problem has no solution.

3.8. The function $1 - |\mathbf{x}|^2/|\mathbf{x}_0 - \mathbf{x}|^n$ ($|\mathbf{x}_0| = 1$) is harmonic inside the unit sphere $|\mathbf{x}| < 1$ in $\mathbb{R}^n$ and vanishes at all points of the sphere except $\mathbf{x}_0$. *Hint*: See the discussion on Poisson's kernel.

3.9. Refine the argument in Theorem 3.7 to show that $u \in C^{1,\alpha}(\bar{\Omega})$.

3.10. If the additional hypotheses that Ω satisfies an external sphere condition, and that the boundary values extend into Ω as a function in $C^2(\bar{\Omega})$, are made, the argument used to bound $|\text{grad}\, u|$ can be replaced by an application of the boundary gradient estimate theorem. Carry out the details in showing this.

3.11. The "scaled version" of D is given by the substitution $\mathbf{y} = \lambda\mathbf{x}$ in the inequalities $-|\mathbf{X}|^{1+\alpha} < x_3 < 2$, that is, $-\lambda^{-\alpha}|\mathbf{Y}| < y_3 < 2\lambda$. Denote this domain by D_λ. In our proof of Lemma 3.3 we have used a translate of D_λ, called D', to derive the inequality $U_r(\mathbf{Y}) \le U_r(0, 0, k|\mathbf{Y}|^{1+\alpha})$. Show that, for $|\mathbf{Y}| \le \varepsilon$, ε fixed, we can choose a fixed point in D_λ to be translated to $\mathbf{Y}$ in this comparison for an appropriate choice of λ. Determine the corresponding value of k.

3.12. Carry out the details in proving (iii) and (iv) in the estimates for G (Theorem 3.9).

3.13. Think through the modifications required to extend Theorem 3.7 to $n \neq 3$.

4.1. Extend to two space dimensions the results obtained for the layer potentials in Section 4.2.

4.2. Consider the vector field $\mathbf{v}(\mathbf{x}) := \text{grad} \int_S \partial E(\mathbf{x}, \mathbf{y})\, \partial n(\mathbf{y}) dS_y$ where S is a smooth open surface. In hydrodynamics, $\mathbf{v}$ describes a *line vortex* along $\gamma = \partial S$. Show that $\mathbf{v}(\mathbf{x})$ diverges as $\mathbf{x}$ approaches γ. *Hint*: See Ref. 15, p. 509.

4.3. Prove that $\mu + g$ has average value zero over $\partial\Omega$ for the solution μ of $(K' - \frac{1}{2}I)\mu = g$. *Hint*: Integrate the equation over $\partial\Omega$ and apply Fubini's theorem and the Gauss solid angle formula.

4.4. Derive (11) from Green's identities for smooth g and explain why this condition is not needed for the exterior Neumann problem ($n > 2$).

4.5. Prove that if the boundary has a conical point $\mathbf{x}_0$, $\bar{\omega}(\mathbf{x}_0)$ in (IR) or (ER) is equal to the solid angle at the conical vertex.

4.6. Prove that (17), (17a) hold. *Hint*: E, grad $E \in L^1_{\text{loc}}(\mathbb{R}^3)$.

4.7. Show that Poincaré's identity holds in Ω^c under the stated asymptotic assumptions on $\mathbf{v}$.

4.8. Prove that Poincaré's identity applied to $\mathbf{v} = \text{grad}\, u$ is equivalent to Green's representation theorem for u. *Hint*: From Stokes' theorems

$$\int_{\partial\Omega} \mathbf{n} \wedge \text{grad}\, u\, dS_y = 0, \qquad \int_{\partial\Omega} \mathbf{n}(\mathbf{y}) \cdot \text{curl}\, \mathbf{v}\, dS_y = 0, \tag{ST}$$

derive the formula

$$\text{curl}_x \int_{\partial\Omega} E(\mathbf{x}, \mathbf{y})\mathbf{n}(\mathbf{y}) \wedge \text{grad}\, u(\mathbf{y}) dS_y = \text{grad}_x \int_{\partial\Omega} u(\mathbf{y}) \frac{\partial E(\mathbf{x}, \mathbf{y})}{\partial \mathbf{n}(\mathbf{y})}\, dS_y. \tag{EQ}$$

This identity expresses the *equivalence of double layers and vortex layers* distributed on a closed surface $S = \partial\Omega$ in aerodynamics (Ref. 15), or of permanent magnets and current loops in electromagnetism.

4.9. Prove the relation

$$\int_S \mathbf{x} \wedge \mathbf{v}\, dS = \int_S \mathbf{x}\mathbf{n} \cdot \operatorname{curl} \mathbf{v}\, dS$$

for an arbitrary smooth closed compact surface S in $\mathbb{R}^3$ (see Ref. 16, p. 668).

4.10. Show that if div $\mathbf{v}$, curl $\mathbf{v} = O(|\mathbf{x}|^{-4})$ as $|\mathbf{x}| \to \infty$, the asymptotic behavior in $\mathbb{R}^3$

$$\mathbf{v}(\mathbf{x}) = \frac{|\mathbf{x}|^{-3}}{4\pi}\left\{\mathbf{x}\left[\left[\int_{\partial\Omega} \mathbf{n} \cdot \mathbf{v}\, dS + \int_{\Omega^c} \operatorname{div} \mathbf{v}\, d\mathbf{y}\right]\right]\right\} + O(|\mathbf{x}|^{-3})$$

follows from (P3) applied to an exterior domain Ω^c and from Exercise 4.9.

4.11. Show that if div $\mathbf{v}, \omega = O(|\mathbf{x}|^{-3})$, (P2) applied to an exterior domain Ω^c yields

$$\mathbf{v}(\mathbf{x}) = \frac{|\mathbf{x}|^{-2}}{2\pi}\left\{\mathbf{x}\left[\left[\int_{\partial\Omega} \mathbf{n} \cdot \mathbf{v}\, dS + \int_{\Omega^c} \operatorname{div} \mathbf{v}\, d\mathbf{y}\right] - \mathbf{x} \wedge \mathbf{k}\left[\left[\int_{\partial\Omega} \mathbf{t} \cdot \mathbf{v}\, dS\right.\right.\right.\right.$$
$$\left.\left.\left.\left. + \int_{\Omega^c} \omega\, d\mathbf{y}\right]\right]\right\} + O(|\mathbf{x}|^{-2}), \qquad |\mathbf{x}| \to \infty.$$

4.12. Show that the operator $(\partial/\partial s)V$ is a Cauchy singular operator on $\partial\Omega$, and that $(\partial/\partial s)V\mu_0 = 0$, where μ_0 is a Robin density for $\partial\Omega$.

4.13. Discuss the first boundary integral equation (13) for a harmonic vector field in an exterior two-dimensional domain Ω^c with normal trace $\mathbf{n} \cdot \mathbf{v}$ assigned on the boundary:

$$\left(K' + \frac{1}{2}I\right)\mathbf{t} \cdot \mathbf{v} = g := -\frac{\partial}{\partial s} V\mathbf{n} \cdot \mathbf{v}.$$

Hint: This integral equation coincides with that obtained for the *interior* Neumann problem from the layer Ansatz. As the orthogonality condition $\int_{\partial\Omega} g ds = 0$ is automatically satisfied, the tangential trace $\mathbf{t} \cdot \mathbf{v}$ is determined up to a term $\kappa\mu_0/\int_{\partial\Omega} \mu_0$, where κ is the circulation of $\mathbf{v}$ around $\partial\Omega$ (compare this with the results at the end of Section 4.3).

4.14. Discuss the first boundary integral equation (13) for a harmonic vector field in an interior two-dimensional domain Ω with normal trace $\mathbf{n} \cdot \mathbf{v}$ assigned on the boundary:

$$\left(K' - \frac{1}{2}I\right)\mathbf{t} \cdot \mathbf{v} = g := \frac{\partial}{\partial s} V\mathbf{n} \cdot \mathbf{v}.$$

Hint: This integral equation coincides with that obtained for the *exterior* Neumann problem from the layer Ansatz. Hence, there exists a unique solution $\mathbf{t} \cdot \mathbf{v}$ for every $\mathbf{n} \cdot \mathbf{v}$, even if the compatibility condition $\int_{\partial\Omega} \mathbf{n} \cdot \mathbf{v}\, ds = 0$ is not satisfied. However, by decomposing $\mathbf{n} \cdot \mathbf{v}$ in the form

$$\mathbf{n} \cdot \mathbf{v} = v_0 + \frac{\int_{\partial\Omega} \mathbf{n} \cdot \mathbf{v}\, ds}{\int_{\partial\Omega} \mu_0\, ds} \mu_0,$$

we see from Exercise 4.12 that the solution depends only on the part v_0 with zero average on the boundary, $\int_{\partial\Omega} v_0\, ds = 0$.

4.15. Construct a harmonic vector field in the exterior of a circle in $\mathbb{R}^2$ of radius R satisfying $\mathbf{n} \cdot \mathbf{v} = 0$ on the circumference. *Hint*: $\mathbf{t} \cdot \mathbf{v} = \kappa/R$, $\mathbf{v} = \kappa \operatorname{grad} \theta$, $\theta = \tan^{-1}(x_2/x_1)$ the angular coordinate, κ a constant. Note that $u = \theta$ is multivalued, while $\mathbf{v}$ is single valued.

4.16. Construct a harmonic vector field in the exterior of a torus in $\mathbb{R}^3$ satisfying $\mathbf{n} \cdot \mathbf{v} = 0$ on the torus (a torus is a surface of topological genus $p = 1$, see Ref. 17).

4.17. Show that if $\partial\Omega$ has genus $p = 0$ (a spheroidal surface in $\mathbb{R}^3$), there exist no nonzero harmonic vector fields $\mathbf{v}$ with $\mathbf{n} \cdot \mathbf{v} = 0$.

4.18. Construct the Green function $G(\mathbf{x}, \mathbf{y}) = E(\mathbf{x}, \mathbf{y}) + h(\mathbf{y})$ for the Neumann problem. *Hint*: Use the Gauss solid angle formula (9) to show that the appropriate boundary condition is $\partial G(\mathbf{x}, \mathbf{y})/\partial n = -1/|\partial\Omega|$.

5.1. The function $u = a$ is the variational solution of the Dirichlet problem for the ball $|\mathbf{x}| < 1$ in $\mathbb{R}^3$ with $u = a$ for $|\mathbf{x}| = 1$, as well as for the punctured ball $0 < |\mathbf{x}| < 1$ with the additional boundary condition $u = b$ for $\mathbf{x} = 0$ (see Exercise 3.7). Thus, the boundary condition at $\mathbf{x} = 0$ has no influence on the variational solution. *Hint*: Changing the value at one point does not change an $H^1(\Omega)$ function.

5.2. Prove that if $\mathbb{Q}(u) = \inf\{\mathbb{Q}(v) : v \in H_0^1(\Omega)\}$, then u satisfies (VP). *Hint*: The function $\mathbb{Q}(u + tv)$ has a minimum at $t = 0$ for fixed $v \in H_0^1(\Omega)$. The equation $\mathbb{Q}'(u + tv)|_{t=0} = 0$ coincides with (VP).

5.3. Prove Theorem 5.3 using the Lax–Milgram theorem (Theorem 1.8 of Chapter 8).

5.4. Prove (23). *Hint*: $[u_n, v] \to [z_1, v]$ yields $[u_n, z_1] \to \|z_1\|^2$ and $|[u_{n+p}, z_1]| \leq (\mu_1 + p^{-1})\|z_1\|$ implies the assertion.

5.5. If $u_n \to z_1$ weakly and $\|u_n\| \to \|z_1\|$, then $\|u_n - z_1\|^2 = [u_n - z_1, u_n - z_1] = \|u_n\|^2 - 2(u_n, z_1) + \|z_1\|^2 \to 0$, hence $u_n \to z_1$ strongly.

5.6. Show, using Theorems 5.8 and 5.9, that there are no eigenvalues other than those in the sequence $\{\lambda_n\}$.

5.7. We may think of $\{\lambda_n\}$ as reciprocals of the eigenvalues of the Green operator G mapping L^2 into itself. Show that G is a compact operator. Using Theorems 5.6 and 5.7 show that each λ_n^{-1} has finite multiplicity as an eigenvalue of G.

5.8. Show that the first eigenvalue λ_1 is simple (i.e., the eigenspace of λ_1^{-1} as an eigenvalue of G is one dimensional) and that the eigenfunction z_1 has one sign in Ω. *Hint*: Suppose z_1 changes sign in Ω. Let $w_1 := \max\{z_1, 0\}$, $w_2 := \min\{z_1, 0\}$. Then $w_1, w_2 \in H_0^1(\Omega)$, and if we choose c_1, c_2 so that $\|c_1 w_1 + c_2 w_2\|_{L^2}^2 = 1$, a direct calculation shows that $\mathbb{D}(c_1 w_1 + c_2 w_2) = \lambda_1$. In particular, if $\int_\Omega w_1^2\, dx > 0$, we can take $c_2 = 0$. Show, if $\int_\Omega w_2^2\, dx > 0$ also, that $\inf \mathbb{D}(w)$ over $w \in H_0^1(\Omega_1)$, $\int_{\Omega_1} w^2\, dx = 1$, where $\Omega_1 = \{\mathbf{x} \in \Omega : z_1(\mathbf{x}) > 0\}$, is strictly larger than λ_1. In other words, $\lambda_1(\Omega_1) > \lambda_1(\Omega)$.

6.1. Carry out the details in the derivation of the capacity of two touching spheres, in particular the computation of $\mathcal{Q}$.

6.2. Prove that $c(A)$ in (28) does not depend on S.

6.3. If A is a plane region bounded by smooth arcs, prove that $c(A) \geq c|A|^{1/2}$, $c > 0$ a constant, $|A|$ the area of A. *Hint*: Compare the conductor potential of A with the single layer potential with unit density over A.

7.1. (*The multiplier method of Jacobi*). By considering Dirichlet integrals of functions $\eta u = \varphi$ where u is an eigenfunction positive in A, $\Delta u + \lambda u = 0$, show that

$$\int_{\mathrm{A}} |\mathrm{grad}\ \varphi|^2 \geq \lambda \int_{\mathrm{A}} \varphi^2$$

with equality sign only if $\eta = \text{const}$. Deduce that λ is the lowest eigenvalue.

7.2. Show by direct calculations (using special functions) that there is a positive radially symmetric eigenfunction.

7.3. Prove (SF). *Hint*: Because $\operatorname{div} \mathbf{v} = 0$, $\psi(P) = \int_A^P u\, dy - v\, dx$ is well defined as a (possibly many-valued) function of position, P. Compare this with the Poincaré identity (P2) in Section 4.5.

7.4. Show that γ_1 is a C^1 curve.

7.5. The theory given in Section 7.2 does not apply to the configuration in Fig. 15 because of the corners on the boundary. (a) If two smooth arcs meet at z_0 in an angle, use a fractional power $\zeta = (z - z_0)^\alpha$ to "straighten" the corner and give compatibility conditions on boundary data for a harmonic function that imply that the gradient is continuous up to z_0. (b) Extend the conclusion of (a) to a solution of $\Delta u = \omega = \text{const}$ at a corner. (A smooth solution of $\Delta u = \omega$ can be subtracted to yield a harmonic Dirichlet problem.)

References

1. HELLWIG, G., *Partial Differential Equations*, Teubner, Stuttgart, Germany, 1977.
2. KELLOGG, O. D., *Foundation of Potential Theory*, Springer, Berlin, Germany, 1929.
3. FOLLAND, G. B., *Introduction to Partial Differential Equations*, Princeton University Press, Princeton, New Jersey, 1976.
4. COLTON, D., and KRESS, R., *Integral Equations Methods in Scattering Theory*, John Wiley, New York, New York, 1983.
5. KRESS, R., *Linear Integral Equations*, Springer-Verlag, Berlin, Germany, 1989.
6. MIKHLIN, S. G., *Mathematical Physics, an Advanced Course*, North-Holland, Amsterdam, Netherlands, 1970.
7. BURAGO, Y. D., MAZ'YA, V. G., and SAPOZHNIKOVA, V. D., *On the Theory of Simple and Double Layer Potential for Domains with Irregular Boundaries*, Boundary Value Problems and Integral Equations, Edited by V. I. Smirnov, Consultants Bureau, New York, New York, 1968.
8. GUNTER, N. M., *Die Potentialtheorie und ihre Anwendung auf Grundaufgaben der Mathematischen Physik*, Teubner, Stuttgart, Germany, 1957.
9. SMIRNOV, V. I., *A Course in Higher Mathematics*, Vol. IV, Pergamon Press, Oxford, England, 1964.
10. JASWON, M. A., and SYMM, G. T., *Integral Equation Methods in Potential Theory and Elastostatics*, Academic Press, New York, New York, 1977.
11. SERRIN, J., *Mathematical Principles of Classical Fluid Mechanics*, Handbuch der Physik IX, Springer, Berlin, Germany, 1959.
12. BASSANINI, P., CASCIOLA, C., LANCIA, M. R., and PIVA, R., *A Boundary Integral Formulation for the Kinetic Field in Aerodynamics. Part I*, European Journal of Mechanics B/Fluids, Vol. 10, pp. 605–627, 1991.
13. BASSANINI, P., CASCIOLA, C., LANCIA, M. R., and PIVA, R., *A Boundary Integral Formulation for the Kinetic Field in Aerodynamics. Part II*, European Journal of Mechanics B/Fluids, Vol. 11, pp. 69–92, 1992.
14. WEYL, H., *The Method of Orthogonal Projections in Potential Theory*, Duke Mathematics Journal, Vol. 7, pp. 411–444, 1940.
15. BATCHELOR, G. K., *Fluid Dynamics*, Cambridge University Press, Cambridge, England, 1991.
16. MILNE-TOMSON, L. M., *Theoretical Hydrodynamics*, Macmillan & Co., London, England, 1968.
17. MARTENSEN, E., *Potentialtheorie*, Teubner, Stuttgart, Germany, 1968.

5

Elliptic Partial Differential Equations of Second Order

We study in this chapter a class of partial differential equations that generalize and are to a large extent represented by Laplace's equation. These are the elliptic partial differential equations of second order. A linear partial differential operator L defined by

$$Lu := a_{ij}(\mathbf{x})D_{ij}u + b_i(\mathbf{x})D_iu + c(\mathbf{x})u$$

is elliptic on $\Omega \subset \mathbb{R}^n$ if the symmetric matrix $[a_{ij}]$ is positive definite for each $\mathbf{x} \in \Omega$. We have used the notation D_iu, $D_{ij}u$ for partial derivatives with respect to x_i and x_i, x_j and the summation convention on repeated indices is used. A nonlinear operator Q,

$$Q(u) := a_{ij}(\mathbf{x}, u, \mathbf{D}u)D_{ij}u + b(\mathbf{x}, u, \mathbf{D}u)$$

[$\mathbf{D}u = (D_1u, \ldots, D_nu)$], is elliptic on a subset of $\mathbb{R}^n \times \mathbb{R} \times \mathbb{R}^n$] if $[a_{ij}(\mathbf{x}, u, \mathbf{p})]$ is positive definite for all $(\mathbf{x}, u, \mathbf{p})$ in this set. Operators of this form are called quasilinear. In all of our examples the domain of the coefficients of the operator Q will be $\Omega \times \mathbb{R} \times \mathbb{R}^n$ for Ω a domain in $\mathbb{R}^n$. The function u will be in $C^2(\Omega)$ unless explicitly stated otherwise.

We will begin by establishing maximum principles for L and associated comparison principles for Q. These are an important part of the general theory for elliptic equations. We also consider equations in two variables that arise in geometric problems, in particular the minimal surface equation and the equation of prescribed mean curvature and capillary surfaces. The rest of the chapter is concerned with equations with discontinuous coefficients and nonlinear Dirichlet problems.

1. Maximum Principle

We first observe that if L is elliptic, there is a linear transformation of the coordinates in $\mathbb{R}^n$ such that the principal part $L_0u = a_{ij}D_{ij}u$ becomes the Laplacian at a fixed point of Ω. In fact, if $\mathbb{C}$ is an orthogonal matrix diagonalizing $\mathbb{A} = [a_{ij}]$ at $\mathbf{x} \in \Omega$, $\mathbb{C}\mathbb{A}\mathbb{C}^{\mathrm{T}} = \mathbb{D}$, $\mathbb{D} = \mathrm{diag}(\lambda_1, \ldots, \lambda_n)$, then the orthogonal transformation $\mathbf{y} = \mathbb{C}\mathbf{x}$ transforms L_0 into $\lambda_i u_{y_i y_i}$, as easily follows using the chain rule. The eigenvalues λ_i of $\mathbb{A}$ are positive so that a scaling transformation $z_i = y_i/\sqrt{\lambda_i}$ accomplishes the desired transformation.

We introduce the further notation $L_1 u = L_0 u + b_i D_i u$. Then an elementary observation shows that if $L_1 u > 0$ in Ω, u cannot attain a relative maximum at an interior point of Ω. In fact, at such a point, in the transformed coordinates u_{z_i} and $u_{z_i z_i} \leq 0$ contradict $L_1 u > 0$.

We will write $\lambda = \lambda(\mathbf{x})$ for the minimum of the eigenvalues of $\mathbb{A}$ so that $a_{ij}\xi_i\xi_i > \lambda|\xi|^2$ for any $\xi \in \mathbb{R}^n$. We will assume that there is a positive constant b_0 such that $|b_i|/\lambda < b_0$. We can then prove the following weak maximum principle for L_1.

Theorem 1.1. Suppose that $u \in C^2(\Omega) \cap C(\bar{\Omega})$, u is bounded, and $L_1 u > 0$ on Ω. Then $\sup_{\Omega} u = \sup_{\partial\Omega} u$.

Proof. Observe that

$$L_1 e^{\gamma x_1} = (\gamma^2 a_{11} + \gamma b_1) e^{\gamma x_1} \geq \lambda(\gamma^2 - \gamma b_0) e^{\gamma x_1}$$

and the last function is positive in Ω if $\gamma > b_0$. For any $\varepsilon > 0$, $L_1(u + \varepsilon e^{\gamma x_1}) > 0$ in Ω so $u + \varepsilon e^{\gamma x_1}$ cannot attain its supremum over $\bar{\Omega}$ at an interior point. Letting $\varepsilon \to 0$ we obtain the result □

The analogous statement holds with $L_1 u \leq 0$ and $\inf_{\Omega} u = \inf_{\partial\Omega} u$. In the most common applications Ω is bounded, so the hypothesis that u is bounded is redundant.

We say that L is *uniformly elliptic* in Ω if the ratio of the largest to the smallest eigenvalue of $\mathbb{A}$ is bounded in Ω.

Theorem 1.2. Suppose that L_1 is uniformly elliptic on Ω, that $L_1 u \geq 0$, and that $\mathbf{x}_0 \in \partial\Omega$ is such that Ω satisfies the interior sphere condition at $\mathbf{x}_0$, u is continuous at $\mathbf{x}_0$, and $u(\mathbf{x}) < u(\mathbf{x}_0)$ for $\mathbf{x} \in \Omega$. Then, if the normal derivative at $\mathbf{x}_0$ exists, $\partial u(\mathbf{x}_0)/\partial n > 0$.

Proof. Choose a ball $B = B(\mathbf{y}, B) \subset \Omega$ with $\mathbf{x}_0 \in \partial B$. Let $0 < \rho < R$ and consider the function $V(\mathbf{x}) := e^{-\alpha r^2} - e^{-\alpha R^2}$ for $r = |\mathbf{x} - \mathbf{y}|$, and $\rho < r < R$. We have

$$\begin{aligned} L_1 V &= e^{-\alpha r^2}(4\alpha^2 a_{ij}(x_i - y_i)(x_j - y_j) - 2\alpha(a_{ii} + b_i(x_i - y_i))) \\ &\geq e^{-\alpha r^2}(4\alpha^2 \lambda r^2 - 2\alpha(a_{ii} + r\lambda b_0)) \end{aligned}$$

so $\alpha > 0$ may be chosen so that $L_1 V \geq 0$ for $\rho < r < R$. (Uniform ellipticity is invoked in order to deal with the spur a_{ii}.) As $u - u(\mathbf{x}_0) < 0$ on $r = \rho$, we can choose $\varepsilon > 0$ sufficiently small that $u - u(\mathbf{x}_0) + \varepsilon V < 0$ for $r = \rho$. As $u - u(\mathbf{x}_0) + \varepsilon V = u - u(\mathbf{x}_0) < 0$ for $r = R$, Theorem 1.1 implies that $u - u(\mathbf{x}_0) + \varepsilon V \leq 0$ for $\rho < r < R$. Then

$$\frac{\partial u}{\partial n}(\mathbf{x}_0) \geq -\varepsilon \frac{\partial V}{\partial n}(\mathbf{x}_0) = -\varepsilon V'(R) > 0. \qquad \square$$

The interior sphere condition cannot be relaxed as the harmonic function $-x_1 x_2$ in the first quadrant of the (x_1, x_2)-plane shows. If $\partial u/\partial n$ is not known to exist, the same argument shows that

$$\varliminf_{\mathbf{x} \to \mathbf{x}_0} \frac{u(\mathbf{x}_0) - u(\mathbf{x})}{|\mathbf{x}_0 - \mathbf{x}|} > 0.$$

Corollary 1.1 (Strong maximum principle). Suppose that L_1 is uniformly elliptic, $L_1 u \geq 0$, and u takes on its maximum at an interior point. Then u is constant.

Proof. Suppose that u is not constant and takes on its maximum M at an interior point of Ω. Consider the set $\Omega_0 := \{u < M\} \subset \Omega$. Our assumptions imply that $\Omega_0 \neq \emptyset$ and $\partial\Omega_0 \cap \Omega \neq \emptyset$. Choose $\mathbf{y} \in \Omega_0$ such that $\text{dist}(\mathbf{y}, \partial\Omega_0) < \text{dist}(\mathbf{y}, \partial\Omega)$. There is a largest ball centered at $\mathbf{y}$ that is contained in Ω_0. Theorem 1.2 implies that $\mathbf{D}u \neq 0$ at a point on the boundary of this ball where $u = M$, and this contradicts the fact that $\mathbf{D}u = 0$ at an interior maximum. $\square$

If $c \leq 0$, we can prove a similar statement for L.

Theorem 1.3. Suppose that, in addition to the hypotheses of Theorem 1.2, $c \leq 0, c/\lambda$ is bounded, and $u(\mathbf{x}_0) \geq 0$. Then , if $Lu \geq 0$, the conclusion of Theorem 1.2 holds. Also under these hypotheses a nonnegative maximum cannot be taken on at an interior point unless u is constant.

The proof is left as an exercise. The condition $c \leq 0$ cannot be relaxed as the existence of eigenfunctions for positive eigenvalues μ of $\Delta u + \mu u = 0$ in Ω, $u = 0$ on $\partial\Omega$ shows.

We should remark here that no hypothesis has been made about the regularity of the coefficients of L. We now state a comparison theorem for quasilinear operators.

Theorem 1.4. Suppose that $u, v \in C^2(\Omega) \cap C(\bar{\Omega})$, $Q(u) \geq Q(v)$ in Ω, and $u \leq v$ on $\partial\Omega$. Further, we assume that a_{ij} does not depend on u, $b(\mathbf{x}, u, \mathbf{p})$ is nonincreasing as a function of u for each fixed $\mathbf{x}$ and $\mathbf{p}$, and a_{ij}, b are continuously differentiable with respect to $\mathbf{p}$ on $\Omega \times \mathbb{R} \times \mathbb{R}^n$. Then $u \leq v$ in Ω. If $u = v$ at some point in Ω, then $u \equiv v$.

Proof. We have, setting $w = v - v$,

$$\begin{aligned} 0 \leq Q(u) - Q(v) &= a_{ij}(\mathbf{x}, \mathbf{D}u)D_{ij}w + [a_{ij}(\mathbf{x}, \mathbf{D}u) - a_{ij}(\mathbf{x}, \mathbf{D}v)]D_{ij}v \\ &\quad + b(\mathbf{x}, u, \mathbf{D}u) - b(\mathbf{x}, u, \mathbf{D}v) + b(\mathbf{x}, u, \mathbf{D}v) - b(\mathbf{x}, v, \mathbf{D}v) \\ &= a_{ij}(\mathbf{x}, \mathbf{D}u(\mathbf{x}))D_{ij}w + [a_{ij/k}(\mathbf{x}, \bar{\mathbf{p}})D_{ij}v + b_{/k}(\mathbf{x}, u, \bar{\mathbf{p}})]D_k w \\ &\quad + b(\mathbf{x}, u, \mathbf{D}v) - b(\mathbf{x}, v, \mathbf{D}v), \end{aligned}$$

where $\bar{\mathbf{p}}$ denotes some intermediate value (varying from function to function) obtained by applying the mean value theorem, and $a_{ij/k}$, $b_{/k}$ denote the partial derivatives with respect to p_k. We can write this inequality in the form

$$L_1 w \equiv \tilde{a}_{ij}(\mathbf{x})D_{ij}w + \tilde{b}_i(\mathbf{x})D_i w \geq b(\mathbf{x}, v, \mathbf{D}v) - b(\mathbf{x}, u, \mathbf{D}v),$$

where $\tilde{a}_{ij}(\mathbf{x}) := a_{ij}(\mathbf{x}, \mathbf{D}u(\mathbf{x}))$ and $\tilde{b}_i(\mathbf{x}) := a_{jk/i}(\mathbf{x}, \bar{\mathbf{p}})D_{jk}v + b_{/i}(\mathbf{x}, u, \bar{\mathbf{p}})$. If $\Omega_0 := \{w > 0\}$, then, because b is nonincreasing in u, $L_1 w \geq 0$ on Ω_0. Because $w \leq 0$ on $\partial\Omega$, the weak maximum principle implies that $\Omega_0 = \emptyset$. The last conclusion follows from the strong maximum principle. □

The hypothesis that a_{ij} do not depend on u cannot be relaxed. To see this we will consider an equation of the form

$$\left(\delta_{ij} + g(r, u)\frac{x_i}{r}\frac{x_j}{r}\right)D_{ij}u = 0.$$

The eigenvalues of a_{ij} here are 1 with multiplicity $(n - 1)$ and $1 + g$ with multiplicity 1. (This is easily seen by considering the eigenvector equations.) If $a \leq 1 + g \leq b$ for positive constants a and b, the equation is uniformly elliptic. We will consider solutions that are functions only of $r = |\mathbf{x}|$ on the annular region $1 < |\mathbf{x}| < 2$. For $u = u(r)$ the equation becomes

$$u'' + \frac{n-1}{r(1+g)}u' = 0.$$

If we write $f(r, u) = (n - 1)/r(1 + g)$, then f is bounded above and below by positive constants for $1 < r < 2$ if and only if $1 + g$ is. We will find a function f

such that $u'' + f(r, u)u' = 0$ has two solutions w and v on $1 \le r \le 2$ that agree at the endpoints and satisfy $w > v$ for $1 < r < 2$. The function f and the solutions w and v will be analytic. First, we find a polynomial V such that $V(1) = 0$, $V(2) = 1$, and $V' > 0$, $V'' < 0$ on $[1;2]$. [A quadratic $V(r) = a(r-1)^2 + (1-a)(r-1)$ with $-1 < a < 0$ will do.] Then we find another polynomial Z with $Z(1) = Z(2) = 0$, $Z > 0$ on $[1,2]$ and $Z'(1) = V'(1)$, $Z'(2) = -V'(2)$, $Z''(1) = V''(1)$, $Z''(2) = -V''(2)$. [If V is the above quadratic, one choice is $Z = (r-1)(2-r)h(r)$ with h the cubic determined by the boundary conditions, a being further restricted to $a \in (-1, -1/3]$.] Then we set $W = V + tZ$. For t positive and sufficiently small, $W' > 0$ and $W'' < 0$ on $[1,2]$. [The smaller of $\inf(V'/\sup -Z')$ and $\sup V''/\inf(-Z'')$ will suffice.] For $1 \le r \le 2$ and $V(r) \le u \le W(r)$ we define

$$f(r, u) = \frac{u - V}{W - V}\left(\frac{V''}{V'} - \frac{W''}{W'}\right) - \frac{V''}{V'}.$$

Observe that $V''/V' - W''/W' = 0$ at the endpoints, and, as $W' \ne V'$ at the endpoints, f is analytic on its domain. We leave as an exercise to show that the definition of f can be extended to all values of u in such a way that f is infinitely differentiable and Q is uniformly elliptic.

We now present a comparison theorem for elliptic operators that is useful in geometric problems. It requires an inequality on only a part of the boundary.

We consider operators that do not explicitly depend on u,

$$Q_1 u = a_{ij}(\mathbf{x}, \mathbf{D}u)D_{ij}u + b(\mathbf{x}, \mathbf{D}u).$$

If N is a neighborhood of a boundary point of Ω, $N \cap \partial\Omega$ is called a boundary neighborhood of this point. For $\Gamma \subset \partial\Omega$ a point is an interior point if there is a boundary neighborhood of it contained in Γ. Suppose that $\Gamma \subset \partial\Omega$ is such that each interior point is the endpoint of a line segment inside Ω. We denote the derivative along such a line segment by D_s. We will use a variant of the result given in Theorem 1.4 as a lemma.

Lemma 1.1. Suppose that $\omega \in C^2(\Omega)$ satisfies $Q_1\omega \le 0$. Then for any $u \in C^2(\Omega)$ such that $Q_1 u \ge 0$ and $\limsup(u - w) \le 0$ approaching any point of $\partial\Omega$, $u - \omega \le 0$ in Ω.

Proof. Setting $v = u - \omega$, we suppose that $k = \sup_\Omega v > 0$. There must be an interior point where this supremum is taken on and which has a neighborhood N in which $0 < v \le k$, v not identically equal to k. Subtracting $Q_1\omega$ from $Q_1 u$ and applying the reasoning used in the proof of Theorem 1.4 we find that

$L_1 v \geq 0$ for a linear elliptic operator satisfying the hypotheses of Corollary 1.1 on N, so that $v \equiv k$ on N. This contradiction proves the lemma. □

The lemma requires only $Q_1\omega \leq Q_1 u$ of course. We have stated it in this form because, in our applications, ω will be given in advance and results will be derived for a variety of functions u. Our principal result requires further hypotheses on ω.

Theorem 1.5. Suppose that $\Gamma \subset \partial\Omega$ satisfies the segment condition described above. Suppose $\omega \in C^2(\Omega) \cap C(\bar{\Omega})$ satisfies $Q_1\omega \leq 0$, and $D_s\omega$ approaches ∞ as interior points of Γ are approached from inside Ω. If $u \in C^2(\Omega) \cap C^1(\bar{\Omega})$ satisfies $Q_1 u \geq 0$, and $u - \omega \leq 0$ on a set $\Gamma_1 \supset \partial\Omega \setminus \Gamma$, then $u \leq \omega$ in Ω and lim sup$(u - \omega) < 0$ for approach to an interior point of Γ along the interior segment.

Proof. First we observe that lim sup$(u - \omega) \leq 0$approaching an arbitrary point of Γ implies that $u - \omega \leq 0$ on Ω by the lemma. Suppose then that $u - \omega$ takes on positive values on Γ. If the maximum over $\bar{\Omega}$ is not taken on at a point of Γ, subtraction of the constant $\sup_\Gamma(u - \omega)$ from $u - \omega$ contradicts the lemma. (The operator Q_1 is invariant under addition of a constant.) But $D_s(u - \omega) \to -\infty$ approaching points of Γ implies that it is impossible for $u - \omega$ to take on a maximum at a point of Γ. □

We also need the following extension of Lemma 1.1 to the more general operators Q.

Lemma 1.2. Suppose Q and ω have the property that $Q(\omega + b) \leq 0$ for all positive constants b. Then, if $Qu \geq 0$ and lim sup$(u - \omega) \leq 0$ for approach to a boundary point of Ω, $u \leq \omega$ in Ω.

Proof. We give only the part of the proof that differs from that given above. Because

$$a_{ij}(\mathbf{x}, \omega + b, \mathbf{D}\omega)D_{ij}\omega + b(\mathbf{x}, \omega + b, \mathbf{D}\omega) \leq 0$$

for all $b > 0$, this also holds if $b = b(\mathbf{x})$. In particular, we can set $b = v = u - \omega$ to get

$$a_{ij}(\mathbf{x}, u, \mathbf{D}\omega)D_{ij}\omega + b(\mathbf{x}, u, \mathbf{D}\omega) \leq 0.$$

Proceeding as before, we get a contradiction to the strong maximum principle. □

This result applies in particular to Q for which a_{ij} is independent of u and b is nonincreasing with respect to u. This extended lemma enables us to prove the following simple, but useful result.

Theorem 1.6. Suppose that $\omega \in C^2(\Omega) \cap C(\bar{\Omega})$, $Q(\omega + b) \leq 0$ for all $b > 0$, and $\partial\omega/\partial n = \infty$ on $\partial\Omega$. Then, if $u \in C^2(\Omega) \cap C(\bar{\Omega})$ and $Qu \geq 0, u \leq \omega$ on Ω.

Proof. Suppose that $u - \omega > 0$ somewhere on $\bar{\Omega}$. If $\max_{\bar{\Omega}}(u - \omega) > \max_{\partial\Omega}(u - \omega) = b > 0$, the function $u - (\omega + b)$ satisfies the hypotheses of the extended lemma and we obtain a contradiction. On the other hand, $\partial\omega/\partial n = \infty$, that is, the limit of $D_s\omega$ along the normal direction approaches infinity from inside Ω, implies that a maximum cannot be attained at a point of $\partial\Omega$ as before. □

2. Applications of the Maximum Principle

We now consider equations in two variables. The notations will be changed to suit this special context.

Our first applications are to the minimal surface equation

$$Mu := (1 + u_y^2)u_{xx} - 2u)_x u_y u_{xy} + (1 + u_x^2)u_{yy} = 0.$$

If $A = u_x/W,\ B = u_y/W,\ W = (1 + u_x^2 + u_y^2)^{1/2}$, this can also be written as

$$A_x + B_y \equiv \left(\frac{u_x}{W}\right)_x + \left(\frac{u_y}{W}\right)_y = 0$$

and this equation arises formally from the problem of minimizing the surface area $\int W dx dy$ of a graph $z = u(x, y)$. Our results will be obtained by considering the function $\omega_0 \equiv \omega_0(r; a) = -a \cosh^{-1}(r/a)$ defined for $r \geq a$. A direct calculation shows that $M\omega_0 = 0$ and that $\partial\omega_0/\partial r|_{r=a} = -\infty$.

Theorem 2.1. Suppose that $u \in C^2(\mathscr{A})$, $\mathscr{A} = \{a < r \leq b\}$, $Mu = 0$, and $m \leq u \leq M$ for $r = b$. Then

$$\omega_0(b; a) - \omega_0(r; a) + m \leq u \leq \omega_0(r; a) - \omega_0(b; a) + M$$

on $\mathscr{A}$. If equality is attained at an interior point, it holds throughout $\mathscr{A}$.

Proof. Let $a \leq a^* < b$, and apply Theorem 1.5 to compare $\omega = \omega_0(r; a^*) - \omega_0(b; a^*) + M$ with u. This implies $u \leq \omega$ on $r = a^*$, and $u \leq \omega$ for $a^* < r < b$. Let $a^* \to a$. The final statement follows from the maximum principle as in Lemma 1.2. The left-hand inequality is obtained by considering $-u$. □

If u is a solution of $Mu = 0$ in a punctured disk $0 < r \leq b$, then $\lim_{a \to 0} \omega_0(r; a) = 0$ (r fixed) implies that $m \leq u \leq M$ if u satisfies these inequalities for $r = b$. This illustrates the contrasting behavior of solutions of $Mu = 0$ and harmonic functions. Further illustrations will be given in the exercises.

We will need in what follows an extension of Theorem 2.1 to domains contained in an annulus $\mathscr{A}$.

Theorem 2.2. Suppose $u \in C^2(\Omega)$, $\Omega \subset \mathscr{A}$, $Mu = 0$, and $\limsup u(\mathbf{x}) \leq \omega_0(r; a) - \omega_0(b; a) + M$ as $\mathbf{x}$ approaches $\partial\Omega \cap \{r > a\}$. Then

$$u(\mathbf{x}) \leq \omega_0(r; a) - \omega_0(b; a) + M$$

on Ω.

We need only apply Theorem 1.5 at points of $\partial\Omega \cap \{r = a\}$.

A point $\mathbf{y} \in \partial\Omega$ is said to be a *point of concavity* of Ω if there exists a circle C through $\mathbf{y}$ and a neighborhood N of $\mathbf{y}$ in $\mathbb{R}^2$ such that the intersection of N with the exterior of C is a subset of Ω. We call C a circle of inner contact at $\mathbf{y}$ (Fig. 1).

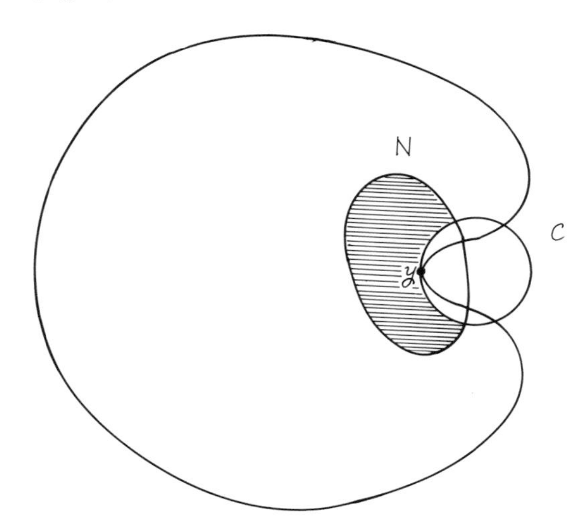

Fig. 1. Points of concavity on the boundary of a domain.

Theorem 2.3. If $u \in C^2(\Omega) \cap C(\bar{\Omega})$ is a solution of $Mu = 0$ in Ω and $\mathbf{y} \in \partial\Omega$ is a point of concavity of Ω with circle of inner contact C, then $u(\mathbf{y})$ is bounded above by a number depending only on the values of u on $\partial\Omega$ exterior to C.

Proof. Suppose that the radius of C is a and Ω is contained in a larger circle of radius b with the same center. If lim sup $u \leq M$approaching all points of $\partial\Omega$ exterior to C, then the Theorem 2.2 implies that $u(\mathbf{x}) \leq \omega(r; a) - \omega(b; a) + M$ on the part of Ω exterior to C. This yields

$$\lim_{\mathbf{x}\to\mathbf{y}} u(\mathbf{x}) \leq M - \omega(b; a). \qquad \square$$

We will consider in some detail later the *nonparametric Plateau problem*, i.e., the problem of finding a solution of $Mu = 0$ in Ω with prescribed values on $\partial\Omega$. It follows easily from Theorem 2.3 that for a domain that has point of concavity there are continuous boundary values for which this problem has no solution. We need only prescribe boundary values that exceed the bound implied by Theorem 2.3 in a neighborhood on the point of concavity. The significance of this is better understood in light of the following result.

Theorem 2.4. Ω is convex if and only if there are no points of concavity on $\partial\Omega$.

A proof is sketched in the exercises.

Consider the equation for surfaces of prescribed mean curvature,

$$A_x + B_y \equiv \left(\frac{u_x}{W}\right)_x + \left(\frac{u_y}{W}\right)_y = 2H, \tag{1}$$

where H is mean curvature, i.e., the average of the principal normal curvatures, of the graph of $u(x, y)$. For minimal surfaces $H \equiv 0$.

We begin with a preliminary result about vector fields. Suppose that $A^2 + B^2 < 1$ on Ω and $A_x + B_y \geq 2\Lambda > 0$ where Λ is a constant. The divergence theorem then implies that $2\Lambda|\Omega| \leq |\partial\Omega|$, where $|\Omega|$ is the area of Ω and $|\partial\Omega|$ is the length of $\partial\Omega$. Similarly, if $A_x + B_y \leq -2\Lambda < 0$, the same inequality holds. Suppose now that $A = u_x/W, B = u_y/W$. Then if $|H(x, y)| \geq \Lambda > 0$ in Ω,

$$|\Omega|/\partial\Omega| \leq 1/2\Lambda. \tag{2}$$

In particular, if Ω is a circle of radius a, then $a \leq \Lambda^{-1}$. If Ω is an annulus $a < r < b$, then $b - a \leq \Lambda^{-1}$. If we think of Ω as fixed, these inequalities impose an upper bound on how large the mean curvature of a surface over Ω can be. Suppose, in particular, that $H = H_0 > 0$ is constant. In the circle $r \le H_0^{-1}$, $u_0 = -(H_0^{-2} - r^2)^{1/2}$, a hemispherical cap of radius H_0^{-1}, is a solution of the constant mean curvature equation. The inequality (2) is sharp in this case. If $H_0 < 0$, $-u_0$ provides an analogous example.

Consider the annulus $a < r < b$, and suppose that $H_0 > 0$. If $u = u(r)$,

$$\operatorname{div}\left(\frac{u_x}{W}, \frac{u_u}{W}\right) = \frac{1}{r}\left[\frac{ru_r}{\sqrt{1+u_r^2}}\right]_r,$$

so solutions of (1) are obtained by solving

$$\frac{u_r}{\sqrt{1+u_r^2}} = H_0 r + \beta r^{-1}. \tag{3}$$

The constant β is at our disposal in seeking useful solutions. The situation differs according to whether $\beta < 0$ or $\beta > 0$. We will consider $\beta > 0$. The other case is dealt with in the exercises. We need to determine an interval on the positive r-axis on which the right-hand side in (3) has magnitude less than one. If $4\beta H_0 \ge 1$, there is no such interval, so we assume $4\beta H_0 < 1$. Then, for

$$(1 - \sqrt{1 - 4\beta H_0})/2H_0 < r < (1 + \sqrt{1 - 4\beta H_0})/2H_0,$$

we have $0 < H_0 r + \beta r^{-1} < 1$, and (3) can be solved by quadrature. The resulting solution, denoted u_2 is determined up to an additive constant. Analysis of (3) shows that u_2 is finite at the endpoints of the interval, but $\partial u_2/\partial r = \infty$ there. For the resulting solution of the mean curvature equation $\partial u_2/\partial n = \infty$ on the outer circumference and $\partial u_2/\partial n = -\infty$ on the inner one, where $\mathbf{n}$ is the outer normal direction. We normalize u_2 to be zero on the outer circumference. If $\beta \to 0$, the annulus approaches the disk of radius H_0^{-1} and u_2 approaches u_0. We will use u_2 to prove the following uniqueness theorem.

Theorem 2.5. Suppose that $(u_x/W)_x + (u_y/W)_y \ge 2H_0 > 0$ in the disk $x^2 + y^2 < H_0^{-2}$. Then

$$\left(\frac{u_x}{W}\right)_x + \left(\frac{u_y}{W}\right)_y \equiv 2H_0$$

on this disk and $u = u_0 + \text{const}$.

Proof. For $0 < \beta < 1/4H_0$, let $u^* = u_2 + c(\beta)$ where $c(\beta)$ is chosen so that $u^* = \max u$ on the inner circumference. Theorem 1.5 implies that $u^* \ge u$ on the entire annulus. Letting $\beta \to 0$ we obtain $u \le u_0 + u(0,0) - H_0^{-1}$ as $c(\beta) \to$

$u(0,0) - H_0^{-1}$. Theorem 1.4 implies that either $u < u_0 + u(0,0) - H_0^{-1}$ in the disk or $u = u_0 + u(0,0) - H_0^{-1}$ there. As equality holds at the origin it must hold everywhere. □

The conclusion shows that hemispheres of radius R are the only graphs with mean curvature $1/R$ that can be extended to the full disk of radius R. This remarkable theorem required no hypothesis about the behavior of u near the boundary.

Consider the equation for the height $u(x, y)$ of a fluid in equilibrium in a cylindrical tube with cross section Ω under the action of gravity and surface tension. The partial differential equation is

$$\left(\frac{u_x}{W}\right)_x + \left(\frac{u_y}{W}\right)_y = Ku,$$

where $K > 0$ is called the capillarity constant and the boundary condition to be satisfied on $\partial\Omega$ is

$$\mathbf{T}u \cdot \mathbf{n} = \cos\gamma, \tag{BC}$$

where $\mathbf{T}u = \text{grad } u/(1 + |\text{grad } u|^2)^{1/2}$, $\mathbf{n}$ is the exterior normal to $\partial\Omega$, and γ is a constant, characterized by the fluid and the material of the walls, called the *contact angle*. We assume that $0 < \gamma < \pi/2$. We write $Qu = \text{div } \mathbf{T}u - Ku = Mu - Ku$. Recall that if $v = -\sqrt{a^2 - r^2}$ in the disk $r < \delta$, then $Mv = 2/a$ there and $\partial v/\partial n = \infty$ on $r = a$. Our first result gives bounds for solutions of $Qu = 0$ using v as a comparison function.

Theorem 2.6. If $Qu = 0$ in Ω and $B_\delta \subset \Omega$, then $u < \delta + 2/K\delta$ in B_δ.

Proof. Choose $\delta' < \delta$ and let

$$v' = -(\delta'^2 - r^2)^{1/2} + 2K/\delta' + \delta'$$

on $B_{\delta'}$, where the origin has been chosen at the center of B_δ. Then $Mv' = 2/\delta' = Kv'(0,0) \leq Kv'$ so that $Qv' \leq 0$ in $B_{\delta'}$. Applying Theorem 1.6 we find that $u < v' < \delta' + 2/K\delta'$ on $B_{\delta'}$. Letting $\delta' \to \delta$ proves the assertion. □

This theorem implies a uniform bound on solutions of $Qu = 0$ in domains Ω that satisfy an interior sphere condition with uniform radius. Note that no use has been made of boundary conditions on u. If $\partial\Omega$ fails to have a normal at a point, the boundary condition (BC) cannot be imposed. We now present a theorem that can be applied to this situation. The significant point is that no assumption is made about growth of u and v near $\mathbf{x}_0$.

Theorem 2.7. Suppose that $u, v \in C^2(\Omega) \cap C^1(\bar{\Omega}\backslash\{\mathbf{x}_0\})$, $\mathbf{x}_0 \in \partial\Omega$, $Qu \geq Qv$ in Ω, and $\mathbf{T}u \cdot \mathbf{n} \leq \mathbf{T}v \cdot \mathbf{n}$ on $\partial\Omega\backslash\{\mathbf{x}_0\}$. Then $u \leq v$ in Ω.

Proof. For $\varepsilon > 0$ sufficiently small let B_ε be a disk centered at $\mathbf{x}_0$ of radius ε and $\Omega_\varepsilon = \Omega\backslash\bar{B}_\in$. If $u - v > 0$ somewhere in Ω, we can consider, for an arbitrary $M > 0$, the function

$$w := \begin{cases} 0, & u - v \leq 0, \\ u - v, & 0 < u - v < M, \\ M, & M \leq u - v. \end{cases}$$

Then

$$\begin{aligned} 0 &\leq \int_{\Omega_\varepsilon} w(Qu - Qv)dx \\ &= -\int_{\Omega_\varepsilon} [\operatorname{grad} w \cdot (\mathbf{T}u - \mathbf{T}v) + Kw^2]\, dx + \int_{\partial\Omega_\varepsilon} w(\mathbf{T}u - \mathbf{T}v) \cdot \mathbf{n}\, ds. \end{aligned}$$

On the part of $\partial\Omega_\varepsilon \subset \partial\Omega$ we have $(\mathbf{T}u - \mathbf{T}v) \cdot \mathbf{n} \leq 0$. As $|\mathbf{T}u \cdot \mathbf{n}| < 1$ in Ω, $\omega(\mathbf{T}u - \mathbf{T}v) \cdot \mathbf{n} \leq 2M$ on $\partial B_\varepsilon \cap \Omega$, and

$$\int_{\Omega_\varepsilon} \operatorname{grad} w \cdot (\mathbf{T}u - \mathbf{T}v)dx + K\int_{\Omega_\varepsilon} w^2 dx \leq 4\pi\varepsilon M.$$

Letting $\varepsilon \to 0$, the left-hand integrals taken over Ω are nonpositive. From Exercise 2.10 the integrand in the first integral is nonnegative, so both integrands vanish on Ω. Because M is arbitrary, the conclusion follows. ☐

This conclusion can be extended easily to domains with a finite number of corners. An interesting example to which this theorem can be applied is a wedge with sufficiently large opening.

Theorem 2.8. Suppose that $Qu = 0$ in a wedge Ω with opening 2α and $\mathbf{T}u \cdot \mathbf{n} = \cos\gamma$ on the sides of the wedge with $\alpha + \gamma \geq \pi/2$. Then, if B_δ is a disk passing through the vertex of the wedge with its diameter along the centerline of the wedge,

$$u \leq \delta + 2/K\delta$$

on $\Omega \cap B_\delta$.

Proof. A lower hemisphere over B_δ makes the constant contact angle $\pi/2 - \alpha$ along the sides of the wedge (Exercise 2.9). On $\Omega \cap \partial B_\delta$ the hemisphere attaches vertically. Using the upward translation of the hemisphere introduced in the proof of Theorem 2.6, the conclusion follows from Theorem 2.7 □

The amazing fact is that if $\alpha + \gamma < \pi/2$, this theorem is no longer true. In fact, a solution must grow like $1/r$ near the vertex if r is the distance to the vertex. In order to prove this we need the following extension of Theorem 2.7 whose proof is left as an exercise.

Theorem 2.9. Suppose that $u, v \in C^2(\Omega) \cap C^1(\bar{\Omega}\backslash\{\mathbf{x}_0\})$, $\mathbf{x}_0 \in \partial\Omega$, $Qu \geq Qv$ in Ω, and $\mathbf{T}u \cdot \mathbf{n} \leq \mathbf{T}v \cdot \mathbf{n}$ on Σ_1, $u \leq v$ on Σ_2 where $\Sigma_1 \cup \Sigma_2 = \partial\Omega\backslash\{\mathbf{x}_0\}$. Then $u \leq v$ in Ω.

Let Ω be the wedge as above except that $\alpha + \gamma < \pi/2$. Let

$$v = \frac{\cos\theta - \sqrt{k^2 - \sin^2\theta}}{kKr} = \frac{f(\theta)}{r},$$

where $k = \sin\alpha/\cos\gamma$. Using polar coordinates a short calculation shows that

$$\mathbf{T}v = (-f(\theta)\mathbf{c}_r + f'(\theta)\mathbf{c}_\theta)/\sqrt{r^4 + |f(\theta)|^2 + |f'(\theta)|^2},$$

where $\mathbf{c}_r$, $\mathbf{c}_\theta$ are the unit vectors in that coordinate system. As $\mathbf{n} = \mathbf{c}_\theta$ on the sides of the wedge, $kKf(\alpha) = \cos\alpha - \tan\alpha \sin\alpha = \cos(\alpha + \gamma)/\cos\gamma$, and $f'(\alpha) = \cot\gamma f(\alpha)$,

$$kK\mathbf{T}v \cdot \mathbf{n} = \frac{\cos(\alpha + \gamma)}{\sin\gamma}\left(r^4 + \frac{\cos^2(\alpha + \gamma)}{(kK)^2\cos^2\gamma\,\sin^2\gamma}\right)^{-1/2}$$

there. If the last term is expanded using a binomial series, we obtain

$$\mathbf{T}v \cdot \mathbf{n} = \cos\gamma(1 - \zeta), \tag{4}$$

where $\zeta = O(r^4)$ and ζ is positive for r sufficiently small. In a similar way we can show that $Qv = \eta = O(r^3)$. The demonstration is left as an exercise.

Theorem 2.10. Suppose that $\alpha + \gamma < \pi/2$. Then, if $Qu = 0$ in Ω_δ (defined as Ω_ε in Theorem 2.7 for $\varepsilon = \delta$) and $\mathbf{T}u \cdot \mathbf{n} = \cos \gamma$ on the sides, $|u - v| = O(1)$ on Ω_δ.

Proof. If suffices to show this for δ sufficiently small. First we choose δ so that ζ in (4) is positive. Then, using Theorem 2.6, we can get a bound $|u| < M$ on $\partial B_\delta \cap \Omega$. Letting $w = v - b$ we have $Qw = \eta + Kb$. Choose $b > 0$ so that $w < u$ on $\partial B_\delta \cap \Omega$, and then further restrict δ so that $\eta + Kb$ is positive on Ω_δ. Theorem 2.9 then implies that $u \geq w$, i.e., $u - v \geq -b$. For the other half of the proof, let $\gamma' < \gamma$ and let v' be obtained from the formula for v by replacing γ by γ'. Then

$$\mathbf{T}v' \cdot \mathbf{n} = \cos \gamma'(1 - O(r^4))$$

and, with possibly another restriction on δ, $\mathbf{T}v' \cdot \mathbf{n} > \cos \gamma$. Choose $A > 0$ so that $w = v' + A > u$ on $\partial B_\delta \cap \Omega$, and restrict δ again so that $\eta - KA < 0$ in Ω_δ. Then, as before, $u \leq v' + A$ in Ω_δ. A further argument is required in order to show that a similar estimate holds with v' replaced by v. This is left as an exercise. □

3. Equations with Discontinuous Coefficients

It is useful for both theoretical and practical reasons to introduce a generalization of the classical solutions of elliptic equations that we have considered in Sections 1 and 2. We have already met this idea in discussing the variational solutions of the Dirichlet problem for Laplace's equation that arose from Dirichlet's principle in Section 5 of Chapter 4.

Suppose, for example, that we need to consider a steady-state heat conduction problem in which the conductivity jumps discontinuously on an interior surface in the domain, i.e.,

$$k(\mathbf{x}) = \begin{cases} k_1, & \mathbf{x} \in \Omega_1, \\ k_2, & \mathbf{x} \in \Omega_2, \end{cases}$$

where $\bar{\Omega}_2 \subset \Omega$, $\Omega_1 = \Omega \backslash \bar{\Omega}_2$. Then we are led to the variational problem

$$\min \int_\Omega k(\mathbf{x}) |\text{grad } u|^2 d\mathbf{x}, \tag{V}$$

where the minimum is taken over functions $u \in H^1(\Omega)$ such that $u - \varphi \in H_0^1(\Omega)$, where φ is a function in $H^1(\Omega)$ whose trace is the required boundary function on

$\partial\Omega$. If u is a solution, Weyl's lemma implies that $u \in C^2(\Omega_1 \cup \Omega_2)$ and $\Delta u = 0$ there. If u is sufficiently regular up to $\partial\Omega_2$, use of appropriate variations shows that

$$u^+ = u^-, \qquad k_1 \frac{\partial u^-}{\partial n} = k_2 \frac{\partial u^+}{\partial n},$$

where $+$ denotes the limit taken from Ω_2, $-$ the limit from Ω_1, and $\mathbf{n}$ is the exterior normal to Ω_2. These conditions describe what is sometimes called an *interface* or *transmission* problem (see Chapter 2), and the variational problem (V) is a natural way to treat this problem theoretically and numerically. For our purposes here, however, the crucial thing to observe is that a minimizer is a solution of the equation

$$\operatorname{div}(k(\mathbf{x}) \operatorname{grad} u) = 0$$

in the sense of distributions, i.e.

$$\int_\Omega k(\mathbf{x}) \operatorname{grad} u \cdot \operatorname{grad} \varphi \, d\mathbf{x} = 0$$

for all $\varphi \in C_0^\infty(\Omega)$.

Another example is given by the problem of minimizing

$$\int_\Omega f(\operatorname{grad} u) d\mathbf{x}$$

for $u \in H^1(\Omega)$ such that $u = \varphi$ on $\partial\Omega$, $\Omega \subset \mathbb{R}^n$. A necessary condition on u is that

$$\int_\Omega f_{p_i}(\operatorname{grad} u)\varphi_{x_i} \, d\mathbf{x} = 0$$

for all test functions φ, so that u satisfies $(f_{p_i}(\mathbf{p}))_{x_i} = 0$, $\mathbf{p} := \operatorname{grad} u$, in the sense of distributions. [Special cases of this problem will arise in Section 4. In particular, if $f = (1 + \operatorname{grad}^2 u)^{1/2}$, this is the nonparametric least area problem.]

We are led to define a class of differential operators, *linear elliptic divergence structure* operators, by

$$Lu := -(a_{ij}(\mathbf{x})u_{x_i})_{x_j}.$$

We assume that $a_{ij} \in L^\infty(\Omega)$, and

$$a_{ij}\xi_i\xi_j \geq \nu|\xi|^2 \tag{E}$$

for $\nu > 0$ and all $\mathbf{x} \in \Omega$. (We do not assume a_{ij} is symmetric. If is not necessary for the general theory, and we will give an important application to a case in which a_{ij} is not symmetric.)

If $u \in H^1_{\text{loc}}(\Omega)[= W^{1,2}_{\text{loc}}(\Omega)]$ and $f_i \in L^1_{\text{loc}}(\Omega)$, $i = 1, \ldots, n$, we say u is a (weak) solution of

$$Lu = (f_i)_{x_i} \equiv \text{div } \mathbf{f}$$

in Ω if

$$\int_\Omega a_{ij}(\mathbf{x}) u_{x_i} \varphi_{x_j} \, d\mathbf{x} = \int_\Omega f_i \varphi_{x_i} \, d\mathbf{x}$$

for all $\varphi \in C^1_0(\Omega)$. In this way sense can be made of the differential operator L without assuming any regularity of the coefficients a_{ij}. We define $b : H^1_0(\Omega) \times H^1_0(\Omega) \to \mathbb{R}$ by

$$b(u, v) = \int_\Omega a_{ij}(\mathbf{x}) u_{x_i} v_{x_j} \, d\mathbf{x}.$$

Then (E) immediately implies that

$$b(u, u) \geq \nu \|u\|^2. \tag{C}$$

[Recall that we may take $(\int_\Omega |\text{grad } u|^2 d\mathbf{x})^{1/2}$ as the norm on $H^1_0(\Omega)$.] As $a_{ij} \in L^\infty(\Omega)$, the Cauchy–Schwarz inequality implies

$$|b(u, v)| \leq C \|u\| \, \|v\|. \tag{B}$$

Suppose that $f_i \in L^2(\Omega)$. Then $u \in H^1_0(\Omega)$ is a solution of $Lu = (f_i)_{x_i}$ in Ω if and only if

$$b(u, \varphi) = F(\varphi) := \int_\Omega f_i \varphi_{x_i} \, d\mathbf{x} \tag{P}$$

for all $\varphi \in C^1_0(\Omega)$. As (B) holds, this condition also holds for all $\varphi \in H^1_0(\Omega)$. It immediately follows that there exists a unique solution of (P) as b satisfies the hypotheses of the Lax–Milgram theorem (Theorem 1.8 of Chapter 8). If we wish to consider inhomogenous boundary values u_0, where (the lifting) $u_0 \in H^1(\Omega)$, $\tilde{u} = u - u_0$ is a solution of $b(\tilde{u}, \varphi) = \tilde{F}(\varphi) := \int_\Omega \tilde{f}_i \varphi_{x_i} \, d\mathbf{x}$ where $\tilde{f}_i = f_i - (u_0)_{x_i}$, and existence and uniqueness follows from the homogeneous case.

In the rest of this section we will prove the celebrated De Giorgi–Nash–Moser theorem, which says that solutions of $Lu = 0$ are Hölder continuous on interior subdomains. The proof follows Moser (Ref. 1) and Evans (Ref. 2).

Suppose then that $u \in L^1_{\text{loc}}(\Omega)$ and $Lu = 0$ in Ω, i.e.

$$\int_\Omega a_{ij}(\mathbf{x})u_{x_i}, \varphi_{x_j}\, d\mathbf{x} = 0 \tag{5}$$

for all $\varphi \in C_0^1(\Omega)$. We say that $v \in H^1_{\text{loc}}(\Omega)$ is a *subsolution* of $Lu = 0$ if

$$\int_\Omega a_{ij}(\mathbf{x})v_{x_i}\varphi_{xj}\, d\mathbf{x} \leq 0 \tag{6}$$

for all $\varphi \in C_0^1(\Omega)$, $\varphi \geq 0$, i.e., $Lv \geq 0$ in this weak sense. The next theorem contains the heart of the proof of our main theorem.

Theorem 3.1. Suppose that v is a nonnegative subsolution of (5). If $\mathbf{x}_0 \in \Omega$ and $B_R = B(\mathbf{x}_0, R) \subset \Omega$, then there is a constant C depending only on n, ν, and $\max\|a_{ij}\|_{L^\infty(\Omega)}$ such that

$$\max_{B_{R/2}} v \leq C\left[\frac{1}{R^n}\int_{B_R} v^2 d\mathbf{x}\right]^{1/2}.$$

The idea of the proof is to estimate a sequence of L^p norms over balls $B(\mathbf{x}_0, R_k)$ lying between $B(\mathbf{x}_0, R/2)$ and $B(\mathbf{x}_0, R)$ starting with $p = 2$, and recursively estimate each norm in terms of the previous one. We need to recall the following result from analysis (Ref. 3, p. 163).

Proposition 3.1. If $\mathcal{O}$ is a bounded domain and $f \in L^p(\mathcal{O})$ for some $p \geq 1$, then

$$\lim_{p\to\infty}\left[\frac{1}{|\mathcal{O}|}\int_{\mathcal{O}} |f|^p d\mathbf{x}\right]^{1/p} = \|f\|_{L^\infty(\mathcal{O})}.$$

Proof of Theorem 3.1. Suppose that $p \geq 2$ and $\zeta \in C_0^\infty(\Omega)$, $0 \leq \zeta \leq 1$. The idea of the start of the proof is to choose $\varphi = \zeta^2 v^{p-1}$ as a test function in (6). Because this function is not in $C_0^1(\Omega)$, some preliminary modifications must be made, however, First, we observe that we can use any function $\varphi \in H_0^1(\Omega)$ whose support is bounded away from $\partial\Omega$; we need only take limits as in our observation that (P) holds for all $\varphi \in H_0^1(\Omega)$. A more serious difficulty arises, however, from the fact that $v^{p-1} \notin H^1_{\text{loc}}(\Omega)$ for $p > 2$ unless v is bounded, and this is an essential part of the conclusion of the theorem! Therefore, we define

$$h(u) = \begin{cases} u^{p-1}, & 0 \le u \le M, \\ au + b, & u > M, \end{cases}$$

with a, b determined in such a way that $h \in C^1(\mathbb{R}^+)$ $[a = (p-1)M^{p-2}$, $b = M^{p-1}(2-p)]$. Because h satisfies a uniform Lipschitz condition, Proposition 3.10 of Chapter 8 implies that $h(v) \in H^1_{\text{loc}}(\Omega)$ for any $p \ge 2, M > 0$. (We will suppress the dependence of h on p and M in our notation.) We note for future reference that $uh'(u) = (p-1)h(u) \ge 0$ for $0 \le u \le M$. We take $\varphi = \zeta^2 h(v)$ in (6). It follows that

$$\int_\Omega a_{ij} v_{x_i} v_{x_j} h'(v)\zeta^2 d\mathbf{x} \le -2 \int_\Omega a_{ij} v_{x_i} \zeta_{x_j} \zeta h(v) d\mathbf{x}.$$

Then (E) and $2ab \le \varepsilon b^2 + a^2/\varepsilon$ for an appropriate ε imply

$$\begin{aligned} v \int_\Omega |\text{grad } v|^2 h'(v)\zeta^2 \, d\mathbf{x} &\le \frac{C}{p-1} \int_\Omega \zeta v h'(v) \text{ grad } v \cdot \text{grad } \zeta \, d\mathbf{x} \\ &\le \varepsilon \int_\Omega |\text{grad } v|^2 h'(v)\zeta^2 \, d\mathbf{x} + \frac{C}{\varepsilon(p-1)^2} \int_\Omega |\text{grad } \zeta|^2 v^2 h'(v) d\mathbf{x}, \end{aligned}$$

where C denotes a constant depending only on n, ν, and $\max \|a_{ij}\|_{L^\infty(\Omega)}$. Setting $\varepsilon = v/2$, we get

$$\int_\Omega |\text{grad } v|^2 h'(v)\zeta^2 d\mathbf{x} \le \frac{C}{(p-1)^2} \int |\text{grad } \zeta|^2 v^2 h'(v) d\mathbf{x}.$$

For notational convenience we define $H(u) = (u^2 h'(u))^{1/2}[= (p-1)^{1/2]} u^{p/2}$ for $u \le M]$. Because

$$|\text{grad } v|^2 v^{p-2} = \frac{4}{p^2} |\text{grad}(v^{p/2})|^2 \ge \frac{1}{(p-1)^2} |\text{grad}(v^{p/2})|^2,$$

we have $|\text{grad } H(v)|^2 \le (p-1)^2 |\text{grad } v|^2 h'(v)$ and

$$\int_\Omega |\text{grad } H(v)|^2 \zeta^2 d\mathbf{x} < C \int_\Omega |\text{grad } \zeta|^2 v^2 h'(v) d\mathbf{x}.$$

Then $|\text{grad}(\zeta H(v))|^2 \leq 2(|\text{grad } H(v)|^2\zeta^2 + (H(v))^2|\text{grad } \zeta|^2)$ implies

$$\int_\Omega |\text{grad}(\zeta H(v))|^2 d\mathbf{x} \leq C\int_\Omega |\text{grad } \zeta|^2 v^2 h'(v) d\mathbf{x}.$$

Let $w = \min[v, M]$. If we apply the Sobolev inequality to the left-hand integral in the last inequality, we obtain

$$\left(\int_\Omega (\zeta H(v))^{2^*} d\mathbf{x}\right)^{2/2^*} \leq C\int_\Omega |\text{grad } \zeta|^2 v^2 h'(v) d\mathbf{x}.$$

Because $2^*/2 := n/(n-2) > 1$ if $n > 2$ and may be taken > 1 if $n = 2$, $(p-1)^{2^*/2} > 1$ and the last inequality implies that

$$\left(\int_\Omega (\zeta w^{p/2})^{2^*} d\mathbf{x}\right)^{2/2^*} \leq C\int_\Omega |\text{grad } \zeta|^2 w^p \, d\mathbf{x}. \tag{7}$$

If p is such that the right-hand side is finite when w is replaced by v, then so is the left-hand side. (Apply Lebesgue's theorem.) We will use this idea recursively in what follows. Let

$$R_k = \frac{R}{2}(1 + 2^{-k}), \qquad k = 0, 1, 2, \ldots.$$

Choose $\zeta = \zeta_k$ so that

$$\zeta = \begin{cases} 1, & \text{on } B_{k+1} = B(\mathbf{x}_0, R_{k+1}), \\ 0, & \text{on } B_k^c = B(\mathbf{x}_0, R_k)^c, \end{cases}$$

with $0 \leq \zeta \leq 1, \zeta \in C_0^\infty(B_k)$. If

$$p_k = 2\left(\frac{n}{n-2}\right)^k, \qquad k = 0, 1, 2, \ldots,$$

then $p_0 = 2$ and $p_k \to \infty$ as $k \to \infty$. We may assume $|\text{grad } \zeta| \leq 2/(R_k - R_{k+1}) = 2^{k+3}/R$. It follows from taking p_kth roots of (7) that

$$\left(\int_{B_{k+1}} v^{p_{k+1}} \, d\mathbf{x}\right)^{1/p_{k+1}} \leq \frac{C^{1/p_k} 4^{k/p_k}}{R^{2/p_k}} \left(\int_{B_k} v^{p_k} \, d\mathbf{x}\right)^{1/p_k} \tag{8}$$

Setting

$$a_k = \|v\|_{L^{p_k}(B_k)}, \qquad \gamma_k = C^{1/p_k} 4^{k/p_k} / R^{2/p_k},$$

(8) says $a_{k+1} \leq \gamma_k a_k$ and, hence, $a_{n+1} \leq \gamma_0 \cdots \gamma_n a_0$, i.e.,

$$a_{n+1} \leq C^{s_1} 4^{s_2} R^{-2s_1} a_0, \tag{9}$$

where $s_1 = \sum_{k=0}^{n} 1/p_k$, $s_2 = \sum_{k=0}^{n} k/p_k$. We can write

$$\sum_{k=0}^{\infty} \frac{1}{p_k} = \frac{1}{2} \sum_{k=0}^{\infty} \left(\frac{n-2}{n}\right)^k = n/4, \qquad \sum_{k=0}^{\infty} \frac{k}{p_k} = \frac{n^2}{4} - \frac{n}{4}.$$

The theorem follows from taking limits in (9) and from the Proposition. □

Lemma 3.1. Suppose that $\phi \in C^1(R)$, ϕ' satisfies a uniform Lipschitz condition, and $\phi'' \geq 0$ a.e. Then if u is solution of (5), $\phi(u)$ is a subsolution.

Proof. The assumption of a uniform Lipschitz condition implies that $\phi'(u) \in H^1_{\text{loc}}(\Omega)$ (Proposition 3.10 of Chapter 8). Let $\varphi = \eta\phi'(u)$ in (5) where $\eta \in C_0^1(\Omega)$, $\eta \geq 0$. Then

$$\int_\Omega a_{ij} u_{x_i} u_{x_j} \eta \phi''(u) d\mathbf{x} + \int_\Omega a_{ij} u_{x_i} \eta_{x_j} \phi'(u) d\mathbf{x} = 0$$

and

$$\int_\Omega a_{ij} (\phi(u))_{x_i} \eta_{x_j}\, d\mathbf{x} = -\int_\Omega a_{ij} u_{x_i} u_{x_j} \eta \phi''(u) d\mathbf{x}.$$

The assumed convexity of ϕ, (E), and $\eta \geq 0$ imply that the last integral is nonnegative and (6) is satisfied. □

We need the following corollary to Theorem 3.1.

Theorem 3.2. If u is a solution of (5), $\mathbf{x}_0 \in \Omega$, $B_R = B(\mathbf{x}_0, R) \subset \Omega$,

$$\max_{B_{R/2}} |u| \leq C \left[\frac{1}{R^n} \int_{B_r} u^2 d\mathbf{x} \right]^{1/2}$$

Proof. The result follows formally by setting $v = |u|$ in Theorem 3.1. In order to make this rigorous we need to consider $\phi_\delta(u)$ defined by $\phi_\delta(0) = 0$ and

$$\phi'_\delta(u) = \begin{cases} 1, & u > \delta, \\ u/\delta, & -\delta \le u \le \delta, \\ -1, & u < -\delta. \end{cases}$$

$v = \phi_\delta(u)$ is a subsolution by Lemma 3.1 and Theorem 3.1 applies. As the conclusion of Theorem 3.1 contains no derivatives, we can let δ go to zero to obtain the result. □

If $\mathbf{x}_0 \in \Omega$ and $R_0 > 0$ is such that $B_{R_0} \subset \Omega$, we define, for a function u,

$$\omega(\mathbf{x}_0, R) = \sup_{B_R} u - \inf_{B_R} u$$

for $0 < R \le R_0$. Then u is Hölder continuous, with exponent γ, $0 < \gamma < 1$, at $\mathbf{x}_0$, if and only if there is a constant $C > 0$ such that

$$\omega(\mathbf{x}_0, R) \le CR^\gamma \tag{10}$$

for all $0 < R \le R_0$. Our next lemma gives a condition on a nonnegative real-valued function $\omega(R)$ that implies (10).

Lemma 3.2. Suppose that $\omega(R) \ge 0$ and there is η, $0 < \eta < 1$, such that

$$\omega(R/4) \le \eta\omega(R) \tag{11}$$

for $0 < R \le R_0$. Then there is $C = C(\eta, \sup \omega) > 0$ and $\gamma = \gamma(\eta)$, $0 < \gamma < 1$, such that

$$\omega(R) \le CR^\gamma. \tag{12}$$

Proof. Choose $\eta < a < 1$, and let $4^\gamma\eta = a$,

$$M = \sup_{R_0/4 \le R \le R_0} (\omega(R)/R^\gamma). \tag{13}$$

Then ω satisfies (12) for $R_0/4 \le R_0$ with $C = M$. For $R_0/4^2 < R < R_0/4$, (11) and (13) imply

$$\omega(R) \le \eta\omega(4R) \le \eta M(4R)^\gamma = aMR^\gamma.$$

Continuing recursively, for $R_0/4^{n+1} \le R \le R_0/4^n$,

$$\omega(R) \le a^n M R^\gamma.$$

The result follows. □

The next lemma may be thought of as a version of the Poincaré inequality. The proof is sketched in the exercises.

Lemma 3.3 Suppose that $u \in H^1(B_r)$ and $u \equiv 0$ on $N \subset B_r = B(\mathbf{x}_0, R)$ with $|N| > 0$. Then there exists $C = C(n)$ such that

$$\int_{B_R} u^2 d\mathbf{x} \le C(R^n/|N|)^2 R^2 \int_{B_R} |\text{grad } u|^2 d\mathbf{x}.$$

Theorem 3.3 Assume that u is a solution of (5) in $B_{2R} = B(\mathbf{x}_0, 2R) \subset \Omega$, $0 \le u \le 1$, and

$$|\{\mathbf{x} \in B_R : u(\mathbf{x}) \ge 1/2\}| \ge \tfrac{1}{2}|B_R|.$$

Then there is a $c = c(n, v \max\|a_{ij}\|_{L^\infty(\Omega)}) > 0$ such that

$$\min_{B_{R/2}} \ge c.$$

Proof. Let $0 < \varepsilon < 1$, and define

$$\phi(u) = \max\{-\ln(2u + \varepsilon), 0\}.$$

Note that ϕ is convex, ϕ is C^∞ on $[0, \infty) \setminus \{1 - \varepsilon)/2\}$, and $\phi'' = (\phi')^2$, $\phi \ge 0$ there, $\phi \equiv 0$ for $u > (1 - \varepsilon)/2$. Let $\zeta \in C^\infty(\Omega), 0 \le \zeta \le 1, \zeta \equiv 1$ on B_R, supp $\zeta = B_{2R}$. We want to take $\phi'(u)\zeta^2$ as a test function in (5). For this to be correct the corner at $u = (1 - \varepsilon)/2$ must be smoothed as in the proof of Theorem 3.2. For a smoothed ϕ,

$$0 = \int_\Omega a_{ij} u_{x_i} u_{x_j} \zeta^2 \phi''(u)\, d\mathbf{x} + 2 \int_\Omega a_{ij} u_{x_i} \zeta_{x_j} \phi'(u)\zeta\, d\mathbf{x} \tag{14}$$

and the limiting operation to obtain this for ϕ works as before. In a similar way, if $v = \phi(u)$, we can write

$$\int_\Omega a_{ij} v_{x_i} v_{x_j} \zeta^2\, d\mathbf{x} = \int_\Omega a_{ij} u_{x_i} u_{x_j} (\phi')^2 \zeta^2\, d\mathbf{x}.$$

If we use $\phi'' = (\phi')^2$ and (14), this implies

$$\int_\Omega a_{ij} v_{x_i} v_{x_j} \zeta^2 d\mathbf{x} = -2 \int_\Omega a_{ij} v_{x_i} \zeta_{x_j} \zeta \, d\mathbf{x}.$$

Then (E) and $2ab \leq \delta b^2 + a^2/\delta$ for an appropriate δ yield

$$\nu \int_\Omega |\text{grad } v|^2 \zeta^2 \, d\mathbf{x} \leq \frac{1}{2} \nu \int_\Omega |\text{grad } v|^2 \zeta^2 d\mathbf{x} + C \int_\Omega |\text{grad } \zeta|^2 d\mathbf{x}.$$

As $|\text{grad } \zeta| \leq 2/R$, this implies

$$\int_{B_R} |\text{grad } v|^2 d\mathbf{x} \leq C R^{n-2}. \tag{15}$$

From Theorem 3.1 we have

$$\max_{B_{R/2}} v^2 = \frac{C}{R^n} \int_{B_R} v^2 d\mathbf{x}. \tag{16}$$

Letting $N = \{\mathbf{x} \in B_R : u(\mathbf{x}) \geq 1/2\}$, $v = \phi(u) \equiv 0$ on N as $\phi \equiv 0$ for $u \geq 1/2$, and our main hypothesis is $|N| \geq \frac{1}{2} |B_R|$. Hence, Lemma 3.3 yields

$$\int_{B_R} v^2 \, d\mathbf{x} \leq C R^2 \int_{B_R} |\text{grad } v|^2 d\mathbf{x}. \tag{17}$$

The inequalities (15)–(17) combine to give

$$\max_{B_{R/2}} v^2 \leq C.$$

The constant C does not depend on ε. Therefore, $-\ln(2u(\mathbf{x}) + \varepsilon) \leq C$, that is,

$$u(\mathbf{x}) + \tfrac{1}{2}\varepsilon \geq \tfrac{1}{2} e^{-C}$$

On $B_{R/2}$. Letting $\varepsilon \to 0$ the assertion follows. □

We come to our main result.

Theorem 3.4. If u is a solution of (5) on Ω and $\bar{\Omega}' \subset \Omega$, there are $C > 0$ and $0 < \gamma < 1$ such that

$$|u(\mathbf{x}) - u(\mathbf{y})| \leq C|\mathbf{x} - \mathbf{y}|^\gamma$$

for $\mathbf{x}, \mathbf{y} \in \Omega'$. γ depends on ν, $\max \|a_{ij}\|_{L^\infty(\Omega)}$, n, $\|u\|_{L^2(\Omega')}$ and C depends on these quantities and on $\text{dist}(\Omega', \partial\Omega)$.

Proof. Let $\mathbf{x}_0 \in \Omega'$ and

$$\omega(R) = \max_{B_R} u - \min_{B_R} u$$

for $0 < R \leq R_0 = \frac{1}{4}\text{dist}(\mathbf{x}_0, \partial\Omega)$. Note that if $\tilde{u} = a(u + b)$, then the corresponding ω is unaffected by b and scales linearly with a. Suppose for the moment that

$$\max_{B_R} u = 1, \qquad \min_{B_R} u = 0. \tag{18}$$

(Theorem 3.2 implies that u is bounded in Ω'.) Both u and $1 - u$ are solutions of (5) and one or the other must satisfy the hypothesis of Theorem 3.3 on $B_{R/2}$. Suppose first that

$$|\{\mathbf{x} \in B_{R/2} : u(\mathbf{x}) \geq 1/2\}| \leq \tfrac{1}{2}|B_{R/2}|.$$

Then by Theorem 3.3, $\min_{B_{R/4}} u \geq c$, and

$$\omega(R/4) \leq \max_{B_R} u - c \leq 1 - c = \eta\omega(R),$$

where $\eta = 1 - c$. If

$$|\{\mathbf{x} \in B_{R/2} : 1 - u(\mathbf{x}) \geq 1/2\}| \geq \tfrac{1}{2}|B_{R/2}|,$$

then $1 - u(\mathbf{x}) \geq c > 0$ on $B_{R/4}$, $\max_{B_{R/4}} u \leq 1 - c$, and

$$\omega(R/4) \leq 1 - c - \min_{B_R} u = 1 - c - \eta\omega(R),$$

as before. Now we observe that these results hold without the translation and scaling that led to (18). Lemma 3.2 implies existence of C and $0 \leq \gamma < 1$ such that

$$\omega(R) \leq C(R/R_0)^\gamma$$

for $0 < R \leq R_0$. If $\mathbf{x}, \mathbf{y} \in \Omega'$, $|\mathbf{x} - \mathbf{y}| = R \leq R_0$, then

$$|u(\mathbf{x}) - u(\mathbf{y})| \leq CR_0^{-\gamma}|\mathbf{x} - \mathbf{y}|^{\gamma} := C|\mathbf{x} - \mathbf{y}|^{\gamma}.$$

For $|\mathbf{x} - \mathbf{y}| > R_0$,

$$|u(\mathbf{x}) - u(\mathbf{y})| \leq 2 \max_{\Omega'}|u| \leq C(\Omega')\|u\|_{L^2(\Omega')}$$

by Theorem 3.2, and

$$|u(\mathbf{x}) - u(\mathbf{y})| \leq (C(\Omega')\|u\|_{L^2(\Omega')}R_0^{-\gamma})|\mathbf{x} - \mathbf{y}|^{\gamma}.$$

4. Nonlinear Elliptic Equations

In this section we investigate representative aspects of the theory of nonlinear elliptic equations. In Section 4.1 we give an example of the application of *monotone operators* to nonlinear elliptic problems. These operators arise naturally, for example, as gradients of convex functionals. In Section 4.2 we study the Dirichlet problem for the minimal surface equation that lies just beyond the reach of monotone operator theory. We have seen in Section 2 that convexity of the domain is a necessary condition for this problem to be well posed. We will show here that convexity is also sufficient. This is an example of an elliptic problem in which curvature conditions are required for existence, and the methods used are representative of those appropriate for more general problems.

4.1. Monotone Operators. We begin with some finite-dimensional examples that exhibit the basic idea. Suppose that $f \in C^1[a, b]$ and $f(x_0) = \min f(x)$. Then the conditions $f'(x_0) = 0$ if $a < x_0 < b$, $f'(x_0) \geq 0$ if $x_0 = a$, $f'(x_0) \leq 0$ if $x_0 = b$ can be summarized by

$$f'(x_0)(x - x_0) \geq 0, \qquad x \in [a, b].$$

If $f \in C^1(K)$, K a closed, convex set in $\mathbb{R}^n$, and $f(\mathbf{x}_0) = \min_K f(\mathbf{x})$, then the one-dimensional case applied to a line segment connecting $\mathbf{x}_0$ and $\mathbf{x}$ implies that

$$\operatorname{grad} f(\mathbf{x}_0) \cdot (\mathbf{x} - \mathbf{x}_0) \geq 0, \qquad \mathbf{x} \in K,$$

or, writing $\mathbf{F} = \text{grad}\, f$, $\mathbf{F}: K \to \mathbb{R}^n$,

$$\mathbf{F}(\mathbf{x}_0) \cdot (\mathbf{x} - \mathbf{x}_0) \geq 0, \qquad \mathbf{x} \in K.$$

This is an example of a *variational inequality*. If K is compact, the existence of such an $\mathbf{x}_0$ is immediate. More generally, if $K \subset \mathbb{R}^n$ is compact and convex and $\mathbf{F}: K \to \mathbb{R}^n$ is continuous, there is an $\mathbf{x} \in K$ such that

$$\mathbf{F}(\mathbf{x}) \cdot (\mathbf{y} - \mathbf{x}) \geq 0 \tag{19}$$

for all $\mathbf{y} \in K$. If K is not compact, we can apply this result to any $K_R = K \cap \overline{B(0, R)}$, and it is not difficult to show that a necessary and sufficient condition for existence of a solution of (19) is that the solution $\mathbf{x}_R$ on K_R satisfy $\|\mathbf{x}_R\| < R$ for some $R > 0$. A natural sufficient condition for this to hold is that, for some $\bar{\mathbf{y}} \in K$,

$$(\mathbf{F}(\mathbf{x}) - \mathbf{F}(\bar{\mathbf{y}})) \cdot (\mathbf{x} - \bar{\mathbf{y}})/\|\mathbf{x} - \bar{\mathbf{y}}\| \to +\infty$$

as $\|\mathbf{x}\| \to \infty$, $\mathbf{x} \in K$. It is natural to ask for a sufficient condition for uniqueness of a solution of (19). If $\mathbf{x}, \mathbf{x}'$ are two solutions, (19) implies $(\mathbf{F}(\mathbf{x}) - \mathbf{F}(\mathbf{x}')) \cdot (\mathbf{x} - \mathbf{x}) \leq 0$, so

$$(\mathbf{F}(\mathbf{x}) - \mathbf{F}(\mathbf{y})) \cdot (\mathbf{x} - \mathbf{y}) > 0 \tag{20}$$

for $\mathbf{x}$, $\mathbf{y} \in K$, $\mathbf{x} \neq \mathbf{y}$, is such a condition. If

$$(\mathbf{F}(\mathbf{x}) - \mathbf{F}(\mathbf{y})) \cdot (\mathbf{x} - \mathbf{y}) \geq 0, \tag{21}$$

the mapping $\mathbf{F}$ is *monotone* and *strictly monotone* when (20) holds. If $f \in C^1(K)$ and f is convex, i.e.,

$$f(\mathbf{x}) \geq f(\mathbf{y}) + \text{grad}\, f(\mathbf{y}) \cdot (\mathbf{x} - \mathbf{y})$$

for $\mathbf{x}$, $\mathbf{y} \in K$, then $\mathbf{F} = \text{grad}\, f$ is monotone, and if f is strictly convex, $\mathbf{F}$ is strictly monotone.

It turns out that the natural generalization of the above to infinite-dimensional spaces calls for an operator A mapping a convex set, K, in a reflexive Banach space, X, into its dual space, X'. We denote by $\|\cdot\|$ the norm in X and by $\langle \cdot, \cdot \rangle$ the natural pairing between X' and X. Then $A: K \to X'$ is *monotone* if

$$\langle Au - Av, u - v \rangle \geq 0$$

for all $u, v \in K$. The mapping A is *coercive* if there is a $\omega \in K$ such that

$$\frac{\langle Au - A\omega, u - \omega\rangle}{\|u - \omega\|} - \to \infty$$

as $\|u\| \to \infty, u \in K$. We say that A is *continuous on finite-dimensional subspaces* if the restriction of A to $K \cap M$ is weakly continuous for any finite-dimensional subspace M of X.

Proposition 4.1. Suppose that X is a reflexive Banach space and $K \subset X$ is a nonempty, closed convex set, and that $A : K \to X'$ is monotone, coercive, and continuous on finite-dimensional subspaces. Then there is a solution $u \in K$ of

$$\langle Au, v - u\rangle \geq 0$$

for all $v \in K$. If A is strictly monotone, the solution is unique.

A proof of this proposition will not be given here. It can be found, for example, in Ref. 4. Suppose that $\mathbf{a}:\mathbb{R}^n \to \mathbb{R}^n$, $a = (a_1, \ldots, a_n)$, satisfies

i. Each a_i satisfies a uniform Lipschitz condition on R,
ii. $|\mathbf{a}(\mathbf{y})| \leq K|\mathbf{y}|$,
iii. $(\mathbf{a}(\mathbf{y}) - \mathbf{a}(\mathbf{y}')) \cdot (\mathbf{y} - \mathbf{y}') \geq \nu|\mathbf{y} - \mathbf{y}'|^2$

for $\mathbf{y}$, $\mathbf{y}' \in \mathbb{R}^n$. We call such a vector field *strongly coercive*. If we define A by

$$Au = -(a_i(\text{grad } u))_{x_i}, \tag{22}$$

then A maps $H^1(\Omega)$ into $H^{-1}(\Omega)$. Then, if K is any nonempty closed convex set contained in $H'(\Omega)$, the hypotheses of the Proposition are satisfied. (Exercise 4.11.) We consider, in particular,

$$K = \{u \in H^1(\Omega): u - g \in H_0^1(\Omega)\},$$

where $g \in H^1(\Omega)$. The existence of $u \in K$ such that

$$\langle Au, v - u\rangle = \int_\Omega a_i(\text{grad } u)(v - u)_{x_i}\, d\mathbf{x} \geq 0$$

for all $v \in K$ is guaranteed. If $\varphi \in H_0^1$ is arbitrary, we can always choose v such that $v - u = \varphi$, and as $-\varphi \in H_0^1(\Omega)$ also,

$$\int_\Omega a_i(\text{grad } u)\varphi_{x_i}\, dx = 0$$

We may think of u as a solution of $Au = 0$ in Ω with $u = g$ on $\partial\Omega$ in the weak sense. The special case $\mathbf{a}(\mathbf{x}) = \mathbf{x}$ reduces to the Dirichlet problem for Laplace's equation considered in Section 5 of Chapter 4. The condition SC(iii) may be thought of as an ellipticity condition. If $\mathbf{a} \in C^1(\mathbb{R}^n)$, then $\mathbf{a}(\mathbf{y}) - \mathbf{a}(\mathbf{y}') = \mathbf{a_p} \cdot (\mathbf{y} - \mathbf{y}')$ by the mean value theorem and SC(iii) holds if $(\partial a_i/\partial p_j)y_i y_j \geq c_0|\mathbf{y}|^2$ for any $\mathbf{y} \in \mathbb{R}^n$. This converse is easily seen by integrating this inequality along the segment connecting $\mathbf{y}$, $\mathbf{y}' \in \mathbb{R}^n$.

The condition SC (ii) severely restricts the range of this application. In some situations the structure of the equation points toward a useful generalization. We will consider an example that brings out many of the essential features. Consider the convex function $f(\mathbf{y}) = (1 + |\mathbf{y}|^2)^{\alpha/2}/\alpha$ where $1 \leq \alpha < 2$, and let $\mathbf{a}(\mathbf{y}) = \text{grad} f(\mathbf{y})$, i.e.,

$$a_i(\mathbf{y}) = (1 + |\mathbf{y}|^2)^{(\alpha-2)/2} y_i.$$

It follows immediately that $\mathbf{a} \in C^1(\mathbb{R}^n)$, $|\mathbf{a}(\mathbf{y})| \leq 1 + |y|^{\alpha-1}$, and SC (iii) is satisfied. We define the operator A by (22) again, i.e.,

$$Au = -((1 + |\text{grad } u|^2)^{(\alpha-2)/2} u_{x_i})_{x_i}, \tag{23}$$

but a check of the exponents shows that we may not take $H^1(\Omega)$ as a domain for A. In fact, a natural domain is $W^{1,\alpha}(\Omega)$ and then $A : W^{1,\alpha}(\Omega) \to (W^{1,\alpha}(\Omega))'$. [This follows immediately from $|\mathbf{a}(\mathbf{y})| \leq 1 + |\mathbf{y}|^{\alpha-1}$ as the conjugate exponent of α is $\alpha/(\alpha - 1)$.] We will consider the convex set

$$K = \{u \in W^{1,\alpha}(\Omega) : u - g \in W_0^{1,\alpha}(\Omega)\},$$

where $g \in W^{1,\alpha}(\Omega)$. We take $X = W^{1,\alpha}(\Omega)$. If we can show that the hypotheses of the Proposition are satisfied, existence of a weak solution of the Dirichlet problem will follow as before.

Theorem 4.1. If $1 < \alpha < 2$, there is a unique solution $u \in W^{1,\alpha}(\Omega)$ of the Dirichlet problem

$$-((1 + |\text{grad } u|^2)^{(\alpha-2)/2} u_{x_i})_{x_i} = 0$$

in Ω, $u = g$ on $\partial\Omega$, where $g \in W^{1,\infty}(\Omega)$.

Proof. Because $\alpha > 1$, $W^{1,\alpha}(\Omega)$ is reflexive. It suffices to show that A given in (23) is coercive. We take $w = g$, and, letting $\bar{u} = u - g$, we consider

$$\frac{\langle Au - A\omega, u - \omega\rangle}{\|u - \omega\|} = \frac{\langle Au, \bar{u}\rangle}{\|\bar{u}\|} - \frac{\langle Ag, \bar{u}\rangle}{\|\bar{u}\|} \tag{24}$$

for $\|\bar{u}\| \to \infty$, where $\|\cdot\|$ is the norm in $W^{1,\alpha}(\Omega)$. Because $(\alpha - 2)/2 < 0$, denoting by $\|\cdot\|_\alpha$ the norm in $L^\alpha(\Omega)$,

$$|\langle Ag, \bar{u}\rangle| \leq \int_\Omega |g_{x_i}\bar{u}_{x_i}|d\mathbf{x} \leq \|\text{grad } g\|_{\alpha'}\|\text{grad } \bar{u}\|_\alpha \leq \|\text{grad } g\|_{\alpha'}\|\bar{u}\|$$

and the second term in (24) is bounded. We can write

$$\begin{aligned}\langle Au, \bar{u}\rangle &= \int_\Omega \frac{\text{grad}(\bar{u} + g)\cdot \text{grad } \bar{u}}{(1 + |\text{grad}(\bar{u} + g)|^2)^{(2-\alpha)/2}}\, d\mathbf{x} \\ &= \int_\Omega \frac{|\text{grad}(u)|^2 d\mathbf{x}}{(1 + |\text{grad } u|^2)^{(2-\alpha)/2}} - \int_\Omega \frac{\text{grad }(u)\cdot \text{grad } g\, d\mathbf{x}}{(1 + |\text{grad } u|^2)^{(2-\alpha)/2}}.\end{aligned} \tag{25}$$

The second term in (25) is bounded by

$$\|\text{grad } g\|_{\alpha'}\|u\| \leq \|\text{grad } g\|_{\alpha'}(\|\bar{u}\| + \|g\|).$$

In order to deal with the first term we need the inequality $t^2(1 + t^2)^{(\alpha-2)/2} \geq 2^{(\alpha-2)/2}(t^\alpha - 1)$, which can be verified directly. The first term in (25) is then bounded below by

$$2^{\alpha-2/2}\int_\Omega (|\text{grad}(u)|^\alpha - 1)d\mathbf{x}/\|u\| = 2^{\alpha-2/3}\|u\|^{\alpha-1} - \frac{2^{\alpha-2/2}}{\|u\|}\left(\int_\Omega |u|^\alpha d\mathbf{x} + |\Omega|\right).$$

As the second term is bounded and $\alpha > 1$, coercivity is proven. □

If $\alpha = 1$, the proof of coercivity breaks down, and, further, $W^{1,1}(\Omega)$ is not reflexive. This is not surprising as the equation $Au = 0$, A given in (23), is the minimal surface equation if $\alpha = 1$ and $n = 2$; we will see in Section 4.2 that convexity of Ω is necessary and sufficient for the well-posedness of the Dirichlet problem in this case.

We remark that, in contrast to most of the material of the next subsection, the theory of monotone operators can be applied to higher-order equations and systems.

4.2. Dirichlet Problem for the Minimal Surface Equation. The problem to be discussed in this section arises from the classic problem of finding a surface of least area spanning a curve. If the curve has a simple projection onto a plane, it is natural to seek solutions that are graphs over this plane. Suppose that the plane is the (x, y)-plane and the height of the surface is $u(x, y)$. Then this problem is minimization of

$$A(u) = \int_\Omega (1 + u_x^2 + u_y^2)^{1/2} dx\, dy = \int_\Omega W\, dxdy$$

over functions that have derivatives in an appropriate sense to guarantee finiteness of this integral, and are, say, continuous on $\bar{\Omega}$ with $u = \varphi$, $\varphi \in C(\bar{\Omega})$ a prescribed function, on $\Gamma = \partial\Omega$. Suppose, for example, that we require $u \in W^{1,\infty}(\Omega)$, $\varphi \in W^{1,\infty}(\Omega)$, i.e., u satisfies a uniform Lipschitz condition on Ω. Then u_x and u_y exist a.e. on Ω and are bounded, measurable functions there. If u^* is a solution of this minimum problem and $\zeta \in W^{1,\infty}(\Omega)$, $\zeta = 0$ on $\partial\Omega$, then

$$\frac{d}{dt} A(u^* + t\zeta)\Big|_{t=0} = \int_\Omega (u_x^* \zeta_x + u_y^* \zeta_y) \frac{1}{W^*}\, dx\, dy = 0, \tag{WM}$$

where $W^* = (1 + u_x^{*2} + u_y^{*2})^{1/2}$. We see that u^* is a weak solution of

$$Mu = \left(\frac{u_x}{W}\right)_x + \left(\frac{u_y}{W}\right)_y = 0,$$

the minimal surface equation. If we also knew somehow that $u^* \in C^2(\Omega)$, an integration by parts in (WM) with $\zeta \in C_0^\infty(\Omega)$ would imply that u is a classical solution of $Mu = 0$ in Ω. We will study the Dirichlet problem: find $u \in C(\bar{\Omega}) \cap C^2(\Omega)$ such that $Mu = 0$ in Ω, $u = \varphi$ on $\partial\Omega$. On the face of it this is not equivalent to minimizing area as $A(u)$ need not be finite. Theorem 4.2 below shows otherwise. First we need a result that may be thought of as showing that weak solutions arise only as area minimizers. We introduce the notation

$$A(v, S) = \int_S (1 + v_x^2 + v_y^2)^{1/2} dx\, dy$$

for the area of a graph over a domain S.

Theorem 4.2. If $u \in W^{1,\infty}(\mathscr{R})$ is a weak solution of the minimal surface equation on a domain $\mathscr{R}$, i.e., (WM) holds, then u is area minimizing on $\mathscr{R}$, i.e., if $v \in W^{1,\infty}(\mathscr{R})$ and $v = u$ on $\partial\mathscr{R}$, then $A(u, \mathscr{R}) \leq A(v, \mathscr{R})$.

Proof. Because $\zeta = u - v \in W^{1,\infty}(\mathscr{R})$ and $\zeta = 0$ on $\partial\mathscr{R}$,

$$\int_{\mathscr{R}} \left(\frac{u_x}{W}\zeta_x + \frac{v_y}{W}\zeta_y\right) dxdy = 0.$$

Therefore,

$$\int_{\mathscr{R}} (1 + u_x^2 + u_y^2)^{1/2} dxdy = \int_{\mathscr{R}} \frac{1 + u_x^2 + u_y^2}{W} dxdy = \int_{\mathscr{R}} \frac{1 + u_x v_x + u_y v_y}{W} dxdy$$

and the Cauchy–Schwarz inequality implies

$$A(u, \mathscr{R}) \leq \int_{\mathscr{R}} \frac{(1 + u_x^2 + u_y^2)^{1/2}(1 + v_x^2 + v_y^2)^{1/2}}{W} dxdy = A(v, \mathscr{R}). \qquad \square$$

Theorem 4.3. If $u \in C(\bar{\Omega}) \cap C^2(\Omega)$ is a solution of $Mu = 0$, then $A(u, \Omega) < \infty$.

Proof. Let $\Omega_\delta = \{\mathbf{x} \in \Omega : d = \text{dist}(\mathbf{x}, \partial\Omega) > \delta\}$, and

$$\eta(t) = \begin{cases} 2 - t/\delta, & \delta \leq t < 2\delta, \\ 0, & 2\delta \leq t. \end{cases}$$

Let $v = u\eta(d)$ where $d = d(\mathbf{x})$ is the distance function from $\partial\Omega$. As u is area minimizing on Ω_δ and $v = u$ on $\partial\Omega_\delta$, we have $A(u, \Omega_\delta) \leq A(v, \Omega_\delta)$. The inequality $(1 + a^2 + b^2)^{1/2} \leq \sqrt{1 + a^2} + b$ $(a \geq 0, b \geq 0)$ implies

$$A(v, \Omega_\delta) \leq \int_{\Omega_\delta} (1 + \eta^2(p^2 + q^2))^{1/2} + \int_{\Omega_\delta} |D\eta||u|.$$

Therefore,

$$A(v, \Omega_\delta) \leq A(u, \Omega_\delta \backslash \Omega_{2\delta}) + \delta^{-1}|\Omega_\delta \backslash \Omega_{2\delta}| \max|u|.$$

The second term is bounded, say by c max $|u|$, and then $A(v, \Omega_{2\delta}) \leq c \max|u|$. Letting δ approach zero we obtain the result. $\square$

This is another distinctive feature of the minimal surface equation. In contrast, a solution of Laplace's equation continuous up to the boundary can have infinite Dirichlet integral (see Section 5 of Chapter 4).

Our next result implies an *interior gradient estimate* [inequality (26) below] for classical solutions of $Mu = 0$. We use this result in an essential way below, but the theorem and its elementary proof are of independent interest.

Theorem 4.4. Let u be a positive solution of $Mu = 0$ in $x^2 + y^2 < 1$, with $u(0) = m$. Then

$$|\text{grad } u(0)| \leq c \exp(\pi m/\sqrt{2})$$

with $c < 3$.

Corollary 4.1. If u is a bounded solution of $Mu = 0$ in Ω, then

$$|\text{grad } u(\mathbf{x})| \leq c \exp(\pi M/2d), \tag{26}$$

where $M = \max_\Omega |u|$, and $d = \text{dist}(\mathbf{x}, \partial\Omega)$.

The corollary follows from a scale change applied to a disk contained in Ω.

The proof of Theorem 4.4 makes use of an explicit solution of $Mu = 0$ given by the formula

$$v = c(\ln(\cos((y - b)/c)) - \ln(\cos((x - a)/c))).$$

The graph of this function is known as Scherk's surface. We will restrict the domain of v to be the isosceles right triangle T with vertices (a, b), $(a + \pi c/2,\ b + \pi c/2)$, and $(a + \pi c/2,\ b - \pi c/2)$. We see by inspection that v is positive in T, vanishes on the short sides, and approaches ∞ at the hypotenuse.

We need a lemma also.

Lemma 4.1. Suppose that $Mu = Mv = 0$ in a neighborhood of the origin, $u \neq v$, and grad $\leftarrow u(0) = \text{grad } v(0)$. Then for some $n \geq 2$, $v - u = c + H(x, y) + O(r^{n+1})$, where H is a harmonic polynomial of degree n in a coordinate system affinely related to (x, y), and $r = \sqrt{x^2 + y^2}$.

Proof. Because u and v are real analytic, we can expand them in convergent Taylor series in a neighborhood of the origin. Let $n \geq 2$ be the first index at which their coefficients differ. Then, setting $w = v - u$,

$$(1 + q_0^2)w_{xx} - 2p_0q_0w_{xy} + (1 + p_0^2)w_{yy}$$
$$= (q_0^2 - u_y^2)w_{xx} + (u_y^2 - v_y^2)v_{xx} + \cdots = O(r^{n-1}),$$

where grad $u(0) =$ grad $v(0) = (p_0, \ q_0)$. As $w = c + w_n(x, y) + +O(r^{n+1})$ where w_n is a homogeneous polynomial of degree n, w_n is a solution of

$$(1 + q_0^2)w_{xx} - 2p_0q_0w_{xy} + (1 + p_0^2)w_{yy} = 0.$$

A linear transformation transforms this constant coefficient equation into the Laplace equation. From properties of harmonic polynomials, it follows that the level curves $w = w(0)$ divide a neighborhood of the origin into $2n$ regions in which w is alternatively greater or less than $w(0)$. □

Proof of Theorem 4.4 By rotating the coordinate system we may assume that $\partial u/\partial x > 0$ and $\partial u/\partial y = 0$. Let $b = 0$ in the formula for v. We want to choose $-1 < a < 0$ and $c > 0$ so that the hypotenuse of T is a chord of the unit circle passing through $d \in (0, 1)$, and $v(0) \geq m$. For $\sqrt{2} < \pi c < 2$, the geometric requirement is satisfied by $d = [1 - (\pi c/2)^2]^{1/2}$ and $a = d - \pi c/2$. For πc in this range $v(0)$ varies monotonically from $\sqrt{2}/\pi$ to ∞, and we can certainly satisfy the second requirement with an appropriate choice of c. We will show that

$$|\text{grad}\, u(0)| \leq |\text{grad}\, v(0)|.$$

Suppose that this is not true. Then, using the symmetry of v, $u_x > v_x$, $u_y = v_y = 0$ at the origin.

By translating T to the left we can find a new triangle T' contained in the unit disk (Fig. 2) and a solution v' such that $u < v'$, $u_x = v'_x$, $u_y = v'_y$ at the origin. Let $w = v' - u$. Lemma 4.1 implies that there are (at least) four level curves of w passing through the origin that divide a neighborhood of the origin into at least four subdomains on which w is alternatively bigger and smaller than $w(0) = k$. If $G = \{\mathbf{x} \in T' : \text{dist}(\mathbf{x}, \partial T') \leq \varepsilon, w \neq k\}$, then for ε sufficiently small, G has exactly two components, one adjacent to the hypotenuse where $w > k$, one adjacent to the short sides where $w < k$. It follows that these two components are separated by level lines $w = k$ issuing from the ends of the hypotenuse. As the maximum principle implies that a set where $w > k$ must extend to the boundary, this set has exactly one component. It follows that any two regions near the origin where $w > k$ can be connected by a Jordan arc. Using the same

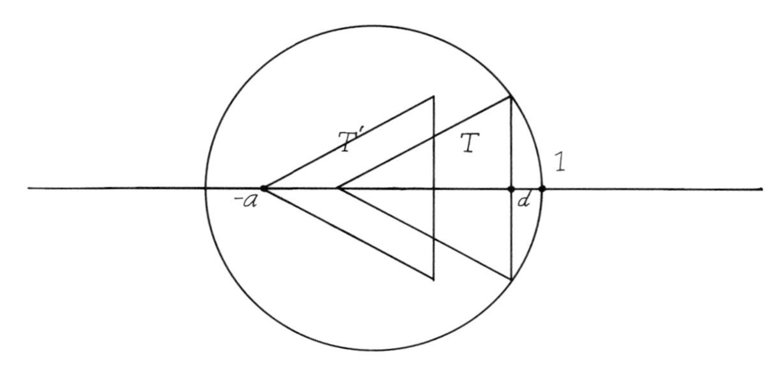

Fig. 2. Triangles for using Scherk's solution as a comparison surface.

argument for the set where $w < k$, we find that there must be two intersecting arcs on which $w > k$ and $w < k$. This contradition proves that

$$|\text{grad } u(0)| \le |\text{grad } v(0)|.$$

Finally, an explicit calculation shows that $v_x(0) = \tan(-a/c)$ and $v(0) = c \ln \sec(-a/c)$. If $u(0) > \sqrt{2}/\pi$, we can choose c so that $u(0) = v(0)$ and the theorem follows with $c = 1$. If $u(0) \le \sqrt{2}/\pi$ we can reach the previous case by adding a positive constant λ to u, and

$$|\text{grad } u(0)| \le e^{\lambda\pi/\sqrt{2}} e^{\pi u(0)/\sqrt{2}}.$$

An upper bound for the required λ is $\sqrt{2}/\pi$. □

We will use a theorem concerning regularity of divergence structure equations. This result will be stated for equations on a domain $\Omega \subset \mathbb{R}^n$ of the form

$$\int A_i(\mathbf{D}u)D_i v = 0 \qquad \text{for all } v \in C_0^1(\Omega), \tag{27}$$

where A_i is bounded, say $|A_i| \le 1$, $A_i(\mathbf{p}) \in C^1(\mathbb{R}^n)$, and

$$\lambda|\xi|^2 \le \frac{\partial A_i}{\partial p_j}\xi_i\xi_j \le |\xi|^2$$

for $\xi \in \mathbb{R}^n$, where λ, $0 \le \lambda \le 1$, is a constant independent of $\mathbf{p}$. The difference operator Δ_i^h is defined by

$$\Delta_i^h w = (w(\mathbf{x} + \mathbf{e}_i h) - w(\mathbf{x}))/h,$$

where $\mathbf{e}_i$ is the unit vector in the ith coordinate direction. We note the property

$$\int x\Delta_i^{-h} g = -\int g\Delta_i^h w$$

of this operator. The following two propositions show how Δ_i^h can be related to weak derivatives.

Proposition 4.2. If $w \in H^1(\Omega)$, $\Omega' \subset \Omega$, and $0 < h < \operatorname{dist}(\Omega', \partial\Omega)$, then

$$\|\Delta_i^h w\|_{L^2(\Omega')} \le \|D_i u\|_{L^2(\Omega)}.$$

Proposition 4.3. If $w \in L^2(\Omega)$, $\|\Delta_i^h w\|_{L^2(\Omega')} \le M$ for all $\Omega' \subset \Omega$ and $0 < h < \operatorname{dist}(\Omega', \partial\Omega)$, then $D_i w$ exists and $\|D_i w\|_{L^2(\Omega)} \le M$.

Proofs are given in Chapter 8.

Theorem 4.5. If $u \in H^1(\Omega)$ is a solution of (27) with A_i satisfying the above hypotheses, then $w \in H^2(\Omega')$ for $\bar{\Omega}' \subset \Omega$.

Proof. Suppose $v \in C_0^1(\Omega)$ and $|2h| < \operatorname{dist}(\partial\Omega, \operatorname{supp} v)$. Then $\Delta_i^{-h} v \in C_0^1(\Omega)$ and

$$0 = \int A_i D_i \Delta_i^{-h} v = \int A_i \Delta_i^{-h} D_i v = -\int \Delta_i^h A_i D_i v.$$

We have

$$\Delta_i^h A_i = h^{-1} \int_0^1 \frac{d}{dt} A_i(\mathbf{D}u(\mathbf{x}) + th\Delta_i^h \mathbf{D}u(\mathbf{x}))dt = \int_0^1 \frac{\partial A_i}{\partial p_j}\, dt \Delta_i^h D_j u := \Theta_{ij} \Delta_i^h D_j u,$$

so that

$$\int \Theta_{ij} \Delta_i^h D_j u D_i v = 0.$$

Note that $|\Theta_{ij}| \leq 1$ and $\Theta_{ij}\xi_i\xi_j \geq \lambda|\xi|^2$. Set $v = \eta^2\Delta_i^h u$ now where $0 \leq \eta \leq 1$ and $\eta \in C_0^1(\Omega)$. Then

$$\int \Theta_{ij}\Delta_i^h D_j u \eta^2 \Delta_i^h D_i u = -2\int \theta_{ij}\eta\Delta_i^h D_j u \Delta_i^h u D_i\eta$$

and

$$\lambda\int \eta^2|\Delta_i^h \mathbf{D}u|^2 \leq 2\ \sup|\mathbf{D}\eta|\left(\int \eta^2|\Delta_i^h \mathbf{D}u|^2\right)^{1/2}\left(\int_{\Omega''}|\Delta_i^h u|^2\right)^{1/2},$$

where $\Omega'' = \text{supp}\ \eta$. Suppose $\eta = 1$ on $\Omega' \subset \Omega''$, $h < \text{dist}(\Omega'', \partial\Omega)$. Then, using Proposition 1 and the inequality $2ab \leq \varepsilon b^2 + a^2/\varepsilon$,

$$\int_{\Omega'}|\Delta_i^h \mathbf{D}u|^2 \leq \int \eta^2|\Delta_i^h u|^2 \leq C(\Omega')\int_{\Omega}|\mathbf{D}u|^2.$$

If $u \in W^{1,\infty}(\Omega)$ is a solution of (WM) with $|u_x|,\ |u_y| \leq K$, that equation satisfies these hypotheses with $\lambda = 1/2\sqrt{1+K^2}$. Therefore, $u \in H^2(\Omega')$ for $\Omega' \subset \Omega$. □

We can also deduce that for any derivative $D^2u = w$ of (27), w is a solution of

$$\int_{\Omega} a_{ij}w_{x_i}\psi_{x_j}\,d\mathbf{x} = 0, \qquad \psi \in C_0^\infty(\Omega), \tag{28}$$

where $a_{ij}(\mathbf{x}) = \partial A_i(\mathbf{p})/\partial p_j$, $\mathbf{p} = \text{grad}\ u(\mathbf{x})$. We need only set $\varphi = D^s\varphi$ in (27) and integrate by parts. We deduce from Theorem 3.4 that

$$|\text{grad}\ u(\mathbf{x}) - \text{grad}\ u(\mathbf{y})| \leq C|\mathbf{x}-\mathbf{y}|^\gamma \tag{29}$$

on Ω' where $0 < \gamma < 1$, γ depends only on λ, n, $|\Omega|$, and $\max_{\Omega'}|\text{grad}\ u|$, and C depends on these quantities and on $\text{dist}(\Omega', \partial\Omega)$. In particular, for a solution of (WM) with $|u_x|$, $|u_y| \leq K$ on Ω we have this estimate with γ depending only on K and C depending on K and $\text{dist}(\Omega', \partial\Omega)$.

Theorem 4.6. If $u \in W^{1,\infty}(\Omega)$ is a solution of (WM) in Ω with $|u_x|$, $|u_y| \leq K$ a.e., then grad u satisfies a uniform Hölder condition (29) on any compact subdomain of Ω. The exponent γ depends only on K, and C depends only on K and on the distance of this compact subdomain from $\partial\Omega$.

We now consider the problem of proving the existence of a minimizer in $W^{1,\infty}(\Omega)$. We denote the norm in $W^{1,\infty}(\Omega)$ by $\|\cdot\|$, i.e.,

$$\|u\| = \lim_{\substack{x,y\in\Omega \\ x\neq y}} \frac{|u(\mathbf{x}) - u(\mathbf{y})|}{|\mathbf{x} - \mathbf{y}|}$$

We denote by $L_K(S, \psi)$ the functions v in $W^{1,\infty}(S)$ such that $\|v\| \leq K$ and $v = \psi$ on ∂S. If the values on ∂S are unrestricted, we write $L_K(S)$. An a priori bound for the Lipschitz constant of an area minimizer will be obtained by using a comparison principle for weak solutions.

If $u \in W^{1,\infty}(\Omega)$ and $W = (1 + u_x^2 + u_y^2)^{1/2}$, we say that u is a *subsolution* of $Mu = 0$ on Ω if

$$\int_\Omega (u_x\zeta_x + u_y\zeta_y)W^{-1}dxdy \leq 0 \tag{30}$$

for all $\zeta \in W^{1,\infty}(\Omega)$, $\zeta \geq 0$, $\zeta = 0$ on $\partial\Omega$. If the inequality is reversed, we say that u is a *supersolution*.

Lemma 4.2. If z is a subsolution and $v \in W^{1,\infty}(\Omega)$, $z \geq v, z = v$ on $\partial\Omega$, then

$$A(z, \Omega) \leq A(v, \Omega). \tag{31}$$

If w is a supersolution, $v \geq w$, $w = v$ on $\partial\Omega$, then

$$A(\omega, \Omega) \leq A(v, \Omega). \tag{32}$$

Proof. We have

$$\begin{aligned}\int_\Omega \sqrt{1 + z_x^2 + z_y^2}\, dxdy &= \int_\Omega W\, dx\, dy = \int_\Omega \frac{1 + z_x^2 + z_y^2}{W}\, dxdy \\ &\leq \int_\Omega \frac{1 + z_xv_x + z_yv_y}{W}\, dxdy \leq \int_\Omega \sqrt{1 + v_x^2 + v_y^2}\, dxdy\end{aligned}$$

as $\zeta = z - v \geq 0$ and (30) holds for z. This establishes (31). In order to establish (32) we set $\zeta = v - w$ in

$$\int_\Omega (w_x\zeta_x + w_y\zeta_y)W^{-1}\, dxdy \leq 0,$$

where $W = \sqrt{1 + w_x^2 + w_y^2}$. □

Lemma 4.3. Suppose that u minimizes A over functions in $W^{1,\infty}(\Omega)$ that coincide with φ on $\partial\Omega$.

a. If w is a supersolution on $S \subset \Omega$ and $w \geq u$ on ∂S, then $w \geq u$ on S.
b. If z is a subsolution on $S \subset \Omega$ and $z \leq u$ on ∂S, then $z \leq u$ on S.

Proof. We will show (a). The proof of (b) is almost identical. We let

$$T = \{\mathbf{x} \in S : w < u\}$$

be nonempty and seek a contradiction. Note that $w = u$ on ∂T. Let $K = \|u\|$. As u minimizes area in $L_K(T, u)$, $A(w, T) \geq A(u, T)$. On the other hand, w is a supersolution and $w \leq u$ on T, $w = u$ on ∂T, so Lemma 4.2 implies $A(w, T) \leq A(u, T)$. Therefore, $A(w, T) = A(u, T)$. As grad $u \neq$ grad w on the nonempty open set T, strict convexity of A implies

$$A\left(\frac{w+u}{2}, T\right) < \frac{1}{2}A(w, T) + \frac{1}{2}A(u, T) = A(u, T).$$

This contradicts $A[(w+u)/2, T] \geq \frac{1}{2}A(u, T)$. □

Lemma 4.4 If u_1 and u_2 are area minimizers, then

$$\sup_{\Omega}|u_1 - u_2| = \sup_{\partial\Omega}|u_1 - u_2|.$$

Proof. We need only observe that area minimizers are solutions of (WM) and that solutions of (WM) are both subsolutions and supersolutions. □

Lemma 4.5. If u is an area minimizer as in Lemma 4.3, then

$$\|u\| = \sup_{\substack{x \in \Omega \\ y \in \partial\Omega}} \frac{|u(\mathbf{x}) - u(\mathbf{y})|}{|\mathbf{x} - \mathbf{y}|}. \tag{33}$$

Proof. If $\mathbf{x}_1, \mathbf{x}_2 \in \Omega$, $\mathbf{x}_1 \neq \mathbf{x}_2$, let $\mathbf{t} = \mathbf{x}_2 - \mathbf{x}_1$ and consider $u_{\mathbf{t}}(\mathbf{x}) := u(\mathbf{x} + \mathbf{t})$. $u_{\mathbf{t}}$ minimizes area on $\Omega_{\mathbf{t}} = \{\mathbf{x} : \mathbf{x} + \mathbf{t} \in \Omega\}$. As both u and $u_{\mathbf{t}}$ minimize area on the nonempty set $\Omega \cap \Omega_{\mathbf{t}}$, Lemma 4.4 implies that

$$|u(\mathbf{x}_1) - u(\mathbf{x}_2)| = |u(\mathbf{x}_1) - u_{\mathbf{t}}(\mathbf{x}_1)| \leq |u(\mathbf{z}) - u_{\mathbf{t}}(\mathbf{z})| = |u(\mathbf{z}) - u(\mathbf{z} + \mathbf{t})|$$

for some $\mathbf{z} \in \partial(\Omega \cap \Omega_t)$. At least one of z and $\mathbf{z} + \mathbf{t}$ belongs to $\partial\Omega$, and

$$|u(\mathbf{x}_1) - u(\mathbf{x}_2)| \leq L|\mathbf{x}_1 - \mathbf{x}_2|,$$

where L is the right-hand side of (33). □

For $t > 0$, we denote

$$\Sigma_t = \{\mathbf{x} \in \Omega : \text{dist}(\mathbf{x}, \partial\Omega) < t\}$$

and $\Gamma_t = \partial\Sigma_t \cap \Omega$.

Definition 4.1. An *upper barrier* relative to φ is a function $v^+ \in W^{1,\infty)}(\Sigma_t)$ for some $t > 0$ such that

i. $v^+ = \varphi$ on $\partial\Omega$ and $v^+ \geq \sup_{\partial\Omega}\varphi$ on Γ_t, and
ii. v^+ is a supersolution in Σ_t.

A *lower barrier* relative to φ is a function $v^- \in W^{1,\infty}(\Sigma_t)$ for some $t > 0$ such that

i. $v^- = \varphi$ on $\partial\Omega$, $v^- \leq \inf_{\partial\Omega}\varphi$ on Γ_t, and
ii. v^- is a subsolution on Σ_t.

Theorem 4.7. Suppose that there exist upper and lower barriers v^+, v^- relative to φ. Then

$$\|u\| \leq Q,$$

where Q is a constant determined by the Lipschitz constants of v^+, v^- on Σ_t, the diameter of Ω, and t

Proof. As constants are both supersolutions and subsolutions, Lemma 4.4 implies $\inf_{\partial\Omega}\varphi \leq u(\mathbf{x}) \leq \sup_{\partial\Omega}\varphi$ on Ω, and, as this implies $v^- \leq u \leq v^+$ on Γ_t, that $v^- \leq u \leq v^+$ on Σ_t. Then, for $\mathbf{x} \in \Sigma_t$, $\mathbf{y} \in \partial\Omega$,

$$u(\mathbf{x}) - u(\mathbf{y}) = u(\mathbf{x}) - v^+(\mathbf{y}) \leq v^+(\mathbf{x}) - v^+(\mathbf{y}) \leq Q|\mathbf{x} - \mathbf{y}|.$$

Similarly,

$$u(\mathbf{x}) - u(\mathbf{y}) \geq v^-(\mathbf{x}) - v^-(\mathbf{y}) \geq -Q|\mathbf{x} - \mathbf{y}|.$$

If $\mathbf{x} \in \Omega\backslash\Sigma_t$, $u(\mathbf{x}) - u(\mathbf{y}) \leq \sup\varphi - u(\mathbf{y})$ and $u(\mathbf{y}) - u(\mathbf{x}) \leq u(\mathbf{y}) - \inf\varphi$, so

$$|u(\mathbf{x}) - u(\mathbf{y})| \leq \max[\sup\varphi - u(\mathbf{y}), u(\mathbf{y}) - \inf\varphi]$$
$$\leq Q\sup_{Z\in\partial\Omega}|\mathbf{z} - \mathbf{y}| \leq \frac{Qd|\mathbf{x} - \mathbf{y}|}{\inf\limits_{z\in\Omega\backslash\Sigma_t}|\mathbf{z} - \mathbf{y}|} = \frac{Qd}{t}|\mathbf{x} - \mathbf{y}|,$$

where $d = \sup\limits_{z\in\partial\Omega}|\mathbf{z} - \mathbf{y}|$. □

First consider the restricted problem of minimizing A in $L_K(\Omega, \varphi)$.

Lemma 4.6. There is a minimum of A on $L_K(\Omega, \varphi)$.

Proof. A minimizing sequence has a subsequence converging uniformly to a function in $L_K(\Omega, \varphi)$. It suffices therefore to prove that A is *lower semi-continuous* on $L_K(\Omega, \varphi)$, i.e., that

$$\|u_n - u\| \to 0 \Rightarrow \liminf A(u_n) \geq A(u).$$

By the convexity of the integrand,

$$A(u_n) - A(u) \geq \int_\Omega \left(\frac{u_x}{W}(u_n - u)_x + \frac{u_y}{W}(u_n - u)_y\right)dxdy,$$

where $W = \sqrt{1 + u_x^2 + u_y^2}$. Consider the first term, the second being similar, and let $\varphi = u_x/W$. Let $\varepsilon > 0$ and choose $\varphi_\varepsilon \in C_0^1(\Omega)$ such that

$$\int_\Omega |\varphi_\varepsilon - \varphi|dxdy < \varepsilon/2K.$$

Then

$$\left|\int_\Omega \varphi(u_n - u)_x\,dxdy\right| \leq \left|\int_\Omega \varphi_\varepsilon(u_n - u)_x\,dxdy\right| + \int_\Omega |\varphi_\varepsilon - \varphi||(u_n - u)_x|dxdy.$$

In the first term integration by parts yields

$$\int_\Omega \varphi_\varepsilon(u_n - u)_x\,dxdy \to 0$$

and $|(u_n - u)_x| \leq 2K$ implies

$$\limsup \int_{\Omega} |\varphi_\varepsilon - \varphi||(u_n - u)_x| dx\, dy \leq \varepsilon.$$

As ε is arbitrary,

$$\liminf (A(u_n) - A(u)) \geq 0. \qquad \square$$

Definition 4.2. A function $\omega \in L_K(\Omega)$ is an $L_K(\Omega)$ supersolution if $A(v, \Omega) \geq A(\omega, \Omega)$ for every $v \in L_K(\Omega, \omega)$ with $v \geq \omega$. A function $z \in L_K(\Omega)$ is an $L_K(\Omega)$ subsolution if $A(v, \Omega) \geq A(z, \Omega)$ for every $v \in L_K(\Omega, z)$ with $v \leq z$.

In other words, we require that the conclusions of Lemma 4.3 hold.

Lemma 4.7. If ω is an $L_K(\Omega)$ supersolution and z an $L_K(\Omega)$ subsolution, and $\omega \geq z$ on $\partial\Omega$, then $\omega \geq z$ on Ω.

The proof is left as an exercise.

We say that "u minimizes area in $L_K(\Omega)$" if u minimizes area over functions in $L_K(\Omega)$ that coincide with u on $\partial\Omega$. If a function minimizes area in $L_K(\Omega)$, it does so in $L_{\tilde{K}}(\tilde{\Omega})$ for any $\tilde{\Omega} \subset \Omega$, $\tilde{K} \leq K$. If u and v minimize area in $L_K(\Omega)$, an immediate consequence of Lemma 4.7 is that $\sup_\Omega |u - v| = \sup_{\partial\Omega} |u - v|$. [Minimizers of area in $L_K(\Omega)$ are automatically both $L_K(\Omega)$ subsolutions and $L_K(\Omega)$ supersolutions.]

Lemma 4.8. If u minimizes area in $L_K(\Omega)$,

$$\|u\| = \sup_{\substack{x \in \Omega \\ y \in \partial\Omega}} \frac{|u(\mathbf{x}) - u(\mathbf{y})|}{|\mathbf{x} - \mathbf{y}|}.$$

The proof is identical to that of Lemma 4.5.

Theorem 4.8. If upper and lower barriers relative to φ exist, the area minimization problem in $W^{1,\infty}(\Omega)$ has a solution.

Proof. Let Q be the constant of Theorem 4.7 and choose $K > Q$. If u_K gives the minimum of area in $L_K(\Omega, \varphi)$, we can mimic the argument in the proof of Theorem 4.7, using Lemmas 4.7 and 4.8, to show that

$$\|u_K\| \leq Q.$$

We will show that $\|u_K\| < K$ implies $u = u_K$ is a solution of our problem. Let $v \in W^{1,\infty}(\Omega)$, $v = \varphi$ on $\partial\Omega$, and for $0 < t < 1$, $v_t = u_K + t(v - u_K)$. Then $A(u_K, \Omega) < A(v_t, \Omega)$ and convexity of the functional A implies

$$A(u_K, \Omega) \le (1 - t)A(u_K, \Omega) + tA(v, \Omega),$$

so that

$$A(u_K, \Omega) \le A(v, \Omega). \qquad \square$$

We emphasize that Theorem 4.7 was not logically necessary for this result. The independent proof of the a priori estimate is of interest on its own merit.

In order to give substance to Theorem 4.8 we must give conditions that guarantee existence of upper and lower barriers. We will investigate this for $\varphi \in C^2(\mathbb{R}^2)$. It is easily seen that $v \in C^2(\Sigma_t) \cap W^{1,\infty}(\Sigma_t)$ is a supersolution if and only if $Mv \le 0$. We can write $Mv = W^{-3}Nv$ where

$$Nv = (1 + v_y^2)v_{xx} - 2v_x v_y v_{xy} + (1 + v_x^2)v_{yy},$$

so it suffices that $Nv \le 0$. Suppose that $v = \varphi + \psi(d)$ where $\psi \in C^2[0, t]$, $\psi(0) = 0$, $\psi' \ge 1$, $\psi'' < 0$, and $d(\mathbf{x})$ is the distance function from $\partial\Omega$. A direct calculation shows that

$$\begin{aligned} Nv = N\varphi &+ (\psi'(1 + |\text{grad } \varphi|^2) + (\psi')^3)\Delta d \\ &+ \psi(2 \text{ grad } \varphi \cdot \text{grad } d\Delta\varphi - (D\varphi)^T D^2 d D\varphi - 2(Dd)^T D^2\varphi D\varphi) \\ &+ \psi^2(\Delta\varphi - (Dd)^T D^2\varphi Dd) + \psi''(1 + |\text{grad } \varphi|^2 - \text{grad } \varphi \cdot \text{grad } d). \end{aligned} \tag{34}$$

In this equation D^2 denotes the Hessian matrix. Then, as $\psi' \ge 1$, $\psi'' < 0$,

$$Nv \le C(\psi')^2 + \psi'' + ((1 + |\text{grad } \varphi|^2)\psi' + (\psi')^3)\Delta d,$$

where $C = C(\varphi) > 0$. In the above we need to assume some regularity of $\partial\Omega$ in order for these formal calculations to make sense. In particular, it can be shown that if $\partial\Omega \in C^2$, there is a $t > 0$ such that $d \in C^2(\Sigma_t)$. Furthermore, if Ω is convex, then $\Delta d \le 0$. We illustrate this in a special case that we will use below, namely, Ω a disk of radius R. If r is a polar coordinate with the center as origin, then $d = R - r$, and $\Delta d = -1/r < 0$. We can then say that $Nv \le C(\psi)^2 + \psi''$. We choose $\psi(d) = (1/C)\ln(1 + \beta d)$. For $0 \le d \le t$,

$$\psi'(d) \ge \frac{\beta}{C(1 + \beta t)}$$

and $\psi(t) = 1/C \ln(1 + \beta t)$. If we take, say, $t = \beta^{-1/2}$ and β big enough, then $\psi' \geq 1$ and $\psi(t) \geq 2 \sup_\Omega |\varphi|$. It follows that v is an upper barrier relative to φ. The construction of a lower barrier is similar. This establishes the existence of a solution of (WM) if Ω is a disk and $\varphi \in C^2(\mathbb{R}^2)$. [To extend this to a convex domain with C^2 boundary, we need only prove the property of $d(\mathbf{x})$ mentioned above.]

The solution we have obtained is actually a classical solution.

Theorem 4.9. u is analytic in Ω.

We need the following lemma whose proof is sketched in the exercises.

Lemma 4.9. Suppose that $\mathscr{R}$ is a rectangle and $a, b \in C(\bar{\mathscr{R}})$ are such that

$$\int_{\mathscr{R}} (a\zeta_x + b\zeta_y) dxdy = 0 \tag{H}$$

for all $\zeta \in C^1(\bar{\mathscr{R}})$ with $\zeta = 0$ on $\partial\mathscr{R}$. Then there exists $\omega \in C^1(\mathscr{R})$ such that $\omega_x = -b$, $\omega_y = a$ in $\mathscr{R}$.

Proof of Theorem 4.9. We will apply Lemma 4.9 to a rectangle $\mathscr{R}$ with $\bar{\mathscr{R}} \subset \Omega$. First, note that (WM) is of the form (H) with $a = p/W$, $b = q/W$ so that there exists ω_1 with $\omega_{1x} = -q/W$, $\omega_{1y} = p/W$. Consider a transformation from points in a (ξ, η)-plane to the (x, y)-plane given by

$$x = \xi + \varepsilon\zeta(\xi, \eta), \qquad y = \eta,$$

where ζ is a C^1 function with compact support. If $\mathrm{M} = \max |\zeta_\xi|$ and $|\varepsilon| < \mathrm{M}^{-1}$, this transformation is one to one, and there is a domain S mapped onto $\mathscr{R}$. We suppose that the support of ζ is a subset of S. Let $u(\xi, \eta)$ be the transplantation to S of u; then the function $u(x, y; \varepsilon)$ is in $\mathscr{L}$. A direct calculation shows that $u_x = u_\xi/(1 + \varepsilon\zeta_\xi)$, $u_y = -\varepsilon u_\xi/(1 + \varepsilon\zeta_\xi) + u_\eta$, and the Jacobian of the transformation is $1 + \varepsilon\zeta_\xi$. If

$$\varphi(\varepsilon) = \int_{\mathscr{R}} \sqrt{1 + u_x^2 + u_y^2}\, dxdy,$$

then

$$\varphi'(0) = \int_{\mathscr{R}} \left(\frac{1 + q^2}{W}\zeta_x - \frac{pq}{W}\zeta_y\right) dxdy = 0$$

$(p = u_x, q = u_y)$ as the transformation reduces to the identity when ε is zero. Applying Lemma 4.9 again yields ω_3 with $\omega_{3x} = pq/W$, $\omega_{3y} = (1+q^2)/W$. If we interchange the roles of x and y, the same reasoning shows that

$$\int_{\mathscr{R}} \left(\frac{1+p^2}{W} \zeta_y - \frac{pq}{W} \zeta_x \right) dxdy = 0$$

and there is ω_2 such that $\omega_{2x} = -(1+p^2)/W$, $\omega_{2y} = -pq/W$. Now introduce the transformation $\alpha = x$, $\beta = \omega_3(x, y)$. As $\beta_y = \omega_{3y} > 0$, this transformation is one to one and maps $\mathscr{R}$ onto a domain $\mathscr{R}^*$ in the $(\alpha, \beta) =$ plane. If $x = \alpha$, $y = y(\alpha, \beta)$, then $y_\alpha = -\omega_{3x}/\omega_{3y} = -pq/(1+q^2)$ and $y_\beta = 1/\omega_{3y} = W/(1+q^2)$. It follows that $x = \alpha$ and $\omega_3(\alpha, y(\alpha, \beta))$ satisfy the Cauchy–Riemann equations $x_\alpha = \omega_{3\beta}$, $x_\beta = -\omega_{3\alpha}$ in $\mathscr{R}^*$. A direct calculation using the chain rule shows that $y(\alpha, \beta)$, $\omega_2(\alpha, y(\alpha, \beta))$ also satisfy $y_\alpha = \omega_{2\beta}$, $y_\beta = -\omega_{2\alpha}$. For example, we have $\omega_{2\beta} = \omega_{2y} y_\beta = (pq/W)[W/(1+q^2)]$. Also, $u(\alpha, \beta) = u(x(\alpha, \beta), y(\alpha, \beta))$ and $\omega_1(\alpha, y(\alpha, \beta))$ satisfy $u_\alpha = \omega_{1\beta}$, $u_\beta = -\omega_{1\alpha}$. As each of $x, y, u, \omega_1, \omega_2, \omega_3$ is therefore a harmonic function of (α, β), they are all real analytic functions of these variables. As ω_3 is analytic, the inverse of the transformation, $\alpha(x, y)$, $\beta(x, y)$, has analytic components (see Ref. 5, Chapter 10), and $u(x, y) = u(\alpha(x, y), \beta(x, y))$ is analytic. □

Suppose we consider arbitrary $\varphi \in C(\partial\Omega)$, Ω a disk. If $\varphi_n \in C^2(\mathbb{R})$, φ_n converging uniformly to φ, and u_n is the corresponding solution of $Mu = 0$ in Ω, then

$$\sup_\Omega |u_n - u_m| \leq \sup_{\partial\Omega} |\varphi_n - \varphi_m|,$$

so $\{u_n\}$ converges uniformly to a function $u \in C(\bar{\Omega})$ with $u = \varphi$ on $\partial\Omega$.

Theorem 4.10. The Dirichlet problem for a disk has a solution for arbitrary $\varphi \in C(\partial\Omega)$.

Proof. Suppose that $\bar{\Omega}' \subset \Omega$. Theorem 4.4 implies that $|\text{grad}\, u_n| \leq C(M, d)$ on Ω' where $d = \text{dist}(\Omega', \partial\Omega)$ and $|u_n|$, $|u| \leq M$ on Ω. We can then bound the Hölder norms of grad u_n, uniformly with respect to n on Ω'. The Arzelà–Ascoli theorem implies that a subsequence of grad u_n converges uniformly on Ω'. It follows easily that $u \in C^1(\Omega')$ and grad u_n converge to grad u. Theorem 4.9 then implies that u is analytic on Ω'. □

We conclude this section with an extension of the conclusion of Theorem 4.10 to an arbitrary convex domain. Suppose that Ω is an arbitrary bounded region. For $u \in C(\bar{\Omega})$ and $B \subset \Omega$ we define $M_B[u]$ to be the function equal to u

off B and equal to the solution of the minimal surface equation in B with boundary values u on ∂B. Theorem 4.10 implies that this solution can always be found. A function $v \in C(\bar{\Omega})$ is *subminimal* if $v \leq M_B[v]$ for any $B \subset \Omega$. If $v \geq M_B[v]$ for any $B \subset \Omega$, u is *superminimal*. The following results are stated without proof. The proofs are completely analogous to the corresponding results for harmonic functions (Chapter 4).

Theorem 4.11. If v is subminimal on Ω and takes its maximum on at an interio point of Ω, then v is constant.

Theorem 4.12. If v_1, v_2 are subminimal, then $\max\{v_1, v_2\}$ is subminimal.

Theorem 4.13. If v is subminimal, then $M_B[v]$ is subminimal for any $B \subset \Omega$.

Suppose that $\varphi \in C(\partial\Omega)$. A *subfunction* for the Dirichlet problem with data φ is a subminimal function on Ω that is not larger than φ on $\partial\Omega$. The set of subfunctions is denoted by $\mathscr{F}$. As any $c \leq \min_{\partial\Omega}\varphi$ is a subfunction, $\mathscr{F}$ is nonempty.

Theorem 4.14. (a) v_1, $v_2 \in \mathscr{F}$ implies $\max\{v_1, v_2\} \in \mathscr{F}$, (b) $v \in \mathscr{F}$ implies $M_B[v] \in \mathscr{F}$ for $B \subset \Omega$.

This is an immediate consequence of Theorems 4.12 and 4.13. As for the Dirichlet problem for Laplace's equation, we define

$$u = \sup_{\mathscr{F}}[v].$$

Theorem 4.15. $Mu = 0$.

Proof. If $B_1 \subset \Omega$, chose $\bar{B} \subset B_1$. We will show $Mu = 0$ on B. If $v_1 \in \mathscr{F}$, we can consider $\mathscr{F}^* = \{\max[v, v_1]: v \in \mathscr{F}\}$. All functions in $\mathscr{F}^*$ are bounded below and $u = \sup_{\mathscr{F}^*} v$. Let $\mathbf{x}_j$ be a sequence of distinct points in B_1 that is dense in B_1. For each j and k we can choose $v_{j,k} \in \mathscr{F}^*$ such that

$$0 \leq u(\mathbf{x}_j) - v_{j,k}(\mathbf{x}_j) \leq 1/k.$$

Then $\max\{v_{1,k}, \ldots, v_{k,k}\}$ satisfies the same inequality at $\mathbf{x}_1, \ldots, \mathbf{x}_k$, and so does $v_k = M_{B_1}[\max\{v_{1,k}, \ldots, v_{k,k}\}]$. It follows that $\lim_{k\to\infty} v_k(\mathbf{x}_j) = u(\mathbf{x}_j)$ for each j. As in the proof of Theorem 4.10, because $\{v_k\}$ is a uniformly bounded family of solutions of the minimal surface equation on B_1 we can give uniform bounds on the magnitude and Hölder norm of grad v_k on B. It follows that v_k converges uniformly to u on B (see Exercise 4.11), and, as in the proof of Theorem 4.10, $Mu = 0$ there. The proof is completed. □

We must now deal with the question as to whether u takes on the boundary values φ. It suffices to find a *barrier* at $\mathbf{y} \in \partial\Omega$ to show that $\lim_{\mathbf{x}\to\mathbf{y}}(\mathbf{x}) = \varphi(\mathbf{y})$.

Definition 4.3. A *local barrier* at $\mathbf{y} \in \partial\Omega$ is a function ω satisfying

a. w is defined and superminimal on $\mathcal{N} \cap \bar{\Omega}$ for some neighborhood $\mathcal{N}$ of $\mathbf{y}$,
b. $w > 0$ on $\mathcal{N} \cap \bar{\Omega}\backslash\{\mathbf{y}\}$, and
c. $w(\mathbf{y}) = 0$.

In (a) we may assume that $\mathcal{N} \subset B$ for a closed disk B centered at $\mathbf{y}$. A *barrier* is a function defined on $\bar{\Omega}$ that satisfies (a), (b), and (c) with $\mathcal{N} \cap \bar{\Omega}$ replaced by $\bar{\Omega}$.

Lemma 4.10. If Ω has a local barrier at $\mathbf{y} \in \partial\Omega$, Ω has a barrier at $\mathbf{y}$.

Proof. If B_1 is a concentric closed subdisk of B and m is the (positive) infimum of w over $\bar{\Omega} \cap (\mathcal{N}\backslash B_1)$, then

$$w_1 = \begin{cases} \min[m, w] & \text{on } B_1 \cap \bar{\Omega} \cap \mathcal{N}, \\ m & \text{on } \bar{\Omega} \cap (B_1 \cap \mathcal{N}^c)' \end{cases}$$

is continuous, and clearly superminimal on each of $\Omega\backslash(B_1 \cap \mathcal{N})^c$ (constant) and $\bar{\Omega} \cap (B_1 \cap \mathcal{N})$ (minimum of two superminimal functions). As a "locally superminimal" function must be superminimal (Exercise 4.12), w_1 is a barrier. □

Theorem 4.16. If Ω has a barrier at $\mathbf{y}$, then $\lim_{\mathbf{x}\to\mathbf{y}} u(\mathbf{x}) = \varphi(\mathbf{y})$.

Proof. As for harmonic functions, consider the two functions $\varphi(\mathbf{y}) - \varepsilon - kw(\mathbf{x})$ and $\varphi(\mathbf{y}) + \varepsilon + Kw(\mathbf{x})$ for k, K sufficiently large. □

Theorem 4.17. If $\partial\Omega$ is locally convex at $\mathbf{y}$, i.e., $B \cap \Omega$ is convex for some disk centered at $\mathbf{y}$, then Ω has a local barrier at $\mathbf{y}$.

Proof. There is a line through $\mathbf{y}$ such that $\partial\Omega \cap B$ lies on the left of this line. Take a line perpendicular to this line as the x-axis and choose the origin inside $\Omega \cap B$ (Fig. 3).

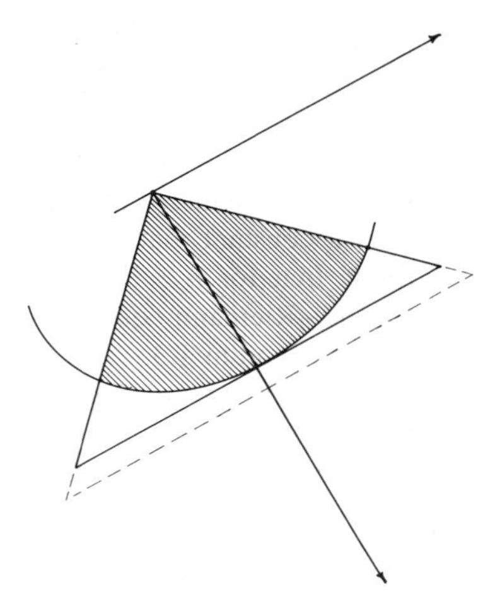

Fig. 3. Scherk's solution provides a barrier at locally convex boundary points.

The Scherk solution used in the proof of Theorem 4.4 with $a = b = 0$ and c chosen so that the hypotenuse lies to the right of the line through **y** can be used to define a local barrier. In fact, if the value of v at **y** is M, then $M - v$ is local barrier. This implies the assertion made earlier that the Dirichlet problem for $Mu = 0$ has a solution for all continuous boundary values f if and only if Ω is convex. □

The Perron method provides a candidate for a solution of the Dirichlet problem if Ω is nonconvex. It is interesting to ask about the behaviour of this "generalized solution" at nonconvex boundary points. In the particular case in which Ω is a quadrilateral with a reentrant corner, complete information is available (Refs. 6, 7).

We have made a considerable effort in studying the nonparametric Plateau problem of finding a minimal surface spanning a given curve. This problem is of great intrinsic and historical interest. On the other hand, most of the techniques used are of wider use in the study of second-order elliptic equations. We mention in particular that these techniques can be applied to n variables, and the method of seeking solutions of variational problems in $L_K(\Omega)$ and using estimates for the Lipschitz constant is generally useful (Ref. 3, Chapter 10). The applications of an interior gradient estimate, and the construction of barriers using conditions on boundary curvatures are also widely used.

Exercises

1.1. (Reduction to canonic form). A smooth transformation of variables $\mathbf{y} = \mathbf{y}(\mathbf{x})$, with Jacobian matrix $\mathbb{J}$ invertible in a domain $D \subset \Omega$, transforms the operator $L_0 u = a_{ij}(\mathbf{x})u_{x_i x_j}$ into a second-order operator with principal part defined by the matrix $\bar{\mathbb{A}} = \mathbb{J}\mathbb{A}\mathbb{J}^{\mathrm{T}}$, $\mathbb{A} = [a_{ij}(\mathbf{x})]$. Write the differential equations for $\mathbf{y}(\mathbf{x})$ required in order that $\bar{\mathbb{A}}$ be the unit matrix. Hence, deduce that the principal part of a second-order elliptic operator can in principle be reduced to a Laplacian in the whole of D only if $n \leq 2$, unless $\mathbb{A}$ is a constant matrix (and then $\mathbf{y} = \mathbb{C}\mathbf{x}$, see beginning of this section). Show that for $n = 2$ and $a_{ij} = \delta_{ij}$, the equations for $\mathbf{y}(\mathbf{x})$ reduce to the Cauchy–Riemann equations.

1.2. Prove Theorem 1.3.

1.3. In Theorem 1.4 we need only assume that the operator $Q(u)$ is elliptic with respect to the function u in order to prove this theorem, that is, the linear equation with coefficients $\tilde{a}_{ij}(\mathbf{x})$ and $\tilde{b}_i(\mathbf{x})$ satisfies the hypotheses of Theorem 1.1. Show that if $Q(u) > Q(v)$, $u < v$ in Ω.

1.4. Carry out the details in finding the polynomials V and Z in the counter-example given above.

1.5. Show that $u = (1 - (x^2 + y^2))/((1 - x)^2 + y^2)$ is harmonic in the unit disk. As $u = 0$ for $x^2 + y^2 = 1$, $(x, y) \neq (1, 0)$, the maximum principle does not apply. Why?

1.6. If the solution of $\Delta u = -1$ in Ω, $u = 0$ on $\partial\Omega$, has constant normal derivative c on $\partial\Omega$, then Ω is a ball of radius nc and $u = (n^2c^2 - r^2)/2n$. The steps are:

a. If r is the distance from a fixed point, $\Delta(r\partial u/\partial r) = r\partial/\partial r(\Delta u) + 2\Delta u = -2$.

b. Apply Green's identity to $r\partial u/\partial r$ and $-u$ to show that

$$\int_\Omega (2u - r\partial u/\partial r) d\mathbf{x} = nc^2\mathbf{V},$$

where $\mathbf{V}$ is the volume of Ω. Then $\int_\Omega r\partial u/\partial r\, d\mathbf{x} = -n\int_\Omega u\, d\mathbf{x}$ implies that $(n+2)\int_\Omega u\, d\mathbf{x} = nc^2\mathbf{V}$.

c. Show that $(\Delta u)^2 = 1$ implies $\Delta(|\text{grad } u|^2 + 2u/n) \geq 0$.

d. As $|\text{grad } u|^2 + 2u/n = c^2$ on $\partial\Omega$, either this function is constant on Ω or

$$|\text{grad } u|^2 + 2u/n < c^2 \qquad \text{in } \Omega.$$

Integration of this inequality over Ω implies

$$(1 + 2/n)\int_\Omega u\, d\mathbf{x} < c^2\mathbf{V},$$

which contradicts the result of (b).

e. Use the constancy of $|\text{grad}\, u|^2 + 2u/n$ and $\Delta u = -1$ to show that $D_{ij}u = -\delta_{ij}/n$.

f. Show that $u = (A - r^2)/2n$ for some constant A with a suitable choice of origin.

2.1. A minimal surface of revolution generated by a curve $y = y(x)$, $\alpha \leq x \leq \beta$, rotated around the x-axis is an extremal for

$$2\pi \int_{\alpha}^{\beta} y\sqrt{1 + y'^2}\, dx.$$

Letting $f(x, y, y') = y\sqrt{1 + y'^2}$, the Euler–Lagrange equation for this problem is $f_y - (d/dx) f_{y'} = 0$. Show that $y = d \cosh[x - c/d]$, $d > 0$ is a solution. The curve generating this surface is called a *catenary*.

2.2. Discuss the possibilities for finding c, d so that $y(\alpha) = A$, $y(\beta) = B$ for given $A, B > 0$. *Hint:* First impose $y(\alpha) = A$ and consider the resulting one-parameter family of catenoids. There exists a curve of right endpoints on which there is a unique solution. Above this curve there are two solutions, below it none.

2.3. Prove the extended theorem 1 (Theorem 2.2).

2.4. If Ω has a point of concavity $\mathbf{y}$, show, using the tangent line to the circle of inner contact at Ω, that Ω cannot be convex.

2.5. If Ω is not convex, and $\mathbf{x}_1$, $\mathbf{x}_2$ are points in Ω such that the segment connecting them is not contained in Ω, show there must be a point of concavity "between" $\mathbf{x}_1$ and $\mathbf{x}_2$. *Hint:* Consider a curve joining $\mathbf{x}_1$ and $\mathbf{x}_2$ in Ω and line segments joining $\mathbf{x}_1$ to points of this curve. Construct a circular arc that contacts $\partial\Omega$.

2.6. Suppose that Ω is bounded by the line segments connecting $A = (-1, 0)$, $B = (0, -1)$, $C = (1, 0)$, and $D = (0, -k)$, $0 < k < 1$, in that order. Let $\mathscr{C}$ be the circular arc through A, D, and C. Find a condition on k that guarantees that this arc lies inside $\bar{\Omega}$. Suppose that $Mu = 0$ in Ω, $u \in C(\bar{\Omega})$, $u = 0$ on ABC, and u rises linearly on AD and CD to $u(D) = h > 0$. Show that there is an $h_0 > 0$ depending only on Ω such that $h \leq h_0$.

2.7. For the annular constant mean curvature surface with $\beta < 0$ a solution exists for

$$(-1 + \sqrt{1 - 4\beta H_0})/2H_0 < r < (1 + \sqrt{1 - 4\beta H_0})/2H_0.$$

If β is chosen so that the inner radius is a given $a > 0$, then the outer radius is $a + H_0^{-1}$. The outward normal derivative of the resulting solution u_1 is ∞ on both circumferences.

2.8. Prove that

$$\left(\frac{u_x}{W}\right)_x + \left(\frac{u_y}{W}\right)_y \geq 2H_0 > 0$$

on $R - H_0^{-1} < r < R$ implies that $u = u_1 + \text{const.}$

2.9. If a wedge with opening angle 2α, $\alpha < \pi/2$, is bounded by the lines $y = \pm(x - a)\tan\alpha$, then the surface of the lower hemisphere $u = -\sqrt{a^2 - r^2}$ makes a constant contact angle $\pi/2 - \alpha$ with the sides of the vertical cylinder over this wedge, i.e., $W^{-1}\text{grad}\, u \cdot \mathbf{n} = \cos(\pi/2 - \alpha)$ along these lines, where $\mathbf{n}$ is the unit normal vector to the line.

2.10. If $\mathbf{f}(\mathbf{p}) = \mathbf{p}/\sqrt{1 + |\mathbf{p}|^2}$ for $\mathbf{p} \in \mathbb{R}^n$, then $(\mathbf{p} - \mathbf{q}) \cdot (\mathbf{f}(\mathbf{p}) - \mathbf{f}(\mathbf{q})) \geq 0$ and equality can occur only if $\mathbf{p} = \mathbf{q}$. *Hint:* If $\Phi(t) = (\mathbf{p} - \mathbf{q}) \cdot (\mathbf{f}(\mathbf{q} + t(\mathbf{p} - \mathbf{q})) - \mathbf{f}(\mathbf{q}))$, then

$$\Phi'(t) = \frac{|\mathbf{p} - \mathbf{q}|^2 - (\mathbf{q} \cdot \mathbf{p})^2 + |\mathbf{p}|^2|\mathbf{q}|^2}{(1 + |\mathbf{q} + t(\mathbf{p} - \mathbf{q})|^2)^{3/2}}.$$

2.11. The capillary problem discussed above assumes that the fluid does not rise to the top of the tube. If the top of the tube is at height a, the boundary conditions become

a. $u = a$ and $\mathbf{T}u \cdot \mathbf{n} \leq \cos\gamma$, or
b. $u \leq a$ and $\mathbf{T}u \cdot \mathbf{n} = \cos\gamma$.

Prove uniqueness for this problem. *Hint:* Consider two different solutions u, w. We may assume that there is a constant $k > 0$ such that $u < v$ in Ω and $u = v$ at a boundary point, where $v = w + k$. Distinguish two cases and use the boundary point lemma in one.

2.12. Prove the extended theorem 6 (Theorem 2.9).

2.13. Show that $Qv = O(r^3)$ for $v = (\cos\theta - \sqrt{k^2 - \sin^2\theta})/kKr$, $k = \sin\alpha/\cos\gamma$. *Hint:* The point is to show there is cancellation of terms, up to r^4, in those terms of Qv that have a factor $1/r$.

2.14. Complete the proof of Theorem 2.10. *Hint:* Begin by choosing γ' so that

$$\cos\gamma + C\delta^4 < \cos\gamma' < \cos\gamma + 2C\delta^4$$

for some constant C (with a further possible restriction on δ). This implies that $v(\gamma') < v(\gamma) + C_0\delta^4/r$ for some C_0 independent of δ. In particular, for $\delta^2 \leq r \leq \delta$, $v(\gamma') < v(\gamma) + C_0\delta^2$, and $u < v + A + C_0^2$ there. Now choose

$\gamma_1, \gamma' < \gamma_1 < \gamma$, such that $v(\gamma_1) \leq v(\gamma) + C_0\delta^8/r$ for $r \leq \delta^2$. Then $v(\gamma_1) \leq v + C_0\delta^4$ for $\delta^4 \leq r \leq \delta^2$. Continue this process to show that

$$u < v + A + C_0\delta^2/(1+\delta^2).$$

2.15. An axially symmetric constant mean curvature surface defined in a neighborhood of the axis of rotation must be part of a sphere. Begin by considering the curve $u(r)$ that generates the surface, r being the distance from the axis of symmetry. Show that the equation for constant mean curvature can be written

$$\frac{1}{r}(r \sin \psi)_r = 2H_0,$$

where ψ is the angle that the tangent to $u(r)$ makes with the horizontal. *Hint:* $\sin \psi = u_r/\sqrt{1+u_r^2}$.

2.16. Assume as known the formulas from the differential geometry of surfaces $[\mathbf{x} = \mathbf{x}(u, v)]$ given by

$$\begin{pmatrix} a_{11} & a_{12} \\ a_{21} & a_{22} \end{pmatrix} = -\begin{pmatrix} e & f \\ f & g \end{pmatrix}\begin{pmatrix} E & F \\ F & G \end{pmatrix}^{-1},$$

where $E = \mathbf{x}_u \cdot \mathbf{x}_u$, $F = \mathbf{x}_u \cdot \mathbf{x}_v$, $G = \mathbf{x}_v \cdot \mathbf{x}_v$ are the coefficients of the first fundamental form, $e = -\mathbf{N}_u \cdot \mathbf{x}_u$, $f = -\mathbf{N}_u \cdot \mathbf{x}_v$, $g = -\mathbf{N}_v \cdot \mathbf{x}_v$ are the coefficients of the second fundamental form, and $\mathbf{N}_u = a_{11}\mathbf{x}_u + a_{12}\mathbf{x}_v$, $\mathbf{N}_v = a_{21}\mathbf{x}_u + a_{22}\mathbf{x}_v$. If $\mathbf{N}$ denotes the unit normal, in the direction $\mathbf{x}_u \times \mathbf{x}_v$, the principal curvatures are the curvatures k_1 and k_2 of normal sections to the surface that are maximum and minimum, and k_1, k_2 are eigenvalues of $\{a_{ij}\}$.

Show that the mean curvature $H := \frac{1}{2}(k_1 + k_2)$ is given by

$$\frac{1}{2}\frac{eG - 2fF + gE}{EG - F^2},$$

and for a surface that is a graph with $\mathbf{x}(u, v) = (u, v, h(u, v))$,

$$H = \frac{(1+h_x^2)h_{yy} - 2h_xh_yh_{xy} + (1+h_y^2)h_{xx}}{(1+h_x^2+h_y^2)^{3/2}} = \frac{1}{2}\left(\left[\frac{h_x}{W}\right]_x + \left[\frac{h_y}{W}\right]_y\right),$$

where $W = (1+h_x^2+h_y^2)^{1/2}$.

3.1. Carry out the details of the proof of Theorem 3.1. for $n = 2$.

3.2. (Proof of Lemma 3.3). Suppose that $\mathbf{x}_0 = 0$, $\mathbf{x} \in B_R$, $\mathbf{y} \in N$ and use polar coordinates centered at $\mathbf{x}$ so that $\mathbf{y} = \mathbf{x} + r\xi$ where $|\xi| = 1$, $r = |\mathbf{x} - \mathbf{y}|$.

a. As

$$u(\mathbf{x}) = -\int_0^r \frac{d}{dt} u(\mathbf{x} + t\xi)dt,$$

we have

$$u(\mathbf{x})|N| \leq \int_N d\mathbf{y} \int_0^r |\text{grad } u(\mathbf{x} + t\xi)|dt \leq \int_{B_R} |\text{grad } u(\mathbf{x} + t\xi)|dtd\mathbf{y}.$$

Writing $d\mathbf{y} = r^{n-1}\, drd\omega$,

$$\begin{aligned} u(\mathbf{x})|N| &\leq \int_0^R drr^{n-1} \int_{|\xi|=1} d\omega \int_0^r |\text{grad } u(\mathbf{x} + t\xi)|dt \\ &\leq \int_0^R drr^{n-1} \int_{|\mathbf{x}-\mathbf{z}|\leq r} d\omega \frac{|\text{grad } u(\mathbf{z})|}{|\mathbf{x} - \mathbf{z}|^{n-1}}\, d\mathbf{z} \leq \frac{R^n}{n} \int_{|\mathbf{z}|\leq R} \frac{|\text{grad } u(\mathbf{z})|}{|\mathbf{x} - \mathbf{z}|^{n-1}}\, d\mathbf{z} \end{aligned}$$

and

$$\int_{B_r} |u(\mathbf{x})|d\mathbf{x} \leq R^n/n|N| \int_{|\mathbf{x}|\leq R} \int_{|\mathbf{z}|\leq R} \frac{|\text{grad } u(\mathbf{z})|}{|\mathbf{x} - \mathbf{z}|^{n-1}}\, d\mathbf{z}|^{n-1}\, d\mathbf{z}d\mathbf{x}.$$

b. The last integral is bounded by

$$\left(\sup_{|\mathbf{z}|\leq R} \int_{|\mathbf{x}|\leq R} \frac{d\mathbf{x}}{|\mathbf{x} - \mathbf{z}|^{n-1}} \right) \int_{B_R} |\text{grad } u(\mathbf{z})|d\mathbf{z}$$

and

$$\int_{|\mathbf{x}|\leq R} \frac{d\mathbf{x}}{|\mathbf{x} - \mathbf{z}|^{n-1}} \leq \int_{|\mathbf{x}|\leq 2R} \frac{d\mathbf{x}}{|\mathbf{x}|^{n-1}} = CR$$

so

$$\int_{B_R} |u(\mathbf{x})|d\mathbf{x} \leq \frac{CR^{n+1}}{|N|} \int_{B_R} |\text{grad } u(\mathbf{z})|d\mathbf{z}.$$

c. Replace u by u^2 in the last inequality and complete the proof.

4.1. If $K \subset \mathbb{R}^n$ is closed and convex, for each $\mathbf{x} \in \mathbb{R}^n$ there is a unique $\mathbf{y} \in K$ such that

$$\|\mathbf{x} - \mathbf{y}\| = \inf_K \|\mathbf{x} - \mathbf{z}\|,$$

where $\|\mathbf{x}\| = \sqrt{\mathbf{x} \cdot \mathbf{x}}$. (This result holds in a Hilbert space, see Exercise 1.8 of chapter 8) We write $\mathbf{y} = P_k\mathbf{x}$, P_k being the *projection* on K.

4.2. Show $\mathbf{y} = P_k\mathbf{x}$ if only if $\mathbf{y} \in K$ and

$$\mathbf{y} \cdot (\mathbf{z} - \mathbf{y}) \geq \mathbf{x} \cdot (\mathbf{z} - \mathbf{y}) \qquad (*)$$

for all $\mathbf{z} \in K$. *Hint:* $(1 - t)\mathbf{y} + t\mathbf{z} = \mathbf{y} + t(\mathbf{z} - \mathbf{y}) \in K$ for $\mathbf{z} \in K$, $t \in [0, 1]$, and $\|\mathbf{x} - \mathbf{y} - t(\mathbf{z} - \mathbf{y})\|^2$ is minimized at $t = 0$ shows necessity of (*). On the other hand, (*) implies $\|\mathbf{x} - \mathbf{y}\|^2 \leq (\mathbf{y} - \mathbf{x}) \cdot (\mathbf{z} - \mathbf{x})$.

4.3. Prove (19) by first showing that $P_k(\mathbf{I} - \mathbf{F})$ is a continuous mapping of K into K, and, hence, has a fixed point by the Brouwer fixed point theorem, and then applying the result of Exercise 4.2.

4.4. Suppose that $Au = -\Delta u + f(u)$ where $f \in C^1(\mathbb{R}) \cap L^\infty(\mathbb{R})$ and $ku \geq f'(u) \geq -c$ where k, $c < 0$. Show that $A: H_0^1(\Omega) \to H^{-1}(\Omega)$ is monotone if $c < (\lambda_1(\Omega))^{-1}$ where λ_1 is the lowest eigenvalue of $-\Delta$ in $H_0^1(\Omega)$.

4.5. Prove that if $\mathbf{a}$ satisfies (SC), then A defined by (22) is monotone, coercive, and continuous on finite-dimensional spaces on $H^1(\Omega)$.

4.6. Suppose that $\psi \in C^1(\bar{\Omega})$, $\psi < 0$ on $\partial\Omega$, and let $K = \{u \in H_0^1(\Omega): u \geq \psi \text{ on } \Omega\}$ where $v \geq 0$ for $v \in H^1(\Omega)$ means there is a sequence of nonnegative functions $v_n \in W^{1,\infty}(\Omega)$ such that $v_n \to v$ in $H^1(\Omega)$. Show that the *obstacle problem*

$$\min_K \int_\Omega |\text{grad } u|^2 d\mathbf{x}$$

has a solution.

4.7. Show that $v = c \ln \{\cos[(y - b)/c]/\cos[(x - a)/c]\}$ is a solution of $Mv = 0$ in the square with side length πc centered at (a, b).

4.8. Prove the assertion about level lines made after the proof of Lemma 4.1.

4.9. If a sequence of functions is uniformly bounded and equicontinuous on a bounded set, and converges on a dense subset, it converges uniformly.

4.10. Prove Theorems 4.11, 4.12, and 4.13.

4.11. Prove that a uniformly bounded family of solutions of the minimal surface equation has a subsequence that is convergent to a solution. The convergence is uniform on compact subdomains.

4.12. Prove that a locally superminimal function is superminimal. *Hint:* Extend the proof of the "strong minimum principle" to locally superminimal functions, and use this result.

4.13. Prove Lemma 4.9 using the following steps.

i. If $f \in C([a,b])$ and $\int_a^b f(x)\varphi'(x)dx = 0$ for all $\varphi \in C([a,b]) \cap C^1(a,b)$ with $\varphi(a) = \varphi(b) = 0$, then f is constant. *Hint:* Observe that $\int_a^b (f(x) - \gamma)\varphi'(x)dx = 0$ for any γ, choose $\gamma = (b-a)^{-1}\int_a^b f(x)dx$ and $\varphi(x) = \int_a^x (f(y) - \gamma)dy$.

ii. If $\mathscr{R} = [a,b] \times [c,d]$ and $A = \int_c^y a(x,z)dz, B = \int_a^x b(z,y)dz$, then existence of ω such that $\omega_x = -b, \omega_y = a$ is equivalent to $A(x,y) + B(x,y) = f(x) + g(x)$.

iii. Choose $l(x)$, $m(y)$ vanishing at a, b and c, d, respectively, $\zeta = lm$, and deduce that

$$\int_{\mathscr{R}} (A+B)l'(x)m'(y)dxdy = 0.$$

Use (i) to show that $H(x,y) = H(a,y) + H(x,c)$ where $H = U + V$.

4.14. Verify the assertion of (WM).

4.15. Generalize Theorem 4.1 to an equation satisfying

$$|\mathbf{a}(\mathbf{p})| \leq c|\mathbf{p}|, \qquad \lambda|\mathbf{y}|^2 \leq \frac{da_i}{dp_j}y_i y_j \leq \nu|\mathbf{y}|^2.$$

Deduce a regularity theorem for solutions of equation $Au = 0$, A defined in (23).

4.16. Prove Lemma 4.3, part (b).

4.17. Prove Lemma 4.7. *Hint:* Consider $T = \{w < z\}$ as in the proof of Lemma 4.3, and $v = \max\{z, w\}$, $v = \min\{z, w\}$, respectively.

4.18. Verify (34).

4.19. Construct a lower barrier relative to $u \in C^2$ on a disk of radius R.

References

1. MOSER, J. *A New Proof of De Giorgi's Theorem Concerning the Regularity Problem for Elliptic Differential Equations*, Communications Pure and Applied Mathematics, Vol. 13, pp. 457–468, 1960.
2. EVANS, L. C., *Bounds for Elliptic Equations*, unpublished lecture notes.
3. GILBARG, D., and TRUDINGER, N. S., *Elliptic Partial Differential Equations of Second Order*, Springer-Verlag, Berlin, Germany, 1983.

4. KINDERLEHRER, D., and STAMPACCHIA, G., *An Introduction to Variational Inequalities and Their Applications*, Academic Press, New York, New York, 1980.
5. DIEUDONNÉ, J., *Foundations of Modern Analysis*, Vol. I, Academic Press, New York, New York, 1960.
6. ELCRAT, A., and LANCASTER, K., *On the Behavior of a Non-parametric Minimal Surface in a Non-convex Quadrilateral*, Archive for Rational Mechanics and Analysis Vol. 94 (3), pp. 209–226, 1986.
7. LANCASTER, K., *Boundary Behaviour of a Non-parametric Minimal Surface in* $\mathbb{R}^3$ *at a Non-convex Point*, Analysis, Vol. 5, pp. 61–69, 1985.

Suggested Further Reading

FINN, R., *Remarks Relevant to Minimal Surfaces and to Surfaces of Prescribed Mean Curvature*, Journal d'Analyse, Vol. 14, pp. 139–160, 1965.

FINN, R., *Equilibrium Capillary Surfaces*, Springer-Verlag, Berlin, Germany, 1986.

GIUSTI, E., *Minimal Surfaces and Functions of Bounded Variation*, Birkhäuser, Boston, Massachusetts, 1984.

MEYERS, N. G., *An Example of Non-uniqueness in the Theory of Quasilinear Elliptic Equations of Second Order*, Archive for Rotational Mechanics and Analysis. Vol. 14, pp. 177–179, 1963.

NITSCHE, J. C. C., *Vorlesungen über Minimalflächen*, Springer-Verlag, Berlin, Germany, 1975.

OSSERMAN, R., *A Survey of Minimal Surfaces*, Dover, New York, New York, 1986.

PROTTER, M. H., and WEINBERGER, H. F., *Maximum Principles in Differential Equations*, Prentice–Hall, Englewood Cliffs, New Jersey, 1967.

RADO, T., *The Problem of Plateau*, Springer-Verlag, Berlin, Germany, 1932.

SERRIN, J., *A Priori Estimates for Solutions of the Minimal Surface Equation*, Archives for Rotational Mechanics and Analysis, Vol. 14, pp. 376–383, 1963.

SERRIN, J., *The Problem of Dirichlet for Quasilinear Elliptic Differential Equations with Many Independent Variables*, Philisophical Transactions of the Royal Society of London, Series A, Vol. 264, pp. 413–196, 1969.

6

Abstract Evolution Equations

In this chapter we reconsider partial differential equations in which there is a distinguished variable t, usually time in physical problems. We might think of such equations as ordinary differential equations in Banach spaces. Consider, for example, the heat equation in a bounded domain Ω. If we let $A = \Delta$, then the equation can be written

$$\frac{du}{dt} = Au, \tag{E}$$

where $u(t)$ is a vector-valued function (Chapter 8). As we will see, if the temperature u vanishes on the boundary of Ω, one reasonable choice of the "state space" is $H_0^1(\Omega)$.

There are several ways of giving a mathematical theory based on this idea. One, which might be described as an "energy-based theory," is given in Section 2. Here we consider a class of equations that are *parabolic*. We will not make use of a general definition of parabolicity in this book. The equations $u_t = Au$ where A is an elliptic operator are the basic examples for us, and we may simply think of these examples when using the term. (In a similar way, the equations $u_{tt} = Au$ where A is an elliptic operator are examples of a class of equations called *hyperbolic*; see also Section 4 of Chapter 1. We consider a large class of hyperbolic equations and systems that are of great practical significance in Chapter 7.)

It turns out that a natural class of initial data for (E) is $L^2(\Omega)$. As the heat equation is instantaneously smoothing, we can expect $u(t)$ for $t > 0$ to be in a more regular class of functions. On the other hand, if we want to think of solutions of (E) as evolving in some fixed Sobolev space, there are difficulties as $A(H^s(\Omega))$ is not contained in $H^s(\Omega)$ for any $s > 0$. The theory in Section 2 deals with this problem in a way closely related to the theory given for elliptic equations in Chapter 5. In special cases the time variable can be "separated" and eigenfunctions of A can be used. This will be illustrated in the special case of the heat equation in Section 1.

1. Solution of the Heat Equation by Eigenfunction Expansions

We will show in this section how the formal separation of variables method that was introduced in Section 5 of Chapter 3 can be set in the framework of Sobolev spaces. Essential use will be made of the variational theory of eigenvalues and eigenfunctions for $-\Delta$ that was developed in Section 5 of Chapter 4.

The problem considered will be

$$\frac{du}{dt} = Au + f, \tag{1}$$

$$u(0) = u_0, \tag{2}$$

where $A = \Delta$. We will consider only the boundary condition $u = 0$ on $\partial\Omega$, and this will be imposed weakly by assuming that $u \in H_0^1(\Omega)$. As $\Delta u \in H^{-1}(\Omega)$ for $u \in H_0^1(\Omega)$, we will assume $u'(t) = du/dt$ and $f(t)$ are elements of $H^{-1}(\Omega)$ for each t. This leaves the question of how to measure the time dependence of solutions. Many choices are possible, but here we will assume $u \in L^2(0, T, H_0^1(\Omega))$, $u' \in L^2(0, T, H^{-1}(\Omega))$ and suppose $f \in L^2(0, T, H^{-1}(\Omega))$, $u_0 \in L^2(\Omega)$ for the data. This yields a Hilbert space structure that will be useful in Section 2.

Suppose that $\{z_k\}$ are the eigenfunctions corresponding to the eigenvalues $\{\lambda_k\}$ of $-\Delta$. We need the following result conserning expansion of functions in Sobolev spaces.

Theorem 1.1 A series $\sum_{n=1}^{\infty} b_n z_n$ represents an element v of $H_0^1(\Omega)$, $L^2(\Omega)$, and $H^{-1}(\Omega)$ if and only if the series

$$\sum_{n=1}^{\infty} \lambda_n |b_n|^2, \qquad \sum_{n=1}^{\infty} |b_n|^2, \qquad \sum_{n=1}^{\infty} |b_n|^2/\lambda_n$$

converge, respectively.

Proof. As $\|z_n\|_{L^2} = 1$, $\|z_n/\sqrt{\lambda_n}\| = 1$, the result for the first two series follows from Parseval's equality (Chapter 8, Exercise 1.13) in L^2 and $H_0^1(\Omega)$ and from the relations

$$\left\| \sum_{n=N}^{N+p} b_n z_n \right\|_{L^2}^2 = \sum_{n=N}^{N+p} |b_n|^2, \qquad \left\| \sum_{n=N}^{N+p} b_n z_n \right\|^2 = \sum_{n=N}^{N+p} \lambda_n |b_n|^2$$

with $b_n = (v, z_n)$. [We recall that $\| \cdot \|$ is the norm in $H_0^1(\Omega)$ defined by the Dirichlet integral.] If $\varphi \in H_0^1(\Omega)$ and $\varphi^{(N)} := \sum_{n=1}^{N} \varphi_n z_n$, where $\varphi_n = (z_n, \varphi)$, and

$F \in H^{-1}(\Omega)$, set $F(z_n) := b_n$. Then, letting $v^{(N)} = \sum_{n=1}^{N} b_n z_n$,

$$F(\varphi) = \sum_{n=1}^{N} \varphi_n b_n = (v^{(N)}, \varphi).$$

As $z_n \in L^2(\Omega)$, $v^{(N)} \in L^2(\Omega)$. The Schwarz inequality yields

$$|F(\varphi)|^2 \leq \sum_{n=1}^{N} \frac{|b_n|^2}{\lambda_n} \sum_{n=1}^{N} \lambda_n \varphi_n^2 \leq \|\varphi\|^2 \sum_{n=1}^{N} \frac{|b_n|^2}{\lambda_n}.$$

Letting $N \to \infty$ we see that

$$v = \sum_{n=1}^{\infty} b_n z_n \in H^{-1}(\Omega) \quad \text{and} \quad \|v\|_{H^{-1}} \leq \left(\sum_{n=1}^{\infty} |b_n|^2/\lambda_n \right)^{1/2}$$

if the third series converges. On the other hand, the function

$$\phi := \sum_{n=1}^{N} b_n z_n / \lambda_n$$

is an element of $H_0^1(\Omega)$ and has norm $\|\phi\|^2 = \sum_{n=1}^{N} |b_n|^2/\lambda_n = |F(\phi)|$. Hence, $\|v\|_{H^{-1}} \geq |F(\phi)|/\|\phi\| = \|\phi\|$. It follows that if $v \in H^{-1}(\Omega)$, the series $\sum_{n=1}^{\infty} |b_n|^2/\lambda_n$ converges, and

$$\|v\|_{H^{-1}}^2 = \sum_{n=1}^{\infty} |b_n|^2/\lambda_n.$$

This completes the proof.

For $g \in L^2(0, T, \mathscr{H})$ where $\mathscr{H}$ is a Hilbert space with orthonormal basis $\{z_k\}$, an application of the dominated convergence theorem implies that

$$g = \sum_{k=1}^{\infty} g_k(t) z_k$$

in $L^2(0, T, \mathscr{H})$ where $g_k(t) = (g(t), z_k)_{\mathscr{H}}$, $0 \leq t \leq T$. Suppose that we write

$$f = \sum_{k=1}^{\infty} f_k(t) z_k, \qquad u_0 = \sum_{k=1}^{\infty} a_k z_k$$

in our problem [the series will then converge in $H^{-1}(\Omega)$ for a.e. t and the second in $L^2(\Omega)$]. We seek a solution in the form

$$u = \sum_{k=1}^{\infty} u_k(t) z_k \tag{3}$$

with the series convergent in $L_2(0, T, H_0^1(\Omega))$. Assuming (3) we obtain, formally,

$$u_k'(t) + \lambda_k u_k(t) = f_k(t), \qquad 0 < t < T,$$
$$u_k(0) = a_k,$$

which are solved by setting

$$u_k(t) = e^{-\lambda_k t} a_k + \int_0^t e^{-\lambda_k (t-s)} f_k(s) ds := v_k(t) + w_k(t). \tag{4}$$

We need to investigate the convergence of the series (3) with $u_k(t)$ given by (4). The series $v := \sum_{k=1}^{\infty} v_k(t) z_k = \sum_{k=1}^{\infty} e^{-\lambda_k t} a_k z_k$ is easily seen to be convergent in $L^2(0, T, H_0^1(\Omega))$ as

$$\int_0^T \|v(t)\|^2 dt = \sum_{k=1}^{\infty} \lambda_k |a_k|^2 \int_0^T e^{-2\lambda_k t} dt \leq \sum_{k=1}^{\infty} \frac{|a_k|^2}{2} (1 - e^{-2\lambda_k T}) \leq C \|u_0\|_{L^2(\Omega)}^2, \tag{5}$$

where $C = C(T)$. Moreover, $v'(t) \in C((0, T], H_0^1(\Omega))$ because of the exponential factors.

Let $w_N(t) = \sum_{k=1}^{N} w_k(t) z_k$. Then

$$\int_0^T \|w_N(t)\|^2 dt = \sum_{k=1}^{N} \lambda_k \int_0^T \left[\int_0^t e^{-\lambda_k (t-s)} f_k(s) ds \right]^2 dt.$$

We recall the following proposition from real analysis.

Proposition 1.1. Suppose $g \in L^1(0, T)$, $h \in L^2(0, T)$, and $k(t) = \int_0^t g(t-s) h(s) ds$. Then

$$\|k\|_{L^2} \leq \|g\|_{L^1} \|h\|_{L^2},$$

where $L^p = L^p(0, T) (p = 1, 2)$.

The proof is a direct application of Hölder's inequality. (A general theorem implying this is given in Chapter 9 of Ref. 1.) Letting $g(s) = e^{-\lambda_k s}$, $h(s) = f_k(s)$, we have

$$\int_0^T \|w_N(t)\|^2 dt \leq \sum_{k=1}^N \lambda_k \left[\int_0^t e^{-\lambda_k t} dt\right]^2 \int_0^T |f_k(s)|^2 ds$$
$$= \sum_{k=1}^N (1 - e^{1-e^{-\lambda_k T}})^2 \lambda_k^{-1} \int_0^T |f_k(s)|^2 ds.$$

If we define $f_N(t) = \sum_{k=1}^N f_k(t) z_k$, this inequality and the previous theorem imply

$$\int_0^t \|w_N(t)\|^2 dt \leq C \int_0^T \|f_N(t)\|^2_{H^{-1}(\Omega)} dt, \tag{6}$$

where, again, $C = C(T)$. We can deduce from (6) that $w = \sum_{k=1}^{\infty} w_k(t) z_k$ is convergent in $L^2(0, T, H_0^1(\Omega))$. As $f_k(t) \in L^2(0, T)$, we can differentiate the integral in the definition of $w_k(t)$. An argument analogous to the above then shows that the series obtained by letting $N \to \infty$ in $w_N'(t)$ converges in $L^2(0, T, H^{-1}(\Omega))$ and the limit is $w'(t)$. Hence, $u'(t) \in L^2(0, T, H^{-1}(\Omega))$. Finally, we can replace w_N and f_N in (6) by w and f, respectively, and (5) and (6) imply that

$$\int_0^T \|u(t)\|^2 dt \leq C\left[\|u_0\|^2_{L^2(\Omega)} + \int_0^T \|f(t)\|^2_{H^{-1}(\Omega)} dt\right],$$

where C depends only on T. This inequality shows uniqueness and continuous dependence in the appropriate norms. We see then that the formal eigenfunction expansion method provides a unique solution in the class of functions considered. In this special case this may be considered an alternative to the theory of Section 2, and motivates the choice of function spaces used there.

We remark that this method can be used for any equation $u_t = Au$ for which an appropriate set of eigenfunctions of A are available.

2. Parabolic Evolution Equations

Here consider a class of parabolic problems of the form $u' = A(t)u$ where the operators $A(t)$ are elliptic for each t. In order to introduce the basic ideas with the minimum technical complications, we consider first the initial value problem for the heat equation,

$$\begin{aligned} u_t - u_{xx} &= f, \qquad (x, t) \in \mathbb{R} \times (0, T), \\ u(x, 0) &= u_0(x), \qquad x \in \mathbb{R}, \end{aligned} \tag{IVP}$$

in this setting. The operator $-\partial^2/\partial x^2$ will be thought of as a bounded linear operator mapping $H^1(\mathbb{R}) = H^1$ into $H^{-1}(\mathbb{R}) = H^{-1}$ and this extends in an obvious way to an operator mapping $L^2(0, T, H^1)$ into $L^2(0, T, H^{-1})$. From the notation $W(0, T)$ in Chapter 8,

$$W(0, T) = \{u \in L^2(0, T, H^1) : u_t \in L^2(0, T, H^{-1})\},$$

and we recall the fact that this is a Hilbert space, with the norm

$$\|u\| = \left(\int_0^T (\|u(t)\|_1^2 + \|u_t(t)\|_{-1}^2) dt \right)^{1/2}$$

in which $C^\infty([0, T], H^1)$ is dense. Further, as stated in Proposition 4.2 of Chapter 8, if $C([0, T], L^2)$ is given the norm $\sup\|u(t)\|_0$, then the natural injection of $C^\infty([0, T], H^1)$ into $C([0, T], L^2)$ can be extended to a continuous injection of $W(0, T)$ into $C([0, T], L^2)$. In particular, functions in $W(0, T)$ may be thought of as being continuous functions on $[0, T]$ taking values in L^2, and functions in $W(0, T)$ have initial values that are taken on continuously *in the L^2 sense*, i.e.,

$$\lim_{t \to 0} \|u(t) - u(0)\|_0 = 0,$$

We can now state the basic existence theorem for (IVP) in this context.

Theorem 2.1. For every $f \in L^2(0, T, H^{-1})$ and $u_0 \in L^2 = L^2(\mathbb{R})$, there is exactly one solution $u \in W(0, T)$ of (IVP), with $u(0) = u_0$ in the L^2 sense. ■

The proof is accomplished using the following proposition.

Proposition 2.1. Let $\mathscr{E}$ be a Hilbert space and $\mathscr{H}$ a subspace of $\mathscr{E}$. Let $a(u, v)$ be a continuous bilinear form defined on $\mathscr{E} \times \mathscr{H}$ such that:

i. For fixed h, $w \to a(w, h)$ is a continuous linear functional on $\mathscr{E}$.
ii. $a(h, h) \geq c\|h\|_{\mathscr{E}}^2$ for $h \in \mathscr{H}$, where $\| \cdot \|_{\mathscr{E}}$ denotes the norm in $\mathscr{E}$.

Then there is a bounded linear operator G from the dual $\mathscr{E}'$ of $\mathscr{E}$ into $\mathscr{E}$ with $\|G\| \leq c^{-1}$ such that $\lambda \in \mathscr{E}'$ implies

$$a(G\lambda, h) = \lambda(h) \qquad \text{for all } h \in \mathscr{H}.$$

Proof. For each $h \in \mathscr{H}$, (i) implies that there is an $Rh \in \mathscr{E}$ such that $a(w, h = (w, Rh)_{\mathscr{E}}$ for all $w \in \mathscr{E}$, where $(,)_{\mathscr{E}}$ denotes the inner product on $\mathscr{E}$. Further, (ii) implies that $\|h\|_{\mathscr{E}} \leq c^{-1}\|Rh\|_{\mathscr{E}}$, hence $R : \mathscr{H} \to \mathscr{E}$ is to a one-to-one linear transformation, and the inverse transformation $Rh \to h$ defined on $R(\mathscr{H})$ (given the norm of $\mathscr{E}$) is continuous. If we extend this transformation to $\overline{R(\mathscr{H})}$ by continuity, and set it to zero on the orthogonal complement of $R(\mathscr{H})$, we obtain an operator $G^* \in B(\mathscr{E})$ with $\|G^*\| \leq c^{-1}$. Let G_1 be the adjoint of G^*. Then

$$a(G_1 w, h) = (G_1 w, Rh)_{\mathscr{E}} = (w, G^* Rh)_{\mathscr{E}} = (w, h)_{\mathscr{E}}.$$

If J is the usual isomorphism of $\mathscr{E}'$ onto $\mathscr{E}$, we can set $G = G_1 J$. □

Proof of Theorem 2.1. We let $\mathscr{E}$ be $L^2(0, T, H^1) \times L^2$, and

$$\mathscr{H} = \{(u, v) \in \mathscr{E} : u \in W(0, T), v = u(0), u(T) = 0\}.$$

The norm in $\mathscr{E}$ is given by

$$\|(u, v)\|^2 = \int_0^T \|u(t)\|_1^2 \, dt + \|v\|_0^2.$$

We define

$$a((w, w_0), (h, h_0)) = \int_0^T \left(-\langle w, h_t \rangle + \int_{\mathbb{R}} w_x h_x \, dx \right) dt,$$

where $\langle\ ,\ \rangle$ denotes the natural duality between H^1 and H^{-1}. We observe that, for $u \in W(0, T)$,

$$2 \int_0^T \langle u, u_t \rangle \, dt = \int_{\mathbb{R}} (|u(x, T)|^2 - |u(x, 0)|^2) dx.$$

For $u \in C^\infty([0, T], H^1)$ this is immediate as the integrand on the left-hand side is just $d\|u\|_0^2/dt$. The density of $C^\infty([0, T], H^1)$ in $W(0, T)$ and Proposition 4.2 of Chapter 8 imply this identity in general. Now

$$\begin{aligned} a((h, h_0), (h, h_0)) &= \int_0^T (-\langle h, h_t\rangle + \|h(t)\|_1^2)dt \\ &= \frac{1}{2}\int_{\mathbb{R}} |h(x, 0)|^2\, dx - \frac{1}{2}\int_{\mathbb{R}} |h(x, T)|^2\, dx + \int_0^T \|h(t)\|_1^2\, dt \\ &\geq \tfrac{1}{2}\|(h, h_0)\|^2, \end{aligned}$$

as $h(x, T) = 0$ and $c = 1/2$. We see that the hypotheses of Proposition 2.1 are satisfied. We let $\lambda \in \mathscr{E}'$ be given by

$$\lambda((h, h_0)) = \int_0^T \langle f, h\rangle dt + \int_{\mathbb{R}} u_0 h_0\, dx.$$

There is, then, a unique $(v, v_0) \in \mathscr{E}$ such that

$$\int_0^T \left(\langle -v, h_t\rangle + \int_{\mathbb{R}} v_x h_x\, dx\right)dt = \int_0^T \langle f, h\rangle dt + \int_{\mathbb{R}} u_0 h(x, 0)dx \tag{7}$$

for all $h \in \mathscr{H}$. We choose first $h \in C_0^\infty([0, T], H^1)$. As $\int_0^T (\langle v, h_t\rangle + \langle v_t, h\rangle)\, dt = 0$ then, and as $\int_{\mathbb{R}} v_x h_x\, dx = -\int_{\mathbb{R}} v_{xx} h\, dx$ in the distributional sense for each t, we have

$$\int_0^T \langle v_t - v_{xx} - f, h\rangle dt = 0.$$

It follows that $v_t - v_{xx} - f = 0$ is an element of H^{-1}. If we now choose a general $h \in \mathscr{H}$ and use

$$\int_0^T \langle -v, h_t\rangle dt = \int_0^T \langle v_t, h\rangle dt + \int_{\mathbb{R}} v(x, 0)h(x, 0)dx,$$

we have

$$\int_{\mathbb{R}} (v(x, 0) - u_0(x))h(x, 0)dx = 0.$$

As $h(x, 0)$ can be an arbitrary element of L_2, the initial condition is satisfied. We have shown the existence of a solution in $L^2(0, T, H^1)$. The conclusion that

$u \in W(0, T)$ follows from the differential equation. The uniqueness question is dealt with by considering

$$b(v, w) := \int_0^T \left(\langle v_t, w \rangle + \int_{\mathbb{R}} v_x w_x \, dx \right) dt$$

on $W(0, T) \times W(0, T)$. In general, we have the identity

$$b(u, u) + \frac{1}{2} \int_{\mathbb{R}} |u(x, 0)|^2 \, dx = \frac{1}{2} \int_{\mathbb{R}} |u(x, T)|^2 \, dx + \int_0^T \|u(t)\|_1^2 \, dt. \tag{8}$$

Suppose that u_1, u_2 are two solutions and $u = u_1 - u_2$. Then

$$b(u, u) = \int_0^T \langle u_t - u_{xx}, u \rangle dt = 0.$$

As $u(x, 0) = 0$, the above identity (8) now implies that

$$\int_0^T \|u(t)\|_1 \, dt = 0.$$

The inequality $\|u_{xx}\|_{-1} \leq \|u\|_1$ and the differential equation then yield $\|u\| = 0$ in $W(0, T)$. □

We remark that the norm of the functional λ defined in the proof is given by

$$\|\lambda\|^2 = \int_0^T \|f(t)\|_{-1}^2 \, dt + \|u_0\|_0^2.$$

It follows from the estimate $\|G\| \leq c^{-1} = 2$ in Proposition 2.1 that

$$\int_0^T \|u(t)\|_1^2 \, dt + \|u_0\|_0^2 \leq 2\left(\int_0^T \|f(t)\|_{-1}^2 \, dt + \|u_0\|_0^2 \right).$$

This shows that the solution depends continuously on the data in the sense that the mapping $(u_0, f) \to u$ from $L^2 \times L^2(0, T, H^{-1})$ to $L^2(0, T, H^1)$ is continuous.

For later work we need to establish an additional identity for the solutions guaranteed by Theorem 2.1.

Theorem 2.2. If $f \in L^2(0, T, H^{-1})$ is of the form $f = f_0 + (f_1)_x$ with $f_0, f_1 \in L^2(0, T, L^2)$, then, for a.e. t,

$$\langle u_t, z\rangle + \int_{\mathbb{R}} u_x z_x \, dx = \int_{\mathbb{R}} (f_0 z - f_1 z_x) dx$$

for all $z \in H^1$.

Proof. Choosing $v = u$ and $h \in C_0^\infty([0, T], H^1)$ as $\varphi(t)z$, where $\varphi \in C_0^\infty([0, T])$, in the identity (7) of the proof of Theorem 2.1, we find

$$\int_0^T \varphi(t)\left(\langle u_t, z\rangle + \int_{\mathbb{R}} u_x z_x \, dx - \langle f, z\rangle\right) dt = 0,$$

and, as φ is arbitrary,

$$\langle u_t, z\rangle + \int_{\mathbb{R}} u_x z_x \, dx - \langle f, z\rangle = 0$$

for a.e. t. Observing that

$$\langle f, z\rangle = \int_{\mathbb{R}} (f_0 z - f_1 z_x) dx,$$

the result follows. □

We also need the following result, which improves the properties of the solution if the data are improved.

Theorem 2.3. If $f \in L^2(0, T, L^2)$ and $u_0 \in H^1$, then $u \in L^2(0, T, H^2)$ and $u_t \in L^2(0, T, L^2)$.

Proof. Formally, $v = u_x$ is a solution of $v_t - v_{xx} = f_x$ with $v(0) = u_{0x}$ and $u = \int_{-\infty}^x v \, dx$. Applying Theorem 2.1 to v implies the result. □

We can also apply this method to the initial–boundary value problem on a finite interval. Suppose for simplicity that we are considering the problem

$$\begin{aligned} u_t - u_{xx} &= f, \qquad (x, t) \in (0, 1) \times (0, T), \\ u(x, 0) &= u_0(x), \qquad x \in (0, 1), \end{aligned} \tag{IBVP}$$

with the boundary conditions $u(0, t) = u(1, t) = 0, t > 0$. We will impose the boundary conditions in a weak sense by simply requiring that $u \in H_0^1(0, 1)$ for fixed t. More precisely, let

$$W_0(0, T) = \{u \in L^2(0, T, H_0^1(I)) : u_t \in L^2(0, T, H^{-1}(I))\}, \qquad I = (0, 1).$$

Then we have

Theorem 2.4. For every $f \in L^2(0, T, H^{-1}(I))$ and $u_0 \in L^2(I)$, there is exactly one solution $u \in W_0(0, T)$ of (IBVP).

In the proof we apply Proposition 2.1 with $\mathscr{E} = L^2(0, T, H_0^1(I)) \times L^2(I)$ and $\mathscr{H}$ the subspace of pairs (v, v_0) with $v(0) = v_0$ and $v(T) = 0$. The rest is almost identical to the proof of Theorem 2.1. The details are left as an exercise.

Suppose now that Ω is a bounded domain in $\mathbb{R}^n$. We will show how Theorem 2.1 extends to a general second-order parabolic initial boundary value problem:

a. $u_t = A(t)u + f, (\mathbf{x}, t) \in \Omega \times (0, T)$,
b. $u(\mathbf{x}, 0) = u_0(\mathbf{x}), \mathbf{x} \in \Omega$,
c. $u(\mathbf{x}, t) = g(\mathbf{x}, t), (\mathbf{x}, t) \in \partial\Omega \times [0, T]$,

where

$$A(t)u = \sum_{i,j=1}^{n} (a_{ij}(\mathbf{x}, t)u_{x_i})x_j$$

and the functions $a_{ij} \in L^\infty(\Omega \times [0, T]), a_{ij} = a_{ji}$, satisfy

$$\sum_{i,j=1}^{n} a_{ij}\xi_i\xi_j \geq c_0|\xi|^2$$

for all $\xi \in \mathbb{R}^n$. It is straightforward to show that $u \to A(t)u$ defines a bounded linear operator $L^2(0, T, H^1(\Omega)) \to L^2(0, T, H^{-1}(\Omega))$. We will assume that g is the trace of a function, again denoted by g, in $H^1(\Omega)$ for almost every t, and the boundary condition (c) will be imposed in the form $u - g \in H_0^1(\Omega)$ for almost every t.

In the definition of $W(0, T)$ we replace H^1 and H^{-1} by $H^{-1}(\Omega)$ and $H^{-1}(\Omega)$, and we define

$$W_0(0, T) = \{u \in W(0, T) : w \in H_0^1(\Omega), t - \text{a.e.}\}.$$

The boundary condition can then be stated: $u - g \in W_0(0, T)$. The proofs of Propositions 4.1 and 4.2 of Chapter 8 go over to this situation unchanged.

We will use a technical assumption about the (lifting of the) boundary values g:

$$\|g(\cdot, t) - g(\cdot, 0)\|_{L^2(\Omega)} \to 0 \qquad \text{as } t \to 0. \tag{g}$$

We can state the following theorem.

Theorem 2.5. Under the stated hypotheses on $A(t)$, for every $f \in L^2(0, T, H^{-1}(\Omega))$, $u_0 \in L^2(\Omega)$ and $g \in L^2(0, T, H^1(\Omega))$ satisfying (g) and $g_t \in L^2(0, T, H^{-1}(\Omega))$, there is a unique solution $u \in L^2(0, T, H^1(\Omega))$ of (a), (b), (c). The initial and boundary conditions are interpreted in the form

b′. $\|u(\cdot, t) - u_0\|_{L^2(\Omega)} \to 0$ as $t \to 0$,
c′. $u - g \in W_0(0, T)$.

Proof. The proof begins by transforming to homogeneous boundary values using the substitution $U = u - g$. The hypotheses made on g are crucial here. We now define

$$a((w, w_0), (h, h_0)) = \int_0^T \left(-\langle w, h_t \rangle + \int_\Omega \sum_{i,j=1}^n a_{ij} w_{x_i} h_{x_j} \, d\mathbf{x} \right) dt$$

and

$$b(v, w) = \int_0^T \left(\langle v_t, w \rangle + \int_\Omega \sum_{i,j=1}^n a_{ij} v_{x_i} w_{x_j} \, d\mathbf{x} \right) dt.$$

The proof is carried out now exactly as for Theorem 2.1. □

We remark that these ideas can be generalized to an abstract situation where H^1, L^2, and H^{-1} are replaced by Hilbert spaces $V \subset H \subset V'$, V dense in H, and a quadratic form $q(t, u, v) = -(A(t)u, v)$ defined on $\mathbb{R} \times V \times V$. If

$$q(t, u, u) \geq c_1 \|u\|_V^2 - c_2 \|u\|_H^2, \qquad c_1, c_2 > 0, \tag{G}$$

then an existence theorem can be given for

$$\frac{du}{dt} = A(t)u + f(t), \qquad u(0) = u_0.$$

This allows for the possibility of considering higher-order operators $A(t)$ and systems of equations.

3. Nonlinear Initial Value Problem

We now consider an initial value problem for a nonlinear equation that will have an important application in Chapter 7. The equation is

$$u_t + f(u)_x = \mu u_{xx} \tag{9}$$

($u, x \in \mathbb{R}, t > 0$) with initial condition

$$u(x, 0) = u_0(x) \tag{10}$$

($x \in \mathbb{R}$). The parameter μ is positive and a crucial aspect of the results obtained is control of estimates on u with respect to μ.

Theorem 3.1. Assume that $f \in C^1$ satisfies the uniform Lipschitz condition

$$|f(u) - f(v)| \leq M|u - v| \tag{UL}$$

and $u_0 \in L^2 \equiv L^2(\mathbb{R})$. Then there is a unique solution $u \in W(0, T)$ of (9), (10).

Proof. (See Ref. 2). The substitution $v = e^{-\lambda t}u$, $\lambda > 0$, transforms (9) into

$$v_t - \mu v_{xx} + \lambda v = -(e^{-\lambda t} f(e^{\lambda t} v))_x \equiv g_{1x}.$$

For $v \in L^2(0, T, L^2)$, let $g = g_{1x}$ be the right-hand side of this equation. Without loss of generality $f(0) = 0$, and $|g_1| \leq M|v|$ implies that $g_1 \in L^2(0, T, L^2)$, $g \in L^2(0, T, H^{-1})$:

$$\int_0^T \|g\|_{-1}^2 \, dt \equiv \int_0^T \|g_{1x}\|_{-1}^2 \, dt \leq \int_0^T \|g_1\|_0^2 \, dt \leq M^2 \int_0^T \|v\|_0^2 \, dt < \infty.$$

Theorem 2.1 implies that there is a unique solution $w \in W(0, T)$ of

$$w_t - \mu w_{xx} + \lambda w = g, \qquad w(x, 0) = u_0(x),$$

and by Theorem 2.2,

$$\langle w_t, z\rangle + \int_{\mathbb{R}} (\mu w_x z_x + \lambda w x) = -\int_{\mathbb{R}} g_1 z_x \, dx \tag{11}$$

for all $z \in H^1$ and for a.e. t. We claim that the mapping F_λ that takes v to w is a contraction on $L^2(0, T, L^2)$ for λ sufficiently large. For $w_i = F_\lambda(v_i)$, $i = 1, 2$, and $w = w_1 - w_2$, (11) implies

$$\langle w_t, z\rangle + \int_{\mathbb{R}} (\mu w_x z_x + \lambda w z) dx = e^{-\lambda t} \int_{\mathbb{R}} (f(e^{\lambda t} v_1) - f(e^{\lambda t} v_2)) z_x \, dx$$

for $z \in H^1$, for a.e. t. Setting $z = w(\cdot, t)$ and integrating from 0 to t, we obtain

$$\frac{1}{2}\|w(t)\|_0^2 + \int_0^t (\mu \|w\|_1^2 + \lambda \|w\|_0^2 \, d\tau \leq M \int_0^t \int_{\mathbb{R}} |v_1 - v_2\|w_x| dx d\tau \tag{12}$$

The Cauchy–Schwarz inequality and $ab \leq (\varepsilon/2)a^2 + (1/2\varepsilon)b^2$ imply

$$\int_{\mathbb{R}} |v_1 - v_2\|w_x| dx \leq \frac{1}{2\varepsilon}\|v_1 - v_2\|^2 + \frac{\varepsilon}{2}\|w\|_1^2.$$

Using this on the right-hand side of (12) with $\varepsilon = 2\mu/M$ in conjunction with Gronwall's lemma yields

$$\lambda \int_0^t \|w\|_0^2 \, d\tau \leq \frac{M^2}{4\mu} \int_0^t \|v_1 - v_2\|_0^2 \, d\tau.$$

Letting t approach T,

$$\|w_1 - w_2\| \leq \frac{M^2}{4\mu\lambda} \|_1 - v_2\|,$$

where $\|\cdot\|$ is the norm in $L^2(0, T, L^2)$. □

We need the following proposition, which contains results from Propositions 3.10 and 4.3 and Exercise 3.16 of Chapter 8.

Proposition 3.1.

a. If G satisfies a Lipschitz condition and G' is continuous except at a finite number of points, then $G \circ v \in H^1$ for $v \in H^1$.

b. $|v|_x = \operatorname{sgn}(v)v_x$ and $v \to |v|$ is continuous on H^1.
c. If $v \in W(0, T)$, $v_+ = \max\{v, 0\} \in L^2(0, T, H^1) \cap C([0, T], L^2)$ and

$$2\int_{t_1}^{t_2} \langle v_t, v_+\rangle dt = \|v_+(t_2)\|_0^2 - \|v_+(t_1)\|_0^2.$$

Theorem 3.2. If $f \in C^1$ and $u_0 \in L^2 \cap L^\infty$, there is a unique solution of (9), (10) in $W(0, T) \cap L^\infty(\mathbb{R} \times [0, T])$ and $\|u(t)\|_\infty \leq \|u_0\|_\infty$ for a.e. t.

Proof. We need to show that the assumption (UL) can be dispensed with. Let $\Psi(u) = \Psi(|u|) \in C^\infty(\mathbb{R})$ with

$$\Psi = \begin{cases} 1, & |u| \leq \|u_0\|_\infty, \\ 0, & |u| \geq \|u_0\|_\infty + 1, \end{cases}$$

and $\tilde{f}(u) = \Psi(u)f(u)$. Then $\tilde{f}$ satisfies (UL) and Theorem 3.1 implies the existence of a solution of (9), (10) with f replaced by $\tilde{f}$, which we will denote by u. Let $v = u - \|u_0\|_\infty$. We will show $v_+ = 0$. First, observe that $v_+ \in H^1$ for a.e. t. As $v_t - \mu v_{xx} = -\tilde{f}'(u)v_x$ and

$$\langle v_t, z\rangle + \mu \int_{\mathbb{R}} v_x z_x \, dx = -\int_{\mathbb{R}} \tilde{f}'(u)v_x z \, dx$$

for all $z \in H^1$, if we take $z = v_+$, and integrate from 0 to t [noting Proposition 3.1, (c) and $v_+(0) = 0$],

$$\frac{1}{2}\|v_+(t)\|_0^2 + \mu \int_0^t \int_{\mathbb{R}} v_x v_{+x} \, dx d\tau = -\int_0^t \int_{\mathbb{R}} \tilde{f}'(u)v_x v_+ \, dx d\tau.$$

Proposition 3.1, (b) implies $\int_{\mathbb{R}} v_x v_{+x} \, dx = \int_{\mathbb{R}} (v_{+x})^2 \, dx$. Applying the Cauchy–Schwarz inequality and $ab \leq \varepsilon a^2 + (1/4\varepsilon)b^2$ we obtain

$$\frac{1}{2}\|v_+(t)\|_0^2 + (\mu - \varepsilon M)\int_0^t \|v_+\|_1^2 d\tau \leq \frac{M}{4\varepsilon}\int_0^t \|v_+\|_0^2 \, d\tau.$$

Setting $\varepsilon = \mu/M$,

$$\|v_+(t)\|_0^2 \leq \frac{M^2}{2\mu}\int_0^t \|v_+\|_0^2 \, d\tau,$$

and Gronwall's inequality implies that $\|v_+(t)\| = 0$ t—a.e. Similarly, we can show $(-u - \|u_0\|_\infty)_+ = 0$ t—a.e. and $\|u\|_\infty \leq \|u_0\|_\infty$ follows. As $\tilde{f}(u) = f(u)$ for $|u| \leq \|u_0\|_\infty$, u is a solution of the original problem.

To prove uniqueness, suppose u_1 and u_2 are two solutions, and truncate f outside $\max\{\|u_i\|_{L^\infty(\mathbb{R}\times[0,T])}\}$ as above. □

We now derive (in Theorem 3.3) some a priori estimates that will be required in Chapter 7 in order to deal with the limit $\mu \to 0$ in (9). We need a sequence of lemmas.

Lemma 3.1. (a) If $f \in C^1$ and $u_0 \in H^1$, then the solution u of (9), (10) obtained in Theorem 3.2 has the further properties $u \in L^2(0, T, H^2) \cap C([0, T], H^1)$, $u_t \in L^2(0, T, L^2)$.

(b) If, in addition, $f \in C^2$ and $u_0 \in H^2$, then $u \in L^2(0, T, H^3) \cap C([0, T], H^2)$, and $u_t \in L^2(0, T, H^1) \cap C([0, T], L^2)$.

Proof. (a) As $-f'(u)u_x \in L^2(0, T, L^2)$, Theorem 2.3 implies that $u \in L^2(0, T, H^2)$ and $u_t \in L^2(0, T, L^2)$. In addition, the proof of that theorem shows that $u_x \in C([0, T], L^2)$.

(b) Setting $v = u_t$, we have

$$v_t - \mu v_{xx} = -(f(u)_t)_x \equiv -(f'(u)u_t)_x,$$

and $f'(u)u_t \in L^2(0, T, L^2)$ [*part*(a)], $v(x, 0) = \mu u_{0xx} - (f(u_0))_x \in L^2$ imply that $v \in L^2(0, T, H^1) \cap C([0, T], L^2)$. We note that $u \in L^2(0, T, H^2)$ implies that $u_{xx} \in L^2([0, T] \times \mathbb{R})$ and $u_t \in L^2(0, T, H^1)$ implies that $u_{xt} \in L^2([0, T] \times \mathbb{R})$; hence, $u_x \in H^1([0, T] \times \mathbb{R})$ and the Sobolev embedding theorem (Section 3 of Chapter 8) says that $u_x^2 \in L^2([0, T] \times \mathbb{R})$. Differentiating (9) with respect to x yields

$$(u_x)_t - \mu(u_x)_{xx} = -f''(u)u_x^2 + f'(u)u_{xx}.$$

The right-hand side is in $L^2([0, T] \times \mathbb{R})$ [$f''(u)$ and $f'(u)$ are in $L^\infty([0, T] \times \mathbb{R})$] and $u_{0x} \in H^1$, so $u_x \in L^2(0, T, H^2) \cap C([0, T], H^1)$ as above, i.e., $u \in L^2(0, T, H^3) \cap C([0, T], H^2)$. □

Lemma 3.2. We can choose a polynomial $p(\rho)$ such that

$$\chi(\rho) := \begin{cases} 1, & \rho \leq 1/2, \\ p(\rho), & 1/2 \leq \rho \leq 1, \\ e^{-p}, & 1 \leq \rho, \end{cases}$$

is in $C^2([0, \infty))$ and $\chi(\rho) \geq 0$ for $\rho \geq 0$. Further, if $\Psi_R(x) = \chi(|x|/R)$, then

$$|\Psi_{R_x}| \leq C\Psi_R/R, \qquad |\Psi_{Rxx}| \leq C\Psi_R/R^2$$

for a constant C independent of R.

Proof. There are many choices for $p(\rho)$. After finding such a p we can choose C such that $|\chi'(\rho), |\chi''(\rho)| \leq C\chi(\rho)$ on $[\frac{1}{2}, 1]$, and hence everywhere. As $\Psi'_R(x) = \chi'(|x|/R)\ \mathrm{sgn}(x)/R$, $\Psi''_R(x) = \chi''(|x|/R)/R^2$, the rest is immediate. □

Lemma 3.3 If $v, \psi \in H^2$ and $\psi \geq 0$ then

$$\int_{\mathbb{R}} v_{xx}\ \mathrm{sgn}(v)\psi dx \leq \int_{\mathbb{R}} |v|\psi_{xx}\, dx.$$

Proof. Consider the piecewise linear function $s_\theta(x)$ approximating $\mathrm{sgn}(x)$,

$$s_\theta = \begin{cases} -1, & x < -\theta, \\ x/\theta, & -\theta \leq x \leq \theta, \\ 1, & x > \theta. \end{cases}$$

As $s'_\theta(x) \geq 0$ (for a.e. x) we have

$$\begin{aligned} \int_{\mathbb{R}} v_{xx} s_\theta(v(x))\psi dx &= -\int_{\mathbb{R}} v_x \psi_x s_\theta(v(x))dx - \int_{\mathbb{R}} v_x s'_\theta(v(x)) v_x \psi dx \\ &\leq -\int_{\mathbb{R}} v_x \psi_x s_\theta(v(x))dx. \end{aligned}$$

Letting $\theta \to 0$ and using Lebesgue's dominated convergence theorem and Proposition 3.1, (b),

$$\int_{\mathbb{R}} v_{xx}\ \mathrm{sgn}(v)\psi\ dx \leq -\int_{\mathbb{R}} v_x\psi_x\ \mathrm{sgn}(v)dx = -\int_{\mathbb{R}} |v|_x \psi_x\ dx = \int_{\mathbb{R}} |v|\psi_{xx}\ dx. \quad □$$

Theorem 3.3 If $f \in C^2$ and $u_0 \in H^2 \cap L^1$, and $M = \sup\{|f'(u)| : |u| \leq \|u_0\|_\infty\}$, then the solution u of (9), (10) satisfies

i. $\int_{\mathbb{R}} |u_x(x, t)|\, dx \leq \int_{\mathbb{R}} |u_{0x}| dx$,
ii. $\int_{\mathbb{R}} |u_t(x, t)|\, dx \leq M \int_{\mathbb{R}} |u_{0x}|\, dx + \mu \int_{\mathbb{R}} |u_{0xx}| dx$,
iii. $\int_{\mathbb{R}} |u(x, t)|\, dx \leq \int_{\mathbb{R}} |u_0|\, dx + MT|_{\mathbb{R}} |u_{0x}| dx$.

Proof. If we differentiate (9) with respect to x and multiply by $\operatorname{sgn}(u_x)$, we obtain, using Proposition 3.1, (b),

$$\begin{aligned}
0 &= \operatorname{sgn}(u_x)(u_{xt} + f''(u)u_x^2 + f'(u)u_{xx} - \mu u_{xxx}) \\
&= (|x_x|)_t + f''(u)u_x|u_x| + f'(u)|u_x|_x - \mu u_{xxx} \operatorname{sgn}(u_x) \\
&= (|u_x|)_t + (f'(u)|u_x|)_x - \mu u_{xxx} \operatorname{sgn}(u_x).
\end{aligned}$$

If this equation is multiplied by Ψ_R and integrated over $\mathbb{R}$, we obtain

$$\begin{aligned}
\frac{d}{dt}\int_{\mathbb{R}} |u_x|\Psi_R \, dx &= \int_{\mathbb{R}} f'(u)|u_x|\Psi_{Rx} \, dx + \mu \int_{\mathbb{R}} u_{xxx} \operatorname{sgn}(u_x)\Psi_r \, dx \\
&\leq \frac{CM}{R}\int_{\mathbb{R}} |u_x|\Psi_{\mathbb{R}} \, dx + \mu \int_{\mathbb{R}} |u_x|\Psi_{Rxx} \, dx \\
&\leq \frac{C}{R}\left(M + \frac{\mu}{R}\right)\int_{\mathbb{R}} |u_x|\Psi_{\mathbb{R}} \, dx,
\end{aligned}$$

where Lemmas 3.2 and 3.3 have been used. If we assume $R \geq 1$, and let $F(t) := \int_{\mathbb{R}} |u_x|\Psi_{\mathbb{R}} \, dx$, this becomes

$$0 \leq F(t) \leq F(0) + \frac{K}{R}\int_0^t F(s)ds$$

$(K > 0)$. Gronwall's inequality implies

$$F(t) \leq F(0)e^{Kt/R},$$

and if we let $R \to \infty$ and use Lebesgue's dominated convergence theorem, conclusion (i) follows.

In a similar way, we can differentiate (0) with respect to t and multiply by $\operatorname{sgn}(u_t)\Psi_{\mathbb{R}}(x)$ to get

$$0 = (|u_t|_t + (f'(u)|u_t|)_x - \mu u_{txx} \operatorname{sgn}(u_t))\Psi_R$$

and

$$\frac{d}{dt}\int_{\mathbb{R}} |u_t|\Psi_{\mathbb{R}} \, dx \leq \frac{C}{R}\left(M + \frac{\mu}{R}\right)\int_{\mathbb{R}} |u_t|\Psi_{\mathbb{R}} \, dx.$$

This implies

$$\int_{\mathbb{R}} |u_t(x, t)| dx \leq \int_{\mathbb{R}} |u_t(x, 0)| dx,$$

as before. As $u_t(x, 0) = -f'(u_0)u_{0x} + \mu u_{0xx}$, conclusion (ii) follows.
Finally, multiplication of (9) by sgn(u) yields

$$\begin{aligned}\frac{d}{dt}\int \mathbb{R}|u|\Psi_R\, dx &= -\int_{\mathbb{R}} f'(u)u_x \operatorname{sgn}(u)\Psi_R\, dx + \mu \int_{\mathbb{R}} u_{xx} \operatorname{sgn}(u)\Psi_R\, dx \\ &\leq M \int_{\mathbb{R}} |u_x|\Psi_R\, dx + \mu \frac{C}{R^2}\int_{\mathbb{R}} |u|\Psi_R\, dx\end{aligned}$$

Integrating and letting $R \to \infty$,

$$\int_{\mathbb{R}} |u(x, t)| dx \leq \int_{\mathbb{R}} |u_0(x)| dx + M \int_0^t \int_{\mathbb{R}} |u_x(x, \tau)| dx\, d\tau,$$

and conclusion (iii) follows from conclusion (i). □

We have illustrated several important techniques in the above. In Theorem 3.1 a fixed point argument is used to obtain a solution. In Theorem 3.2 and a priori estimate is used to extend the scope of the existence theorem. In Lemma 3.1 additional regularity is obtained using the differential equation and added hypotheses on the data. In Theorem 3.3 integral norm estimates of the solution are obtained by differentiating the equation and using appropriate "test functions."

Exercises

1.1 Prove that the initial conditions are taken on in the sense that

$$\|u(t) - u_0\|_{L^2(\Omega)} \to 0 \qquad \text{as } t \to 0.$$

1.2. Consider the initial value problem

$$u_{tt} = Au,$$

$$u(0) = u_0, \qquad u_t(0) = u_1,$$

where $A = \Delta$, $u_0 \in H_0^1(\Omega)$, $u_1 \in L^2(\Omega)$. Assume a solution of the form (3), and show that there is a unique $u \in C([0, T], H_0^1(\Omega))$ with $u_t \in C([0, T],$

$L^2(\Omega))$ and $u_{tt} \in C([0, T], H^{-1}(\Omega))$. *Hint*: First show $\|u(t)\| \le \|u_0\| + C\|u_1\|_{L^2(\Omega)}$, and then consider $\sum_{n=1}^{\infty} \lambda_n |u_n(t) - u_n(s)|^2$.

2.1. Prove Theorem 2.4.

2.2. Give the details in proving Theorem 2.5.

2.3. In Theorem 2.5 we can add lower-order terms and obtain the same result. In particular, we can replace $A(t)$ by $A(t) + B(t)$ where $B(t) = \sum_i b_i u_{x_i} + cu$ and $b_i, c \in L^\infty(\Omega \times [0, T])$. The transformation to homogeneous boundary conditions now becomes $U = e^{-\tau t}(u - g)$ where τ is to be chosen. The equation for U is $U_t + (\tau + A(t) + B(t))U = F$, $F = e^{-\tau t}(f - g_t - (A + B)g)$, and we are led to consider the quadratic form

$$q(t, u, v; \tau) = \tau \int_\Omega uv d\mathbf{x} + \int_\Omega \left(\sum_{i,j} a_{ij} u_{x_i} v_{x_j} + \sum_i b_i u_{x_i} v + cuv \right) d\mathbf{x}$$

on $H^1(\Omega) \times H^1(\Omega)$. Show that τ can be chosen so that

$$q(t, u, u; \tau) \ge C\|u\|^2_{H^1(\Omega)}$$

and carry out the details of the extension of Theorem 2.5.

2.4. Let V, H, V' be a triplet of Hilbert spaces with $V \subset H \subset V'$, V dense in H. We may use the Riesz map J (Section 1 of Chapter 8) to identify H and H': This means that for every $u \in H$ we identify u and Ju, writing

$${}_{H'}\langle u, v \rangle_H = (u, v)_H$$

for every $v \in H$. This implies the inclusion $H = H' \subset V'$, i.e., ${}_{V'}\langle u, v \rangle_V = (u, v)_H$ for every $u \in H$, $v \in V$. One might think of using the Riesz theorem (Section 1 of Chapter 8) to enforce the further identification $V = V'$ (remember that two infinite-dimensional sets A, B with A a proper subset of B may be isomorphic). Show that this further identification is not allowed unless $V = H$ with equal norms. *Hint*: ${}_{V'}\langle u, v \rangle_V = (u, v)_V$ for every $u, v \in V$ implies $(u, v)_V = (u, v)_H$ and as V is dense in H, $V = H$.

3.1. (Gronwall's inequality). If f is continuous and nonnegative with

$$f(t) \le C + K \int_0^t f(s) ds, \qquad C, K \ge 0,$$

on $0 \le t \le a$, then $f(t) \le Ce^{Kt}$ there. *Hint*: Let $F(t) = C + K \int_0^t f(s) ds$ and show $F(t) \le Ce^{Kt}$.

References

1. WHEEDEN, R. L., and ZYGMUND, A., *Measure and Integral*, Marcel Dekker, New York, New York, 1977.
2. GODLEWSKI, E., and RAVIART, P. A., *Hyperbolic Systems of Conservation Laws*, SMAI No. 3/4, Paris, France, 1990–91.

Suggested Further Reading

FRIEDMAN, A., *Partial Differential Equations of Parabolic Type*, Prentice–Hall, Englewood Cliffs, New Jersey, 1964.

LADYZHENSKAYA, O. A., SOLONNIKOV, V, A., and URAL'CEVA, N. N., *Linear and Quasilinear Equations of Parabolic Type*, American Mathematical Society Translations of Mathematical Monographs, 1968.

TREVES, F., *Basic Linear Partial Differential Equations*, Academic Press, New York, New York, 1975.

7

Hyperbolic Systems of Conservation Laws in One Space Variable

1. Introduction

Here we consider systems of partial differential equations that can be written in the form

$$\mathbf{u}_t + \mathbf{f}(\mathbf{u})_x = 0, \tag{1}$$

where $\mathbf{u}(x, t) \in \mathbb{R}^m$, x is a real variable, $t > 0$, and $\mathbf{f}$ is a smooth function defined on a domain $G \subset \mathbb{R}^m$ and taking values in $\mathbb{R}^m$. More particularly, we will study the Cauchy problem for (1) in which $\mathbf{u}$ is sought for $t > 0$ such that

$$\mathbf{u}(x, 0) = \mathbf{u}^0(x). \tag{2}$$

We will always assume $\mathbf{u}^0 \in L^\infty(\mathbb{R})$ and our attention will be focused on solutions that are also bounded functions. If $\mathbf{u}$ is differentiable, (1) can be written

$$\mathbf{u}_t + \mathbb{A}(\mathbf{u})\mathbf{u}_x = 0, \tag{3}$$

where $\mathbb{A} = \partial\mathbf{f}/\partial\mathbf{u}$ is the Jacobian matrix of $\mathbf{f}$. We are concerned with *hyperbolic* systems.

Definition 1.1. The system (3) is hyberbolic if, for each $\mathbf{u} \in G$, $\mathbb{A}(\mathbf{u})$ has m real eigenvalues $\lambda_1 \leq \lambda_2 \leq \cdots \leq \lambda_m$ and a full set of (right) eigenvectors $\boldsymbol{r}_k$, $\mathbb{A}\boldsymbol{r}_k = \lambda_k \boldsymbol{r}_k$.

There are always, then, a full set of (left) eigenvectors $\boldsymbol{l}_k$. $\mathbb{A}^T\boldsymbol{l}_k \equiv \boldsymbol{l}_k\,\mathbb{A} = \lambda_k \boldsymbol{l}_k$, which form a biorthogonal set with $\boldsymbol{r}_k$ (Ref. 1, p. 333), i.e.,

$$\boldsymbol{l}_j(\mathbf{u}) \cdot \boldsymbol{r}_k(\mathbf{u}) = 0 \qquad \text{for } j \neq k \quad (\forall \mathbf{u} \in G).$$

We will always assume that $\lambda_k(\mathbf{u})$, $\boldsymbol{l}_k(\mathbf{u})$, and $\mathbf{r}_k(\mathbf{u})$ are smooth functions on G. In the special case in which the eigenvalues are distinct, the system is said to be *strictly hyperbolic.*

An important role is played by the characteristic curves given, in implicit form, by $\varphi_k(x, t) = c$, where $\varphi = \varphi_k$ is a solution of

$$\varphi_t + \lambda_k(\mathbf{u})\varphi_x = 0$$

with $\varphi_x \neq 0$. They are given in explicit form by $x = x(t) + c$, with $x(t)$ a solution of

$$\frac{dx}{dt} = \lambda_k(\mathbf{u}) \equiv \lambda_k(\mathbf{u}(x(t), t))$$

for every $k = 1, \ldots, m$ (for fixed k, we speak of a k-characteristic curve). Thus, the characteristic curves form m one-parameter families that are distinct for strictly hyperbolic systems. We have met this idea in Chapter 1 for the special case of a constant-coefficient system, and the ideas introduced there will play an important role in further developments here. As the equation that we are dealing with here is nonlinear, these curves must be thought of as arising from a solution $\mathbf{u}$ of (1), presumably already known, i.e., characteristic curves are not a priori known.

An important structural property of hyperbolic systems of the form (3), strictly related to the notion of characteristics, is the existence of simple waves and Riemann invariants.

Definition 1.2 (Simple waves). A solution of (3) of the form

$$\mathbf{u}(x, t) = \mathbf{U}(\varphi(x, t))$$

where φ is a single scalar function of (x, t), and $\mathbf{U}$ is a vector-valued function of one scalar variable, is called a simple wave.

Substituting in (3) we find $\mathbf{U}'\varphi_t + \mathbb{A}(\mathbf{U})\mathbf{U}'\varphi_x = 0$, where $\mathbf{U}' = d\mathbf{U}/d\varphi$. Hence, $\mathbf{U}'(\varphi)$ must coincide with a right eigenvector of $\mathbb{A}(\mathbf{U})$ corresponding to the eigenvalue $-\varphi_t/\varphi_x$ and we arrive at the two equations for $\mathbf{U}$ and φ,

$$\begin{aligned} \mathbf{U}'(\varphi) &= \boldsymbol{r}_k(\mathbf{U}(\varphi)), \\ \varphi_t + \lambda_k(\mathbf{U}(\varphi))\varphi_x &= 0, \end{aligned} \tag{4}$$

for some integer $k, 1 \leq k \leq m$. Thus, the function φ is constant along k-characteristics, the curves $\varphi(x, t) = c$ are k-characteristic curves corresponding to the *constant* solution $\mathbf{u} = \mathbf{U}(c)$, and hence they are straight lines (whose slope, in general, varies with c).

As the $\mathbf{r}_k$ form a full set, a hyperbolic system is characterized by the existence of m distinct families of simple waves (e.g., a detailed proof is given in Ref. 2). When $\varphi = x/t$, the simple wave is called *centered* (at the origin). For a k-centered wave $\lambda_k(\mathbf{U}(x/t)) = x/t$ and the characteristics $x = ct$ form a bundle centered at the origin.

A simple wave can also be defined in terms of the *Riemann invariants*. We illustrate this concept (formally) for 2×2 systems. Left multiplying the system (3) by $\boldsymbol{l}_i(\mathbf{u})$, and using the fact that $\boldsymbol{l}_i \cdot \mathbb{A} = \lambda_i \boldsymbol{l}_i(\mathbf{u})$ yields

$$\boldsymbol{l}_i(\mathbf{u}) \cdot (\mathbf{u}_t + \lambda_i(\mathbf{u})\mathbf{u}_x) = 0, \qquad i = 1, \ldots, m$$

[this is called the "characteristic form" for (3)]. If $m = 2$, the differential form $\boldsymbol{l}_j(\mathbf{u}) \cdot d\mathbf{u}$ is exact or can be made so by introducing a suitable integrating factor (which amounts to choosing a suitable normalizing factor for $\boldsymbol{l}_j$). This implies the (local) existence of functions $z_1(\mathbf{u}), z_2(\mathbf{u})$, called Riemann invariants, such that

$$\text{grad}_u\, z_j(\mathbf{u}) = \boldsymbol{l}_j(\mathbf{u}), \qquad j = 1, 2.$$

As the $\boldsymbol{l}_j$ form a full set, this vector function $z = (z_1(\mathbf{u}), z_2(\mathbf{u}))$ can be (locally) inverted to yield $\mathbf{u} = \mathbf{u}(z)$, and the system (3) reduces to the "diagonal form" for $z_j(x, t) = z_j(\mathbf{u}(x, t))$,

$$\frac{\partial z_j}{\partial t} + \overline{\lambda}_j(z) \frac{\partial z_j}{\partial x} = 0, \qquad j = 1, 2, \tag{5}$$

where $\overline{\lambda}_j(z) := \lambda_j(\mathbf{u}(z))$. We thus see that *j-Riemann invariant z_j is a function of* $\mathbf{u}$ *that is constant along the j-characteristic lines*. We have met this concept in connection with linear hyperbolic systems in Section 4 of Chapter 2: For linear systems, $\boldsymbol{l}_j$ are constant and z_j exist for every m as linear functions of $\mathbf{u}, z_j := \boldsymbol{l}_j \cdot \mathbf{u}$. In $\mathbf{u}$-space, the Riemann invariants may be viewed as a convenient choice of coordinates that diagonalizes the matrix $\mathbb{A}(\mathbf{u})$. A Riemann invariant z_j satisfies the relation

$$R_k z_j := \boldsymbol{r}_k \cdot \text{grad}_u\, z_j = 0 \qquad \text{for } j \neq k.$$

(This relation can be used to extend the definition of Riemann invariants to systems with $m > 2$ dependent variables; see Exercises 1.1, 1.2.)

A simple wave can be defined as a solution $\mathbf{u}$ such that either z_1 or z_2 is constant (everywhere). The proof of this is left as an exercise (Exercise 1.3).

Two further important concepts are given in the following definitions.

Definition 1.3 (Genuine nonlinearity). The kth eigenvalue is said to be genuinely nonlinear if

$$\boldsymbol{r}_k(\mathbf{u}) \cdot \operatorname{grad}_{\mathbf{u}} \lambda_k(\mathbf{u}) \neq 0 \qquad \forall \mathbf{u} \in G. \tag{6}$$

The system is called genuinely nonlinear if (6) holds for all $k = 1, \ldots, m$. For a 2×2 system in diagonal form (5) this condition reduces to $\partial\bar{\lambda}_j/\partial z_j \neq 0$ ($j = 1, 2$).

Definition 1.4 (Linear degeneracy). The kth eigenvalue is called linearly degenerate if

$$\boldsymbol{r}_k(\mathbf{u}) \cdot \operatorname{grad}_{\mathbf{u}} \lambda_k(\mathbf{u}) = 0 \qquad \forall \mathbf{u} \in G. \tag{7}$$

The system is said to be linearly degenerate if (7) holds for all $k = 1, \ldots, m$.

For example, a linear system is linearly degenerate, as all λ_k are constant. A 2×2 system in diagonal form (5) is linearly degenerate if $\bar{\lambda}_1$ does not depend on z_1 and $\bar{\lambda}_2$ does not depend on z_2; in other words, $\lambda_1(\mathbf{u})$ is a 2-Riemann invariant and $\lambda_2(\mathbf{u})$ is a 1-Riemann invariant.

We will need later on to define normalized eigenvectors. For genuinely nonlinear systems this will always be done by setting

$$\hat{\boldsymbol{r}}_k(\mathbf{u}) = \boldsymbol{r}_k(\mathbf{u})/\boldsymbol{r}_k(\mathbf{u}) \cdot \operatorname{grad}_{\mathbf{u}} \lambda_k(\mathbf{u}), \qquad \hat{\boldsymbol{l}}_k(\mathbf{u}) = \boldsymbol{l}_k(\mathbf{u})/\boldsymbol{l}_k(\mathbf{u}) \cdot \hat{\boldsymbol{r}}_k(\mathbf{u})$$

for $\mathbf{u}$ in any compact subset K of G (so that the denominators are bounded away from zero). With this choice the relations

$$\hat{\boldsymbol{r}}_k(\mathbf{u}) \cdot \operatorname{grad}_{\mathbf{u}} \lambda_k(\mathbf{u}) = 1, \qquad \hat{\boldsymbol{r}}_k(\mathbf{u}) \cdot \hat{\boldsymbol{l}}_j(\mathbf{u}) = \delta_{jk} \qquad \forall \mathbf{u} \in K$$

hold, and any vector $\boldsymbol{v} \in \mathbb{R}^m$ can be represented as $\boldsymbol{v} = \sum_j \boldsymbol{v} \cdot \hat{\boldsymbol{l}}_j \hat{\boldsymbol{r}}_j = \sum_j \boldsymbol{v} \cdot \hat{\boldsymbol{r}}_j \hat{\boldsymbol{l}}_j$.

Examples

1.1 (p-systems). A p-system is the 2×2 system in divergence form

$$u_t - v_x = 0, \qquad v_t - (p(u))_x = 0$$

[$p(u)$ a given function] obtained from the nonlinear wave equation

$$w_{tt} = p(w_x)_x$$

by the substitution $u = w_x$, $v = w_t$. Here $\mathbf{u} = (u, v)$, the system is (strictly) hyperbolic for $p'(u) > 0$, and setting $c(u) := \sqrt{p'(u)}$ we have

$$\lambda_1 = -c(u), \quad \boldsymbol{r}_1 = (1, c(u)), \quad \boldsymbol{l}_1 = (c(u), 1), \quad \boldsymbol{r}_1 \cdot \operatorname{grad}_u \lambda_1 = -c'(u),$$

$$\lambda_2 = c(u), \quad \boldsymbol{r}_2 = (1, -c(u)), \quad \boldsymbol{l}_2 = (c(u), -1), \quad \boldsymbol{r}_2 \cdot \operatorname{grad}_u \lambda_2 = c'(u).$$

Hence, the system is genuinely nonlinear if $p''(u) \neq 0$. A global system of Riemann invariants is

$$z_1 = v + \int_a^u c(\alpha)d\alpha, \qquad z_2 = v - \int_a^u c(\alpha)d\alpha \tag{8}$$

(a is any convenient real number). If we set $P(u) := \int_a^u c(\alpha)d\alpha$, then, as $c(u) > 0$, (8) can be inverted to yield

$$v = \tfrac{1}{2}(z_1 + z_2), \qquad u = P^{-1}\big(\tfrac{1}{2}(z_1 - z_2)\big),$$

and thus $\bar{\lambda}_j(z) := \lambda_j(P^{-1}(\frac{1}{2}(z_1 - z_2)))$ $(j = 1, 2)$.

1.2 (Barotropic gas). A gas is called barotropic if its pressure is a function of density alone, $p = p(\rho)$. Such a gas is clearly insentropic (Ref. 3). Suppose the gas flow is inviscid and one-dimensional with (scalar) velocity $u(x, t)$ along the x-axis. Then (see Chapter 1) the flow is governed by the 2×2 system in divergence form

$$\begin{aligned} \rho_t + (\rho u)_x &= 0, \\ (\rho u)_t + (\rho u^2 + p)_x &= 0, \end{aligned} \tag{9}$$

where the "equation of state" $p = p(\rho)$ is given, with $dp/d\rho > 0$. It follows $p_x = c^2(\rho)\rho_x$, where $c = \sqrt{dp/d\rho}$ is the speed of sound. The phase space is $G = \{\rho > 0, u \in \mathbb{R}\}$. If $\mathbf{u} = (\rho, u)$, then the system is strictly hyperbolic with

$$\begin{aligned} &\lambda_1 = u - c, \quad \boldsymbol{r}_1 = (-\rho, c), \quad \boldsymbol{l}_1 = (-c, \rho), \quad \boldsymbol{r}_1 \cdot \operatorname{grad}_u \lambda_1 = c + \rho c'(\rho), \\ &\lambda_2 = u + c, \quad \boldsymbol{r}_2 = (\rho, c), \quad \boldsymbol{l}_2 = (c, \rho), \quad \boldsymbol{r}_2 \cdot \operatorname{grad}_u \lambda_2 = c - \rho c'(\rho), \end{aligned} \tag{10}$$

and a system of Riemann invariants is

$$z_1 = u - \int c(\rho)\rho^{-1}\,d\rho, \qquad z_2 = u + \int c(\rho)\rho^{-1}\,d\rho.$$

For a perfect gas with constant specific heats $p = A\rho^\gamma (A > 0, 1 < \gamma$ are constants, see Chapter 1), hence $c = \sqrt{\gamma A \rho^{\gamma-1}}$, and

$$z_1 = u - 2c/(\gamma - 1), \qquad z_2 = u + 2c/(\gamma - 1). \tag{11}$$

These can be inverted to give

$$c = \frac{\gamma - 1}{4}(z_2 - z_1), \qquad u = \frac{1}{2}(z_1 + z_2),$$

hence

$$\bar{\lambda}_1(z) = \frac{1+\gamma}{4} z_1 + \frac{3-\gamma}{4} z_2, \qquad \bar{\lambda}_2(z) = \frac{1+\gamma}{4} z_2 + \frac{3-\gamma}{4} z_1,$$

and the diagonal form

$$\frac{\partial z_1}{\partial t} + \left(\frac{1+\gamma}{4} z_1 + \frac{3-\gamma}{4} z_2\right)\frac{\partial z_1}{\partial x} = 0,$$
$$\frac{\partial z_2}{\partial t} + \left(\frac{1+\gamma}{4} z_2 + \frac{3-\gamma}{4} z_1\right)\frac{\partial z_2}{\partial x} = 0$$

reduces to two uncoupled Burgers' equations for $\gamma = 3$ (a value appropriate for some explosives). Note that the system would be linearly degenerate if γ were allowed to take the (nonphysical) value $\gamma = -1$: This corresponds to the so-called *Chaplygin gas*, which, although fictitious, has certain interesting mathematical properties. The common range of values of γ is $1 < \gamma \leq 3$, but values as high as 7 arise in special applications.

1.3 (Acoustics). By linearizing system (9) near the constant state $\rho = \rho_0$, $u = 0$ we obtain the 2×2 linear system of acoustics

$$\begin{aligned} \rho_t + \rho_0 u_x &= 0, \\ \rho_0 u_t + c_0^2 \rho_x &= 0, \end{aligned} \tag{12}$$

where $c_0 = c(\rho_0)$. The eigenvalues are $\lambda_j = \mp c_0$, the eigenvectors are obtained by replacing ρ, c by ρ_0, c_0 in (10), and a system of Riemann invariants $\boldsymbol{l}_j \cdot \mathbf{u}$ is

$$z_1 = u - c_0\rho/\rho_0, \qquad z_2 = u + c_0\rho/\rho_0.$$

1.4 (Euler gas dynamics equations). The Euler equations for one-dimensional inviscid gas flow are (e.g., Courant and Friedrichs, Ref. 4)

$$\begin{aligned} \rho_t + (\rho u)_x &= 0, \\ (\rho u)_t + (\rho u^2 + p)_x &= 0, \\ (\rho W)_t + \{u(\rho W + p)\}_x &= 0, \end{aligned} \tag{13}$$

where $W = (e + \frac{1}{2}u^2)$ is the total energy, with e the internal energy, per unit mass. These equations express conservation of mass, linear momentum, and energy, respectively. The pressure p can be written in terms of any two thermodynamic variables, say $p = p(\rho, e)$ (equation of state). Then choosing

$$\mathbf{u} = (\rho, \rho u, \rho W)$$

as dependent variable, varying in the open set $G = \{\rho > 0, \rho u \in \mathbb{R}, \rho W > 0\}$ (the "phase space") in $\mathbb{R}^3$, we can easily reduce (13) to a system of conservation laws (1).

Other choices of dependent variables are legitimate. We may replace $\rho, u, \rho W$ by u and any two thermodynamic variables. For instance, by manipulations and using thermodynamic relations, (13) yields

$$S_t + uS_x = 0, \tag{14}$$

which states that the entropy per unit, mass, S, is constant along streamlines (due to the absence of viscous dissipation). By combining with the first equation (13), this can be written as a conservation law,

$$(\rho S)_t + (\rho u S)_x = 0, \tag{15}$$

for the entropy per unit volume, ρS. Similarly, by combining the first two equations (13), one finds the equation for u,

$$u_t + uu_x + \rho^{-1}p_x = 0. \tag{16}$$

Thus, one may choose the new dependent variable $\mathbf{u} = (\rho, u, S)$ and replace the last two equations (13) by (14) and (16). One advantage of using equation (14) is that we see immediately that if the initial entropy $S(x, 0) = S_0$ is constant, the flow remains isentropic, $S(x, t) = S_0$, at least so long as it remains smooth. It is interesting for later purposes to remark that $U = -\rho S$ is a *convex* function of ρ, ρu, ρW (Exercise 1.5).

Consider for simplicity the case of a perfect gas with constant specific heats c_p, c_v. Then $p = R\rho T$, T the absolute temperature, and thermodynamics tells us that

$$p = \rho^\gamma \exp(S/c_v), \qquad \gamma = c_p/c_v > 1.$$

Choosing the variables (ρ, u, S) one finds that the system is *strictly hyperbolic* with

$$\begin{aligned} \lambda_1 &= u - c, & \boldsymbol{r}_1 &= (\rho, -c, 0), & \boldsymbol{l}_1 &= (-c, \rho, -p_S/c), \\ \lambda_2 &= u, & \boldsymbol{r}_2 &= (p_S, 0, -c^2), & \boldsymbol{l}_2 &= (0, 0, 1), \\ \lambda_3 &= u + c, & \boldsymbol{r}_3 &= (\rho, c, 0), & \boldsymbol{l}_3 &= (c, \rho, p_S/c), \end{aligned}$$

where $p_S = \partial p/\partial S \equiv p/c_v$, and $c = \sqrt{\partial p/\partial \rho} \equiv \sqrt{\gamma p/\rho}$ is the local sound speed. Moreover,

$$\boldsymbol{r}_k(\mathbf{u}) \cdot \operatorname{grad}_{\mathbf{u}} \lambda_k(\mathbf{u}) = \begin{cases} -c - \rho\partial c/\partial\rho = -(\gamma + 1)c/2, & k = 1, \\ 0, & k = 2, \\ c + \rho\partial c/\partial\rho = (\gamma + 1)c/2, & k = 3. \end{cases}$$

Therefore, the eigenvalues λ_1, λ_3 are *genuinely nonlinear*, whereas λ_2 is *linearly degenerate*.

A crucial remark must be made here. The substitution of $\mathbf{u} = (\rho, \rho u, \rho W)$ with any other set of dependent variables and of system (13) with the corresponding variants is perfectly legitimate and all of these variants are equivalent *so long as* $\mathbf{u}$ *is smooth*. However, it is well known from experiments that gas dynamic flows may develop *shock* fronts, across which the solution $\mathbf{u}$, and in particular the entropy S, has a jump discontinuity (the entropy actually *increases* in passing through a shock). If this happens, the equivalence ceases to hold and one must refer to $\mathbf{u} = (\rho, \rho u, \rho W)$ as the correct conservative variables. [We will see later that the conservation law (15) is no longer valid in the presence of shocks, as it turns out to be incompatible with the discontinuity jump in the entropy.] Discontinuous solutions of conservation laws can indeed be defined, but a smooth change of the dependent variables does not preserve such solutions.

In conclusion, one should envisage $\mathbf{u} = (\rho, \rho u, \rho W)$ as the "correct" dependent variable and regard (15) as an additional conservation law satisfied by smooth solutions of the Euler system (13). In the presence of shocks, this conservation law is to be replaced by an entropy condition stating that S increases in passing through a shock. We will return to this important point later on.

1.5 (Scalar equations). A great deal can be learned by studying the special case, $m = 1$, of a scalar equation. We will make an extensive sutdy of this case in Section 3 of this chapter. We make a few simple remarks now. In the case of a scalar equation

$$u_t + f(u)_x = 0,$$

the characteristic curves form a single one-parameter family with equation $dx/dt = f'(u)$; for each fixed u we have a line with slope $1/f'(u)$ in the (x,t)-plane. We may think of the solution of the Cauchy problem (1), (2) as being generated by characteristic lines $x - x_0 = f'(u_0)t$ where $u_0 = u_0(x_0)$ (a precise mathematical statement will be given in Section 3), with u constant, $u = u_0$, along these lines. We will see that this geometric construction is equivalent to $u(x, t)$ being a solution of (1), (2).

The prototype of the scalar equation is the special case

$$u_t + uu_x = 0,$$

where $f(u) = u^2/2$, so that the characteristics are just lines with slope $1/u$. We can see the construction most simply in this case, and also see potential pitfalls. If $u_0(x)$ is smooth, the lines $x - x_0 = u_0 t$, $u = u_0$ in (x, t, u)-space generate a smooth surface $u(x, t)$ solving (1), (2) in some neighborhood of the initial line, but this procedure must break down if two of the lines $x - x_0 = u_0 t$ cross at some $t > 0$. It is not difficult to see that this must happen if $u_0(x)$ is not monotonic (more precisely, if it is a decreasing function of x). This is illustrated in Fig. 1.

The unique smooth solution that exists for some time must become discontinuous in general. The understanding of these discontinuities and continuation past the time they develop for an appropriate generalized or "weak" solution is the main goal of this chapter. We begin, however, with an existence theorem, local in time, for the Cauchy problem (1), (2) for hyperbolic systems in one space variable. As a by-product of the proof given in Section 2 we obtain an alternative theorem that characterizes finite time breakdown.

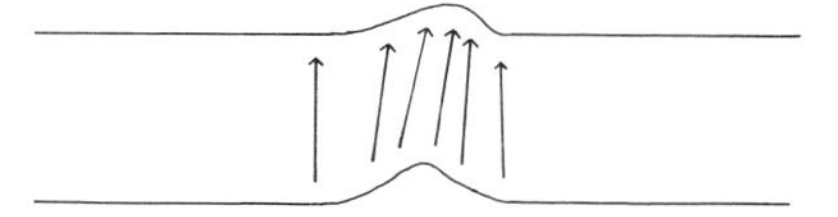

Fig. 1. Crossing of characteristics leads to a breakdown of the solution.

Although these results have been essentially known for a long time (e.g., Douglis, Ref. 5; Hartman and Winter, Ref. 6; Cesari, Ref. 7), we believe the following is a clarification of existing presentation in the literature. We have avoided systems of conservation laws in several space variables because more sophisticated techniques are required. The reader is referred to Majda (Ref. 2) for more information.

2. Local Existence Theorem for a.e. and Smooth Solutions of the Cauchy Problem

We consider the Cauchy problem for a quasilinear hyperbolic system in two independent variables of the form

$$\mathbf{u}_t + \mathbb{A}(\mathbf{u})\mathbf{u}_x = 0, \qquad x \in \mathbb{R}, t > 0, \tag{17}$$

$$\mathbf{u}(x, 0) = \mathbf{u}^0(x), \qquad x \in \mathbb{R}, \tag{18}$$

where $\mathbf{u}, \mathbf{u}^0 \in \mathbb{R}^m$, $\mathbb{A}$ is an $m \times m$ matrix (a given function of $\mathbf{u}$), and $\mathbf{u}^0(x)$ is a given vector function. Let $G \subset \mathbb{R}^m$, G an open connected set, denote the "phase space" for $\mathbf{u}$, and $\boldsymbol{v}_0 \in G$ a constant vector such that

$$Q_\omega := \{\mathbf{u} \in \mathbb{R}^m\colon |\mathbf{u} - \boldsymbol{v}_0|_\infty \leq \omega\} \subset G$$

for some $\omega > 0$. (Here $|\mathbf{u}|_\infty := \max_i |\boldsymbol{u}_i|$.) If $G = \mathbb{R}^m$, then ω is arbitrary. By defining a shifted dependent variable $\tilde{\mathbf{u}} = \mathbf{u} - \boldsymbol{v}_0$, we can always suppose that $\boldsymbol{v}_0 = 0$.

We assume that the system is *hyperbolic*, so that $\mathbb{A}(\mathbf{u})$ has, globally in G, m real eigenvalues $\lambda_k(\mathbf{u})(\mathbf{u} \in G)$ and full sets of left and right eigenvectors $\boldsymbol{l}_k(\mathbf{u}), \boldsymbol{r}_k(\mathbf{u}), k = 1, \ldots, m,$ which form biorthogonal bases of $\mathbb{R}^m(\forall \mathbf{u} \in G)$. Multiplying the system (1) on the left by $\boldsymbol{l}_i(\mathbf{u})$ yields

$$\boldsymbol{l}_i(\mathbf{u}) \cdot (\mathbf{u}_t + \lambda_i(\mathbf{u})\mathbf{u}_x) = 0 \qquad (i = 1, \ldots, m)$$

(see Section 1). This *characteristic form* can be written componentwise as

$$\sum_{j=1}^{m} \Lambda_{ij}(\mathbf{u}(x, t))\left[\frac{\partial u_j}{\partial t} + \lambda_i(\mathbf{u})\frac{\partial u_j}{\partial x}\right] = 0 \qquad (i = 1, \ldots, m), \tag{17a}$$

where Λ_{ij} is the jth component of $\boldsymbol{l}_i(\mathbf{u})$. The matrix $\Lambda = [\Lambda_{ij}]$ is nonsingular in G, and we denote by $[\alpha_{ji}(\mathbf{u})]$ the inverse matrix of $[\Lambda_{ij}(\mathbf{u})]$ in G (the indices are transposed for later convenience).

We first consider the Cauchy problem (17a), (18) under the assumptions

i. $\mathbb{A}(\mathbf{u}), \lambda_k(\mathbf{u}), \boldsymbol{l}_k(u) \in C^1(G)$,
ii. $\mathbf{u}^0(x)$ is bounded and Lipschitz continuous in $\mathbb{R}$:

$$|\mathbf{u}^0(x)|_\infty \leq \omega_0 \quad (\forall x \in \mathbb{R}), \qquad |\mathbf{u}^0(x) - \mathbf{u}^0(x')|_\infty \leq \Lambda_0 |x - x'| \quad (\forall x, x' \in \mathbb{R}).$$

Let us denote by $\xi = g_k(\tau; t, x)$ the characteristic curve of the kth family, parametrized by $\tau \geq 0$:

$$C_k\colon \frac{d}{d\tau} g_k(\tau; t, x) = \lambda_k(\mathbf{u}(b_k, \tau))\Big|_{b_k = g_k(\tau; t, x)} \tag{19}$$

passing through the point (x, t), i.e., $g_k(t; t, x) = x$ [Of course, $g_k(\tau; t, x)$ also depends on $\mathbf{u}$.] The characteristic form (17a) expresses the fact that

$$\sum_{j=1}^{m} \Lambda_{ij}(\mathbf{u}) \frac{du_j}{d\tau}\bigg|_{C_i} = 0 \qquad (i = 1, \ldots, m). \tag{20}$$

Proceeding formally (for the moment), we can rewrite (20) in the form

$$\frac{d}{d\tau}\left(\sum_{j=1}^{m} \Lambda_{ij}(\mathbf{u}) u_j\right) - \sum_{j=1}^{m}\left(\frac{d}{d\tau}\Lambda_{ij}(\mathbf{u})\right) u_j = 0 \qquad (i = 1, \ldots, m)$$

and integrate along the ith characteristric C_i for τ between 0 and t, to obtain

$$\sum_{j=1}^{m} \Lambda_{ij}(\mathbf{u}(x, t)) u_j(x, t) = \sum_{j=1}^{m} \Lambda_{ij}(\mathbf{u}^0(a_i)) u_j^0(a_i)$$
$$+ \sum_{j=1}^{m} \int_0^t \frac{d}{d\tau}(\Lambda_{ij}(\mathbf{u}(b_i, \tau))) u_j(b_i, \tau) d\tau,$$

where

$$a_i := g_i(0; t, x), \qquad b_i := g_i(\tau; t, x).$$

By applying the inverse matrix $\Lambda^{-1} = [\alpha_{ji}(\mathbf{u}(x, t))]$ we find

$$u_i(x, t) = \sum_{s=1}^{m} \alpha_{si}(\mathbf{u}(x, t)) \sum_{j=1}^{m} \left[\Lambda_{sj}(\mathbf{u}^0(a_s))u_j^0(a_s) + \int_0^t \left(\frac{d}{d\tau} \Lambda_{sj}(\mathbf{u}(b_s, \tau)) \right) u_j(b_s, \tau) d\tau \right]. \tag{21}$$

Similarly, letting $h_k(\tau; t, x) = g_k(\tau; t, x) - x$, (19) can be rewritten in intergral form

$$h_k(\tau; t, x) = -\int_\tau^t \lambda_k(\mathbf{u}(b'_k, \tau'))d\tau', \tag{22}$$

where $b'_k = g_k(\tau'; t, x)$. We put

$$\mathbf{h} = \mathbf{h}(\tau; t, x) = (h_1, \ldots, h_m), \qquad \boldsymbol{g} = (h_1 + x, \ldots, h_m + x).$$

If $\mathbf{u}(x, t)$ is a C^1-solution of (17a) and (18), with $\mathbf{u}^0 \in C^1$, all above passages are justified, $\mathbf{u}$ is a fixed point of the pair of integral transformations (21), (22), and vice versa, every C^1 fixed point of (21), (22) yields a smooth solution of (17a), (18). Suppose now $\mathbf{u}$ is a fixed point of (21), (22), bounded and Lipschitz continuous in (x, t). Then $\mathbf{u}_x$, $\mathbf{u}_t$ exist almost everywhere and by applying the Chain Rule Differentiation Lemma of real analysis, we shall prove below that the above passages can still be justified (in an a.e. sense), and $\mathbf{u}(x, \tau)$ turns out to be a solution of (20) in the sense of Carathéodory, i.e., almost everywhere in (a_i, τ) $(i = 1, \ldots, m)$. Finally, it will be shown that $\mathbf{u}(x, t)$ is a solution of (17a), (18) a.e. in (x, t), and every such solution can be obtained in this way.

Remark 2.1. Two particular cases are worth noticing. For *linear* systems (with constant coefficients), $\mathbb{A}(\mathbf{u}) = \mathbb{A}^0$, $\lambda_k(\mathbf{u}) = \lambda_k^0$, $\boldsymbol{l}_k(\mathbf{u}) = \boldsymbol{l}_k^0$, $\Lambda_{sj} = \Lambda_{sj}^0$ and $\alpha_{ji} = \alpha_{ji}^0$ are constant, and the characteristics are straight lines, $g_s(\tau; t, x) = x - \lambda_k^0(t - \tau)$. The solution (*global* in time) of the Cauchy problem is given by

$$u_i(x, t) = \sum_{s=1}^{m} \alpha_{si}^0 \sum_{j=1}^{m} \Lambda_{sj}^0 u_j^0(a_s) \qquad (i = 1, \ldots, m)$$

$(a_s = x - \lambda_s^0 t)$ or, in vector form,

$$\mathbf{u}(x, t) = \sum_{s=1}^{m} \boldsymbol{r}_s^0 Z_s(x - \lambda_s^0 t), \tag{23}$$

where $Z_s(x) = \sum_{j=1}^{m} \Lambda_{sj}^0 \boldsymbol{u}_j^0(x)$ and $\boldsymbol{r}_s^0$ is the vector with components α_{si}^0. (See Section 4 of Chapter 2.) The solution clearly satisfies the global L^∞ estimate

$$\|\mathbf{u}(\cdot\,, t)\|_{L^\infty(\mathbb{R})} \leq c\|\mathbf{u}^0\|_{L^\infty(\mathbb{R})}, \tag{24}$$

where $c > 0$ is a fixed constant.

For *diagonal* systems, that is, $\Lambda_{ij}(\mathbf{u}) = \delta_{ij}$, the solution is

$$u_i(x, t) = u_i^0(g_i(0; t, x)) \qquad (i = 1, \ldots, m),$$

where $g_k(\tau; t, x)$ is the fixed point of the map (22), or

$$g_k(\tau; t, x) = x - \int_\tau^t \lambda_k(\mathbf{u}^0(\boldsymbol{g}(0; \tau', g_k(\tau'; t, x))))d\tau'. \tag{25}$$

Hence, again, the L^∞ estimate (24) holds, at least locally, with $c = 1$.

It is clear that if $\mathbf{u}^0(x)$ is bounded and Lipschitz continuous (23) is a solution of (17a) for a.e. a_s $(s = 1, \ldots, m)$ and a.e. $t > 0$, hence a solution of (17) [with $\mathbb{A}(\mathbf{u}) = \mathbb{A}^0$] a.e. in (x, t) (or more precisely almost everywhere along almost all characteristic lines). The same is true in the diagonal case provided (25) admits a fixed point $\boldsymbol{g}$. As we will see, this in general is true only locally in time.

Returning to the general case, we are led to consider the pair of nonlinear integral transformations

$$\mathbf{U} = T_1[\mathbf{u}, \mathbf{h}], \qquad \mathbf{H} = T_2[\mathbf{u}, \mathbf{h}], \tag{TR}$$

defined componentwise by (a manipulation of) (21) and (22):

$$U_i(x, t) = u_i^0(x) + \sum_{s=1}^{m} \alpha_{si}(\mathbf{u}(x, t))\{D_{s1}[\mathbf{h}] + D_{s2}[\mathbf{u}, \mathbf{h}]\}, \tag{TR$_1$}$$

$$H_i(\tau; t, x) = -\int_\tau^t \lambda_i(\mathbf{u}(b_i', \tau'))d\tau' \tag{TR$_2$}$$

$(i = 1, \ldots, m)$, where

$$
\begin{aligned}
D_{s1}[\mathbf{h}] &= \sum_{j=1}^{m} \Lambda_{sj}(\mathbf{u}^0(a_s))[u_j^0(a_s) - u_j^0(x)], \\
D_{s2}[\mathbf{u}, \mathbf{h}] &= \sum_{j=1}^{m} \int_0^t \left(\frac{d}{d\tau}\Lambda_{sj}(\mathbf{u}(b_s, \tau))\right)[u_j(b_s, \tau) - u_j^0(x)]d\tau
\end{aligned}
\tag{26}
$$

$(s = 1, \ldots, m)$. We have used the definitions of a_i, b_i above, and $g_s = h_s + x$.

The maps (TR) can be viewed as a nonlinear transformation

$$
W = T[w]
$$

defined on pairs $w = (\mathbf{u}, \mathbf{h})$ in the product of Banach spaces for $\mathbf{u}$ and $\mathbf{h}$ to be determined below in such a way that $W = (\mathbf{U}, \mathbf{H})$ belongs to the same product space. We shall then find conditions under which T is a contraction in a suitable weighted norm and apply the contraction mapping principle to prove existence of a unique fixed point $\overline{w} = T[\overline{w}]$, or $\overline{\mathbf{u}} = T_1[\overline{\mathbf{u}}, \overline{\mathbf{h}}]$, $\overline{\mathbf{h}} = T_2[\overline{\mathbf{u}}, \overline{\mathbf{h}}]$.

Let $a > 0$ be an arbitrary positive number, and $D_a = \mathbb{R} \times [0, a]$, $\Delta_a = [0, a]^2 \times \mathbb{R}$. Consider the Banach spaces $S_1 = C^0 \cap L^\infty(D_a)$, $S_2 = C^0 \cap L^\infty(\Delta_a)$, with norms

$$
\|\mathbf{u}\|_{S_1} := \max_i \sup_{D_a} |u_i(x, t)|, \qquad \|h\|_{S_2} := \max_i \sup_{\Delta_a} |h_i(\tau; t, x)|.
$$

Let B_1 be the closed (convex) set of functions in S_1 satisfying $\mathbf{u}(x, 0) = \mathbf{u}^0(x)$ and

$$
\begin{aligned}
|\mathbf{u}(x, t)|_\infty &\leq \omega, \\
|\mathbf{u}(x, t) - \mathbf{u}(x', t)|_\infty &\leq \Lambda|x - x'|, \\
|\mathbf{u}(x, t) - \mathbf{u}(x, t')|_\infty &\leq Q|t - t'|
\end{aligned}
\tag{27}
$$

(for $0 \leq t, t' \leq a$) where Λ, Q are positive constants. Let B_2 be the closed (convex) set of functions in S_2 satisfying $\mathbf{h}(t; t, x) = 0$ and

$$
\begin{aligned}
|\mathbf{h}(\tau; t, x) - \mathbf{h}(\tau; t, x')|_\infty &\leq p_1|x - x'|, \\
|\mathbf{h}(\tau; t, x) - \mathbf{h}(\tau; t', x)|_\infty &\leq p_2|t - t'|, \\
|\mathbf{h}(\tau; t, x) - \mathbf{h}(\overline{\tau}; t, x)|_\infty &\leq p_3|\tau - \overline{\tau}|
\end{aligned}
\tag{28}
$$

$t, t', \tau, \bar{\tau} \in [0, a]$) where p_i ($i = 1, 2, 3$) are positive constants; for reasons to be explained later, we also take

$$0 < p_1 < 1.$$

Note that $\mathbf{h} \in B_2$ is bounded as

$$|\mathbf{h}(\tau; t, x)|_\infty = |\mathbf{h}(\tau; t, x) - \mathbf{h}(t; t, x)|_\infty \leq p_3|\tau - t| \leq 2ap_3.$$

From the first inequality (27) and from assumption (i) we see that there exist constants $C_i = C_i(\omega) > 0$ such that

$$\begin{aligned} |\Lambda_{ij}(\mathbf{u})| &\leq C_1, \qquad & |\Lambda_{ij}(\mathbf{u}) - \Lambda_{ij}(\mathbf{u}')| &\leq C_2|\mathbf{u} - \mathbf{u}'|_\infty, \\ |\alpha_{ij}(\mathbf{u})| &\leq C_3, \qquad & |\alpha_{ij}(\mathbf{u}) - \alpha_{ij}(\mathbf{u}')| &\leq C_4|\mathbf{u} - \mathbf{u}'|_\infty, \\ |\lambda_i(\mathbf{u})| &\leq C_5, \qquad & |\lambda_i(\mathbf{u}) - \lambda_i(\mathbf{u}')| &\leq C_6|\mathbf{u} - \mathbf{u}'|_\infty \end{aligned} \tag{29}$$

for all $\mathbf{u}, \mathbf{u}' \in Q_\omega \subset G$.

We proceed now along the steps (I)–(IV). These will be the essential ingredients of the proof of a local existence theorem.

(I) *Under suitable conditions* T_2: $B_1 \times B_2 \to B_2$.

From the definition of H_i and previous relations we have

$$H_i(t; t, x) = 0; \qquad |H_i(\tau; t, x)| \leq C_5|t - \tau| \leq 2aC_5 < \infty;$$

$$|H_i(\tau; t, x) - H_i(\tau; t, x')| \leq C_6\Lambda(1 + p_1)|t - \tau||x - x'| \leq 2aC_6\Lambda(1 + p_1)|x - x'|;$$

$$\begin{aligned} |H_i(\tau; t, x) - H_i(\tau; t', x)| &\leq C_5|t - t'| + C_6\Lambda p_2|t - t'||t - \tau| \\ &\leq 2a(C_5 + C_6\Lambda p_2)|t - t'|; \end{aligned}$$

$$|H_i(\tau; t, x) - H_i(\bar{\tau}; t, x)| \leq C_5|\tau - \bar{\tau}|.$$

Therefore, by taking

$$p_2 = p_3 = C_5$$

and a sufficiently small, so that

(a_1) $2aC_6\Lambda \leq p_1/(1 + p_1) < 1$,
(a_2) $2a(1 + C_6\Lambda) \leq 1$,

we have that $T_2\colon B_1 \times B_2 \to B_2$.

(II) *Under suitable conditions* $T_1\colon B_1 \times B_2 \to B_1$.

From the inequalities (29) and from (27), (28) we see that the derivative $d\Lambda_{sj}/d\tau$ in (26) exists a.e. and

$$\frac{d}{d\tau}\Lambda_{sj}(\mathbf{u}(b_s, \tau)) = \sum_{i=1}^{m} \frac{\partial \Lambda_{sj}}{\partial u_i}\left(\frac{\partial u_i}{\partial \tau} + \frac{\partial u_i}{\partial x}\frac{\partial b_s}{\partial \tau}\right)$$

(a.e.). Hence,

$$\left|\frac{d}{d\tau}\Lambda_{sj}(\mathbf{u}(b_s, \tau))\right| \le mC_2(Q + C_5\Lambda),$$

amd by previous relations and the fact that $g_s(t; t, x) = x$, $\mathbf{u}(x, 0) = \mathbf{u}^0(x)$ and $0 \le t \le a$, we find

$$|D_{s2}| \le \tfrac{1}{2}m^2C_2(Q + C_5\Lambda)^2a^2, \qquad |D_{s1}| \le m\Lambda_0C_1C_5a$$

(remember that $p_2 = p_3 = C_5$). Then

$$U_i(x, 0) = u_i^0(x),$$

$$|\mathbf{U}(x, t)|_\infty \le \omega_0 + m^2C_3[C_1\Lambda_0C_5 + \tfrac{1}{2}mC_2(Q + C_5\Lambda)^2a]a, \tag{30}$$

and

$$\begin{aligned}|\mathbf{U}(x, t) - \mathbf{U}(x', t)|_\infty \le {} & \{\Lambda_0 + m^2C_4\Lambda[C_1\Lambda_0C_5 + \tfrac{1}{2}mC_2(Q + C_5\Lambda)^2a]a \\ & + m^2C_3(1 + p_1)\Lambda_0[C_2C_5a + C_1]\}|x - x'| \\ & + mC_3|D_{s2}[\mathbf{u}, \mathbf{h}](x, t) - D_{s2}[\mathbf{u}, \mathbf{h}](x', t)|.\end{aligned}$$

The evaluation of the last term above can be carried out from (26) as follows. Let

$$\beta_s' = g_s(\tau; t, x'), \qquad \alpha_s' = g_s(0; t, x'),$$
$$\Lambda_{sj} = \Lambda_{sj}(\mathbf{u}(b_s, \tau)), \qquad \Lambda_{sj}' = \Lambda_{sj}(\mathbf{u}(\beta_s', \tau)).$$

Then

$$D_{s2}[\mathbf{u}, \mathbf{h}](x, t) - D_{s2}[\mathbf{u}, \mathbf{h}](x', t) = \sum_{j=1}^{m} \int_0^t \left(\frac{d}{d\tau}\Lambda_{sj}\right)[u_j(b_s, \tau) - u_j^0(x)]d\tau$$

$$- \sum_{j=1}^{m} \int_0^t \left(\frac{d}{d\tau}\Lambda'_{sj}\right)[u_j(\beta'_s, \tau) - u_j^0(x')]d\tau$$

$$= \sum_{j=1}^{m} \int_0^t \left(\frac{d}{d\tau}\right)\Lambda_{sj} - \Lambda'_{sj})[u_j(b_s, \tau) - u_j^0(x)]d\tau$$

$$+ \sum_{j=1}^{m} \int_0^t [u_j(b_s, \tau) - u_j^0(x) - u_j(\beta'_s, \tau)$$

$$+ u_j^0(x')]\frac{d\Lambda'_{sj}}{d\tau}\, d\tau,$$

and, integrating by parts in the first term, which is legitimate because we have Lipschitz functions,

$$\sum_{j=1}^{m} \big(\Lambda_{sj}(\mathbf{u}(x, t)) - \Lambda_{sj}(\mathbf{u}(x', t))\big)[u_j(x, t) - u_j^0(x)]$$

$$- \sum_{j=1}^{m} \big(\Lambda_{sj}(\mathbf{u}^0(a_s)) - \Lambda_{sj}(\mathbf{u}^0(\alpha'_s))\big)[u_j^0(a_s) - u_j^0(x)]$$

$$- \sum_{j=1}^{m} \int_0^t (\Lambda_{sj} - \Lambda'_{sj})\frac{du_j(b_s, \tau)}{d\tau}\, d\tau$$

$$+ \sum_{j=1}^{m} \int_0^t [u_j(b_s, \tau) - u_j^0(x) - u_j(\beta'_s, \tau) + u_j^0(x')]\frac{d\Lambda'_{sj}}{d\tau}\, d\tau.$$

From (27) and (28) we have (for a.e. τ)

$$\left|\frac{d}{d\tau}u_j(b_s, \tau)\right| = \left|\frac{\partial u_j}{\partial \tau} + \frac{\partial u_j}{\partial x}\frac{\partial b_s}{\partial \tau}\right| \leq Q + C_5\Lambda,$$

so that

$$|D_{s2}[\mathbf{u}, \mathbf{h}](x, t) - D_{s2}[\mathbf{u}, \mathbf{h}](x', t)|$$
$$\leq mC_2\big\{Q\Lambda + C_5\Lambda_0^2(1 + p_1) + (Q + C_5\Lambda)[\Lambda(1 + p_1)(1 + m) + m\Lambda_0]\big\}t|x - x'|$$

$(t \leq a)$. Summarizing

$$|\mathbf{U}(x, t) - \mathbf{U}(x', t)|_\infty \leq \{\Lambda_0[1 + m^2 C_1 C_3 (1 + p_1)] + O_1(a)\}|x - x'|, \qquad (31)$$

where

$$\begin{aligned} O_1(a) = m\Big\{ & mC_4\Lambda\Big[C_1C_5\Lambda_0 + \frac{m}{2}C_2(Q + C_5\Lambda)^2 a\Big] \\ & + mC_2C_3[C_5\Lambda_0(1 + p_1) + Q\Lambda] \\ & + mC_3[C_2C_5\Lambda_0^2(1 + p_1) + mC_2(Q + C_5\Lambda)(\Lambda_0 + \Lambda(1 + p_1))] \\ & + (Q + C_5\Lambda)C_2\Lambda(1 + p_1)\Big\}a. \end{aligned}$$

Finally, we have to evaluate $|\mathbf{U}(x, t) - \mathbf{U}(x, t')|_\infty$. Again, the delicate term is the one involving $d\Lambda_{sj}/d\tau$ in (26) for D_{s2}. Letting

$$\beta_s' = g_s(\tau; t', x), \qquad \alpha_s' = g_s(0; t', x),$$

and $\Lambda_{sj} = \Lambda_{sj}(\mathbf{u}(b_s, \tau))$, $\Lambda_{sj}' = \Lambda_{sj}(\mathbf{u}(\beta_s', \tau))$, we have

$$\begin{aligned} D_{s2}[\mathbf{u}, \mathbf{h}](x, t) - D_{s2}[\mathbf{u}, \mathbf{h}](x, t') = & \sum_{j=1}^{m} \int_0^t \left(\frac{d\Lambda_{sj}}{d\tau} - \frac{d\Lambda_{sj}'}{d\tau}\right)[u_j(b_s, \tau) - u_j^0(x)]d\tau \\ & - \sum_{j=1}^{m} \int_t^{t'} [u_j(\beta_s', \tau) - u_j^0(x)]\frac{d\Lambda_{sj}'}{d\tau}\, d\tau \\ & - \sum_{j=1}^{m} \int_0^t [u_j(\beta_s', \tau) - u_j(b_s, \tau)]\frac{d\Lambda_{sj}'}{d\tau}\, d\tau. \end{aligned}$$

Integrating by parts in the first integral yields

$$\begin{aligned} \sum_{j=1}^{m} \Big\{ & \Big(\Lambda_{sj}(\mathbf{u}(x, t)) - \Lambda_{sj}(\mathbf{u}(g_s(t; t', x), t))\Big)[u_j(x, t) - u_j^0(x)] \\ & - \Big(\Lambda_{sj}(\mathbf{u}_0(a_s)) - \Lambda_{sj}(\mathbf{u}_0(\alpha_s'))\Big)[u_j^0(a_s) - u_j^0(x)] \\ & - \int_0^t (\Lambda_{sj} - \Lambda_{sj}')\frac{du_j(b_s, \tau)}{d\tau}\, d\tau\Big\}. \end{aligned}$$

Hence,

$$|D_{s2}[\mathbf{u}, \mathbf{h}](x, t) - D_{s2}[\mathbf{u}, \mathbf{h}](x, t')| \leq O_2(a)|t - t'|,$$

where

$$O_2(a) = mC_2\{C_5(Q\Lambda + C_5\Lambda_0^2) + (Q + C_5\Lambda)[C_5\Lambda(1 + m) + m(Q + 2C_5\Lambda)]\}a,$$

and we finally find

$$|\mathbf{U}(x, t) - \mathbf{U}(x, t')|_\infty \leq [m^2C_1C_3C_5\Lambda_0 + O_3(a)]|t - t'|, \tag{32}$$

where

$$O_3(a) = m^2\left\{C_4Q\left[C_1C_5\Lambda_0 + \frac{m}{2}C_2(Q + C_5\Lambda)^2a\right] + C_2C_3C_5^2\Lambda_0^2\right\}a + mC_3O_2(a).$$

From inequalities (30), (31), (32) and from (27) we see that $T_1\colon B_1 \times B_2 \to B_1$ if the following conditions are satisfied:

$$(\mathrm{a}_3) \qquad \omega_0 + O_4(a) \leq \omega,$$

$$(\mathrm{a}_4) \qquad \Lambda_0[1 + m^2C_1C_3(1 + p_1)] + O_1(a) \leq \Lambda,$$

$$(\mathrm{a}_5) \qquad m^2C_1C_3C_5\Lambda_0 + O_3(a) \leq Q,$$

where

$$O_4(a) = m^2\left\{C_1C_3C_5\Lambda_0 + \frac{m}{2}C_2C_3(Q + C_5\Lambda)^2a\right\}a.$$

Conditions (a_3)–(a_5) are satisfied for a sufficiently small provided

$$\omega_0 < \omega, \qquad Q_\omega \subset G, \tag{33}$$

$$\Lambda_0[1 + m^2C_1C_3(1 + p_1)] < \Lambda, \tag{34}$$

$$m^2C_1C_3C_5\Lambda_0 < Q, \tag{35}$$

as $O_i(a)$ $(i = 1, \ldots, 4)$, although dependent on Q and Λ, tend to zero as $a \to 0$.

Note that (34) and (35) can always be satisfied by taking Λ and Q large enough, while (33) places a restriction on ω_0 when G is bounded in at least one coordinate direction.

(III) *The operators* T_i: $B_1 \times B_2 \to B_i$ $(i = 1, 2)$ *are Lipschitz continuous.*

We begin with the operator T_2. For all $\mathbf{u}, \tilde{\mathbf{u}}$ in B_1, $\mathbf{h}$ and $\tilde{\mathbf{h}}$ in B_2, and corresponding H and $\tilde{\mathbf{H}}$ in B_2, we find

$$H_i(\tau; t, x) - \tilde{H}_i(\tau; t, x) = -\int_\tau^t \Big[\lambda_i(\mathbf{u}(b_i', \tau')) - \lambda_i(\tilde{\mathbf{u}}(\tilde{b}_i', \tau'))\Big] d\tau', \qquad 0 \le t, \ \tau \le a,$$

where $\tilde{b}_i' = \tilde{g}_i(\tau'; t, x)$. Hence, using (28) and (29), we find

$$\|\mathbf{H} - \tilde{\mathbf{H}}\|_{S_2} \equiv \|T_2[\mathbf{u}, \mathbf{h}] - T_2[\tilde{\mathbf{u}}, \tilde{\mathbf{h}}]\|_{S_2} \le k_3 \|\mathbf{h} - \tilde{\mathbf{h}}\|_{S_2} + k_4 \|\mathbf{u} - \tilde{\mathbf{u}}\|_{S_1}, \tag{36}$$

where $k_3 = C_6 \Lambda a$, $k_4 = C_6 a$.

We turn now to the operator T_1: For all $\mathbf{u}, \tilde{\mathbf{u}}$ and corresponding $\mathbf{U}, \tilde{\mathbf{U}}$ in B_1, $\mathbf{h}$ and $\tilde{\mathbf{h}}$ in B_2, we denote by $\alpha_{si} = \alpha_{si}(\mathbf{u}(x, t))$, $\tilde{\alpha}_{si} = \alpha_{si}(\tilde{\mathbf{u}}(x, t))$, $\tilde{a}_i = \tilde{g}_i(0; t, x)$, $\Lambda_{sj} = \Lambda_{sj}(\mathbf{u}(b_s, \tau))$, $\tilde{\Lambda}_{sj} = \Lambda_{sj}(\tilde{\mathbf{u}}(\tilde{b}_s, \tau))$. Then from the definition of $\mathbf{U}$ we find

$$\begin{aligned} U_i(x, t) - \tilde{U}_i(x, t) = \sum_{s=1}^{m} \{&(\alpha_{si} - \tilde{\alpha}_{si})[D_{s1}[\mathbf{h}] + D_{s2}[\mathbf{u}, \mathbf{h}]] \\ &+ \tilde{\alpha}_{si}[D_{s1}[\mathbf{h}] + D_{s2}[\mathbf{u}, \mathbf{h}] - D_{s1}[\tilde{\mathbf{h}}] - D_{s2}[\tilde{\mathbf{u}}, \tilde{\mathbf{h}}]]\}. \end{aligned}$$

Integration by parts in (26) then implies

$$\begin{aligned} |D_{s1}[\mathbf{h}] - D_{s1}[\tilde{\mathbf{h}}]| = \Bigg| \sum_{j=1}^{m} \{&\Lambda_{sj}(\mathbf{u}^0(\tilde{a}_s))[u_j^0(a_s) - u_j^0(\tilde{a}_s)] \\ &+ [\Lambda_{sj}(\mathbf{u}^0(a_s)) - \Lambda_{sj}(\mathbf{u}^0(\tilde{a}_s))][u_j^0(a_s) - u_j^0(x)]\} \Bigg| \\ \le{}& m\Lambda_0(\Lambda_0 C_2 C_5 a + C_1)\|\mathbf{h} - \tilde{\mathbf{h}}\|_{S_2} \end{aligned}$$

and

$$
\begin{aligned}
|D_{s2}[\mathbf{u}, \mathbf{h}] - D_{s2}[\tilde{\mathbf{u}}, \tilde{\mathbf{h}}]| = \Bigg| & \sum_{j=1}^{m} \int_0^t \frac{d}{d\tau}(\Lambda_{sj} - \tilde{\Lambda}_{sj})[u_j(b_s, \tau) - u_j^0(x)]d\tau \\
& + \sum_{j=1}^{m} \int_0^t \left(\frac{d}{d\tau}\tilde{\Lambda}_{sj}\right)[u_j(b_s, \tau) - \tilde{u}_j(\tilde{b}_s, \tau)]d\tau \Bigg| \\
\le & \left| \sum_{j=1}^{m} (\Lambda_{sj}(\mathbf{u}(x, t)) - \Lambda_{sj}(\tilde{\mathbf{u}}(x, t)))[u_j(x, t) - u_j^0(x)] \right| \\
& + \left| \sum_{j=1}^{m} (\Lambda_{sj}(\mathbf{u}^0(a_s)) - \Lambda_{sj}(\mathbf{u}^0(\tilde{a}_s)))[u_j^0(a_s) - u_j^0(x)] \right| \\
& + \left| \sum_{j=1}^{m} \int_0^t (\Lambda_{sj} - \tilde{\Lambda}_{sj}) \frac{du_j(b_s, \tau)}{d\tau}\, d\tau \right| \\
& + \left| \sum_{j=1}^{m} \int_0^t \frac{d\tilde{\Lambda}_{sj}}{d\tau}\, [u_j(b_s, \tau) - \tilde{u}_j(\tilde{b}_s, \tau)]d\tau \right|.
\end{aligned}
$$

Then

$$
\begin{aligned}
|D_{s2}[\mathbf{u}, \mathbf{h}] - D_{s2}[\tilde{\mathbf{u}}, \tilde{\mathbf{h}}]| \le\; & maC_2[Q + (m+1)(Q + C_5\Lambda)]\|\mathbf{u} - \tilde{\mathbf{u}}\|_{S_1} \\
& + maC_2\Lambda[\Lambda_0 C_5 + (Q + C_5\Lambda)(m+1)]\|\mathbf{h} - \tilde{\mathbf{h}}\|_{S_2}
\end{aligned}
$$

and

$$
\begin{aligned}
|U_i(x, t) - \tilde{U}_i(x, t)| \le\; & m^2\left[\Lambda_0 C_1 C_5 a + \frac{m}{2} C_2(Q + C_5\Lambda)^2 a^2\right] C_4\|\mathbf{u} - \tilde{\mathbf{u}}\|_{S_1} \\
& + m^2 C_3[C_1\Lambda_0 + C_2 C_5 \Lambda_0(\Lambda_0 + \Lambda)a + (Q + C_5\Lambda)C_2\Lambda a(1 + m)]\|\mathbf{h} - \tilde{\mathbf{h}}\|_{S_2} \\
& + m^2 C_3[QC_2 a + (Q + C_5\Lambda)C_2 a(1 + m)]\|\mathbf{u} - \tilde{\mathbf{u}}\|_{S_1}.
\end{aligned}
$$

Summarizing,

$$
\|\mathbf{U} - \tilde{\mathbf{U}}\|_{S_1} \equiv \|T_1[\mathbf{u}, \mathbf{h}] - T_1[\tilde{\mathbf{u}}, \tilde{\mathbf{h}}]\|_{S_1} \le k_1\|\mathbf{h} - \tilde{\mathbf{h}}\|_{S_2} + k_2\|\mathbf{u} - \tilde{\mathbf{u}}\|_{S_1}, \tag{37}
$$

where $k_1 = m^2 C_1 C_3 \Lambda_0 + O_5(a)$, $k_2 = O_6(a)$, with

$$O_5(a) = m^2 C_3[C_2 C_5 \Lambda_0(\Lambda_0 + \Lambda) + (Q + C_5\Lambda)C_2\Lambda(1 + m)]a,$$

$$O_6(a) = m^2 \left\{ C_1 C_4 C_5 \Lambda_0 + C_2 C_3 Q + C_2[C_3(1 + m) + \frac{m}{2} C_4(Q + C_5\Lambda)a] \cdot (Q + C_5\Lambda) \right\} a.$$

For small a, k_2, k_3, and k_4 are small, but k_1 need not be.

(IV) *The operator* $T: B_1 \times B_2 \to B_1 \times B_2$ *is a contraction in a suitable norm.*

We choose the weighted product norm

$$\|w\|_S = \|\mathbf{u}\|_{S_1} + \alpha\|\mathbf{h}\|_{S_2}$$

for the element $w = (\mathbf{u}, \mathbf{h})$ in the Banach space $S = B_1 \times B_2$, with $\alpha > 0$ to be fixed below. Then from (36) and (37) we obtain (for every $w, \tilde{w}$ in S):

$$\begin{aligned} \|W - \tilde{W}\|_S &\equiv \|T[w] - T[\tilde{w}]\|_S = \|\mathbf{U} - \tilde{\mathbf{U}}\|_{S_1} + \alpha\|\mathbf{H} - \tilde{\mathbf{H}}\|_{S_2} \\ &\leq (k_1 + \alpha k_3)\|\mathbf{h} - \tilde{\mathbf{h}}\|_{S_2} + (\alpha k_4 + k_2)\|\mathbf{u} - \tilde{\mathbf{u}}\|_{S_1} \leq k\|w - \tilde{w}\|_S \end{aligned}$$

with $0 \leq k < 1$, if α is chosen so that

$$\begin{aligned} \alpha k_4 + k_2 &\leq k, \\ k_1 + \alpha k_3 &\leq \alpha k. \end{aligned}$$

As $k_2, k_3, k_4 = O(a)$, $k_1 = m^2 C_1 C_3 \Lambda_0 + O(a)$, we can always select α such that this is true; for instance, we may take $\alpha = 1 + 2m^2 C_1 C_3 \Lambda_0$, $\frac{1}{2} \leq k < 1$, and require that a is sufficiently small, so that

$$(\mathrm{a}_6) \qquad O_6(a) + C_6\alpha a \leq k,$$

$$(\mathrm{a}_7) \qquad O_5(a) + \alpha C_6 \Lambda a \leq k + (2k - 1)m^2 C_1 C_3 \Lambda_0.$$

In this way, T will be a contraction in the weighted norm, with contraction constant $k \in [\frac{1}{2}, 1)$. We recall that the constant p_1 is arbitrary in the interval $(0,1)$, p_2 and p_3 are fixed according to $p_2 = p_3 = C_5$, and the constants C_i in (29) are known once ω has been fixed (with $Q_\omega \subset G$).

We are now able to prove our main theorem.

Theorem 2.1. Let $\mathbf{u}^0(x)$ satisfy conditions (ii) with $\omega_0 > 0$ such that $Q_{\omega_0} \subset G$, and let system (17) be hyperbolic and fulfill assumptions (i). Then there is a unique a.e. solution $\overline{\mathbf{u}}(x, t)$ of the Cauchy problem (17), (18) satisfying (27) in the strip $D_a = \mathbb{R} \times [0, a]$, where the constants ω, Q, Λ are chosen according to inequalities (33), (34), (35), and $a > 0$ satisfies (a_1)–(a_7). This solution depends continuously on $\mathbf{u}^0(\mathbf{x})$ in the sense of the sup norm in S_1.

Proof. Under the stated assumptions, the map $T\colon S \to S$ has a unique fixed point $\overline{w} = (\overline{\mathbf{u}}, \overline{\mathbf{h}})$ that satisfies

$$\overline{h}_k(\tau; t, x) = -\int_\tau^t \lambda_k(\overline{\mathbf{u}}(\overline{b}_k', \tau'))d\tau',$$

$$\overline{u}_i(x, t) = u_i^0(x) + \sum_{s=1}^m \alpha_{si}(\overline{\mathbf{u}}(x, t)\{D_{s1}[\overline{\mathbf{h}}] + D_{s2}[\overline{\mathbf{u}}, \overline{\mathbf{h}}]\}$$

($i = 1, \ldots, m$; $\overline{b}_k, \overline{b}_k', \overline{a}_k$ the quantities corresponding to $\overline{\mathbf{h}}$).Thus, $\xi = \overline{g}_k(\tau; t, x)$ yields the kth characteristic curve corresponding to the fixed point $\overline{\mathbf{u}}(x, t)$, and

$$\overline{u}_i(x, t) = \sum_{s=1}^m \alpha_{si}(\overline{\mathbf{u}}(x, t)) \sum_{j=1}^m$$
$$\times \left[\Lambda_{sj}(\mathbf{u}^0(\overline{a}_s))u_j^0(\overline{a}_s) + \int_0^t \left(\frac{d}{d\tau}\Lambda_{sj}(\overline{\mathbf{u}}(\overline{b}_s, \tau))\right)\overline{u}_j(\overline{b}_s, \tau)d\tau\right],$$

which is equivalent to the previous expression for $\overline{u}_i$. Integration by parts in this expression yields

$$\sum_{s=1}^m \alpha_{si}(\overline{\mathbf{u}}(x, t)) \sum_{j=1}^m \int_0^t \Lambda_{sj}(\overline{\mathbf{u}}(\overline{b}_s, \tau))\frac{d}{d\tau}\overline{u}_j(\overline{b}_s, \tau)d\tau = 0 \qquad (i = 1, \ldots, m),$$

and, as the matrix $[\alpha_{si}]$ is nonsingular,

$$\sum_{j=1}^m \int_0^t \Lambda_{ij}(\overline{\mathbf{u}}(\overline{b}_i, \tau))\frac{d}{d\tau}\,\overline{u}_j(\overline{b}_i, \tau)d\tau = 0. \tag{38}$$

As $0 < p_1 < 1$, from (28) we see that $\overline{b}i = \overline{g}_i(\tau; t, x)$ satisfies

$$(1 - p_1)|x - x'| \leq |\overline{g}_i(\tau; t, x) - \overline{g}_i(\tau; t, x')| \leq (1 + p_1)|x - x'| \tag{39}$$

for any $t, \tau \in [0, a]$. Moreover, from the uniqueness of the fixed point $\overline{\mathbf{h}}$ we find that $\overline{g}_i$ satisfies the group relation

$$\overline{g}_s(\tau; t, g_s(t; t', x)) = \overline{g}_s(\tau; t', x) \tag{G}$$

(Exercise 2.1), and this implies that the symmetric relations

$$\overline{a}_i = \overline{g}_i(0; t, x) \Leftrightarrow x = \overline{g}_i(t; 0, \overline{a}_i) \qquad (i = 1, \ldots, m) \tag{40}$$

hold for $0 \leq t \leq a$ and $x \in \mathbb{R}$. By force of (39) these relations yield *homeomorphisms* of the slab D_a in (x, t)-space onto slabs D_a in $(\overline{a}_i, t)$-space *that preserve sets of Lebesgue measure zero* ["property (N)"]. Taking $x = \overline{g}_i(t; 0, \overline{a}_i)$ in (38), we thus obtain, for every $\overline{a}_i \in \mathbb{R}$,

$$\sum_{j=1}^{m} \int_0^t \Lambda_{ij}(\overline{\mathbf{u}}(\overline{g}_i(\tau; 0, \overline{a}_i), \tau)) \frac{d}{d\tau} \overline{u}_j(\overline{g}_i(\tau; 0, \overline{a}_i), \tau) d\tau = 0,$$

and differentiating with respect to t,

$$\sum_{j=1}^{m} \Lambda_{ij}(\overline{\mathbf{u}}(\overline{g}_i(t; 0, \overline{a}_i), t)) \frac{d}{dt} \overline{u}_j(\overline{g}_i(t; 0, \overline{a}_i), t) = 0,$$

an explicit form of equation (20). Applying the chain rule,

$$\sum_{j=1}^{m} \Lambda_{ij}(\overline{\mathbf{u}}) \left(\frac{\partial}{\partial t} + \lambda_i(\overline{\mathbf{u}}) \frac{\partial}{\partial x} \right) \overline{u}_j = 0 \qquad (i = 1, \ldots, m)$$

for almost every $(\overline{a}_i, t) \in D_a$, where $\overline{\mathbf{u}} = \overline{\mathbf{u}}(\overline{g}_i(t; 0, \overline{a}_i), t)$. Finally, letting $\overline{a}_i = \overline{g}_i(0; t, x)$ gives

$$\sum_{j=1}^{m} \Lambda_{ij}(\overline{\mathbf{u}}(x, t)) \left(\frac{\partial \overline{u}_j(x, t)}{\partial t} + \lambda_i(\overline{\mathbf{u}}(x, t)) \frac{\partial \overline{u}_j(x, t)}{\partial x} \right) = 0 \qquad (i = 1, \ldots, m)$$

for almost every $(x, t) \in D_a$, because of "property (N)" of the map (40).

Thus, $\overline{\mathbf{u}}(x, t)$ is a solution of (17a), (18) and hence of (17), (18). The previous steps can be reversed and so $\mathbf{u}$ is a solution of (17a), (18) if and only if it is the fixed point of T.

To prove that the solution depends continuously on $\mathbf{u}^0(x)$ we may argue as follows. Let $\mathbf{u} = \overline{\mathbf{u}}[\mathbf{u}^0]$, $\mathbf{u}' = \overline{\mathbf{u}}[\mathbf{u}'^0]$ be the solutions corresponding to initial data

$\mathbf{u}^0$, $\mathbf{u}'^0$, and let $\mathbf{h}$, $\mathbf{h}'$ be the corresponding fixed points of (TR). Then, proceeding as in the derivation of (36), (37) we obtain

$$\|\mathbf{u}-\mathbf{u}'\|_{S_1} \leq k_1\|\mathbf{h}-\mathbf{h}'\|_{S_2} + k_2\|\mathbf{u}-\mathbf{u}'\|_{S_1} + \|\mathbf{u}^0-\mathbf{u}'^0\|_{S_1},$$
$$\|\mathbf{h}-\mathbf{h}'\|_{S_2} \leq k_3\|\mathbf{h}-\mathbf{h}'\|_{S_2} + k_4\|\mathbf{u}-\mathbf{u}'\|_{S_1}.$$

From step (IV) we get $k_2 \leq k, k_3 \leq k$ so that

$$\begin{aligned}\|\mathbf{u}-\mathbf{u}'\|_{S_1} &+ \alpha\|\mathbf{h}-\mathbf{h}'\|_{S_2} \\ &\leq (k_1+\alpha k_3)\|\mathbf{h}-\mathbf{h}'\|_{S_2} + (k_2+\alpha k_4)\|\mathbf{u}-\mathbf{u}'\|_{S_1} + \|\mathbf{u}^0-\mathbf{u}'^0\|_{S_1} \\ &\leq \alpha k\|\mathbf{h}-\mathbf{h}'\|_{S_2} + k\|\mathbf{u}-\mathbf{u}'\|_{S_1} + \|\mathbf{u}^0-\mathbf{u}'^0\|_{S_1}\end{aligned}$$

and

$$\|\mathbf{u}-\mathbf{u}'\|_{S_1} + \alpha\|\mathbf{h}-\mathbf{h}'\|_{S_2} \leq \frac{1}{1-k}\|\mathbf{u}^0-\mathbf{u}'^0\|_{S_1}.$$

On the other hand,

$$\|\mathbf{h}-\mathbf{h}'\|_{S_2} \geq \frac{1-k_2}{k_1}\|\mathbf{u}-\mathbf{u}'\|_{S_1} - \frac{1}{k_1}\|\mathbf{u}^0-\mathbf{u}'^0\|_{S_1},$$

and by combining the last two inequalities we finally find the bound

$$\|\mathbf{u}-\mathbf{u}'\|_{S_1} \leq \frac{k_1+\alpha(1-k)}{k_1+\alpha(1-k_2)}\,\frac{1}{1-k}\|\mathbf{u}^0-\mathbf{u}'^0\|_{S_1},$$

which completes the proof of Theorem 2.1. □

Note that the functional class B_1 is locally persistent for a.e. solutions of (17).

Remark 2.2. The solution $\bar{\mathbf{u}}(x, t)$ satisfies the local L^∞ estimate

$$\|\bar{\mathbf{u}}(x,t)\|_{S_1} \leq c\|\mathbf{u}^0\|_{S_1}, \qquad 0 \leq t \leq a,$$

for some $c = c(\omega, \Lambda, Q, a)$, with c a fixed constant for diagonal systems and scalar conservation laws (Exercise 2.2).

After the a.e. solution has been constructed on the interval $0 \leq t \leq a$, we may restart the procedure from the initial time $t = a$ taking $\bar{\mathbf{u}}(x, a)$ as new initial value, and so forth. By inspection of the proof of Theorem 2.1 we see that the solution can be continued as long as $\mathbf{u} \in G$ and a remains positive. We assert that one of the three following mutually exclusive cases necessarily arises.

(G) *The a.e. solution exists globally in time.*

(S) *There exists a finite* $T^* > 0$ *such that* $\Lambda \to \infty$ *and/or* $Q \to \infty$ *as* $t \to T^*$, *while* $\mathbf{u}$ *remains in a compact subset* $Q_{\bar{\omega}} \subset G$ *for* $0 \leq t < T^*$. This corresponds to formation of a *shock.*

(B) *For every compact set* $K \subset G$ *there is a sequence* (x_j, t_j) *such that* $t_j \uparrow T^*$ *and* $\mathbf{u}(x_j, t_j) \notin K$. This corresponds to *blow-up* of the solution.

Alternative (B) can be ruled out in case a *uniform* L^∞ estimate can be proved to hold for $0 \leq t < T^*$. This is the case for diagonal systems and scalar conservation laws. When $T^* < \infty$, the interval $t \in [0, T^*)$ is the *maximal interval of existence* of the a.e. solution.

To prove the assertion that these are the only possibilities we need to examine the proof of Theorem 2.1 carefully, and in particular the dependence of the constants $O_i(a) = O_i(a, \Lambda, Q)$ on a, Λ, and Q. These constants must be made small by choosing a small. The conditions (a_4) and (a_5) require choosing Λ and Q sufficiently large that (34), (35) hold and then choosing a small. The rest of the conditions (a_i) require choosing $O_i(a)$ small enough and these constants contain products of powers of a, Λ, Q and of the constants $C_i(\omega)$. If existence has been proven on an interval $[0, t_1]$, it is always possible to extend to an interval $[0, t_1 + a]$ by using $\mathbf{u}(x, t_1)$ as initial data at $t = t_1$. The process can then be continued. A maximal interval of existence is encountered only if the new values of a in the sequence of continuations tend to zero. If ω remains bounded, then either Λ or Q must become unbounded; if Λ, Q, and ω remain bounded, this cannot happen. We have thus proven the following theorem.

Theorem 2.2 (Continuation). Suppose there is a maximal interval of existence for (17), (18), $[0, T^*)$. Then either (B) or (S) hold.

It is possible to prove a regularity theorem showing that the solution $\bar{\mathbf{u}}(x, t)$ is a classical solution under additional hypotheses.

Theorem 2.3. Suppose, in addition to the hypotheses of Theorem 2.1, $\mathbb{A}(\mathbf{u}) \in C^2(G)$ and $\mathbf{u}^0$ is differentiable with

$$\left|\frac{d\mathbf{u}^0(x)}{dx}\right| \leq \omega_1, \qquad \left|\frac{d\mathbf{u}^0(x)}{dx} - \frac{d\mathbf{u}^0(x')}{dx'}\right| \leq \Lambda_1 |x - x'| \qquad (\forall x, x' \in \mathbb{R}).$$

Then the solution $\bar{\mathbf{u}}(x, t)$ has Lipschitz continuous first derivatives $\bar{\mathbf{u}}_x, \bar{\mathbf{u}}_t$ in a strip $(x, t) \in D_{a'}$, with $a' > 0$.

Proof. The proof is left to the reader as an extended, guided exercise.

i. Let $\boldsymbol{v}(x, t)$ be a solution of the Cauchy problem [obtained formally by differentiating (17), (18) with respect to x, and replacing u by z and $\mathbf{z}_x$ by $\boldsymbol{v} = (v_1, \ldots, v_m)$],

$$\begin{aligned} \boldsymbol{v}_t + \mathbb{A}(\mathbf{z})\boldsymbol{v}_x &= -\sum_{k=1}^{m} \mathbb{J}_k \boldsymbol{v} v_k, \qquad && (x, t) \in \tilde{D}_a, \\ \boldsymbol{v}(x, 0) &= d\mathbf{u}^0(x)/dx, && |x - x_0| \leq C_5 a \end{aligned}$$

(where $x_0 \in \mathbb{R}$ is arbitrary), and let $\mathbf{z}(x, t)$ be a fixed point of the integral transformation (which implies $\mathbf{z}_x = \boldsymbol{v}$),

$$\boldsymbol{Z}(x, t) = \int_{x_0}^{x} \boldsymbol{v}(y, t)dy + \mathbf{u}^0(x_0) - \int_0^t \mathbb{A}(\mathbf{z}(x_0, \tau))\boldsymbol{v}(x_0, \tau)d\tau, \qquad (x, t) \in \tilde{D}_a. \quad (*)$$

Here $\tilde{D}_a = \{(x, t)\colon 0 \leq t \leq a, |x - x_0| \leq C_5(a - t)\}$, and $\mathbb{J}_k$ are the matrices $\partial\mathbb{A}(\mathbf{z})/\partial z_k (k = 1, \ldots, m)$. Define $\tilde{\Delta}_a = \{(\tau; t, x)\colon 0 \leq \tau \leq t, (x, t) \in \tilde{D}_a\}$, $\tilde{B}_i$ the analogues of $B_i (i = 1, 2)$ with D_a, Δ_a replaced by $\tilde{D}_a, \tilde{\Delta}_a$, and $\tilde{B}_3$ the analogue of $\tilde{B}_1$ with constants $\tilde{\omega} = \Lambda, \tilde{\Lambda}, \tilde{Q}$.

ii. Take $\mathbf{z}$ in $\tilde{B}_1$ [with $\mathbf{z}(x, 0) = \mathbf{u}^0(x)$], $\mathbf{h}$ in $\tilde{B}_2$, and $\boldsymbol{v}$ in $\tilde{B}_3$ [with $\boldsymbol{v}(x, 0) = d\mathbf{u}^0(x)/dx$]. Show that the solution triplet $(\mathbf{z}, h, \boldsymbol{v})$ is a fixed point of the map $\tilde{T}\colon \tilde{B}_1 \times \tilde{B}_2 \times \tilde{B}_3 \to \tilde{B}_1 \times \tilde{B}_2 \times \tilde{B}_3$, $\tilde{T}(\mathbf{z}, \mathbf{h}, \boldsymbol{v}) = (\mathbf{Z}, \mathbf{H}, \boldsymbol{V})$, defined by $(*)$ and

$$H_i(\tau; t, x) = -\int_\tau^t \lambda_i(\mathbf{z}(b_i', \tau'))d\tau',$$

$$V_i(x, t) = du_i^0(x)/dx + \sum_{s=1}^{m} \alpha_{si}(\mathbf{z}(x, t)) \sum_{l=1}^{3} D_{sl}[\mathbf{z}, \mathbf{h}, \boldsymbol{v}],$$

where (in vector form)

$$\boldsymbol{D}_1 = \Lambda(\mathbf{u}^0(a_s))d[\mathbf{u}^0(a_s) - \mathbf{u}^0(x)]/dx,$$

$$\boldsymbol{D}_2 = \int_0^t \left(\frac{d}{d\tau}\Lambda(\mathbf{z}(b_s, \tau))\right)[\boldsymbol{v}(b_s, \tau) - d\mathbf{u}^0(x)/dx]d\tau,$$

$$\boldsymbol{D}_3 = -\int_0^t \Lambda(\mathbf{z}(b_s, \tau)) \sum_{k=1}^m \mathbb{J}_k(\mathbf{z}(b_s, \tau))\boldsymbol{v}(b_s, \tau)v_k(b_s, \tau)d\tau.$$

iii. By taking a weighted norm for $\tilde{B}_1 \times \tilde{B}_2 \times \tilde{B}_3$ as in step (IV) above, show that $\tilde{T}$ has a unique fixed point $(\tilde{\mathbf{z}}, \tilde{\mathbf{h}}, \tilde{\boldsymbol{v}})$ if $a \leq a'$, for some $a' > 0$. Note that the triangle $\tilde{D}_{a'}$ is the domain of determinacy for the interval $|x - x_0| \leq C_5 a'$ on the line of initial data.

iv. Show that $\tilde{\mathbf{z}} = \bar{\mathbf{u}}$ and therefore, from (*) and (17), that $\bar{\mathbf{u}}_x = \tilde{\boldsymbol{v}}$ and $\bar{\mathbf{u}}_t$ are bounded (by Λ and Q, respectively) and Lipschitz continuous in $\tilde{D}_{a'}$. As x_0 is arbitrary, this proves the assertion. □

For this classical solution, alternative (S) corresponds to blow-up of the first derivatives.

The present approach can be extended to prove local existence and uniqueness for quasilinear hyperbolic systems in a slab $\{0 < x < a, t \in \mathbb{R}\}$, with impedance boundary conditions at $x = 0, a$ of the type considered in Section 3 of Chapter 2 (see Refs. 7, 8).

3. Scalar Equations

We now consider in detail the Cauchy problem for a scalar conservation law,

$$u_t + f(u)_x = 0 \qquad (x \in \mathbb{R}, t > 0), \tag{41}$$

$$u(x, 0) = u_0(x) \qquad (x \in \mathbb{R}), \tag{42}$$

in which $f \in C^2(G)$, G is an open interval $(-R, R)$ on the real line, and u_0 takes values in G.

The general existence theorem proven in the last section implies existence of an a.e. solution in some neighborhood of the initial line, if u_0 is a Lipschitz continuous function, and the continuation theorem indicates the two ways in which this solution may break down in finite time. In the scalar case, if u_0 is differentiable, we can give a construction that is much simpler, and that gives more precise information on the nature of the solution. If we write (41) in the form

$$u_t + a(u)u_x = 0, \tag{41a}$$

with $f' = a$, then, as the surface $u = u(x, t)$ has $u_x, u_t, -1$ as components of a normal vector, the curves in (x, t, u)-space satisfying

$$\frac{dx}{ds} = a(u), \qquad \frac{dt}{ds} = 1, \qquad \frac{du}{ds} = 0$$

are tangent to this surface. Setting $s = t$ we find that $dx/dt = a(u)$, and, if $x(0) = x_0$, the projection of these curves onto the (x, t)-plane are the characteristics, $u = \text{const} = u_0(x_0)$ along these (planar) curves, and they are lines with slope $a(u) = a(u_0(x_0))$. We may anternatively think of the construction of the solution surface as arising, on the one hand, geometrically from the lines $x = x_0 + ta(u_0(x_0))$ carrying the value $u_0(x_0)$, or, on the other hand, by solving the equation

$$F(x, t, u) = u - u_0(x - ta(u)) = 0$$

for $u = u(x, t)$. The latter can be done, by the implicit function theorem, so long as $F_u = 1 + tu_0'(x - ta(u))a'(u) \neq 0$, and then

$$u_x = u_0'/(1 + tu_0'a'), \qquad u_t = -u_0'a/(1 + tu_0'a'). \tag{43}$$

[Direct substitution shows that u is a solution of (41a) then.] The geometric construction is equivalent to the implicit function argument and works up to the first time at which characteristics cross. If two characteristics cross, they necessarily carry different values of u. Assume, for simplicity, that (41) is genuinely nonlinear, say $a'(u) = f''(u) > 0$ on G. Then if $u_0' \geq 0$, the solution exists for all time, but if $u_0' < 0$ on an interval, the maximal interval of existence is $[0, T^*)$, where

$$T^* = \left[\sup_{x_0 \in \mathbb{R}} (-u_0'(x_0) f''(u_0(x_0))) \right]^{-1}. \tag{44}$$

Furthermore, u_x and u_t become infinite (at some point x) as $t \uparrow T^*$.

As $\|u(x, t)\|_\infty \leq \|u_0(x)\|_\infty$ (for $0 \leq t < T^*$) by construction, a qualitative result of this sort could be derived from the continuation theorem in the previous section (Theorem 2.2). If we consider initial data in the form of a smooth bump (Fig. 2) and the equation $u_t + uu_x = 0$, the geometric construction of the solution gives a surface that "breaks" and is no longer a graph (Fig. 3).

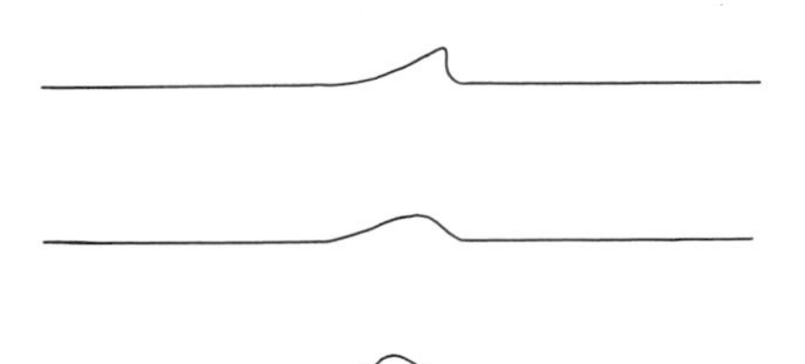

Fig. 2. The steepening of the solution.

All of this suggests that we allow a "generalized" solution that has a jump discontinuity for $t > T^*$. The development of this discontinuity at $t = T^*$ is also natural in view of the fact that $\|u_x\|_\infty + \|u_t\|_\infty \to \infty$ as $t \uparrow T^*$.

The local solution constructed above can be written as

$$u(x, t) = u_0(x_0),$$

where $x_0 = x_0(x, t)$ is obtained by inverting the equation of characteristic lines $x = x_0 + ta(u_0(x_0))$ for fixed t. This can be done so long as there is a unique charateristic line passing through the point (x, t), that is, by applying the implicit function theorem, for all $t < T^*$, where T^* coincides with (44). Therefore, characteristic lines cannot cross unless $t \geq T^*$.

3.1. Weak Solutions. If (41) is thought of as a conservation law for a physical system, as described in Chapter 1, then an integral form for this conservation law is as appropriate as (41), and will be easier to work with in the present situation. In view of the fact that discontinuities may develop anyway, we assume at the outset that $u_0(x) \in L^\infty(\mathbb{R})$ and seek solutions in $L^\infty(\mathbb{R} \times [0, \infty))$. The concept of generalized solution that we use, called here a *weak solution*, is obtained by multiplying (41) by an appropriate test function, integrating, and applying integration by parts to remove derivatives of u from the equation.

Definition 3.1. A weak solution of (41), (42) is a function $u \in L^\infty(\mathbb{R} \times [0, \infty))$ such that

$$\int_{\mathbb{R}} \int_0^\infty \big(\psi_t u + \psi_x f(u)\big) dx\, dt + \int_{\mathbb{R}} \psi(x, 0) u_0(x) dx = 0 \tag{W}$$

for every $\psi \in C_0^1(\mathbb{R} \times [0, \infty))$.

Fig. 3. The solution has "broken."

If the discontinuities of u are confined to smooth curves, we can derive a very important result.

Theorem 3.1. Suppose u is a piecewise C^1 weak solution of (41), (42), and Σ, given by $\varphi(x, t) = 0$ (with grad $\varphi \neq 0$) is a C^1 discontinuity curve for u. Then the Rankine–Hugoniot relation

$$\varphi_t[u] + \varphi_x[f(u)] = 0$$

holds along Σ, where $[u] = u_l - u_r$, $[f(u)] = f(u_l) - f(u_r)$, u_l, u_r being the limits from the left and the right along Σ of u.

Proof. Choose a point in $\mathbb{R} \times (0, \infty)$ on Σ and a ψ with support in a neighborhood Q of this point also contained in $\mathbb{R} \times [0, \infty)$. Then

$$0 = \int_Q \big(\psi_t u + \psi_x f(u)\big) dxdt = \int_{Q_l} \cdots + \int_{Q_r} \cdots,$$

where Q_l and Q_r are the parts of Q to the left and right of Σ, respectively. Integrating by parts in these integrals we obtain

$$0 = \int_{Q_l \cup Q_r} \psi(u_t + f(u)_x) dxdt + \int_\Sigma \psi\big(\varphi_t[u] + \varphi_x[f(u)]\big)(\varphi_x^2 + \varphi_t^2)^{-1/2}\, ds.$$

If we first choose ψ so that its support is Q_l or Q_r we see that ψ arbitrary implies $u_t + f(u)_x = 0$, i.e., the weak solution satisfies the equation off of Σ. Then a similar argument shows that

$$\varphi_t[u] + \varphi_x[f(u)] = 0$$

on Σ. $\square$

If $[u] \neq 0$, then $\varphi_x \neq 0$, the curve Σ has the form $x = x(t)$, and this relation takes the simpler form

$$s[u] = [f(u)] \tag{RH}$$

(or $s = \Delta f / \Delta u$), where $s = dx/dt$. The discontinuity is called a *shock* and s is the *shock speed*. The shock speed is then the slope of the chord subtended by $(u_l, f(u_l))$ and $(u_r, f(u_r))$ on the graph of f. It will be important for later purposes to observe that $f''(u) \neq 0$, i.e., (41) genuinely nonlinear, implies that the shock speed is strictly between the characteristic speeds from the left and right.

Another useful observation is that the jump relations (RH) are symmetric in u_l and u_r, that is, $s = \Delta f/\Delta u$ implies $s(u_l, u_r) = s(u_r, u_l)$. In this sense we may say that shocks are "reversible."

The following remarks are of help in clarifying the sense in which a weak solution assumes its initial value.

Remarks

3.1. Every u such that

$$\int_{\mathbb{R}} \int_0^{\infty} (\psi_t u + \psi_x f(u)) dx\, dt = 0 \qquad \forall \psi \in C_0^1(\mathbb{R} \times (0, \infty)), \tag{W$_1$}$$

$$|u(x, t)| \leq M, \qquad u(x, t) \to u_0(x) \qquad \text{a.e. as } t \to 0, \tag{W$_2$}$$

is a weak solution, i.e., satisfies (W) (Exercise 3.3).

3.2. Every weak solution $u(x, \tau)$ converges weakly to $u_0(x)$ for a.e. $\tau \to 0$ (Exercise 3.4.).

There are significant differences between weak and a.e. solutions of (41). For one thing a smooth change of the dependent variable does not preserve weak solutions. If u is transformed to $U(u)$ and $F'(u) = U'(u)f'(u)$, then $U_t + F_x = U'(u)(u_t + f'(u)u_x)$ which vanishes in regions where u is a C^1 solution of (41). If the transformation is invertible we may think of $U_t + F(U)_x = 0$ as a conservation law for U, and, if shocks develop, weak solutions do not in general coincide. We illustrate this with *Burgers' equation*,

$$u_t + uu_x = u_t + \left(\tfrac{1}{2}u^2\right)_x = 0,$$

for which (RH) becomes $s = \frac{1}{2}(u_l + u_r)$. If $U(u) = u^2/2$ and $F(u) = u^3/3$, the jump relation for U, expressed in terms of u, is

$$s' = \Delta F/\Delta U = \frac{2(u_l^2 + u_l u_r + u_r^2)}{3(u_l + u_r)},$$

hence $s' \neq s$. For these functions $U, F, s\Delta U - \Delta F = -(u_l - u_r)^3/12$, where $s = (u_l + u_r)/2$, so that the inequality

$$s\Delta U - \Delta F \leq 0$$

holds for $u_l \geq u_r$.

Another new feature is that *weak solutions are not unique*. We illustrate this with the example

$$u_t + \left(\tfrac{1}{2}u^2\right)_x = 0, \qquad t > 0,$$

$$u(x, 0) = \begin{cases} u_l, & x < 0, \\ u_r, & x > 0, \end{cases}$$

with $u_l < u_r$. One obvious solution is the "rarefaction shock,"

$$u(x, t) = \begin{cases} u_l, & x < st, \\ u_r, & x > st, \end{cases}$$

with $s = \frac{1}{2}(u_l + u_r)$ (Fig. 4).

There is another, the "rarefaction fan" given, for $t > 0$, by

$$u(x, t) = \begin{cases} u_l & x \leq u_l t, \\ x/t & u_l t \leq x \leq u_r t, \\ u_r & x \geq u_r t \end{cases}$$

(see Fig. 5).

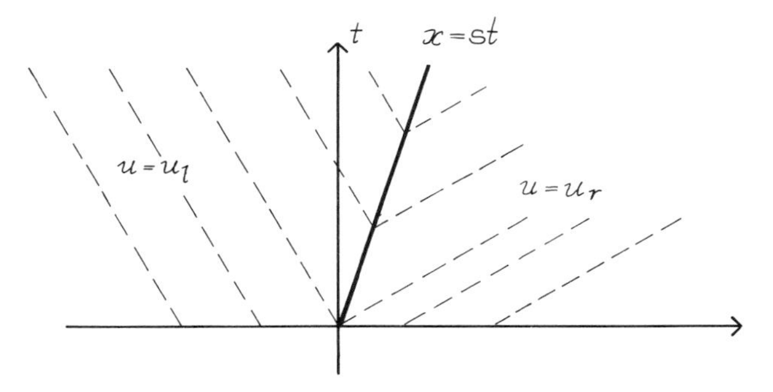

Fig. 4. A rarefaction shock.

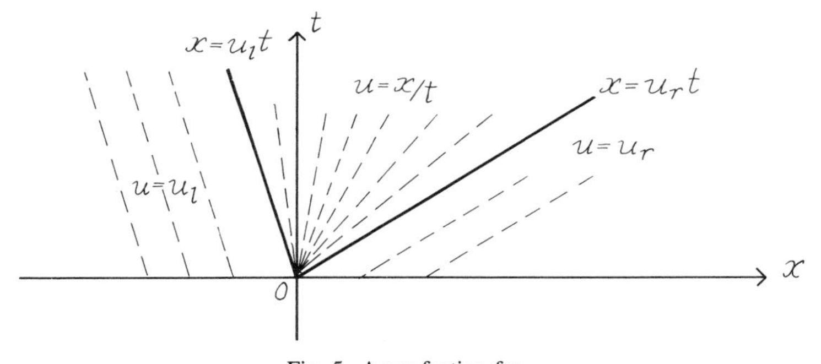

Fig. 5. A rarefaction fan.

This function is bounded, continuous, piecewise C^1 and satisfies the differential equation a.e. It is not difficult to see that this implies that it is a weak solution.

The nonuniqueness of weak solutions requires introducing another condition in order to single out a solution. The first study of problems with shocks was in gas dynamics and in that physical situation the appropriate condition was that the physical quantity called "entropy" should increase in passing through a shock. Shocks should be "irreversible." It turns out that the mathematical device that we will introduce is analogous to entropy in gas dynamics. It might be called a mathematical notion of entropy, and takes a relatively simple form for a scalar equation. A convex function $U(u)$ will be called an *entropy*, and then the function $F(u) = \int U'(u)f'(u)du$ will be the corresponding entropy flux, (U, F) being an entropy–entropy flux pair. (The assumption of convexity is explained shortly.) An example is the pair $(u^2/2, u^3/3)$ considered above for $f = u^2/2$ in relation to a change of dependent variables. A particular pair that will play an important role in what follows is

$$U(u) = |u - k|, \qquad F(u) = \operatorname{sgn}(u - k)(f(u) - f(k)),$$

where $k \in \mathbb{R}$ is fixed but arbitrary. This pair was first considered by Kruzhkov (Ref. 9) and we will refer to it simply as the Kruzhkov entropy.

The significance of a pair (U, F) for weak solutions of (41) is most easily understood by considering a related parabolic problem in which (41) is replaced by

$$u_t + f(u)_x = \mu u_{xx}, \tag{V}$$

where $\mu > 0$, and we will consider solutions of (V), (42) for μ tending to zero. We will consider the limit for $\mu \to 0$ in giving a proof of an existence theorem, due to Kruzhkov, in what follows.

For now, however, we give only an intuitive argument for the introduction of an "entropy condition" for weak solutions of (41). Suppose we multiply (V) by $U'(u)$, obtaining

$$U(u)_t + F(u)_x = \mu U'(u)u_{xx} = \mu(U'(u)u_x)_x - \mu U''(u)u_x^2.$$

Then multiplying by an arbitrary nonnegative test function ψ and integrating we obtain

$$\int_{\mathbb{R}} \int_0^{\infty} (\psi_t U + \psi_x F)dxdt = \mu \int_{\mathbb{R}} \int_0^{\infty} U'(u)u_x\psi_x \, dxdt + \mu \int_{\mathbb{R}} \int_0^{\infty} U''(u)u_x^2\psi \, dxdt.$$

Of course, function $u = u_\mu$ also depends on μ as it is a solution of (V). The second integral can be integrated by parts once more to obtain

$$-\mu \int_{\mathbb{R}} \int_0^\infty U(u)\psi_{xx}\, dxdt$$

if $\psi \in C_0^2$. We see then that if $u = u_\mu$ is bounded, this integral goes to zero as $\mu \to 0$, and an approximation argument then shows that this is true for the original integral also if $\psi \in C_0^1$. The integral

$$\mu \int_{\mathbb{R}} \int_0^\infty U''(u)u_x^2\psi\, dxdt$$

is more difficult, however. In fact, there is no reason to expect that it goes to zero with μ, but it is, by virtue of the convexity of U and the nonnegativity of ψ, nonnegative and we obtain in the limit

$$\int_{\mathbb{R}} \int_0^\infty (\psi_t U + \psi_x F) dx\, dt \geq 0 \qquad \forall \psi \in C_0^1(\mathbb{R} \times (0, \infty)), \psi \geq 0, \tag{LK}$$

or $U_t + F_x \leq 0$ in the weak sense. This is known as the *Lax–Kruzhkov entropy condition*, abbreviated (LK). An unsatisfying aspect of this condition is that (U, F) are arbitrary. A partial ameliorization of this is given in the following theorem.

Theorem 3.2. Suppose that $f'' > 0$, and u is a piecwise C^1 weak solution of (41). Then if u satisfies (LK) for one pair (U, F), U strictly convex, it satisfies (LK) for any pair.

Proof. We can argue as in the establishment of (RH) to deduce that

$$s[U] \leq [F]$$

on an "admissible" shock. Then, using (RH), this implies

$$s(u_r, u_l)\big((U(u_r) - U(u_l)\big) - \big(F(u_r) - F(u_l)\big) \leq 0$$

along the shock, where $s(v, u_l) = [f(v) - f(u_l)]/(v - u_l)$. Fix u_l and define

$$E_U(v) = s(v, u_l)\big((U(v) - U(u_l)\big) - \big(F(v) - F(u_l)\big).$$

Then $E_u(u_r) \leq 0$ and $E_u(u_l) = 0$. For $v \neq u_l$, $U'f' = F'$ implies

$$-E'_U(v) = (f(u_l) - f(v) - (u_l - v)f'(v))\frac{U(u_l) - U(v) - U'(v)(u_l - v)}{(v - u_l)^2},$$

and

$$\big(f(u_l) - f(v) - (u_l - v)f'(v)\big) > 0, \qquad U(u_l) - U(v) - U(v)(u_l - v) \geq 0$$

show that $E'_U(v) \leq 0$ for U convex and $E'_U(v) < 0$ for U strictly convex. Then, for U strictly convex $E_U(u_r) > 0$ if and only if $u_r < u_l$, and then for another convex entropy $\tilde{U}$, $E_{\tilde{U}}(u_r) \geq 0$. □

It is perhaps worth remarking that, for f strictly convex ($f'' > 0$), a shock satisfies (LK) if and only if $u_r < u_l$. Then, in the example of nonuniqueness that we have given, (LK) eliminates the rarefaction shock.

A weak solution of (41) that arises by letting $\mu \to 0$ in (V) is called a *viscosity solution* in analogy to the fluid dynamic situation where shocks are thought of as smoothed on a small enough scale by viscosity. The mathematics involved in carrying out the limit is substantial, however, as we will illustrate in presenting a proof of Kruzhkov's existence theorem (Ref. 9), and a condition that guarantees uniqueness without small-scale diffusion is desirable. We will see in what follows that (LK) is such a condition.

There is a special case of (V) that has the remarkable property that it can be transformed into a linear equation by a change of the dependent variable. The equation (Burgers–Hopf equation, Ref. 10)

$$u_t + uu_x = \mu u_{xx}$$

transforms to $\varphi_t = \mu\varphi_{xx}$ if $u = -2\mu\varphi_x/\varphi$, and, if $u(x, 0) = u_0(x)$,

$$\varphi(x, 0) = \varphi_0(x) \equiv \exp\left(-\frac{1}{2\mu}\int_0^x u_0(y)dy\right) := \exp(-U_0(x)/2\mu).$$

As u_0 is bounded, $\varphi_0(x)$ satisfies the growth condition $\varphi_0(x) = o(e^{Ax^2})$ as $|x| \to \infty (A > 0)$. Then, letting

$$U(x, y, t) := U_0(y) + (x - y)^2/2t, \qquad G(x, y, t) := \frac{\exp(-U(x, y, t)/2\mu)}{\int_{\mathbb{R}} \exp(-U(x, y, t)/2\mu)dy},$$

from the results of Chapter 3 we have

$$\varphi(x, t) = (4\pi\mu t)^{-1/2} \int_{\mathbb{R}} \exp\left[-\frac{U(x, y, t)}{2\mu}\right] dy$$

and

$$u(x, t) = \int_{\mathbb{R}} G(x, y, t) \frac{x - y}{t} dy.$$

From this formula it follows that

$$\begin{aligned} &\text{i. } u(x, t) \to u_0(x) \text{ at continuity points of } u_0(x), \\ &\text{ii. } u(x, t) \to u_0(x) \text{ weakly (in the distributional sense)} \end{aligned} \tag{45}$$

(Exercise 3.6).

3.2. Kruzhkov's Theorem. We proceed now to prove a theorem on uniqueness of solutions of (41), (42) satisfying (LK).

Let $T > 0$ (arbitrary) and π_T denote the closed band

$$\pi_T = \{(x, t): x \in \mathbb{R},\ 0 \le t \le T\}.$$

Let $u_0(x)$ be a (measurable) bounded function, $|u_0(x)| \le M_0$, and $f(u) \in C^1(\mathbb{R})$.

Definition 3.2. A bounded measurable function $u(x, t)$ is called an *entropy solution* of (41), (42) in the band π_T if:

i. For any constant k and any smooth test function $\psi^+(x, t) \ge 0$ with compact support contained in the interior of of π_T,

$$\int_{\pi_T} \{|u(x, t) - k|\psi_t^+ + \operatorname{sgn}[u(x, t) - k] \times [f(u(x, t) - f(k)]\psi_x^+\} dxdt \ge 0. \tag{46}$$

ii. There exists a set $\mathcal{N}$ of zero measure on $[0, T]$ such that for any $t \notin \mathcal{N}$, $u(x, t)$ is defined for a.e. $x \in \mathbb{R}$, and for any closed interval $K_r = \{|x| \leq r\}$,

$$\lim_{\substack{t \to 0 \\ t \in [0,T] \setminus \mathcal{N}}} \int_{K_r} |u(x, t) - u_0(x)| dx = 0. \tag{47}$$

As will be shown later, the inequality (46) implies that u is a weak solution satisfying the (LK) entropy condition (see Exercise 3.8).

We introduce some notation and terminology. We let $\omega(\sigma)$ designate modulus of continuity type functions:

$$\omega(\sigma) \in C([0, \infty)), \qquad \omega(0) = 0, \qquad \omega(\sigma) \text{ nondecreasing},$$

and $v_h(x)(h > 0)$ the mean functions or mollifications of $v(x) \in L^1_{\mathrm{loc}}(\mathbb{R})$ via mollifiers $\delta(x)$ such that

$$\delta(x) \in C^\infty(\mathbb{R}), \qquad \delta(x) \geq 0, \qquad \delta(x) = 0 \text{ for } |x| \geq 1, \qquad \int_{\mathbb{R}} \delta(x) dx = 1. \tag{48}$$

Then the limit relation

$$v^h(x) \equiv \frac{1}{h} \int_{\mathbb{R}} \delta\left(\frac{x-y}{h}\right) v(y) dy \to v(x) \qquad h \downarrow 0 \tag{49}$$

holds a.e. in $\mathbb{R}$, i.e., at all of the Lebesgue points of $v(x)$. [In case of functions $v(x)$ defined only on a subset Ω of $\mathbb{R}$, $v(x)$ will be continued by zero outside Ω.] We always denote by K_R the closed interval $K_R = \{|x| \leq R\}$, and by

$$\mathcal{K} = \left\{(x, t)\colon x \in K_{R-Nt},\ 0 \leq t \leq T_0 := \min[T, R/N]\right\}$$

the characteristic cone with

$$N = N_M := \max[|f'(u)|\colon |u| \leq M] \tag{50}$$

for any $R > 0, M > 0$ (Fig. 6).

Finally, we let $S_\tau = \{x\colon |x| \leq R - N\tau\}$ denote the cross section of $\mathcal{K}$ at $t = \tau, \tau \in [0, T_0]$. We will need in what follows a few auxiliary lemmas.

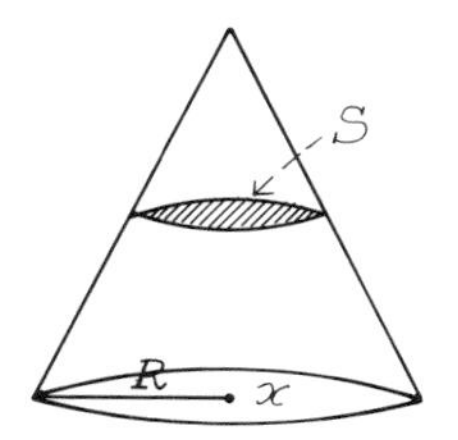

Fig. 6. A characteristic cone.

Lemma 3.1. Let the function $v(x)$ be integrable in the interval $K_{r+2\rho}, r > 0, \rho > 0$, where

$$J_s(v, \Delta x) := \int_{K_s} |v(x + \Delta x) - v(x)| dx \leq \omega_s(|\Delta x|)$$

for $|\Delta x| \leq \rho$ and $0 \leq s \leq r + \rho$. Then for $h \leq \rho$,

$$J_r(v^h, \Delta x) \leq \omega_{r+h}(|\Delta x|), \tag{51}$$

$$\int_{K_r} ||v| - v(\text{sgn } v)^h| dx \leq 2\omega_r(h). \tag{52}$$

Proof. The inequality (51) follows from (48), 49):

$$\begin{aligned} J_r(v^h, \Delta x) &= \int_{K_r} |v^h(x + \Delta x) - v^h(x)| dx \\ &= \frac{1}{h} \int_{K_r} dx \left| \int_{\mathbb{R}} \left[\delta\left(\frac{x + \Delta x - y}{h} \right) - \delta\left(\frac{x - y}{h} \right) \right] v(y) dy \right| \\ &\leq \int_{\mathbb{R}} \delta(z) \int_{K_r} |v(x + \Delta x - hz) - v(x - hz)| dx dz \leq \omega_{r+h}(|\Delta x|). \end{aligned}$$

To prove (52), it suffices to note that

$$\big| |v(x)| - v(x) \text{sgn}\, v(y) \big| = \big| |v(x)| - |v(y)| - [v(x) - v(y)] \,\text{sgn}\, v(y) \big| \leq 2|v(x) - v(y)|,$$

and consequently,

$$\int_{K_r} ||v(x)| - v(x)[\operatorname{sgn} v(x)]^h| dx$$

$$= \frac{1}{h}\int_{K_r} dx \left| \int_{\mathbb{R}} \delta\left(\frac{x-y}{h}\right)[|v(x)| - v(x)\operatorname{sgn} v(y)]dy \right|$$

$$\leq 2\int_{\mathbb{R}} \delta(z) \int_{K_r} |v(x) - v(x-hz)| dxdz \leq 2\omega_r(h). \qquad \square$$

Lemma 3.2. Let $v(x, t)$ be bounded and measurable in some cylinder $Q = K_r \times [0, T]$. If for some $\rho \in (0, \min[r, T])$ and any number h, $0 < h < \rho$, we set

$$V_h := h^{-2} \int_{Q^2_{\rho,h}} |v(x, t) - v(y, \tau)| dxdtdyd\tau, \tag{53}$$

where

$$Q^2_{\rho,h}\colon\ |t-\tau| \leq 2h,\ 2\rho \leq t+\tau \leq 2(T-\rho),\ |x-y| \leq 2h,\ |x+y| \leq 2(r-\rho),$$

then

$$\lim_{h\to 0} V_h = 0.$$

Proof. After substituting $t + \tau = 2\alpha, t - \tau = 2\beta, x + y = 2\eta, x - y = 2\xi$, we have $\rho \leq \alpha \leq T - \rho$, $|\eta| \leq r - \rho$, $|\beta| \leq h$, $|\xi| \leq h$, and

$$V_h = 4\int G_h(\alpha, \eta) d\alpha\, d\eta,$$

$$G_h(\alpha, \eta) = h^{-2} \int |v(\alpha+\beta, \eta+\xi) - v(\alpha-\beta, \eta-\xi)| d\beta\, d\xi.$$

As almost all points (α, η) of the cylinder $Q_\rho = [\rho, T-\rho] \times K_{r-\rho}$ are Lebesgue points of the function $v(\alpha, \eta)$ and as

$$|v(\alpha+\beta, \eta+\xi) - v(\alpha-\beta, \eta-\xi)|$$
$$\leq |v(\alpha+\beta, \eta+\xi) - v(\alpha, \eta)| + |v(\alpha, \eta) - v(\alpha-\beta, \eta-\xi)|,$$

it follows that $G_h(\alpha, \eta) \to 0$ as $h \to 0$ a.e. in Q_ρ. As $|G_h(\alpha, \eta)| \leq 8 \sup|v|$, the dominated convergence theorem then completes the proof. $\square$

Lemma 3.3. If the function $F(u)$ satisfies a Lipschitz condition on an interval $[-M, M]$ with constant L, then the function $H(u, v) := \text{sgn}(u - v)[F(u) - F(v)]$ also satisfies a Lipschitz condition in u and v with the same constant L.

The proof of this lemma is immediate.

Uniqueness of the entropy solution of (41), (42) is a consequence of the following result on stability of solutions with respect to changes of initial data in the L^1-norm.

Theorem 3.3. Let $u(x, t)$, $v(x, t)$ be entropy solutions of (41), (42) with initial functions $u_0(x)$ and $v_0(x)$, respectively, and $|u(x, t)| \leq M$, $|v(x, t)| \leq M$ a.e. in the cylinder $K_R \times [0, T]$. Then for almost all $t \in [0, T_0]$,

$$\int_{S_t} |u(x, t) - v(x, t)| dx \leq \int_{S_0} |u_0(x) - v_0(x)| dx, \tag{54}$$

where $T_0 = \min[T, R/N]$, and $R > 0, M > 0$ are arbitrary [see (50)].

Proof. Take a smooth function $g(x, t; y, \tau) \geq 0$ and set in inequality (46) $k = v(y, \tau)$, $\psi^+(x, t) = g(x, t; y, \tau)$ for fixed (y, τ) in π_T; for this we require that the support of g is compact and contained in the interior of $\pi_T^2 = \pi_T \times \pi_T$ (see below). Then integrating (46) over π_T [in the variables (y, τ)] yields

$$\int_{\pi_T^2} \{|u - v| g_t + \text{sgn}\,[u - v]\,[f(u) - f(v)] g_x\} dx dt dy d\tau \geq 0$$

$[u = u(x, t), v = v(y, \tau)]$. In exactly the same way, starting from the inequality (46) for the function $v = v(y, \tau)$ and taking $k = u = u(x, t)$, $\psi^+(y, \tau) = g(x, t; y, \tau)$ for fixed (x, t), integrating over π_T [in the variables (x, t)], we find

$$\int_{\pi_T^2} \{|v - u| g_\tau + \text{sgn}\,[v - u]\,[f(v) - f(u)] g_y\} dy d\tau dx dt \geq 0.$$

Combining these and rearranging we obtain the inequality

$$\int_{\pi_T^2} \{|u - v|(g_t + g_\tau) + \text{sgn}\,[u - v]\,[f(u) - f(v)](g_x + g_y)\} dx dt dy d\tau \geq 0 \tag{55}$$

valid for any smooth function $g(x, t; y, \tau) \geq 0$ with support inside π_T^2. Hence, we may choose g as follows. Let $\psi^+(x, t)$ be a test function from Definition 3.2

such that $\psi^+(x, t) \equiv 0$ outside some cylinder $(x, t) \in K_{r-2\rho} \times [\rho, T - 2\rho]$, $0 < 2\rho < \min[T, r]$, and let

$$g = \psi^+\Big(\frac{x+y}{2}, \frac{t+\tau}{2}\Big)\lambda_h\Big(\frac{x-y}{2}, \frac{t-\tau}{2}\Big), \qquad h \leq \rho,$$

where $\lambda_h(x, t) := \delta_h(x)\delta_h(t)$, $\delta_h(x) = h^{-1}\delta(x/h)$. Note that, from (48), $\delta_h(x) \geq 0$, $\delta_h(x) \equiv 0$ for $|x| \geq h$, $|\delta_h(x)| \leq \text{const}/h$, and $\int_{\mathbb{R}} \delta_h(x)dx = 1$; hence, $\delta_h(x)$ is delta-shaped at the point $x = 0$ as $h \downarrow 0$ ["delta-approximate" family, see (49) and Chapter 8]. With this choice we have

$$g_t + g_\tau = \psi_t^+\lambda_h, \qquad g_x + g_y = \psi_x^+\lambda_h,$$

where λ_h is delta-shaped at the point $(x, t) = (y, \tau)$ as $h \downarrow 0$. Let us show that: *in the limit* $h \downarrow 0$ (55) *yields the inequality*

$$\int_{\pi_T} \{|u - v|\psi_t^+ + \operatorname{sgn}[u - v]\,[f(u) - f(v)]\psi_x^+\}dxdt \geq 0, \tag{56}$$

where the argument of both u and v is (x, t). In other words, (46) holds with k replaced by $v(x, t)$.

We proceed now to prove (56). This result can be proved by means of Lemmas 3.2 and 3.3 (the proof is nontrivial as u and v are not continuous). In fact, for this choice of g each of the two terms in the integrand of (55) is of the form

$$P_h(x, t; y, \tau) = F(x, t, y, \tau, u(x, t), v(y, \tau))\lambda_h\Big(\frac{x-y}{2}, \frac{t-\tau}{2}\Big),$$

where by force of Lemma 3.3 the function $F(x, t, y, \tau, u, v)$ is Lipschitz continuous in all of its variables, $P_h \equiv 0$ outside the region

$$|t - \tau| \leq 2h, \quad 2\rho \leq t + \tau \leq 2T - 4\rho, \quad |x - y| \leq 2h, \quad |x + y| \leq 2r - 4\rho,$$

and each integral in (55) can be written as

$$\begin{aligned}\int_{\pi_T^2} P_h\, dxdtdyd\tau &= \int_{\pi_T^2} F(x,t,x,t,u(x,t),v(x,t))\lambda_h\Big(\frac{x-y}{2},\frac{t-\tau}{2}\Big)dxdtdyd\tau \\ &\quad + \int_{\pi_T^2} \{F(x,t,y,\tau,u(x,t),v(y,\tau)) \\ &\quad - F(x,t,x,t,u(x,t),v(x,t))\}\lambda_h(\cdots)dxdtdyd\tau \\ &:= J_2 + J_1(h),\end{aligned}$$

where $|\lambda_h((x-y)/2,(t-\tau)/2)| \le \text{const}/h^2$ [see after (55)], $|x-y|$ and $|t-\tau|$ are of order h, and the integral on π_T^2 is of order h^2. Hence, we find

$$J_1(h) \le c[h + V_h],$$

where V_h is defined in (53), and the constant c does not depend on h. By Lemma 3.2, $J_1(h) \to 0$ as $h \downarrow 0$. The integrand J_2 does not depent on h:

$$J_2 = 4\int_{\pi_T} F(x,t,x,t,u(x,t),v(x,t))dxdt,$$

as $\int_{\pi_T} \lambda_h((x-y)/2,(t-\tau)/2)dyd\tau = 4$ as can be immediately verified. Hence,

$$\lim_{h\to 0} \int_{\pi_T^2} P_h\, dxdtdyd\tau = 4\int_{\pi_T} F(x,t,x,t,u(x,t),v(x,t))dxdt,$$

and (55) implies (56), as asserted. Thus, (56) is proven.

Let $\mathscr{K}$ be a characteristic cone, and let $\mathscr{N}_u$, $\mathscr{N}_v$ be the null sets for u and v in Definition 3.2, (ii). We let $\mathscr{N}_\mu$ designate the set of points on $[0,T]$ that are not Lebesgue points of the bounded measurable function

$$\mu(t) := \int_{S_t} |u(x,t) - v(x,t)|dx. \tag{57}$$

Let $\mathscr{N}_0 = \mathscr{N}_u \cup \mathscr{N}_v \cup \mathscr{N}_\mu$: It is clear that $\mathscr{N}_0$ is a null set in $[0, T]$. We define

$$\alpha_h(\sigma) = \int_{-\infty}^{\sigma} \delta_h(\sigma)d\sigma$$

and take two numbers ρ and $\tau \in (0, T_0)\backslash\mathcal{N}_0$, $\rho < \tau$. In (56) we set

$$\psi^+ = [\alpha_h(t-\rho) - \alpha_h(t-\tau)]\chi(x, t), \qquad h < \min[\rho, T_0 - \tau],$$

where

$$\chi(x, t) = \chi_\varepsilon(x, t) := 1 - \alpha_\varepsilon(\xi + \varepsilon), \qquad \xi = |x| + Nt - R \qquad (\varepsilon > 0).$$

χ is identically zero outside the cone $\mathcal{K}$: $|x| \le R - Nt$, $0 \le t \le T_0$ (it is easy to see that this is a permissible test function). As $\alpha'_\varepsilon(\sigma) = \delta_\varepsilon(\sigma) \ge 0$, and

$$N \ge \left|\frac{f(u) - f(v)}{u - v}\right|$$

[see (50)], for $(x, t) \in \mathcal{K}$ we have

$$0 = \chi_t + N|\chi_x| \ge \chi_t + \frac{f(u) - f(v)}{u - v}\chi_x. \tag{58}$$

With this choice for ψ^+ we obtain from (56):

$$\begin{aligned} 0 \le \int_{\pi_{T_0}} |u - v| \Big\{ & [\delta_h(t-\rho) - \delta_h(t-\tau)]\chi + [\alpha_h(t-\rho) - \alpha_h(t-\tau)]\chi_t \\ & + \operatorname{sgn}(u - v)\frac{f(u) - f(v)}{|u - v|}[\alpha_h(t-\rho) - \alpha_h(t-\tau)]\chi_x \Big\} dxdt \\ = \int_{\pi_{T_0}} & |u - v|[\delta_h(t-\rho) - \delta_h(t-\tau)]\chi \, dx \, dt \\ + \int_{\pi_{T_0}} & [\alpha_h(t-\rho) - \alpha_h(t-\tau)]\left\{\chi_t + \frac{f(u) - f(v)}{u - v}\chi_x\right\}|u - v| dxdt. \end{aligned}$$

From (58) it follows that the second integral is nonpositive. We thus arrive at the inequality

$$\int_{\pi_{T_0}} |u(x, t) - v(x, t)|[\delta_n(t-\rho) - \delta_h(t-\tau)]\chi_\varepsilon(x, t)dxdt \ge 0.$$

Letting ε approach zero we find that

$$\chi_\varepsilon(x, t) = 1 - \alpha_\varepsilon(\xi + \varepsilon) = 1 - \int_{-\infty}^{1+\xi/\varepsilon} \delta(\sigma) d\sigma \to 1 - H(\xi),$$

where $H(\xi)$ is the Heaviside step function. Hence,

$$\int_0^{T_0} \mu(t)[\delta_h(t - \rho) - \delta_h(t - \tau)]dt \geq 0 \tag{59}$$

[see (57)]. As ρ and τ are Lebesgue points of $\mu(t)$, in the limit $h \downarrow 0$ we have

$$\begin{aligned}\left|\int_0^{T_0} \mu(t)\delta_h(t - \sigma)dt - \mu(\sigma)\right| &= \left|\int_0^{T_0} [\mu(t) - \mu(\sigma)]\delta_h(t - \sigma)dt\right| \\ &\leq \frac{C}{h}\int_{\delta-h}^{\sigma+h} |\mu(t) - \mu(\sigma)| \to 0\end{aligned}$$

for $\sigma = \rho$ and $\sigma = \tau$ (the constant C does not depend on h). Thus, (59) yields

$$\mu(\tau) = \int_{S_\tau} |u(x, \tau) - v(x, \tau)|dx \leq \int_{S_\rho} |u(x, \rho) - v(x, \rho)|dx = \mu(\rho) \tag{60}$$

for $\rho < \tau$. Taking into account that

$$|u(x, \rho) - v(x, \rho)| \leq |u(x, \rho) - u_0(x)| + |v(x, \rho) - v_0(x)| + |u_0(x) - v_0(x)|,$$

we now let ρ approach zero over a sequence of points $\{\rho_i\}$ not belonging to $\mathcal{N}_0$. From (60) and (47) we find

$$\begin{aligned}\mu(\tau) \leq & \int_{S_{\rho_i}} |u(x, \rho_i) - u_0(x)|dx + \int_{S_{\rho_i}} |v(x, \rho_i) - v_0(x)|dx \\ &+ \int_{S_{\rho_i}} |u_0(x) - v_0(x)|dx \\ &\to \int_{S_0} |u_0(x) - v_0(x)|dx \qquad \text{as } i \to \infty,\end{aligned}$$

which by the arbitrariness of τ coincides with (54). □

As for any point $(x, t) \in \pi_T$ we can find a characteristic cone $\mathscr{K}$ containing the point (for any $M > 0$), Theorem 3.3 implies *uniqueness* of the entropy solution.

Theorem 3.4. The entropy solution of (41), (42) in the band π_T is unique.

Concerning monotonic dependence of the entropy solutions on the initial data we have the following.

Theorem 3.5. Let $u(x, t)$, $v(x, t)$ be the entropy solutions of (41), (42) in π_T with initial data $u_0(x)$ and $v_0(x)$, respectively. Let $u_0(x) \leq v_0(x)$ a.e. in $\mathbb{R}$. Then $u(x, t) \leq v(x, t)$ a.e. in π_T.

We present a proof of Theorem 3.5 based on previous estimates; an alternative proof can be derived from the vanishing viscosity method (see the existence theorem, Theorem 3.6 below).

We begin by showing that for the solutions u and v the following analogue of estimate (54),

$$\int_{S_t} \Phi(u(x, t) - v(x, t))dx \leq \int_{S_0} \Phi(u_0(x) - v_0(x))dx, \tag{61}$$

holds for $\Phi(\sigma) = \sigma + |\sigma|$ and for any $R > 0, M > 0$. We first prove the following.

Lemma 3.4. The entropy solution $u = u(x, t)$ satisfies the integral identity

$$\int_{\pi_T} (u\psi_t + f(u)\psi_x\}dxdt = 0 \tag{62}$$

for all test functions $\psi(x, t)$ with compact support contained in the interior of π_T.

Proof. Taking first $k > \sup_{\pi_T} u(x, t)$ and then $k < \inf_{\pi_T} u(x, t)$ (this is possible because u is bounded) we find from (46) the two inequalities

$$\int_{\pi_T} (u\psi_t^+ + f\psi_x^+\}dx\,dt \leq 0, \qquad \int_{\pi_T} (u\psi_t^+ + f\,\psi_x^+\}dxdt \geq 0,$$

which together imply (62). □

Proof of Theorem 3.5. A similar integral identity,

$$\int_{\pi_T} (v\psi_\tau + f(v)\psi_y)dyd\tau = 0,$$

follows from the lemma applied to the function $v = v(y, \tau)$ [we use the variables (y, τ) for v in what follows]. By means of these integral identities (and Fubini's theorem) we obtain

$$\begin{aligned}
&\int_{\pi_T^2} \{(u - v)(g_t + g_\tau) + [f(u) - f(v)](g_x + g_y)\}dxdtdyd\tau \\
&\quad = \int_{\pi_T} dyd\tau \int_{\pi_T} [ug_t + f(u)g_x]dxdt - \int_{\pi_T} dxdt \int_{\pi_T} [vg_\tau + f(v)g_y]dyd\tau \\
&\quad + \int_{\pi_T} u\, dxdt \int_{\pi_T} g_\tau\, dyd\tau - \int_{\pi_T} v\, dyd\tau \int_{\pi_T} g_t\, dxdt \\
&\quad + \int_{\pi_T} f(u)dxdt \int_{\pi_T} g_y\, dyd\tau - \int_{\pi_T} f(v)dyd\tau \int_{\pi_T} g_x\, dxdt = 0,
\end{aligned}$$

and adding (55) yields the inequality

$$\int_{\pi_T^2} \Phi(u - v)\left\{g_t + g_\tau + \frac{f(u) - f(v)}{u - v}(g_x + g_y)\right\}dxdtdyd\tau \geq 0,$$

which is the analogue of (55) with $\Phi(u - v)$ in place of $|u - v|$. Hence, proceeding exactly as in the proof of Theorem 3.3 we obtain (61). As $u_0(x) \leq v_0(x)$ a.e. in $\mathbb{R}$, and R is arbitrary, inequality (61) implies that

$$u(x, t) - v(x, t) \leq -|u(x, t) - v(x, t)| \leq 0 \qquad \text{a.e. in } \pi_T,$$

and the theorem is proved. □

The existence of the entropy solution of the problem (41), (42) can be proven using the *vanishing viscosity method* (Ref. 11). We have investigated Cauchy's problem for the parabolic equation

$$u_t + f(u)_x = \mu u_{xx}, \qquad \mu > 0, \tag{63}$$

with initial condition (42) in Section 3 of Chapter 6. We need here an approximation lemma concerned with smoothing the initial data. We recall the notation u_0^h for the mollification of a function u_0 [see (49)].

Lemma 3.5. Suppose that $u_0 \in L^1 \cap L^\infty \cap BV, BV = BV(\mathbb{R})$. Then $u_0^h \in C^\infty \cap H^2$ and

(a). $\displaystyle\int_{\mathbb{R}} |u_0^h| dx \leq \int_{\mathbb{R}} |u_0| dx,$

(b). $\|u_0^h\|_\infty \leq \|u_0\|_\infty,$

(c). $\displaystyle\int_{\mathbb{R}} |u_{0x}^h| dx \leq TV(u_0),$

(d). $\displaystyle\int_{\mathbb{R}} |u_{0xx}^h| dx \leq \frac{C}{h}\, TV(u_0).$

Proof. (a) and (b) are immediate. We recall that

$$TV(u_0) = \sup \int_{\mathbb{R}} u_0 \phi_x \, dx,$$

where the supremum is over all $\phi \in C_0^1(\mathbb{R})$ with $\|\phi\|_\infty \leq 1$. To prove (c), let ϕ be such a test function, and observe

$$\int_{\mathbb{R}} u_0^h \phi_x \, dx = \int_{\mathbb{R}} u_0 * \zeta_h \phi_x \, dx = \int_{\mathbb{R}} u_0 (\phi * \zeta_h)_x \, dx \leq TV(u_0),$$

as $\phi_h = \phi * \zeta_h$ is also in C_0^1 and $\|\phi_h\|_\infty \leq 1$, for ζ_h an even mollifier with compact support. If we take the supremum over all ϕ on the left, we obtain

$$TV(u_0^h) = \int_{\mathbb{R}} |u_{0x}^h| dx$$

and (c) follows. For $\psi \in L^\infty$,

$$\int_{\mathbb{R}} u_{0xx}^h \psi \, dx = \int_{\mathbb{R}} u_0 \zeta_{hxx} * \psi \, dx = \int_{\mathbb{R}} u_0 \psi_{hxx} \, dx.$$

Then, as functions in L^1 define linear functionals on L^∞, and $\psi_{hx} = \zeta_{hx} * \psi$ is in C_0^1,

$$\int_{\mathbb{R}} |u_{0xx}^h| dx = \sup_{\psi \in L^\infty} \int_{\mathbb{R}} u_{0xx}^h \psi \, dx / \|\psi\|_\infty$$
$$= \sup \int_{\mathbb{R}} u_0 (\psi_{hx})_x \, dx / \|\psi\|_\infty \leq TV(u_0) \sup \frac{\|\psi_{hx}\|_\infty}{\|\psi\|_\infty},$$

where

$$\|\psi_{hx}\|_\infty \leq \|\psi\|_\infty \int_{\mathbb{R}} |\zeta_{hx}| dx \leq \frac{C}{h} \|\psi\|_\infty,$$

hence (d) follows. □

Using these results we can now prove the following existence theorem.

Theorem 3.6. Assume $f \in C^2$ and $u_0 \in L^1 \cap L^\infty \cap BV$. Then (41), (42) has an entropy solution $u \in L^\infty([0, T] \times \mathbb{R}) \cap C([0, T], L^1)$ with

(a). $\|u(t)\|_\infty \leq \|u_0\|_\infty$,
(b). $TV(u(t)) \leq TV(u_0)$

for $t \in [0, T]$.

Proof. Consider

$$u_t + (f(u))_x - \mu u_{xx} = 0$$

with $u(x, 0) = u_0^\mu$ (mollified initial data with $h = \mu$). Combining Theorem 3.3 of Chapter 6 and Lemma 3.5, we know that $\{u_\mu\}$ is bounded in $BV(K_n)$, so that by Proposition 3.18 of Chapter 8 there is a subsequence converging in $L^1(K_n)$. We have shown that there is a solution u_μ with

i. $\|u_\mu(t)\|_\infty \leq \|u_0\|_\infty$,
ii. $\int_{\mathbb{R}} |u_{\mu_x}(x, t)| dx \leq TV(u_0)$,
iii. $\int_{\mathbb{R}} |u_{\mu_t}(x, t)| dx \leq CTV(u_0)$,
iv. $\int_{\mathbb{R}} |u_\mu(x, t)| dx \leq \int_{\mathbb{R}} |u_0(x)| dx + CTTV(u_0)$,

with C independent of μ. Suppose that K_n are Lipschitz domains, $K_n \subset K_{n+1}$, with $\cup K_n = \mathbb{R} \times [0, T]$. By again passing to a subsequence and diagonalizing we can assume that u_μ converges a.e. and in $L^1_{\text{loc}}(\mathbb{R} \times [0, T])$ to a function u satisfying (a).

We rederive from (63) the Lax–Kruzhkov entropy condition. Let $u = u_\mu$ be a solution of (63), $U(u)$ an arbitrary twice smooth function of $u \in \mathbb{R}$, strictly convex in $\mathbb{R}$, and $\psi^+(x, t)$ a nonnegative test function with compact support in the interior π_T^0 of π_T. As $U'' \geq 0$, from (63) we find

$$\begin{aligned} 0 = U'(u)[u_t + f'(u)u_x - \mu u_{xx}] &= U(u)_t + F(u)_x - \mu U_{xx} + \mu U'' u_x^2 \\ &\geq U_t + F_x - \mu U_{xx}, \end{aligned} \tag{64}$$

where $F(u) := \int_k^u U'(u) f'(u) du$, k any real constant. Multiplying by ψ^+ and integrating by parts of π_T yields

$$\int_{\pi_T} (U\psi_t^+ + F\psi_x^+ + \mu U \psi_{xx}^+) dxdt \geq 0 \qquad \forall \psi^+ \geq 0, \psi^+ \in C_0^2(\pi_T^0). \tag{65}$$

We can now approximate the Kruzhkov entropy $U(u) = |u - k|$ by twice smooth convex functions $\tilde{U}(u)$, and the corresponding entropy flux $F(u) = \text{sgn}(u - k)[f(u) - f(k)]$ by the smooth approximation $\int_k^u \tilde{U}(u) f'(u) du$. In this way we see that the entropy inequality (65) holds also for the Kruzhkov entropy,

$$\int_{\pi_T} \{|u - k|(\psi_t^+ + \mu \psi_{xx}^+) + \text{sgn}\,(u - k)[f(u) - f(k)]\psi_x^+\} dxdt \geq 0. \tag{66}$$

From (ii), (iii) and (iv) we see that $\{u_\mu\}$ is contained in a bounded subset of $C([0, T], W^{1,1}(\mathbb{R}))$. For any bounded interval J, (iii) implies that

$$\int_J |u_\mu(x, t_2) - u_\mu(x, t_1)| dx \leq C(t_2 - t_1) TV(u_0),$$

so that $\{u_\mu\}$ is uniformly equicontinuous as a set of functions from $[0, T]$ to $L^1(J)$. As $W^{1,1}(J)$ is compactly embedded in $L^1(J)$, we can deduce from Ascoli–Arzela's theorem that $u_\mu \to u$ in $C([0, T], L^1(J))$. With the usual diagonalization argument we see that we may assume $u_\mu \to u$ in $C([0, T], L^1_{\text{loc}})$. Finally, (iv) implies that $u \in L^1$. Letting $\mu \to 0$ we see that u satisfies (46), hence u is an entropy solution. To prove (b), let $\phi \in C_0^1$, $\|\phi\|_\infty \leq 1$. From (ii),

$$\int_{\mathbb{R}} u_\mu \phi_x \, dx = -\int_{\mathbb{R}} u_{\mu_x} \phi \, dx \leq TV(u_0) \|\phi\|_\infty.$$

The mode of convergence of u_μ just established implies that we can take the limit as $\mu \to 0$ in this inequality and (b) follows. □

The theorem [part (a)] can be proven under the sole assumption $u_0 \in L^\infty$ (see Ref. 11).

It is interesting to investigate the *asymptotic behavior* of the viscosity solution as t tends to infinity. Typical results are given in the following examples (Ref. 12).

Examples

3.1. The Burgers equation $u_t + (u^2/2)_x = 0$ with periodic discontinuous initial data

$$u_0(x) = x - p/2 - np, \qquad np < x < (n+1)p$$

$(n \in \mathbb{Z}, p > 0)$ has the viscosity solution

$$u(x,t) = (x - p/2 - np)/(1+t), \qquad np < x < (n+1)p,$$

and hence $u(x,t) = O(1/t)$ as $t \to +\infty$. This solution is called a "sawtooth wave" from the form of its graph (Fig. 7).

3.2. If $u_0(x)$ is a (smooth) function with compact support, the asymptotic behavior of the viscosity solution $u(x,t)$ is given by an "N-wave" (see Section 3.3), and $u(x,t) = O(1/\sqrt{t})$ as $t \to +\infty$.

The decay estimates in these examples are interesting as they show that in the nonlinear case the solution is damped as t increases due to the presence of singularities (admissible shocks), even in the absence of "viscosity."

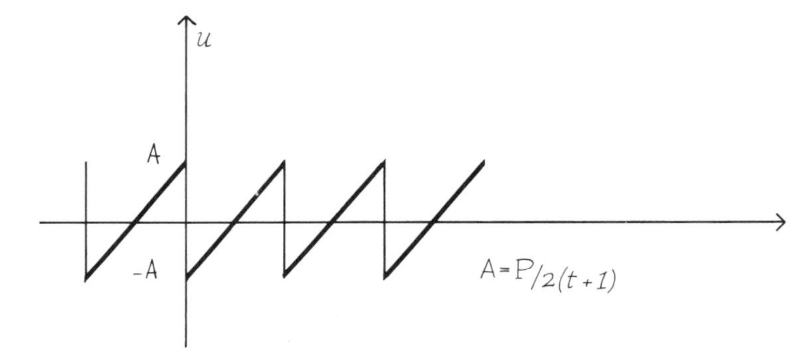

Fig. 7. A decaying sawtooth wave.

There are other entropy conditions that have been introduced and one of particular interest is the *Lax geometric entropy condition*, which for a scalar equation says that characteristics should come *into* a shock (hence, this condition applies to noncharacteristic shocks). For a genuinely nonlinear scalar equation $[f''(u) \neq 0]$ this means that at a point of a shock curve,

$$f'(u_r) < s < f'(u_l), \tag{GE}$$

where $s = s(u_l, u_r)$ is the shock speed. If $f''(u) > 0$, this again implies $u_l > u_r$ and we have seen that this is equivalent to (LK) for a piecewise C^1 solution. The generalization of this condition to (genuinely nonlinear) systems will play an important role later.

3.3. Riemann Problem. Consider the special initial data that define the *Riemann problem*,

$$u_0(x) = \begin{cases} u_l, & x < 0, \\ u_r, & x > 0. \end{cases} \tag{67}$$

Suppose we look for a *self-similar* solution of (41), (42), i.e., $u(x, t) = U(\xi)$, $\xi = x/t$. Then, in regions of smoothness of u,

$$U'(\xi)[f'(U(\xi)) - \xi] = 0.$$

We consider two cases.

(a) For a *linearly degenerate* equation $f'(u) = \lambda_0$, a constant for u in the interval between u_l and u_r. Then $u(x, t) = U(x/t)$ is constant for $x \neq \lambda_0 t$ and the solution is the shock

$$u(x, t) = \begin{cases} u_l, & x < st, \\ u_r, & x > st. \end{cases} \tag{68}$$

where $s = \lambda_0$ ("characteristic shock").

(b) For (41) *genuinely nonlinear* with $f''(u) > 0$ the function U, when smooth, is either constant or a "centered wave," $U(\xi) = b(\xi)$ where $b(\xi)$ is the inverse function of $f'(u)$. As $b(\xi)$ is an increasing function of ξ, a centered wave connecting two constant states u_l, u_r exists only if $u_l < u_r$. [The case of $f''(u) < 0$ is treated similarly by interchanging the roles of u_l and u_r.] We see that if $u_l > u_r$, the unique solution satisfying (LK) is the shock (68), where $s = s(u_l, u_r)$ is given by (RH): This follows from the discussion in Section 3.1

and Kruzhkov's uniqueness theorem. If $u_l < u_r$ then the centered wave or *rarefaction fan* given by

$$u(x,t) = \begin{cases} u_l, & x \le f'(u_l)t, \\ b(x/t), & f'(u_l)t \le x \le f'(u_r)t, \\ u_r, & x \ge f'(u_r)t, \end{cases} \tag{69}$$

is a weak solution, and as it is continuous and piecewise C^1 it satisfies $U_t + F_x = 0$ a.e. for any entropy–entropy flux pair U, F and, hence, satisfies (LK) trivially. Note that a rarefaction fan may be centered at a generic point x_0, $u = b[(x - x_0)/t]$.

It is instructive to investigate the Riemann problem (67) for the viscous equation

$$u_t + f(u)_x = \mu u_{xx}$$

($\mu > 0$) and then take the limit for vanishing viscosity μ. We look for solutions $u = u_\mu(x,t)$ of the form

$$u_\mu = \mathcal{U}(\mu^{-\alpha}(x - st), 2(1-\alpha)t) \equiv \mathcal{U}(\eta, \tau),$$

where α, S are real constants, and the variables $\eta = \mu^{-\alpha}(x - st)$, $\tau = 2(1-\alpha)t$ are introduced for later convenience. As will be seen, this form is not suitable to describe a rarefaction fan in the limit $\mu \downarrow 0$. We obtain

$$2(1-\alpha)\mathcal{U}_\tau + \mu^{-\alpha}(f'(\mathcal{U}) - s)\mathcal{U}_\eta = \mu^{1-2\alpha}\mathcal{U}_{\eta\eta}.$$

Exact solutions of this equation can be easily found when the limiting equation is either genuinely nonlinear or linearly degenerate, by choosing values for α such that μ disappears from the resulting equation.

If $f'(u) = \lambda_0$ is constant for u in the interval between u_l, u_r, setting $\alpha = \frac{1}{2}$, $s = \lambda_0$ leaves only terms of zero order in μ, and we find

$$u_\mu = \mathcal{U}(\eta, \tau), \qquad \eta = (x - st)/\sqrt{\mu}, \qquad \tau = t,$$

where $\mathcal{U}$ is a (bounded) solution to the *Riemann problem for the heat equation*

$$\mathcal{U}_t = \mathcal{U}_{\eta\eta} \qquad t > 0,\ \eta \in \mathbb{R},$$

$$\mathcal{U}(\eta, 0) = u_l + (u_r - u_l)H(\eta), \qquad \eta \in \mathbb{R}.$$

Hence,

$$u_\mu(x,t) = \mathscr{U}(\eta,\tau) = \tfrac{1}{2}(u_l + u_r) + \tfrac{1}{2}(u_l - u_r)\,\mathrm{erf}(\eta/2\sqrt{t}),$$

where erf(x) is the error function. We see that

$$u^\mu(x,t) \to U_l + (u_r - u_l)H(x - st), \qquad u_x \to 0 \text{ as } |x| \to \infty.$$

The "boundary layer" is the region, centered at $x = st(s = \lambda_0)$ and having thickness $\sqrt{\mu t}$, outside which u is essentially constant, $u_x \simeq 0$. Note that, on the half-line $x = st$, $u_\mu(st,t) = \frac{1}{2}(u_l + u_r)$ for every μ. As μ approaches zero, we see that $u_\mu(x,t)$ approaches the characteristic shock (68) for $x \neq st$, while the "natural" value of the limiting solution on the shock ($x = st$) is the average of the limits from the left and the right. As

$$\mu u_{\mu_x}^2 = \frac{(\Delta u)^2}{4\pi t}\exp(-(x - st)^2/2\mu t), \qquad \Delta u := u_l - u_r$$

($s = \lambda_0$) in the limit $\mu \downarrow 0$, we find, for every $t > 0$:

$$\mu(u_{\mu_x})^2 \to \begin{cases} 0, & x \neq st, \\ \dfrac{(\Delta u)^2}{4\pi t}, & x = st, \end{cases} \qquad \text{(a.e.)} \tag{70}$$

$$u_{\mu_x} \to -\Delta u\,\delta(x - st) \qquad \text{(in the distribution sense)},$$

$$\mu \int_{\mathbb{R}} (u_{\mu_x})^2\,dx = (\Delta u)^2 \sqrt{\mu}(8\pi t)^{-1/2} \to 0,$$

so that $\mu(u_{\mu_x})^2 \to 0$ in $L^1(\mathbb{R})$, hence in the sense of distributions in $\mathbb{R}$, for every $t > 0$. The same is true also in the sense of distributions in $\mathbb{R} \times (0, \infty)$, as can be seen from the relation

$$\mu \int_0^T dt \int_{\mathbb{R}} (u_{\mu_x})^2\,dx = (\Delta u)^2 \sqrt{\mu T}(2\pi)^{-1/2} \to 0 \qquad \mu \downarrow 0,$$

valid for every $T > 0$. This result is interesting as it shows that the term $\mu U'' u_x^2$ in (64) tends weakly to zero in this case.

If $f''(u) \neq 0$, choosing $\alpha = 1$ leaves only terms of order μ^{-1}, and

$$u = u_\mu(x,t) = \mathscr{U}(\eta), \qquad \eta := (x - st)/\mu.$$

The solution is then sought in the form of a *traveling wave* $\mathcal{U}((x - st)/\mu)$ (a function of $x - st$ alone) to be determined, together with s, from the conditions

$$\mathcal{U}_{\eta\eta} = (f'(\mathcal{U}) - s)\mathcal{U}_\eta \qquad \eta \in \mathbb{R},$$

$$\mathcal{U}(\eta) \to u_l - \Delta u H(\eta), \qquad \mathcal{U}_\eta(\eta) \to 0 \qquad |\eta| \to \infty.$$

It will be seen that this entails a smoothing of the initial discontinuity over a layer of thickness $O(\mu)$. Integrating the equation gives

$$0 = \int_{\mathbb{R}} \mathcal{U}_{\eta\eta}\, d\eta = \int_{\mathbb{R}} f(\mathcal{U})_\eta\, d\eta - s\int_{\mathbb{R}} \mathcal{U}_\eta\, d\eta = -\Delta f + s\Delta u,$$

and we find that s satisfies (RH). Multiplying the equation by $\mathcal{U}_\eta$ we then find

$$\frac{1}{2}\frac{\partial}{\partial\eta}\mathcal{U}_\eta^2 = (f'(\mathcal{U}) - s)\mathcal{U}_\eta^2.$$

As f' is monotone and $f'(u) \to f'(u_l) - \Delta f' H(\eta)$ as $|\eta| \to \infty$, $\mathcal{U}_\eta^2$ is monotone and infinitesimal in this limit. Thus, $f' < s$ as $\eta \to +\infty$, $f' > s$ as $\eta \to -\infty$, and the geometric entropy condition

$$f'(u_r) < s < f'(u_l), \qquad s = s(u_l, u_r),$$

follows. The solution u_μ is a *viscous shock*: For $\mu > 0$ the transition between u_l and u_r occurs in a layer of thickness $O(\mu)$, which for $\mu \downarrow 0$ collapses to the admissible shock.

For example, in the case of the Burgers equation, $f(u) = u^2/2$, performing the computations we obtain

$$u_\mu = u_r + \frac{\Delta u}{1 + \exp(\eta\Delta u/2)},$$

and $u = \frac{1}{2}(u_l + u_r)$ still holds for $x = st$. Thus, the "natural" value for the viscosity solution on the shock is the average of the limits as in the case when $f(u)$ is affine. We can also obtain

$$\mu(u_{\mu_x})^2 = \frac{(\Delta u)^4}{4\mu}\frac{\exp(\eta\Delta u)}{[1 + \exp(\eta\Delta u/2)]^4} \to c\delta(x - st) \qquad \mu \downarrow 0,$$

where the limit is in the sense of distributions ($c = |\Delta u|^3/12$). Hence, in contrast to the linear case, the term $\mu U'' u_x^2$ in (64) does not vanish in this limit. It follows that

$$\mu \int_{\mathbb{R}} (u_{\mu_x})^2 \, dx = |\Delta u|^3/12,$$

$$\mu \int_0^T dt \int_{\mathbb{R}} (u_{\mu_x})^2 \, dx = T|\Delta u|^3/12.$$

The first equation is of particular interest. It shows that the energy dissipation rate (due to "viscosity") is independent of μ. In contrast to the linear case, the *energy dissipation does not approach zero with* μ. (There is a conjecture of this sort of behavior in the theory of turbulent flows in fluid dynamics known as the *Kolmogorov hypothesis*.) We see, again, that in the nonlinear case energy dissipation can occur (in the presence of singularities) even if the viscosity is zero.

Another contrast with the linear case is that the "viscous shock" here has a constant thickness with time. In the linear case it diffuses like $t^{1/2}$.

Remark 3.3 (N-wave). The Riemann problem with $u_l = u_r = 0$ for the equation $u_t + (u^2/2)_x = 0$ has the one-parameter family of piecewise smooth solutions

$$u(x,t) = \begin{cases} 0, & x < -c\sqrt{t}, \\ x/t, & -c\sqrt{t} < x < c\sqrt{t}, \\ 0, & x > c\sqrt{t}, \end{cases}$$

for $c > 0$. They all satisfy (RH) at the shock curve consisting of the two branches $x = \pm c\sqrt{t}$, but are not bounded in any neighborhood of the origin, and are not self-similar solutions. One speaks of "N-wave" because of the form of the graph for $t > 0$ (Fig. 8).

It can be verified that the N-wave satisfies (GE) (Exercise 3.12). Thus, we have here an example of a.e. solutions for $t > 0$, satisfying the (RH) relation as well as the entropy condition, which are a.e. zero initially. Being unbounded at the origin, they are not weak solutions.

3.4. Wave Interaction. We finally discuss the important phenomenon of the interaction of solutions (shocks and rarefaction fans). We do this by the series of examples below. In case of shocks, one speaks of *shock collision*: When two shocks collide, the two discontinuities merge into one and in order to determine

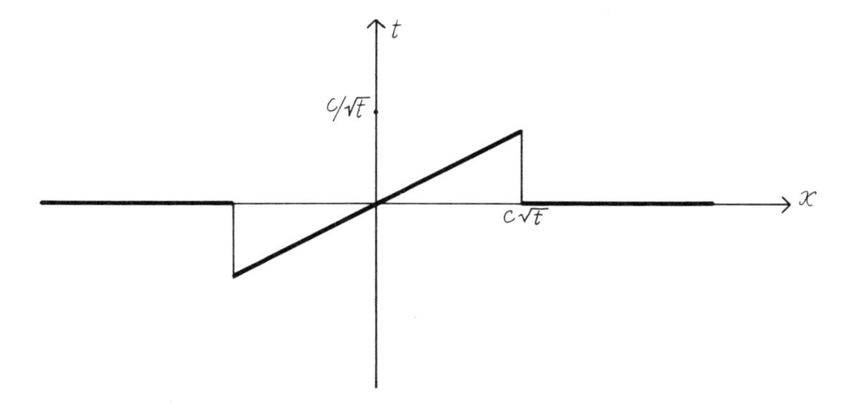

Fig. 8. The N-wave.

what happens beyond this point one has to solve a new Riemann problem. For the scalar equation, the collision of two shocks always gives rise to a shock.

We consider, for simplicity, the (weak entropic) solution of the Cauchy problem for the Burgers equation

$$u_t + (u^2/2)_x = 0$$

with initial data

$$u(x, 0) = \begin{cases} a, & x < 0, \\ b, & 0 < x < 1, \\ c, & 1 < x. \end{cases}$$

Using the entropy $U = u^2$ it is easy to verify, from (LK), that $u_l > u_r$ on all 'admissible" shocks having left and right states u_l and u_r respectively.

i. *Collision of two shocks.* Let $a > b > c$. then the solution for $0 \leq t < \tilde{t} = 2/(a - c)$ consists of the three states a, b, c separated (from left to right) by the two shocks

$$\Sigma: \ x = st, \qquad \Sigma': \ x = 1 + s't,$$

where $s = \frac{1}{2}(a+b)$, $s' = \frac{1}{2}(b+c)$. The two shocks collide for $t = \tilde{t}$, $x = \tilde{x} = (a+b)/(a-c)$. For $t > \tilde{t}$ the solution is found by solving the Riemann problem with initial data

$$u(x, \tilde{t}) = \begin{cases} a, & x < \tilde{x}, \\ c, & x > \tilde{x}, \end{cases}$$

and is given by the two constant states, a, c separated by the shock

$$\Sigma'': \; x = \tilde{x} + s''(t - \tilde{t}),$$

where $s'' = \frac{1}{2}(a+c)$ (Fig. 9).

Conversely, two admissible shocks require $a > b > c$, so that necessarily $s > s'' > s'$. In conclusion, *two consecutive admissible shocks always interact, yielding a shock line with intermediate slope.*

ii. *Collision of a shock with a rarefaction fan.* Suppose $b < a = c$. The solution for small t consists of the three constant states, a, b, a separated (from left to right) by the shock Σ: $x = st$, $s = \frac{1}{2}(a+b)$, and by a rarefaction fan $(x-1)/t$ in the angular sector $bt < x - 1 < at$. As $b < a$, the two waves interact for $t = \tilde{t} = 2/(a-b)$, $x = \tilde{x} = (b+a)/(a-b)$. To find the solution for $t \geq \tilde{t}$ we

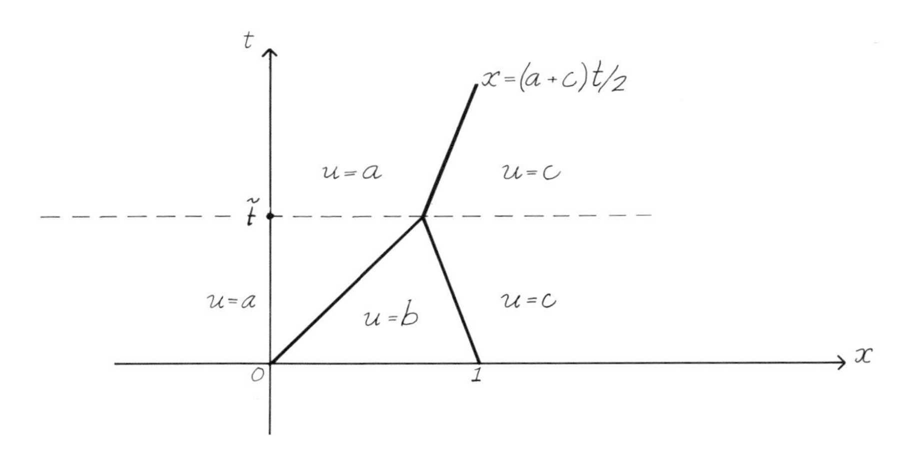

Fig. 9. Shock collision.

need to determine the form of a shock separating the constant state a from the rarefaction fan $(x-1)/t$, using the differential equation

$$\frac{dx}{dt} = \frac{1}{2}\left(a + \frac{x-1}{t}\right), \qquad t > \tilde{t},$$

which follows from the Rankine–Hugoniot relation. The solution satisfying the initial condition $x(\tilde{t}) = \tilde{x}$ is

$$x(t) = 1 + A\sqrt{t} + at, \qquad t \geq \tilde{t},$$

where $A = -\sqrt{2(a-b)}$. Thus, the shock slope $s = dx/dt$ satisfies

$$s(\tilde{t}) = \tfrac{1}{2}(a+b), \qquad s(t) \uparrow a \text{ as } t \to +\infty.$$

The first relation tells us that the shock slope matches continuously with that of Σ, so that *the global shock is a C^1 curve for $t > 0$*. The second relation shows that the shock slope tends asymptotically to the characteristic slope, $dx/dt = a$ (see Fig. 10).

To summarize, *an admissible shock and a rarefaction fan always interact, yielding a global C^1 shock curve.*

iii. *Interaction of two rarefaction fans.* Finally, let $a < b < c$. Then the solution consists of the constant states a, b, c separated by two rarefaction fans

$$u = x/t, \qquad at < x < bt; \qquad u = (x-1)/t, \qquad bt < x - 1 < ct.$$

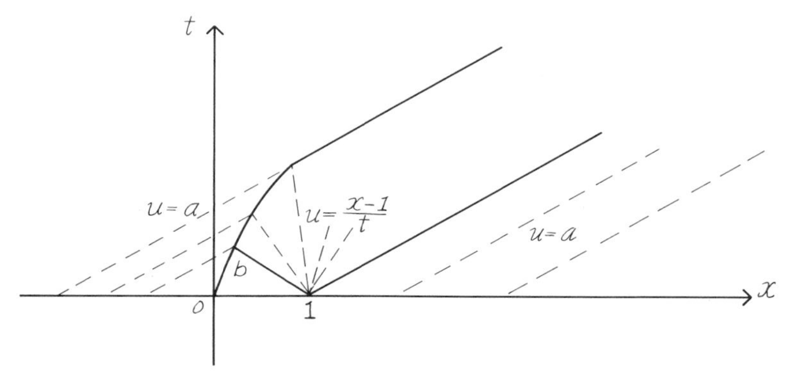

Fig. 10. Collision of a shock with a rarefaction fan.

As the "tail" of the first fan travels with the same speed as the "head" of the second, we see that *two consecutive rarefaction fans never interact* (Fig. 11).

iv. *Approaching waves.* By combining the results of (i)–(iii) we conclude that *two consecutive "waves" approach if and only if at least one of them is an admissible shock. Approaching waves always interact.*

4. Systems in One Space Variable

We return to the study of the Cauchy problem for a hyperbolic system of conservation laws,

$$\mathbf{u}_t + \mathbf{f}(\mathbf{u})_x = 0 \qquad x \in \mathbb{R},\ t > 0, \tag{71}$$

$$\mathbf{u}(x, 0) = \mathbf{u}^0(x) \qquad x \in \mathbb{R}, \tag{72}$$

with $\mathbf{u}^0 \in L^\infty(\mathbb{R})$, where now $\mathbf{u}, \mathbf{u}^0$ takes values in $\mathbb{R}^m$, and $\mathbf{f}$ is a smooth function defined on a domain $G \subset \mathbb{R}^m$ and taking values in $\mathbb{R}^m$. With this in mind we see that the definition of weak solution is formally identical to the scalar case and we carry it over here. If $\mathbf{u}$ is a piecewise smooth weak solution with a discontinuity curve Σ given in the form $x = x(t)$, the Rankine–Hugoniot relation takes the form

$$s[\mathbf{u}] = [\mathbf{f}(\mathbf{u})], \tag{RH}$$

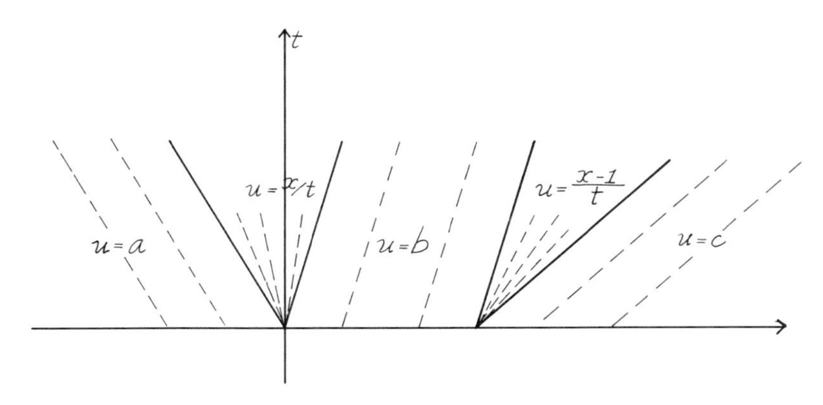

Fig. 11. Noninteraction of rarefaction fans.

where $s = dx/dt$, $[\mathbf{u}] = \mathbf{u}_l - \mathbf{u}_r$, and the proof is the same as in the scalar case. These are now m scalar equations involving the $2m + 1$ scalar quantities $\mathbf{u}_l, \mathbf{u}_r, s$. If the system is linear, so that $\mathbb{A} = \partial\mathbf{f}/\partial\mathbf{u}$ is a constant matrix $\mathbb{A}^0$ with eigenvalues λ_k^0 and left and right eigenvectors $\boldsymbol{l}_k^0, \boldsymbol{r}_k^0$, (RH) becomes $s[\mathbf{u}] = \mathbb{A}^0[\mathbf{u}]$, and hence has the m solutions $s = \lambda_k^0, [\mathbf{u}] = \varepsilon_k \boldsymbol{r}_k^0$ corresponding to a kth characteristic shock, of strength $|\varepsilon_k||\boldsymbol{r}_k^0|$ $(k = 1, \ldots, m)$. For weak shocks, $|[\mathbf{u}]| \to 0$, these are the limiting solutions of (RH) for any $\mathbf{f}$. In other words, infinitesimal shocks and linear shocks are characteristic, as for the scalar law.

Example 4.1 (Rankine–Hugoniot relations for an ideal gas). Consider a piecewise smooth weak solution $\mathbf{u} = (\rho, \rho u, \rho W)$ of the system

$$\begin{aligned}
\rho_t + (\rho u)_x &= 0,\\
(\rho u)_t + (\rho u^2 + p)_x &= 0,\\
(\rho W)_t + \{u(\rho W + p)\}_x &= 0
\end{aligned}$$

[see Section 1, equations (13) and ff.] with $\mathbf{u}_l = (\rho_l, \rho_l u_l, \rho_l W_l)$, $\mathbf{u}_r = (\rho_r, \rho_r u_r, \rho_r W_r)$ the limits from the left and the right, respectively, along a discontinuity curve Σ. They correspond to the constant states ρ_l, u_l, p_l and ρ_r, u_r, p_r, where $p = \rho^\gamma \exp(S/c_v) = (\gamma - 1)\rho e$. Then the Rankine–Hugoniot relations can be written in the form

$$\begin{aligned}
(s - u_l)[\rho] - \rho_r[u] &= 0,\\
[p] - (s - u_l)\rho_l[u] &= 0,\\
(s - u_l)([e] + \tfrac{1}{2}(p_l + p_r)[1/\rho]) &= 0,
\end{aligned}$$

where $e = c_v p/R\rho$, $R = c_p - c_v$ the gas constant. It is not difficult to check that (with the exception of the Chaplygin gas, which does not exist in nature) entropy is always discontinuous across Σ,

$$[S] \neq 0,$$

even when $[u] = [p] = 0$ (contact discontinuity, $(s = u_l = u_r)$. In contrast, the Rankine–Hugoniot relation for the conservation law

$$(\rho S)_t + (\rho u S)_x = 0$$

[see Section 1, equation (15)] is $s[\rho S] = [\rho u S]$, whence $(s - u_r)[S] = 0$, a contradiction (Exercise 4.1). We will see that in the presence of shocks the conservation law for entropy must be replaced by an inequality, $(s - u_r)[S] \geq 0$, ensuring the increase of physical entropy at the passing of a shock (with $s \neq u_r$).

4.1. Entropy Conditions. The generalization of the entropy conditions is more complicated. If $U = U(\mathbf{u})$, it need not be the case that there exists an $F(\mathbf{u})$ such that (71) satisfies the additional conservation law

$$U_t + F_x = \operatorname{grad} U \cdot \mathbf{u}_t + \operatorname{grad} F \cdot \mathbf{u}_x = 0,$$

as multiplying $\mathbf{u}_t + \mathbb{A}\mathbf{u}_x = 0$, $\mathbb{A} = \partial \mathbf{f}/\partial \mathbf{u}$, by grad U shows that this requires

$$\operatorname{grad} U \cdot \mathbb{A} = \operatorname{grad} F. \tag{73}$$

This may be thought of as a system of m partial differential equations for the two functions U and F, and it is, in general, overdetermined. For many important systems such a pair, called an entropy–entropy flux pair, can be found, however, and we will generalize the (LK) criterion to this situation. If we consider, formally, the parabolic system

$$\mathbf{u}_t + \mathbb{A}(\mathbf{u})\mathbf{u}_x = \mu \mathbf{u}_{xx}, \tag{P}$$

then, assuming existence of a pair (U, F) for (71),

$$U_t + F_x = \mu U_{xx} - \mu \mathbf{u}_x \cdot \mathbb{H}\mathbf{u}_x,$$

where $\mathbb{H}$ is the Hessian matrix of $U(\mathbf{u})$. We assume, as in the scalar case, that U is a convex function of $\mathbf{u}$. Then

$$U_t + F_x \leq \mu U_{xx}.$$

All of this holds exactly whenever the functions involved are smooth, presumably everywhere for solutions of (P). If we multiply by a test function ψ, integrate by parts, and take the limit as $\mu \downarrow 0$, assuming $\mathbf{u}$ bounded, so that $\int U\psi_{xx}$ is bounded, we obtain an inequality of the same form as (LK) for a scalar equation.

Definition 4.1. Suppose (71) admits an entropy–entropy flux pair (U, F), with U convex. Then a weak solution satisfies the Lax–Kruzhkov entropy condition, (LK), if

$$\int_{\mathbb{R}} \int_0^{\infty} (\psi_t U + \psi_x F) dx\, dt \geq 0 \qquad \forall \psi \in C_0^1(\mathbb{R} \times (0, \infty)), \psi \geq 0. \qquad \text{(LK)}$$

We can show exactly as before that for a piecewise C^1 weak solution satisfying (LK),

$$s[U] \leq [F]$$

across a shock.

For a strictly hyperbolic and genuinely nonlinear system the geometric entropy condition can be generalized as follows. We say that a shock is an *admissible k-shock* if its speed $s = s_k$ satisfies $\lambda_{k-1}(\mathbf{u}_l) < s < \lambda_k(\mathbf{u}_l)$ and $\lambda_k(\mathbf{u}_r) < s < \lambda_{k+1}(\mathbf{u}_r)$ (we set here $\lambda_0 := -\infty, \lambda_{m+1} := +\infty$). These can be rewritten as

$$\lambda_k(\mathbf{u}_r) < s < \lambda_k(\mathbf{u}_l), \qquad \lambda_{k-1}(\mathbf{u}_l) < s < \lambda_{k+1}(\mathbf{u}_r), \qquad \text{(GE)}$$

where $k = 1, \ldots, m$ and $s = s(\mathbf{u}_l, \mathbf{u}_r)$ satisfies (RH). As for each k

$$\lambda_1(\mathbf{u}_r) < \cdots < \lambda_k(\mathbf{u}_r) < s_k < \lambda_k(\mathbf{u}_l) < \cdots < \lambda_m(\mathbf{u}_l),$$

then k characteristics impinge on Σ from the line of initial data on the $\mathbf{u}_r$ side and $m - k + 1$ on the $\mathbf{u}_l$ side, and the information carried by these $m + 1$ characteristics plus the $m - 1$ relations obtained from (RH) after eliminating s are sufficient to determine the $2m$ values $\mathbf{u}_l, \mathbf{u}_r$. There are m families of admissible k-shocks $s = s_k$, with $s_k < s_{k+1} (k = 1, \ldots, m)$, characterized by having speed s_k intermediate between the left and right characteristic speeds $\lambda_k(\mathbf{u}_l), \lambda_k(\mathbf{u}_r)$. As in the case of the scalar equation, if the system is genuinely nonlinear the admissible shocks are never characteristic, that is, $\lambda_k(\mathbf{u}_l) \neq s(\mathbf{u}_l, \mathbf{u}_r) \neq \lambda_k(\mathbf{u}_r)$. The equivalence of the two entropy conditions in this case will be discussed after solving the Riemann problem.

If instead the kth eigenvalue is linearly degenerate, the admissible k-shock is a characteristic shock, or *contact discontinuity* (this name comes from gas dynamics), defined by

$$\lambda_k(\mathbf{u}_l) = s(\mathbf{u}_l, \mathbf{u}_r) = \lambda_k(\mathbf{u}_r), \qquad \text{(CD)}$$

and (GE) for this index k reduces to (CD). In particular, in the *linear* case, (CD) is satisfied for all k and the entropy condition becomes redundant, in the sense that *all linear shocks are admissible provided they are characteristic.*

4.2. Riemann Problem. Finally, we consider the Riemann problem for a strictly hyperbolic and genuinely nonlinear system

$$\mathbf{u}_t + \mathbf{f}(\mathbf{u})_x = 0 \qquad x \in \mathbb{R},\ t > 0; \qquad \mathbf{u}(x, 0) = \begin{cases} \mathbf{u}_l, & x < 0, \\ \mathbf{u}_r, & x > 0. \end{cases}$$

We will begin our study of this problem by thinking of the left state $\mathbf{u}_l$ as fixed and varying $\mathbf{u}_r$.

We can get insight into this problem by considering a linear problem, $\mathbf{u}_t + \mathbb{A}^0 \mathbf{u}_x = 0$. Then by taking normalized eigenvectors, $\boldsymbol{r}_k = \boldsymbol{r}_k^0$, $\boldsymbol{l}_k = \boldsymbol{l}_k^0/(\boldsymbol{l}_k^0 \cdot \boldsymbol{r}_k)$ so that $\boldsymbol{l}_j \cdot \boldsymbol{r}_k = \delta_{j,k}$, the representation

$$\mathbf{u}_r - \mathbf{u}_l = \sum_{k=1}^{m} \varepsilon_k \boldsymbol{r}_k, \qquad \varepsilon_k = (\mathbf{u}_r - \mathbf{u}_l) \cdot \boldsymbol{l}_k$$

holds, and it is not difficult to see that the solution consists of $m + 1$ constant states, depending linearly on the parameters ε_k:

$$\mathbf{u}_0 = \mathbf{u}_l, \quad \mathbf{u}_j = \mathbf{u}_{j-1} + \varepsilon_j \boldsymbol{r}_j \quad (j = 1, \ldots, m-1), \quad \mathbf{u}_m = \mathbf{u}_{m-1} + \varepsilon_m \boldsymbol{r}_m^0 = \mathbf{u}_r,$$

separated by m characteristic (linear) shocks $x = \lambda_j^0 t$ of strength $|\varepsilon_j|$:

$$\mathbf{u}(x, t) = \begin{cases} \mathbf{u}_l, & x < \lambda_1^0 t, \\ \mathbf{u}_k := \mathbf{u}_l + \sum_{j=1}^{k} \varepsilon_j \boldsymbol{r}_j, & \lambda_k^0 t < x < \lambda_{k+1}^0 t \qquad (k = 1, \ldots, m-1), \\ \mathbf{u}_r, & \lambda_m^0 t < x. \end{cases}$$

Example 4.2 (*Riemann's problem in acoustics*). We consider the Riemann problem for the 2×2 linear system of acoustic equations

$$\rho_t + \rho_0 u_x = 0, \qquad \rho_0 u_t + c_0^2 \rho_x = 0,$$

$$\rho(x, 0) = \begin{cases} a, & x < 0, \\ b, & x > 0, \end{cases} \qquad u(x, 0) = \begin{cases} 0, & x < 0, \\ V, & x > 0 \end{cases}$$

[see (12), Section 1]. Setting $\mathbf{u} = (\rho, u)$, $\mathbf{u}_l = (a, 0)$, $\mathbf{u}_r = (b, V)$, the eigenvalues and normalized eigenvectors are

$$\lambda_1 = -c_0, \qquad \boldsymbol{r}_1 = (-\rho_0, c_0), \qquad \boldsymbol{l}_1 = \left(\frac{-1}{2\rho_0}, \frac{1}{2c_0}\right),$$

$$\lambda_2 = c_0, \qquad \boldsymbol{r}_2 = (\rho_0, c_0), \qquad \boldsymbol{l}_2 = \left(\frac{1}{2\rho_0}, \frac{1}{2c_0}\right).$$

It follows that

$$\varepsilon_1 = (\mathbf{u}_r - \mathbf{u}_l) \cdot \boldsymbol{l}_1 = \frac{a-b}{2\rho_0} + \frac{V}{2c_0}, \qquad \varepsilon_1 \boldsymbol{r}_1 = \left(\frac{b-a}{2} - \frac{V\rho_0}{2c_0}, \frac{a-b}{2\rho_0}\, c_0 + \frac{V}{2}\right),$$

and the solution is

$$\rho(x, t) = \begin{cases} a, & x < -c_0 t, \\ a + \dfrac{b-a}{2} - \dfrac{V\rho_0}{2c_0}, & -c_0 t < x < c_0 t, \\ b, & x > c_0 t, \end{cases}$$

and

$$u(x, t) = \begin{cases} 0, & x < -c_0 t, \\ \dfrac{a-b}{2\rho_0}\, c_0 + \dfrac{V}{2}, & -c_0 t < x < c_0 t, \\ V, & x > c_0 t. \end{cases}$$

The structure of the solution in the nonlinear case is similar, as a consequence of the following results. We recall that, for genuinely nonlinear systems, the normalized eigenvectors are defined as $\tilde{\boldsymbol{r}}_k(\mathbf{u}) = \boldsymbol{r}_k(\mathbf{u})/(\boldsymbol{r}_k(\mathbf{u}) \cdot \operatorname{grad}_{\mathbf{u}} \lambda_k(\mathbf{u}))$, $\tilde{\boldsymbol{l}}_k(\mathbf{u}) = \boldsymbol{l}_k(\mathbf{u})/(\boldsymbol{l}_k(\mathbf{u}) \cdot \tilde{\boldsymbol{r}}_k(\mathbf{u}))$. For the rest of this chapter, by $\boldsymbol{r}_k, \boldsymbol{l}_k$ we will always mean these normalized eigenvectors, so that the relations $\boldsymbol{r}_k(\mathbf{u}) \cdot \operatorname{grad}_{\mathbf{u}} \lambda_k(\mathbf{u}) = 1$, $\boldsymbol{r}_k(\mathbf{u}) \cdot \boldsymbol{l}_j(\mathbf{u}) = \delta_{j,k}$ $(j, k = 1, \ldots, m)$ hold for $\mathbf{u}$ in any compact subset of G.

Theorem 4.1. Let the system be strictly hyperbolic and genuinely nonlinear. There are m smooth one-parameter families of states $\mathbf{u}$ near $\mathbf{u}_l$

$$S_k(\mathbf{u}_l)\colon \qquad \mathbf{u} = \mathbf{u}_k(\varepsilon), \qquad \mathbf{u}(0) = \mathbf{u}_l$$

(*Hugoniot curves*), and smooth functions $s = s_k(\varepsilon)$, defined for $|\varepsilon| < a_k$ ($k = 1, \ldots, m$), such that the Rankine–Hugoniot conditions

$$s(\varepsilon)(\mathbf{u}(\varepsilon) - \mathbf{u}_l) = f(\mathbf{u}(\varepsilon)) - f(\mathbf{u}_l) \tag{74}$$

are satisfied, and the relations

$$s(0) = \lambda_k(\mathbf{u}_l), \qquad \mathbf{u}'(0) = \boldsymbol{r}_k(\mathbf{u}_l), \qquad \mathbf{u}''(0) = \boldsymbol{r}'_k(\mathbf{u})_l), \qquad s'(0) = \tfrac{1}{2} \tag{75}$$

hold by a suitable choice of parametrization. (Primes indicate derivatives with respect to ε.)

Proof. We write

$$\begin{aligned}\mathbf{f}(\mathbf{u}) - \mathbf{f}(\mathbf{u}_l) &= \int_0^1 \frac{d}{dt}\mathbf{f}(\mathbf{u}_l + t(\mathbf{u} - \mathbf{u}_l))dt = \int_0^1 \mathbb{A}(\mathbf{u}_l + t(\mathbf{u} - \mathbf{u}_l))(\mathbf{u} - \mathbf{u}_l)dt \\ &:= \mathbb{G}(\mathbf{u})(\mathbf{u} - \mathbf{u}_l),\end{aligned}$$

where $\mathbb{G}(\mathbf{u})$ approaches $\mathbb{A}(\mathbf{u}_l)$ as $\mathbf{u}$ approaches $\mathbf{u}_l$. Thus, for $\mathbf{u}$ close to $\mathbf{u}_l$, $\mathbb{G}(\mathbf{u})$ must have real and distinct eigenvalues $\mu_k(\mathbf{u})$ and full sets of biorthogonal left and right eigenvectors $\boldsymbol{L}_k(\mathbf{u}), \boldsymbol{R}_k(\mathbf{u})$ ($k = 1, \ldots, m$). The jump conditions (74) become $\mathbb{G}(\mathbf{u})(\mathbf{u} - \mathbf{u}_l) = s(\mathbf{u} - \mathbf{u}_l)$ and therefore if $\mathbf{u} \neq \mathbf{u}_l$ necessarily $s = \mu_k(\mathbf{u})$ and $(\mathbf{u} - \mathbf{u}_l) \propto \boldsymbol{R}_k(\mathbf{u})$ for some index k. It follows that $\mathbf{u}$ is a solution of the nonlinear system of $m - 1$ equations

$$\boldsymbol{L}_j(\mathbf{u}) \cdot (\mathbf{u} - \mathbf{u}_l) = 0 \qquad j \neq k,$$

whose Jacobian matrix at $\mathbf{u} = \mathbf{u}_l$, $[\boldsymbol{l}_j(\mathbf{u}_l)]_{j \neq k}$, has rank $m - 1$. Then by the implicit function theorem (74) defines m distinct smooth curves $\mathbf{u} = \mathbf{u}_k(\varepsilon)$ defined for small $|\varepsilon|$, near $\mathbf{u}_l = \mathbf{u}_k(0)$, and $s = \mu_k(\mathbf{u}_k(\varepsilon))$ ($k = 1, \ldots, m$).

To prove (75), observe first that $(\mathbf{u} - \mathbf{u}_l) \propto \boldsymbol{R}_k(\mathbf{u})$ implies, by continuity as $\varepsilon \to 0$, that $\mathbf{u}'(0) = c\boldsymbol{r}_k(\mathbf{u}_l)$ for some constant $c \neq 0$, and by a change of scale $c\varepsilon$ in the parametrization we can achieve $\mathbf{u}'(0) = \boldsymbol{r}_k(\mathbf{u}_l)$. Differentiating (74) specialized to $\mathbf{u} = \mathbf{u}_k(c\varepsilon)$ with respect to $c\varepsilon$ gives

$$s\mathbf{u}' + s'(\mathbf{u} - \mathbf{u}_l) = \mathbb{A}\mathbf{u}',$$

and at $\varepsilon = 0$ we find $s(0)\mathbf{u}'(0) = \mathbb{A}(\mathbf{u}_l)\mathbf{u}'(0)$, whence $s(0) = \lambda_k(\mathbf{u}_l)$. Differentiating once more yields $s\mathbf{u}'' + 2s'\mathbf{u}' + s''(\mathbf{u} - \mathbf{u}_l) = \mathbb{A}\mathbf{u}'' + \mathbb{A}_\mathbf{u}\mathbf{u}' \cdot \mathbf{u}'$ and setting $\varepsilon = 0$ gives

$$\lambda_k(\mathbf{u}_l)\mathbf{u}''(0) + 2s'(0)\boldsymbol{r}_k(\mathbf{u}_l) = \mathbb{A}(\mathbf{u}_l)\mathbf{u}''(0) + \mathbb{A}_\mathbf{u}(\mathbf{u}_l)\boldsymbol{r}_k(\mathbf{u}_l) \cdot \boldsymbol{r}_k(\mathbf{u}_l). \quad (76)$$

On the other hand, differentiating $\lambda_k(\mathbf{u})\boldsymbol{r}_k(\mathbf{u}) = \mathbb{A}(\mathbf{u})\boldsymbol{r}_k(\mathbf{u})$ for $\mathbf{u} = \mathbf{u}_k(c\varepsilon)$ we obtain $\lambda_k'\boldsymbol{r}_k + \lambda_k\boldsymbol{r}_k' = \mathbb{A}_\mathbf{u}\mathbf{u}' \cdot \boldsymbol{r}_k + \mathbb{A}\boldsymbol{r}_k'$, where $\lambda_k'(\mathbf{u}) \equiv \boldsymbol{r}_k(\mathbf{u}) \cdot \text{grad}_\mathbf{u}\lambda_k(\mathbf{u}) = 1$ for every ε (from the normalization of $\boldsymbol{r}_k$). Setting $\varepsilon = 0$ we then find

$$\lambda_k(\mathbf{u}_l)\boldsymbol{r}_k'(\mathbf{u}_l) + \boldsymbol{r}_k(\mathbf{u}_l) = \mathbb{A}(\mathbf{u}_l)\boldsymbol{r}_k'(\mathbf{u}_l) + \mathbb{A}_\mathbf{u}(\mathbf{u}_l)\boldsymbol{r}_k(\mathbf{u}_l) \cdot \boldsymbol{r}_k(\mathbf{u}_l). \quad (77)$$

Multiplying both (76) and (77) by $\boldsymbol{l}_k(\mathbf{u}_l)$ on the left and subtracting yields

$$\boldsymbol{l}_k \cdot \mathbb{A}(\mathbf{u}''(0) - \boldsymbol{r}_k') = \lambda_k\boldsymbol{l}_k \cdot (\mathbf{u}''(0) - \boldsymbol{r}_k') + \boldsymbol{l}_k \cdot \boldsymbol{r}_k(2s'(0) - 1)$$

(all arguments evaluated at $\varepsilon = 0$), whence $s'(0) = \frac{1}{2}$. Now subtract (77) from (76) to get

$$\lambda_k(\mathbf{u}_l)(\mathbf{u}''(0) - \boldsymbol{r}_k'(\mathbf{u}_l)) = \mathbb{A}(\mathbf{u}_l)(\mathbf{u}''(0) - \boldsymbol{r}_k'(\mathbf{u}_l)).$$

Hence, $\mathbf{u}''(0) - \boldsymbol{r}_k'(\mathbf{u}_l) = b\boldsymbol{r}_k(\mathbf{u}_l)$, where b is a real number. We can again change our parametrization (to $\tilde{\varepsilon} = c\varepsilon - b\varepsilon^2/2$) so as to achieve $\mathbf{u}''(0) = \boldsymbol{r}_k'(\mathbf{u}_l)$. This completes the proof. □

Theorem 4.2. The geometric entropy condition holds along $\mathbf{u}_k(\varepsilon)$ for $|\varepsilon|$ small if and only if $\varepsilon < 0$.

Proof. We write $\lambda_k(\varepsilon) := \lambda_k(\mathbf{u}_\mathbf{k}(\varepsilon))$ and $s(\varepsilon) := s(\mathbf{u}_k(\varepsilon))$. (GE) says then

(a) $\lambda_{k-1}(0) < s(\varepsilon) < \lambda_k(0)$,
(b) $\lambda_k(\varepsilon) < s(\varepsilon) < \lambda_{k+1}(\varepsilon)$.

All statements following will be for ε sufficiently small. Let $\Phi(\varepsilon) := \lambda_k(\varepsilon) - s(\varepsilon)$. Then $\Phi(0) = 0, \Phi'(0) = \lambda_k'(0) - s'(0) = 1 - \frac{1}{2} > 0$ implies that whenever (b) holds $\varepsilon < 0$.

Conversely, if $\varepsilon < 0, \Phi(\varepsilon) < 0$ so $\lambda_k(\varepsilon) < s(\varepsilon)$. As $s'(0) = \frac{1}{2}, \lambda_k(0) = s(0) > s(\varepsilon)$. As $\lim_{\varepsilon \to 0} s(\varepsilon) = \lambda_k(0) > \lambda_{k-1}(0), s(\varepsilon) > \lambda_{k-1}(0)$. Finally, $\lambda_{k+1}(0) > \lambda_k(0) = s(0)$ and $s'(0) = \frac{1}{2}$ implies $s(\varepsilon) < \lambda_{k+1}(\varepsilon)$. □

Therefore, the set of states $\mathbf{u}_r = \mathbf{u}(\varepsilon)$ that can be connected to $\mathbf{u}_l$ by a k-shock satisfying (GE) is the part of the kth Hugoniot curve for $\varepsilon \in (-a_k, 0]$, for some $a_k > 0$, and this requires $\lambda_k(\mathbf{u}_r) < \lambda_k(\mathbf{u}_l)$. From $s(\varepsilon) = s(0) + \varepsilon s'(0) + O(\varepsilon^2) = \lambda_k(\mathbf{u}_l) + \varepsilon/2 + O(\varepsilon^2)$ and $L_k(\mathbf{u}_r) = \lambda_k(\mathbf{u}_l) + \varepsilon + O(\varepsilon^2)$ we find

$$s(\varepsilon) = \tfrac{1}{2}\{\lambda_k(\mathbf{u}_r) + \lambda_k(\mathbf{u}_l)\} + O(\varepsilon^2).$$

This relation was exact for the Burgers equation.

It remains to investigate the case $\lambda_k(\mathbf{u}_r) > \lambda_k(\mathbf{u}_l)$. By direct substitution we see that a function $\mathbf{u} = \mathbf{U}(x/t) \equiv \mathbf{U}(\xi)$ is a nonconstant solution of (71) if and only if $(\mathbb{A}(\mathbf{U}) - \xi\mathbb{I})\mathbf{U}'(\xi) = 0, \mathbf{U}'(\xi) \neq 0$, so that

$$\xi = \lambda_k(\mathbf{U}(\xi)), \qquad \mathbf{U}'(\xi) = \boldsymbol{r}_k(\mathbf{U}(\xi)).$$

As a matter of fact, we should write $\mathbf{U}'(\xi) = g(\xi)\boldsymbol{r}_k(\mathbf{U}(\xi))$, with $g(\xi) \neq 0$ a suitable scale factor, but differentiating the first relation and using the normalization $\boldsymbol{r}_k(\mathbf{u}) \cdot \mathrm{grad}_{\mathbf{u}}\lambda_k(\mathbf{u}) = 1$ for $\boldsymbol{r}_k$, we see that $g(\xi) \equiv 1$.

If we want to find such a solution in an angular sector $\alpha \leq \xi = x/t \leq \beta$, this requires

$$\lambda_k(\mathbf{U}(\alpha)) = \alpha < \beta = \lambda_k(\mathbf{U}(\beta)).$$

If this inequality holds, we can define $\mathbf{u} = \mathbf{U}(x/t)$ for $t \geq 0$ by

$$\mathbf{u}(x/t) = \begin{cases} \mathbf{U}(\alpha), & x \leq \alpha t, \\ \mathbf{U}(x/t), & \alpha t \leq x \leq \beta t, \\ \mathbf{U}(\beta), & \beta t \leq x, \end{cases}$$

and, as we have seen for the scalar equation, this function is a weak solution of (71). If $\mathbf{u}_l = \mathbf{U}(\alpha)$ and $\mathbf{u}_r = \mathbf{U}(\beta)$, this is a solution of the Riemann problem for $\lambda_k(\mathbf{u}_r) > \lambda_k(\mathbf{u}_l)$, which is called a *k-rarefaction fan*. The function $\mathbf{U}(\xi)$ is determined as the solution of

$$\mathbf{U}'(\xi) = \boldsymbol{r}_k(\mathbf{U}(\xi)), \qquad \mathbf{U}(\alpha) = \mathbf{u}_l,$$

where $\alpha = \lambda_k(\mathbf{u}_l)$. This initial value problem can be uniquely solved locally and the solution identically satisfies the condition $\xi = \lambda_k(\mathbf{U}(\xi))$. If we introduce $\varepsilon = \xi - \alpha$ as a parameter, then $d\lambda_k(\varepsilon)/d\varepsilon = 1$, so for $\varepsilon > 0$ we can use $\mathbf{U}(\alpha + \varepsilon)$ to define a rarefaction fan solution of the Riemann problem $\mathbf{u}_r = \mathbf{U}(\alpha + \varepsilon), \varepsilon \in [0, b_k)$ for $\lambda_k(\mathbf{u}_r) > \lambda_k(\mathbf{u}_l)$. This solution is called a *k-centered wave*, and as the $\boldsymbol{r}_k$ form a full set the m k-centered waves for $k = 1, \ldots, m$ are distinct.

If we paste these two families together (for any fixed k), we obtain a family of solutions $\mathbf{u} = \mathbf{u}_k(\varepsilon), \varepsilon \in (-a_k, b_k), \mathbf{u}(0) = \mathbf{u}_l$ that are admissible k-shocks for

$\varepsilon < 0$, and k-centered waves for $\varepsilon > 0$. The above shows that $\mathbf{u}'(0+) = \boldsymbol{r}_k(\mathbf{u}_l)$, and the theorem on Hugoniot curves shows $\mathbf{u}'(0-) = \boldsymbol{r}_k(\mathbf{u}_l)$ also. Direct differentiation of the equation for $\mathbf{U}$ shows $\mathbf{u}''(0+) = \boldsymbol{r}'_k(\mathbf{u}_l)$ so we also have $\mathbf{u}''$ continuous at $\varepsilon = 0$, and, as $\mathbf{u}$ is smooth for $\varepsilon \neq 0$, $\mathbf{u} \in C^2(-a_k, b_k)$.

We are now able to discuss the solution of the Riemann problem. The weak solution constructed below will be global in time, but local in $\mathbf{u}$, in the sense that the distance between the states $\mathbf{u}_l$, $\mathbf{u}_r$ must be of first order in ε.

We look for a self-similar solution $\mathbf{u}(x, t) = \mathbf{u}(x/t)$ (this entails that all possible shocks are rectilinear, $x = st$ with s independent of t), consisting of $m + 1$ constant states $\mathbf{u}_0, \mathbf{u}_1, \ldots, \mathbf{u}_m$ ($\mathbf{u}_0 = \mathbf{u}_l$, $\mathbf{u}_m = \mathbf{u}_r$), with $\mathbf{u}_{k-1}, \mathbf{u}_k (k = 1, \ldots, m)$ connected by either a (rectilinear) k-shock or a k-centered wave. The states $\mathbf{u}_j$ are actually j-parameter families of constant states defined for $j = 1, \ldots, m$ as follows. First, let $\mathbf{u}_1 = \mathbf{u}_1(\mathbf{u}_0, \varepsilon_1)$ be the one-parameter family defined above for $\mathbf{u}_1(\mathbf{u}_0, 0) = \mathbf{u}_0$; then let $\mathbf{u}_2 = \mathbf{u}_2(\mathbf{u}_1, \varepsilon_2)$ be the one-parameter family of states with initial state $\mathbf{u}_2(\mathbf{u}_1, 0) = \mathbf{u}_1$, so that $\mathbf{u}_2 = \mathbf{u}_2(\mathbf{u}_1(\mathbf{u}_0, \varepsilon_1), \varepsilon_2)$. By continuing this process we arrive at an m-parameter family

$$\mathbf{u}_m = \mathbf{u}(\mathbf{u}_0;\ \varepsilon_1, \ldots, \varepsilon_m), \qquad \mathbf{u}(\mathbf{u}_0;\ 0, \ldots, 0) = \mathbf{u}_0.$$

All that is needed now is to show that the ε_k's can be fixed so that $\mathbf{u}_m = \mathbf{u}_r$, at least for any $\mathbf{u}_r$ in a neighborhood of $\mathbf{u}_0$. Letting $\mathbf{u}_m = (u^1, \ldots, u^m)$, the Jacobian of the (u^j) with respect to the (ε_k) at the origin, due to $\mathbf{u}'(0) = \boldsymbol{r}_k(\mathbf{u}_l)$, has the vectors $\boldsymbol{r}_k(\mathbf{u}_l)$ as its columns, and hence does not vanish. Therefore, by the implicit function theorem, a sufficiently small neighborhood of the origin in ε-space, $\varepsilon := \max_{1 \le k \le m} |\varepsilon_k| < \varepsilon_0$ is mapped one to one onto a neighborhood of $\mathbf{u}_l$, and for every $\mathbf{u}_r$ in such a neighborhood it is possible to determine the values $\bar{\varepsilon}_1, \ldots, \bar{\varepsilon}_m$ such that $\mathbf{u}_r = \mathbf{u}(\mathbf{u}_l;\ \bar{\varepsilon}_1, \ldots, \bar{\varepsilon}_m)$. This fixes the states $\mathbf{u}_j$ and the m "waves" connecting them and thus defines the unique (self-similar) solution of the Riemann problem for $\mathbf{u}_r$ in a neighborhood of $\mathbf{u}_l$. A solution for $\mathbf{u}_r$ far from $\mathbf{u}_l$ need not exist (see Smoller, Ref. 13, Chapter 17).

It has been shown that there is a strictly hyperbolic, genuinely nonlinear system that has a convex entropy for which global uniqueness fails in the Riemann problem (Ref. 14).

A consequence of this analysis is the following result on the equivalence of entropy conditions.

Lemma 4.1. Let the system (71) be strictly hyperbolic, genuinely nonlinear, and with an entropy–entropy flux pair (U, F), with U strictly convex. If a solution $\mathbf{u}$ contains a weak shock $x = x(t)$, and $s = dx/dt$, then

$$s[U] < [F] \Leftrightarrow \text{(GE) holds.}$$

Proof. We have just seen that the totality of states $\mathbf{u}_r$ that can be connected to a state $\mathbf{u}_l$ by a weak k-shock forms a one-parameter family $\mathbf{u}(\mathbf{u}_l, \varepsilon)$; this family satisfies (GE) if and only if $\varepsilon < 0$ (and $|\varepsilon| < a_k$, with a_k small enough). Therefore, setting

$$E(\varepsilon) := s(\varepsilon)\{U(\varepsilon) - U(0)\} - \{F(\varepsilon) - F(0)\},$$

all that is needed is to show that $E > 0$ if and only if $-a_k < \varepsilon < 0$ (remember that we have defined the jump $[U]$ as $U_l - U_r$). By differentiating three times with respect to ε and taking (73) into account we find (Exercise 4.6)

$$E(0) = E'(0) = E''(0) = 0, \qquad E'''(0) = -\tfrac{1}{2}\boldsymbol{r}_k(\mathbf{u}_l) \cdot \mathbb{H}(\mathbf{u}_l)\boldsymbol{r}_k(\mathbf{u}_l), \tag{78}$$

where the Hessian matrix $\mathbb{H}$ is positive definite by assumption. This proves the assertion. □

Example 4.3 (*Riemann's problem for all an ideal gas*). We consider here initial data (see Example 4.1)

$$\mathbf{u}(x, 0) = \begin{cases} \mathbf{u}_l = (\rho_l, \rho_l u_l, \rho_l W_l), & x < 0, \\ \mathbf{u}_r = (\rho_r, \rho_r u_r, \rho_r W_r), & x > 0, \end{cases}$$

corresponding to the constant states ρ_l, u_l, p_l and ρ_r, u_r, p_r. We recall that the Euler system is strictly hyperbolic with eigenvalues and eigenvectors

$$\begin{aligned} \lambda_1 &= u - c, & \boldsymbol{r}_1 &= (\rho, -c, 0), & \boldsymbol{l}_1 &= (-c, \rho, -p_S/c), \\ \lambda_2 &= u, & \boldsymbol{r}_2 &= (p_S, 0, -c^2), & \boldsymbol{l}_2 &= (0, 0, 1), \\ \lambda_3 &= u + c, & \boldsymbol{r}_3 &= (\rho, c, 0), & \boldsymbol{l}_3 &= (c, \rho, p_S/c), \end{aligned}$$

and that λ_1, λ_3 are genuinely nonlinear, λ_2 is *linearly degenerate*. The linearly degenerate case is, strictly speaking, not included in the above theory, but is easy to treat directly. The k-shocks and k-centered waves coalesce for $k=2$ into a contact discontinuity (linear shock) with $s = u_l = u_r$, and from the Rankine–Hugoniot relations (obtained in Example 4.1) we immediately find

$$[u] = [p] = 0, \qquad [\rho] \text{ arbitrary}.$$

Let $\rho_r/\rho_l = e^{\varepsilon}$. Then the one-parameter family of contact discontinuities for $k=2$ is given by

$$p_r = p_l, \qquad u_r = u_l, \qquad \rho_r = \rho_l e^{\varepsilon} \qquad \text{for } \varepsilon \in \mathbb{R}.$$

From the equation of state $p = \rho^\gamma \exp(S/c_v)$, $\gamma = c_p/c_v > 1$, we find

$$p_\rho = \gamma p/\rho = c^2 > 0, \qquad p_{\rho\rho} = \gamma(\gamma - 1)p/\rho^2 > 0, \qquad p_S = p/c_v > 0.$$

These inequalities imply (see Ref. 13, p. 346) that the geometric entropy conditions (GE) are valid globally along the shock curves. From the previous lemma applied to the entropy function $U = -\rho S$ and entropy flux $F = -\rho u S$, it follows that

$$(s - u_r)[S] > 0 \tag{79}$$

for admissible k-shocks with $k = 1$ and $k = 3$ (we recall that $-\rho S$ is convex, see Exercise 1.5). As the one-parameter families of admissible shocks and centered waves for $k = 1$ and 3 will be computed explicitly, we need not bother with the normalization of eigenvectors.

We begin with admissible shocks. We want to see what further restrictions are imposed by the entropy conditions (GE) on the jumps satisfying the (RH) relations. For 1-shocks we find

$$s < u_l - c_l, \qquad u_r - c_r < s < u_r \Rightarrow 0 < c_l < u_l, \qquad 0 < u_r < c_r \qquad (k = 1)$$

and u_l, $u_r > s$, so gas particles cross a 1-shock from left to right. For 3-shocks, (GE) yield

$$u_l < s < u_l + c_l, \quad u_r + c_r < s \Rightarrow -c_l < u_l < 0, \quad u_r < -c_r < 0 \qquad (k = 3)$$

and u_l, $u_r < s$, so gas particles cross a 3-shock from right to left. For both families v_l, $v_r \neq 0$, and $s - u_r$, $s - u_l \neq 0$. From (RH) and (79) we then have

$$[\rho] \neq 0, \qquad [u] \neq 0, \qquad [p] \neq 0, \qquad [S] \neq 0.$$

Furthermore, if we denote by an index b resp. a the state of a particle *before* and *after* the shock, then $l = b$, $r = a$ for $k = 1$ and $l = a$, $r = b$ for $k = 3$, and (79) be omes

$$S_a > S_b.$$

Hence, for both shock families ($k = 1$, 3) *the entropy increases at the passing of a shock*, as dictated by the second law of thermodynamics. It follows also that $p_a > p_b$ (admissible shocks are *compressive*), and we may define

$$p_a/p_b = e^{-\varepsilon} \tag{80}$$

for $\varepsilon < 0$.

We now turn to centered waves. From the relation $d\mathbf{u} = \boldsymbol{r}_k \, d\varphi$ [see (4) in Section 1] and the expressions of $\boldsymbol{r}_1$ and $\boldsymbol{r}_3$, we easily recognize that

$$S, u + 2c/(\gamma - 1) \qquad \text{are constant in a 1-centered wave}$$

and

$$S, u - 2c/(\gamma - 1) \qquad \text{are constant in a 3-centered wave}$$

(these quantities are Riemann invariants, see Exercise 1.2 and Ref. 13). As now $p_a < p_b$, a parameter $\varepsilon \geq 0$ can be defined by the formula (80).

Performing the computations, the two one-parameter families of curves for $k = 1, 3$ can be written as follows:

$$\begin{aligned} k = 1{:} &\qquad p_r = p_l e^{-\varepsilon}, \qquad \rho_r = \rho_l f_1(\varepsilon), \qquad u_r = u_l + c_l h_1(\varepsilon), \qquad \varepsilon \in \mathbb{R}, \\ k = 3{:} &\qquad p_r = p_l e^{\varepsilon}, \qquad \rho_r = \rho_l f_3(\varepsilon), \qquad u_r = u_l + c_l h_3(\varepsilon), \qquad \varepsilon \in \mathbb{R}, \end{aligned}$$

where

$$f_1(\varepsilon) := \begin{cases} e^{-\varepsilon/\gamma}, & \varepsilon \geq 0, \\ \dfrac{\beta + e^{\varepsilon}}{1 + \beta e^{\varepsilon}}, & \varepsilon < 0, \end{cases} \qquad f_3(\varepsilon) := 1/f_1(\varepsilon),$$

are C^2 functions with $f_1(0) = f_3(0) = 1$, β and τ are constants defined by

$$\beta := (\gamma + 1)/(\gamma - 1), \qquad \tau := (\gamma - 1)/2\gamma,$$

and

$$h_1(\varepsilon) := \begin{cases} \dfrac{2}{\gamma - 1}(1 - e^{-\tau\varepsilon}), & \varepsilon \geq 0, \\ \dfrac{2\sqrt{\tau}}{\gamma - 1} \dfrac{1 - e^{-\varepsilon}}{\sqrt{1 + \beta e^{-\varepsilon}}}, & \varepsilon < 0, \end{cases}$$

$$h_3(\varepsilon) := \begin{cases} \dfrac{2}{\gamma - 1}(e^{\tau\varepsilon} - 1), & \varepsilon \geq 0, \\ \dfrac{2\sqrt{\tau}}{\gamma - 1} \dfrac{e^{\varepsilon} - 1}{\sqrt{1 + \beta e^{\varepsilon}}}, & \varepsilon < 0, \end{cases}$$

are C^2 functions with $h_1(0) = h_3(0) = 0$.

It is possible to prove that the Riemann problem has a unique (global) solution, obtained by combining the three one-parameter families for $k = 1, 2, 3$, if and only if

$$u_r - u_l < 2(c_l + c_r)/(\gamma - 1), \tag{VS}$$

and that if this condition is violated, a *vacuum state* $\rho = 0$ occurs.

Suppose in particular that $u_r = u_l = 0$. Then the Riemann problem models two gases (or a single gas at different densities and pressures) initially at rest separated by a screen that is removed at time $t = 0$ (*shock tube problem*). Suppose $\rho_r > \rho_l$, then the solution consists of a 1-shock propagating to the left with shock speed $s' < 0$, a 2-contact discontinuity propagating with speed s'', $s' < s'' < 0$, and a centered wave propagating to the right. The two gases do not mix, but remain in contact at the 2-wave, which moves with the common velocity of the gas particles on the left- and right-hand sides (hence the term *contact discontinuity* for this kind of wave).

For details and proofs we refer the reader to Smoller (Ref. 13).

5. Proof of Existence for Weak Solutions of Systems of Conservation Laws

We will consider the Cauchy problem for a system of conservation laws

$$\mathbf{u}_t + \mathbf{f}(\mathbf{u})_x = 0 \qquad x \in \mathbb{R}, t > 0, \tag{81}$$

$$\mathbf{u}(x, 0) = \mathbf{u}^0(x) \qquad x \in \mathbb{R}. \tag{82}$$

Our goal is to prove the following theorem.

Theorem 5.1 (Glimm, Ref. 15). Suppose that the system (81) is strictly hyperbolic and genuinely nonlinear for $\mathbf{u} = (u_1, \ldots, u_m) \in \mathbb{R}^m$. If $\mathbf{u}^0 \in L^\infty \cap BV(\mathbb{R})$, and $TV(\mathbf{u}^0, \mathbb{R}) := \max_i TV(u_i, \mathbb{R})$ is sufficiently small, then there exists a weak solution of (81), (82) for $t \geq 0$.

We begin by sketching the idea of the proof. First the initial data $\mathbf{u}^0(x)$ are approximated by a piecewise constant function $\mathbf{u}_0^\Delta(x)$,

$$\mathbf{u}_0^\Delta(x) = u_m := \mathbf{u}^0((m+1)\Delta x), \qquad \text{for } (m-1)\Delta x < x < (m+1)\Delta x,$$

m even, and then the Riemann problems with data on intervals $(m-1)\Delta x \leq x \leq (m+1)\Delta x$ (m even) are solved for $0 \leq t \leq \Delta t$. We assume that all values of

u lie in a fixed compact neighborhood U of a constant state, and that Δt satisfies the Courant–Friedrichs–Lewy condition

$$\Delta t < \Delta x/M, \qquad M := \max_k \sup_U |\lambda_k(u)|. \tag{CFL}$$

The set U will be precisely determined in the proof. The solutions $\mathbf{u}^\Delta$ of these Riemann problems are made up of shocks and/or centered rarefaction waves, and the CFL condition implies that they do not interact up to time Δt (Fig. 12).

In the next step a random number θ_1 is chosen in $[-1, 1]$ and $\mathbf{u}^\Delta$ is replaced at time $t = \Delta t$ by a piecewise constant function

$$\mathbf{u}^\Delta(x, \Delta t+) = \mathbf{u}^\Delta((m + \theta_1)\Delta x, \Delta t - 0), \qquad (m-1)\Delta x < x < (m+1)\Delta x$$

(m even). Then the process is continued. These piecewise constant data are used to solve another set of Riemann problems on a time interval of length Δt and an independent random number θ_2 is used to get a new set of piecewise constant data. If $\theta = \{\theta_n\}$ is an independent set of random numbers chosen from $[-1, 1]$, then

$$\mathbf{u}^\Delta(x, n\Delta t+) = \mathbf{u}^\Delta((m + \theta_n)\Delta x, n\Delta t - 0), \qquad (m-1)\Delta x < x < (m+1)\Delta x$$

($n = 1, 2, 3, \ldots;\ m + n$ odd). Thus, at each advance by Δt the x-intervals are shifted by Δx, so that a "diamond mesh" is created (Fig. 13).

In this way a family of approximated solutions $\mathbf{u}_\theta^\Delta(x, t)$ is obtained. It will be shown that this family is compact in $L^1_{\text{loc}}(\mathbb{R} \times [0, \infty))$ and that a suitable subsequence converges to a weak solution. Before entering into details, let us see how the method works on a simple example.

Consider the scalar equation $u_t + (u^2/2)_x = 0$ with the initial value $u(x, 0) = -\text{sgn}(x)$. The exact weak (entropy) solution **u** coincides with the

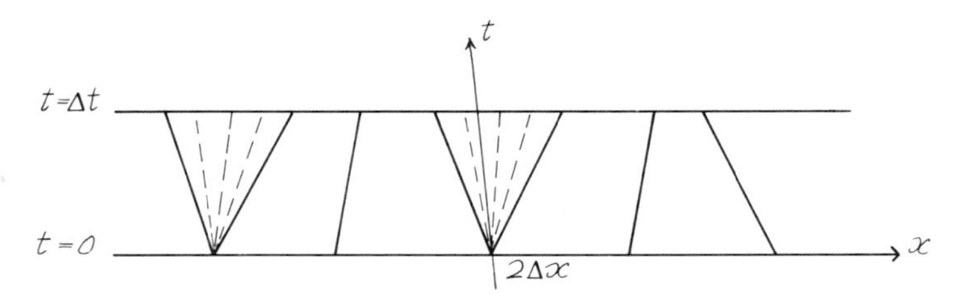

Fig. 12. Solution of Riemann's problem in one time step.

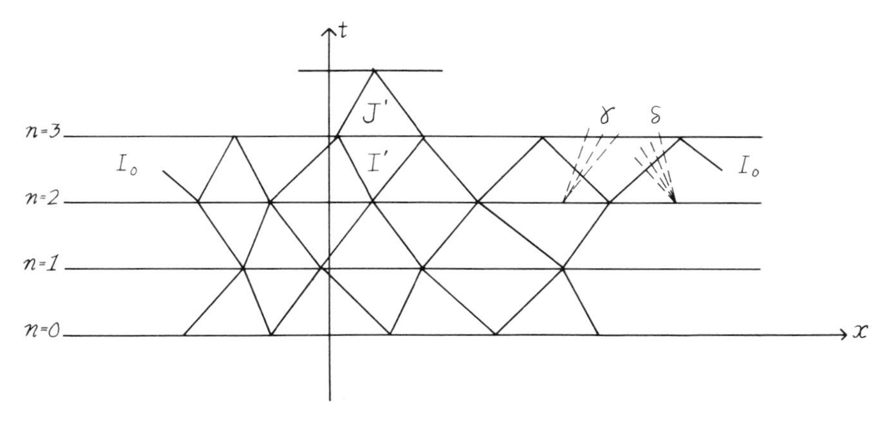

Fig. 13. The diamond mesh.

initial data and gives rise to a fixed shock line $x=0$ ($s=0$). The above procedure gives

$$\mathbf{u}_\theta^\Delta(x,t)=\begin{cases}1, & x<\Gamma_n\Delta t,\\ -1, & x>\Gamma_n\Delta t,\end{cases}$$

where $\Gamma_n=-r\sum_{i=1}^{n}\operatorname{sgn}(\theta_i)$, and the number $r=\Delta x/\Delta t$ is chosen larger than $M=\sup|u|=1$. This corresponds to shifted shock segments $x-\Gamma_n\Delta t=0$ in each horizontal strip $n\Delta t<t<(n+1)\Delta t$, and the probability that the shifted shock is to the left or to the right of the exact position $x=0$ is the same (1/2). Now Γ_n is the sum of n independent random variables $-r\operatorname{sgn}(\theta_i)$ having the same (uniform) distribution on $\{-r\}\cup\{r\}$, average value zero, and variance equal to r. It follows from the Central Limit Theorem of probability theory (Ref. 16) that $\Gamma_n=O(\sqrt{n})$. As $\Delta t=O(1/n)$, this implies that for fixed $t>0$, $\Gamma_n\Delta t=O(n^{-1/2})$. It then follows that $\mathbf{u}_\theta^\Delta\to\mathbf{u}$ as $\Delta x\to 0$, $n\to\infty$ for any $x\neq 0$ and any θ.

The proof of Glimm's theorem requires careful estimates of the dependence of the solution of Riemann's problem on intermediate states. The main technical effort in the proof is in establishing these "interaction estimates." The derivations are given now.

Recall that the solution of Riemann's problem is given in terms of $m+1$ constant states $\mathbf{u}_0=\mathbf{u}_l,\mathbf{u}_1,\ldots,\mathbf{u}_m=\mathbf{u}_r$ together with a "wave system" defined by an m-tuple of parameters $\varepsilon=(\varepsilon_1,\ldots,\varepsilon_m)$. The states $\mathbf{u}_{k-1}$ and $\mathbf{u}_k$ are connected by a k-shock wave or a k-centered wave; in the first case $\varepsilon_k<0$, in the second $\varepsilon_k>0$. Formally, we may think of $|\varepsilon|=\max_k|\varepsilon_k|$ as measuring the variation of the solution of Riemann's problem. A more precise result is given in the following.

Lemma 5.1. Suppose that $\mathbf{u}_j \in U$, $j = 1, \ldots, m$. Then we can choose U so that

$$K_1|\varepsilon_k| \leq |\mathbf{u}_k - \mathbf{u}_{k-1}| \leq K_2|\varepsilon_k|, \qquad k = 1, \ldots, m,$$

for positive constants K_1, K_2 depending only on U.

Proof. As $\mathbf{u}_k = \mathbf{u}_{k-1} + \varepsilon_k \boldsymbol{r}_k(\mathbf{u}_{k-1}) +$ second-order terms, we need only choose U so that the second-order terms are dominated by $\varepsilon_k \boldsymbol{r}_k(\mathbf{u}_{k-1})$ for all k.

In what follows we will always assume that U has been chosen in this way and we will use $|\varepsilon|$ as a measure of the variation of the solution of Riemann's problem. The latter will be denoted by

$$(\mathbf{u}_l, \mathbf{u}_r) = [\mathbf{u}_0, \ldots, \mathbf{u}_m;\ \varepsilon_1, \ldots, \varepsilon_m]$$

in order to exhibit the dependence on intermediate states. For brevity we shall speak of a "wave" $\mathbf{u}_k$ to denote the k-wave $(\mathbf{u}_{k-1}, \mathbf{u}_k)$. If $\mathbf{u}_l$, $\mathbf{u}_c$, $\mathbf{u}_r \in$ U, and

$$(\mathbf{u}_l, \mathbf{u}_c) = [\mathbf{u}'_0, \ldots, \mathbf{u}'_m;\ \gamma_1, \ldots, \gamma_m], \qquad (\mathbf{u}_c, \mathbf{u}_r) = [\mathbf{u}''_0, \ldots, \mathbf{u}''_m;\ \delta_1, \ldots, \delta_m],$$

we will say that "the left wave system γ and the right wave system δ interact to yield ε" and write

$$\gamma + \delta \to \varepsilon = \varepsilon(\gamma, \delta, \mathbf{u}_c).$$

The interaction estimates are concerned with the dependence of ε on γ and δ. In the case of a linear system we have seen that the dependence on the parameters is linear, hence $\varepsilon = \gamma + \delta$. In general, $\varepsilon(\gamma, \delta, \mathbf{u}_c)$ is a C^2 function of γ and δ (Exercise 5.1).

Proposition 5.1. If $|\gamma|$, $|\delta| \to 0$, then

$$\varepsilon_i = \gamma_i + \delta_i + C_i(\mathbf{u}_c) \sum_{k,j} |\gamma_k||\delta_j| + O(|\gamma| + |\delta|)^3 \qquad (i = 1, \ldots, m)$$

with

$$|C_i(\mathbf{u}_c)| \le m^2 \sup_{k,j}\left[\left|\boldsymbol{l}_i(\mathbf{u}_c)\cdot\left(\frac{\partial \boldsymbol{r}_j(\mathbf{u}_c)}{\partial \varepsilon_k} - \frac{\partial \boldsymbol{r}_k(\mathbf{u}_c)}{\partial \varepsilon_j}\right)\right|\right].$$

The proof is straightforward and is given in the Appendix.

We need to refine our knowledge of the dependence of these estimates on the waves $\mathbf{u}_j'$ (to the left) and $\mathbf{u}_k''$ (to the right).

Definition 5.1. The j-wave $\mathbf{u}_j'$ and the k-wave $\mathbf{u}_k''$ are *approaching* if either $j > k$ or $j = k$ and at least one of γ_k, δ_k is negative. We write $\mathbf{u}_j' \mathscr{A} \mathbf{u}_k''$, or $\gamma_j \mathscr{A} \delta_k$.

Equivalently, the "left" wave $\mathbf{u}_j'$ corresponds too a faster characteristic speed, or, in case the characteristic speeds are the same, at least one is a shock. It will be shown in Exercises 5.2–5.4 that approaching waves always interact.

The crucial fact is that the nonlinear terms in Proposition 5.1 arises only from approaching waves.

Proposition 5.2. As $|\gamma| + |\delta| \to 0$,

$$\varepsilon_i = \gamma_i + \delta_i + c_i(\gamma, \delta, \mathbf{u}_c)D(\gamma, \delta)$$

($i = 1, \ldots, m$), where

$$D(\gamma, \delta) = \sum_{\gamma_j \mathscr{A} \delta_k} |\gamma_j||\delta_k|$$

and $|c_i(\gamma, \delta, \mathbf{u}_c)| \le c|C_i(\mathbf{u}_c)|$, $c > 0$ a fixed constant (depending on U).

The proof is long and technical and is given in the Appendix.

Proposition 5.3. If two solutions $\varepsilon = \varepsilon(\gamma, \delta, \mathbf{u}_c)$ and $\varepsilon' = \varepsilon(\gamma', \delta', \mathbf{u}_c')$ are replaced with intermediate transitions, as indicated, the number of approaching waves, $\varepsilon_j \mathscr{A} \varepsilon_k'$, does not decrease for $|\gamma|$, $|\delta|$, $|\gamma'|$, $|\delta'|$ small.

Proof. As

$$\varepsilon_i = \gamma_i + \delta_i + c_i(\gamma, \delta, \mathbf{u}_c)D(\gamma, \delta), \qquad \varepsilon_i' = \gamma_i' + \delta_i' + c_i(\gamma', \delta', \mathbf{u}_c')D(\gamma', \delta'),$$

if $\varepsilon_j \mathscr{A} \varepsilon'_k$ for $j = k$, at least one of $\gamma_j \mathscr{A} \delta'_k$, $\delta_j \mathscr{A} \delta'_k$, $\gamma_j \mathscr{A} \gamma'_k$, $\delta_j \mathscr{A} \gamma'_k$ occurs and the number of approaching waves does not decrease. If $j > k$, then all of them occur and the number of approaching waves increases. □

We are now ready to give a more detailed description of the discrete scheme and proceed to the main part of the proof of Glimm's theorem. Let $Y = \{(m, n) \in \mathbb{Z}^2: m + n \text{ odd}, \ n \geq 0\}$ and

$$A = \prod_{m,n \in Y} R_m \times \{n\Delta t\}, \qquad R_m := [(m-1)\Delta x, (m+1)\Delta x].$$

Every factor in A is a horizontal segment of length $2\Delta x$ in the (x, t)-plane, to which we give the measure 1; A is given the product measure dA. In each R_m we choose a point (a mesh node) $a_{m,n} = m\Delta x + \theta_n \Delta x$, where θ_n is a random number uniformly distributed in $[-1, 1]$ (Fig. 14).

The measure space A depends on Δx, but is isomorphic, through $\theta = \{\theta_n: n = 1, 2, \ldots, \}$ to a denumerable infinity of copies of the interval $[-1, 1]$, a fixed probability space $\Phi = \bigotimes_{n=1}^{\infty} [-1, 1]$, independent of Δx, endowed with the product measure $d\theta = \prod_{n=1}^{\infty} d\theta_n$, $d\theta_n = \frac{1}{2}\, dz$ (dz the Lebesgue measure on $[-1, 1]$), so that $\int_\Phi d\theta = 1$.

The discrete scheme is defined inductively as follows. Suppose $\mathbf{u}^\Delta = \mathbf{u}_\theta^\Delta(x, t)$ is known at the points $a_{m-1,n-1}$ and $a_{m+1,n-1}$ (see the figure). In order to define $\mathbf{u}_\theta^\Delta$ in $a_{m,n}$ we solve the Riemann problem

$$\mathbf{v}_t + \mathbf{f}(\mathbf{v})_x = 0, \qquad x \in R_m, \qquad (n-1)\Delta t \leq t < n\Delta t,$$

with initial data $\mathbf{u}_l$, $\mathbf{u}_r$ defined according to the rule

$$\mathbf{v}(x, (n-1)\Delta t) = \begin{cases} \mathbf{u}_\theta^\Delta(a_{m-1,n-1}), & (m-1)\Delta x \leq x < m\Delta x, \\ \mathbf{u}_\theta^\Delta(a_{m+1,n-1}), & m\Delta x < x \leq (m+1)\Delta x, \end{cases}$$

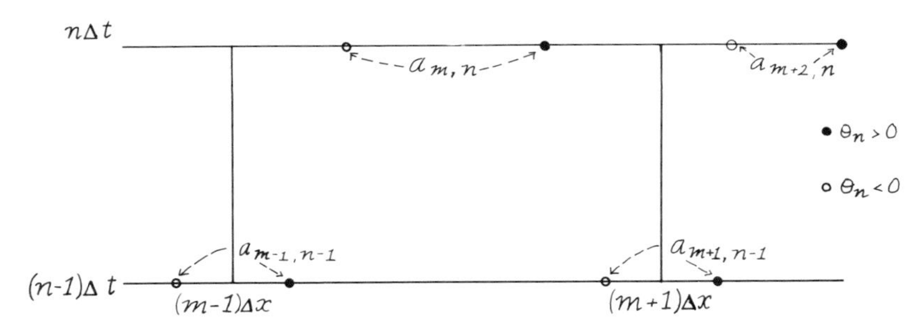

Fig. 14. The random choice.

where $m+n$ is odd, and we set

$$\mathbf{u}^{\Delta}_{\theta}(a_{m,n}) = \mathbf{v}(a_{m,n}).$$

In much of what follows, dropping the subscript θ in our notation will cause no confusion and we will suppress it.

The above Riemann problem is locally solvable. We assume that the values of $\mathbf{u}^0(x)$ lie in a fixed neighborhood V of a constant state $\hat{\mathbf{u}}$, sufficiently small that intermediate states lie in the neighborhood U of Lemma 5.1, which in turn is restricted to be small enough for solvability of Riemann's problem with states in U. [There will be a further restriction on the variation of $\mathbf{u}^0(x)$ in the proof.] We use the neighborhood U to define the CFL condition.

We extend the definition of $\mathbf{u}^{\Delta}$ by defining it to be $\mathbf{v}$ in the rectangle $\bar{R}_m \times [(n-1)\Delta t, n\Delta t)$, and to be the constant value $\mathbf{u}^{\Delta}(a_{m,n})$ for $(m-1)\Delta x < x < (m+1)\Delta x$, $t = n\Delta t$. The CFL condition prevents the shocks and rarefaction waves from entering vertical strips around $x = (m \pm 1)\Delta x$, $(n-1)\Delta t < t \leq n\Delta t (m+n$ odd), where $\mathbf{u}^{\Delta}$ is necessarily constant. The approximate solution $\mathbf{u}^{\Delta}$ of the Cauchy problem is thus composed of "waves" in each cell $R_m \times [(n-1)\Delta t, n\Delta t)$ and has jump discontinuities across the horizontal lines $t = n\Delta t$.

The "mesh lines" I will play a crucial role in what follows. These are defined as being composed of segments joining "from left to right" a set of nodes $a_{m,n}$ for increasing $m \in \mathbb{Z}$, and clearly depend on θ. Let $\hat{O}$ be the (unique) mesh line crossing all nodes on $t=0$ and $t=\Delta t$ once θ_1 is fixed. We require estimates on certain functionals $F(I) = F(\mathbf{u}^{\Delta}|_I)$. The values of the restriction $\mathbf{u}^{\Delta}|_I$ are given by the shocks and rarefaction curves that cross I (Fig. 15).

If α_j is the parameter of a j-wave contained in $\mathbf{u}^{\Delta}|_I$, then

$$L(I) := \sum |\alpha_j|,$$

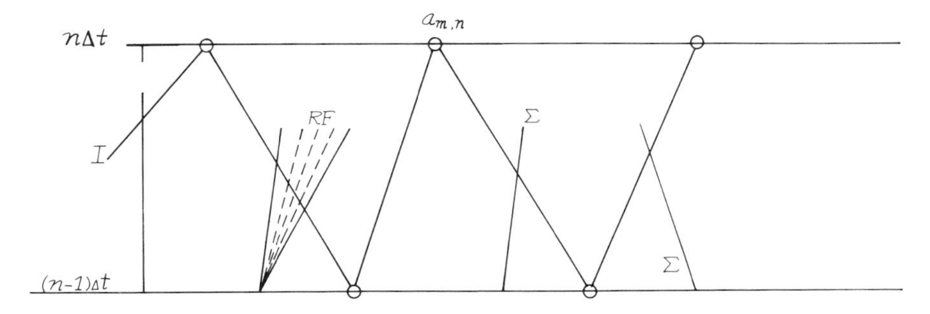

Fig. 15. Shocks and rarefaction fans crossing I.

where the summation extends over all waves that cross I. If β_k is the parameter of a k-wave belonging to a different system,

$$Q(I) := \sum |\alpha_j||\beta_k|,$$

where the summation extends over all α_j, β_k crossing I and such that $\alpha_j \mathscr{A} \beta_k$. By Lemma 5.1, as long as the values of $\mathbf{u}^\Delta$ remain in U, $L(I)$ is a measure of the total variation of $\mathbf{u}^\Delta|_I$. Therefore, $L(\hat{O})$ measures the total variation of $\mathbf{u}_0^\Delta$, the initial piecewise constant approximation of $\mathbf{u}^0(x)$.

Suppose that J is a mesh line obtained from I by replacing one node $a_{m,n}$ by the point $a_{m,n+2}$ so that they differ by a single diamond $ABCD$, where $A = a_{m-1,n+1}$, $B = a_{m,n}$, $C = a_{m+1,n+1}$, $D = a_{m,n+2}$. Let I' be the part of I connecting A, B, C, and J' the part of J connecting A, D, and C. Let $I_0 = I \backslash I'$ so that $I = I_0 \cup I'$ and $J = I_0 \cup J'$(Fig. 16).

Proposition 5.4. If $L(I)$ is sufficiently small, then

i. $Q(I) \geq Q(J)$,
ii. There is a constant k such that $L(I) + kQ(I) \geq L(J) + kQ(J)$.

Proof. Let $\mathbf{u}_l = \mathbf{u}^\Delta(A)$, $\mathbf{u}_c = \mathbf{u}^\Delta(B)$, and $\mathbf{u}_r = \mathbf{u}^\Delta(C)$. Then, if $\mathbf{u}_c' = \mathbf{u}^\Delta(a_{m-2,n})$, $\mathbf{u}_c'' = \mathbf{u}^\Delta(a_{m+2,n})$, $\mathbf{u}_l$ and $\mathbf{u}_r$ depend only on $\mathbf{u}_c$, $\mathbf{u}_c'$, $\mathbf{u}_c''$ through the solution of two Riemann problems at the previous time step. One can pass from $\mathbf{u}_l$ to $\mathbf{u}_r$ either through the intermediate state $\mathbf{u}_c$ via waves, with parameters γ and δ, crossing I, or directly across J' via waves with parameter ε. The γ waves cross I' to the left of $B = a_{m,n}$ and the δ waves cross I' to the right of B. (See Fig. 17.)

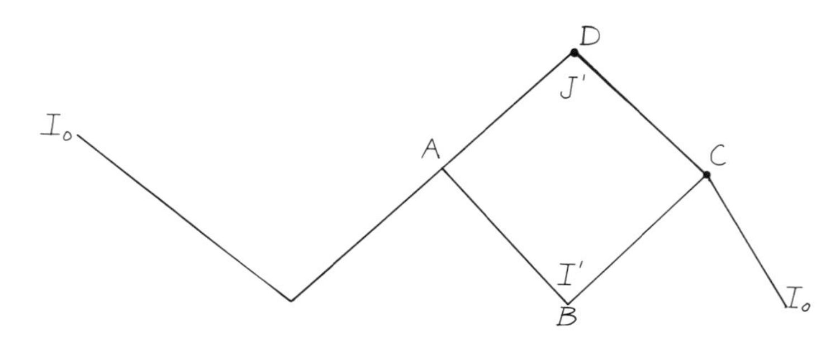

Fig. 16. The modified mesh line J.

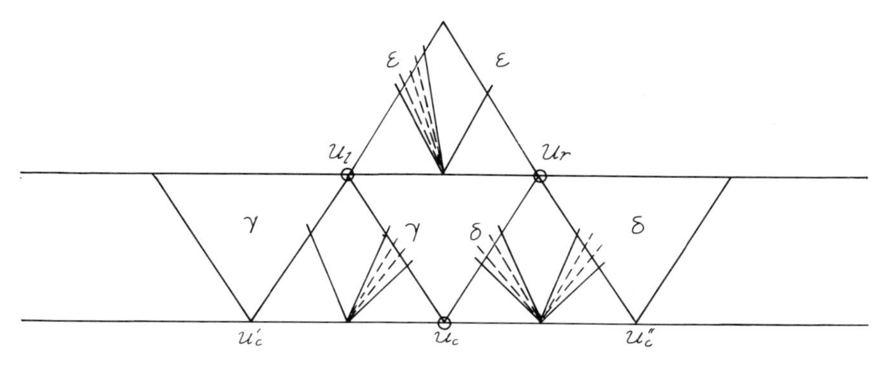

Fig. 17. γ and δ waves crossing I'.

Then

$$L(I) = L(I_0) + L(I') = L(I_0) + \sum_{I'} |\gamma_i| + \sum_{I'} |\delta_i|$$

and

$$L(J) = L(I_0) + L(J') = L(I_0) + \sum_{J'} |\varepsilon_i|.$$

As Proposition 5.2 implies

$$|\varepsilon_i| \leq |\gamma_i| + |\delta_i| + c|C_i(\mathbf{u}_c)|D(\gamma, \delta),$$

we can deduce

$$L(J) \leq L(I) + k_0 Q(I'), \tag{83}$$

where $Q(I') = D(\gamma, \delta)$ and $k_0 = c \max_i |C_i(\mathbf{u}_c)|$. Also,

$$Q(I) = Q(I_0) + Q(I') + Q(I_0, I')$$

and

$$Q(J) = Q(I_0) + Q(I_0, J'),$$

where

$$Q(I_0, I') = \sum |\alpha_j||\beta_k|$$

for $\alpha_j \mathscr{A} \beta_k$, α_j crossing I_0, β_k crossing I', and

$$Q(I_0, J') = \sum |\alpha_j||\varepsilon_i|$$

for $\alpha_j \mathscr{A} \varepsilon_i$, α_j crossing I_0, ε_i crossing J'. Proposition 5.2 implies

$$Q(I_0, J') \leq \sum_{\alpha \mathscr{A} \varepsilon} |\alpha_j||\gamma_i + \delta_i| + k_0 L(I_0) Q(I').$$

Proposition 5.3 shows that the first sum is not decreased if the sum is extended over $\gamma \mathscr{A} \alpha$ or $\delta \mathscr{A} \alpha$ with α crossing I_0, γ and δ crossing I'. Then

$$Q(I_0, J') \leq Q(I_0, I') + k_0 Q(I') L(I_0),$$

so, if

$$L(I_0) \leq 1/2k_0,$$

then

$$Q(I_0, J') \leq Q(I_0, I') + \tfrac{1}{2} Q(I')$$

and

$$Q(J) - Q(I) = Q(I_0, J') - Q(I_0, I') - Q(I') \leq -\tfrac{1}{2} Q(I'),$$

and the first statement is proven. Now (83) implies

$$L(J) + kQ(J) \leq L(I) + k_0 Q(I') + kQ(I) - \frac{k}{2} Q(I') \leq L(I) + kQ(I),$$

provided $k \geq 2k_0$. The proof of the proposition is completed. □

We will apply this result recursively to obtain an estimate for $L(I_n)$, where I_n is the mesh line joining all nodes at the levels $t = n\Delta t$ and $t = (n+1)\Delta t$.

First, we observe that if intermediate states lie in U, we can give a uniform value $k_0 = k_0(\mathrm{U})$ in the argument of Proposition 5.4. Then the argument can be applied one mesh point at a time to show finally that, for $k \geq 2k_0$,

$$L(I_1) + kQ(I_1) \leq L(\hat{O}) + kQ(\hat{O}).$$

Lemma 5.2. There is a $\delta > 0$ such that $L(I_n) > \delta$ implies $\mathbf{u}^{\Delta}(x, n\Delta t) \in \mathrm{V}$.

Proof. We need only show that $L(I_n)$ is equivalent to $TV(\mathbf{u}^\Delta(\cdot, n\Delta t), \mathbb{R})$ as a (semi)norm on the grid functions $\mathbf{u}^\Delta$. As in Lemma 5.1 we can find positive constants K_1, K_2, depending only on U, so that

$$K_1|\varepsilon_k| \leq |\mathbf{u}_k - \mathbf{u}_{k-1}| \leq K_2|\varepsilon_k|, \qquad k = 1, \ldots, m,$$

in the k-wave solution of the Riemann problem at the time level $(n-1)\Delta t$. This proves the lemma. □

Now we choose

$$k = \max\{4k_0, 2/\delta\}.$$

Then $L(\hat{O}) < 1/k$ implies that

$$L(I_1) \leq L(\hat{O}) + kQ(\hat{O}) \leq L(\hat{O}) + k[L(\hat{O})]^2 < 2/k = \min\{1/2k_0, \delta\}.$$

Lemma 5.2 then implies that $\mathbf{u}^\Delta(x, \Delta t) \in$ V, the solution can be continued, and

$$L(I_2) + kQ(I_2) \leq L(I_1) + kQ(I_1) \leq L(\hat{O}) + kQ(\hat{O}),$$

hence $L(I_2) < 2/k = \min\{1/2k_0, \delta\}$. The process can be continued to arbitrary n.

We observe for use in the next result that the oscillation of $\mathbf{u}^\Delta$ at any time level $n\Delta t$ is bounded above by $TV(\mathbf{u}^\Delta(\cdot, n\Delta t), \mathbb{R})$. In particular, as $\lim_{x\to\infty} \mathbf{u}^\Delta(x, t) = \hat{\mathbf{u}}$ for each t,

$$|\mathbf{u}^\Delta(x, t) - \hat{\mathbf{u}}| \leq TV(\mathbf{u}^\Delta(\cdot, t), \mathbb{R}).$$

Proposition 5.5. If $TV(\mathbf{u}^0)$ is sufficiently small, the inequality

$$\int_{\mathbb{R}} |\mathbf{u}^\Delta(x, t) - \mathbf{u}^\Delta(x, t')|dx \leq CTV(\mathbf{u}^0)|t - t'|$$

holds for some positive constant C depending only on U.

Proof. Fix x and t, choose $t' > t + \Delta t$, and let

$$\tilde{\mathbf{u}}(y, t) = \begin{cases} \mathbf{u}^\Delta(y, t), & |y - x| < |t - t'|\Delta x/\Delta t, \\ 0, & \text{otherwise.} \end{cases}$$

If we take $\tilde{\mathbf{u}}$ as initial data for our scheme at $t=0$, the CFL condition implies $\tilde{\mathbf{u}} = \mathbf{u}^\Delta$ in the domain of dependence of (x, t'), so

$$|\mathbf{u}^\Delta(x, t) - \mathbf{u}^\Delta(x, t')| = |\tilde{\mathbf{u}}(x, t) - \tilde{\mathbf{u}}(x, t')| \leq |\tilde{\mathbf{u}}(x, t') - \tilde{\mathbf{u}}_\infty| + |\tilde{\mathbf{u}}(x, t) - \tilde{\mathbf{u}}_\infty|$$
$$\leq TV(\tilde{\mathbf{u}}(\cdot, t'), \mathbb{R}) + TV(\tilde{\mathbf{u}}(\cdot, t), \mathbb{R})$$

($\tilde{\mathbf{u}}_\infty = \hat{\mathbf{u}}$). From the fact that the total variation is constant between the time levels $n\Delta t$ and $(n+1)\Delta t$, the equivalence of L and TV as observed in the proof of Lemma 5.2, and the inequality

$$L(I_n) \leq L(I_m) + kQ(I_m) \leq L(I_m) + k(L(I_m))^2 \leq L(I_m) + \frac{2k}{k} L(I_m) = 3L(I_m) \quad (84)$$

$(m < n)$, we deduce that $TV(\tilde{\mathbf{u}}(\cdot, t'), \mathbb{R}) \leq c(\mathrm{U}) TV(\tilde{\mathbf{u}}(\cdot, t), \mathbb{R})$, and

$$|\mathbf{u}^\Delta(x, t) - \mathbf{u}^\Delta(x, t')| \leq c(\mathrm{U}) TV(\tilde{\mathbf{u}}(\cdot, t), \mathbb{R}) = c(\mathrm{U}) TV(\mathbf{u}^\Delta(\cdot, t), X)$$
$$= c(\mathrm{U}) \int_X |d\mathbf{u}^\Delta(\xi, t)|,$$

where by $c(\mathrm{U})$ we denote a set of positive constants depending only on U, and $X = \{y: |y - x| \leq |t - t'| \Delta x / \Delta t\}$ as indicated in Fig. 18.
Integrating, we obtain

$$\int_{\mathbb{R}} |\mathbf{u}^\Delta(x, t) - \mathbf{u}^\Delta(x, t')| dx \leq c(\mathrm{U}) \int_{\mathbb{R}} dx \int_X |d\mathbf{u}^\Delta(\xi, t)|$$
$$= c(\mathrm{U}) \int_{\mathbb{R}} |d\mathbf{u}^\Delta(\xi, t)| \int_{a_-}^{a_+} dx = c(\mathrm{U}) |t' - t| TV(\mathbf{u}^\Delta(\cdot, t), \mathbb{R}),$$

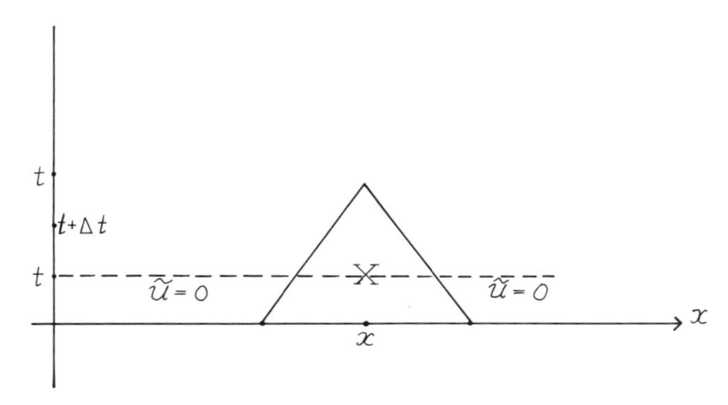

Fig. 18. The domain of integration X.

where $a_- = \xi - K|t' - t|$, $a_+ = \xi + K|t' - t|$. As the argument used in showing (84) implies $TV(\mathbf{u}^\Delta(\cdot, t), \mathbb{R}) \le c' TV(\mathbf{u}_0^\Delta)$ for some constant $c' < 0$ depending only on U, and by construction $TV(\mathbf{u}_0^\Delta) \le TV(\mathbf{u}^0)$, we have

$$TV(\mathbf{u}^\Delta(\cdot, t), \mathbb{R}) \le c'(\mathrm{U}) TV(\mathbf{u}^0, \mathbb{R}) \qquad \forall t > 0, \tag{85}$$

and the result follows. □

Now we set $\Delta = \Delta x$, and observe that we have shown the approximate family $\{\mathbf{u}^\Delta\} \equiv \{\mathbf{u}_\theta^\Delta\}$ satisfies, for $\Delta > 0$ and any fixed $\theta \in \Phi$,

$$\|\mathbf{u}^\Delta\|_{L^\infty} \le M_1, \tag{H1}$$

$$TV(\mathbf{u}^\Delta(\cdot, t), \mathbb{R}) \le M_2, \tag{H2}$$

$$\|\mathbf{u}^\Delta(\cdot, t) - \mathbf{u}^\Delta(\cdot, t')\|_{L^1(\mathbb{R})} \le M_3 |t - t'|. \tag{H3}$$

for every $t, t' \ge 0$, with constants $M_i \ge 0$ depending only on U, but not on Δ, θ, t, t'. We will show that $\{\mathbf{u}^\Delta\}$ is (pre)-compact in L^1_{loc}.

Proposition 5.6. There exists a subsequence $\{\mathbf{u}_\theta^{\Delta_i}\} \subset \{\mathbf{u}_\theta^\Delta\}$ convergent in $L^1_{\mathrm{loc}}(\mathbb{R})$ and in $L^1_{\mathrm{loc}}(\mathbb{R} \times [0, \infty))$ to a function $\mathbf{u} = \mathbf{u}_\theta$ as $\Delta_i \to 0$, with $\|\mathbf{u}\|_{L^\infty} \le M_1$.

Proof. From (H1) and (H2) the set of functions $\{\mathbf{u}^\Delta\}$, considered as functions of x, are uniformly bounded and have uniformly bounded total variation on each bounded interval on any line $t = \mathrm{const.} \ge 0$. By Helly's theorem (bounded sequences of functions of equibounded variation are precompact) a sequence $\{\mathbf{u}^{\Delta_i'}\}$ converges on any bounded interval of this line, and by a standard diagonal process a subsequence $\{\mathbf{u}^{\Delta_i''}\}$ converges at each point of this line. Let $\{t_m\}$ be a countable dense subset of the interval $[0, T]$ $(T > 0)$. By a further diagonal process, we can select a subsequence $\{\mathbf{u}^{\Delta_i}\} \subset \{\mathbf{u}^{\Delta_i''}\}$ that converges at every point of the lines $t = t_m$, $m = 1, 2, \dots,$ as $\Delta_i \to 0$. For fixed $\varepsilon > 0$ we can choose a finite set $\{\tau_s\} \subset \{t_m\}$ with the property that if $t \in [O, T]$, there is a τ_s such that $2C|t - \tau_s| < \varepsilon/2$, where C is the constant in Proposition 5.5. For each compact $K \subset \mathbb{R}$ let

$$I(j, t|k, t') := \int_K |\mathbf{u}^{\Delta_j}(x, t) - \mathbf{u}^{\Delta_k}(x, t')| dx.$$

Then

$$|I(j, t|k, t)| \le |I(j, t|j, \tau_s)| + |I(j, \tau_s|k, \tau_s)| + |I(k, \tau_s|k, t)| := I_1 + I_2 + I_3.$$

In view of the choice of τ_s and the Lebesgue bounded convergence theorem we can choose j, k so large that $I_2 < \varepsilon/2$ for all τ_s, and using Proposition 5.5 we can choose τ_s so that $I_1 + I_3 \leq 2C|t - \tau_s| < \varepsilon/2$. This shows that $\{\mathbf{u}^{\Delta_i}\}$ is a Cauchy sequence in $L^1_{\text{loc}}(\mathbb{R})$, and hence there exists a function $\mathbf{u}(x, t) \in L^1_{\text{loc}}(\mathbb{R})$ such that

$$\int_K |\mathbf{u}^{\Delta_i}(x, t) - \mathbf{u}(x, t)|dx \to 0 \qquad \Delta_i \to 0$$

for every fixed $t \in [O, T]$. Moreover, the convergence $|I(j, t|k, t)| \to 0$ as j, $k \to \infty$ is uniform in t for $0 \leq t \leq T$, so that also

$$\int_0^T dt \int_K |\mathbf{u}^{\Delta_j}(x, t) - \mathbf{u}^{\Delta_k}(x, t)|dx \to 0$$

as j, $k \to \infty$. Hence,

$$\int_0^T dt \int_K |\mathbf{u}^{\Delta_i}(x, t) - \mathbf{u}(x, t)|dx \to 0 \qquad \Delta_i \to 0$$

for every $T > 0$ and every compact K. Finally, as local convergence in L^1 implies pointwise convergence a.e. of a subsequence, and $\|\mathbf{u}^{\Delta_i}\|_{L^\infty} \leq M_1$, the limit function $\mathbf{u}$ (possibly redefined on a set of measure zero) satisfies the same inequality. □

In particular, Proposition 5.6 will be applied below to a countable family $\{\mathbf{u}_\theta^\Delta\}$ with $\{\Delta\}$ a subsequence of $\{2^{-i}\}$.

Corollary 5.1. If $\mathbf{f}(\mathbf{u})$ is continuous, there exists a subsequence of $\{\mathbf{u}^{\Delta_i}\}$ denoted again by $\{\mathbf{u}^{\Delta_i}\}$ such that $\mathbf{f}(\mathbf{u}^{\Delta_i}) \to \mathbf{f}(\mathbf{u})$ in $L^1_{\text{loc}}(\mathbb{R} \times [0, \infty))$ as $\Delta_i \to 0$.

Proof. By Proposition 5.6 we can find a subsequence such that $\mathbf{u}^{\Delta_i} \to \mathbf{u}$ boundedly a.e. Then $\mathbf{f}(\mathbf{u}^{\Delta_i}) \to \mathbf{f}(\mathbf{u})$ boundedly a.e. and hence also in L^1_{loc}. □

Let $\psi \in C_0^1(\mathbb{R} \times [0, \infty))$ be a (vector valued) test function, and let us define the functional

$$\mathscr{L}_\psi(\mathbf{u}, \mathbf{f}(\mathbf{u})) := \int_0^T dt \int_{\mathbb{R}} (\mathbf{u} \cdot \psi_t + \mathbf{f}(\mathbf{u}) \cdot \psi_x)dx + \int_{\mathbb{R}} \psi(x, 0) \cdot \mathbf{u}^0(x)dx.$$

We know that $\mathbf{u}^\Delta$ is a weak solution in every horizontal strip $\mathbb{R} \times [n\Delta t, (n+1)\Delta t]$ so that, for every such ψ,

$$\int_{n\Delta t}^{(n+1)\Delta t} dt \int_{\mathbb{R}} (\mathbf{u}^\Delta \cdot \psi_t + \mathbf{f}(\mathbf{u}^\Delta) \cdot \psi_x) dx + \int_{\mathbb{R}} \psi(x, n\Delta t) \cdot \mathbf{u}^\Delta(x, n\Delta t + 0) dx$$
$$- \int_{\mathbb{R}} \psi(x, (n+1)\Delta t) \cdot \mathbf{u}^\Delta(x, (n+1)\Delta t - 0) dx = 0.$$

Summing over n we obtain

$$\begin{aligned} \mathscr{L}_\psi(\mathbf{u}^\Delta, \mathbf{f}(\mathbf{u}^\Delta)) &\equiv \int_0^\infty dt \int_{\mathbb{R}} (\mathbf{u}^\Delta \cdot \psi_t + \mathbf{f}(\mathbf{u}^\Delta) \cdot \psi_x) dx + \int_{\mathbb{R}} \psi(x, 0) \cdot \mathbf{u}^\Delta(x, 0) dx \\ &= \sum_{n=1}^{\infty} \int_{\mathbb{R}} \psi(x, n\Delta t) \cdot [\mathbf{u}^\Delta]_n \, dx, \end{aligned} \tag{86}$$

where $[\mathbf{u}^\Delta]_n$ denotes the jump of $\mathbf{u}^\Delta$ across the horizontal lines $t = n\Delta t$:

$$\begin{aligned} [\mathbf{u}^\Delta]_n &:= \mathbf{u}^\Delta(x, n\Delta t + 0) - \mathbf{u}^\Delta(x, n\Delta t - 0) \\ &= \mathbf{u}^\Delta((m + \theta_n)\Delta x, n\Delta t - 0) - \mathbf{u}^\Delta(x, n\Delta t - 0), \end{aligned}$$

and the series in (86) is in reality a finite sum due to ψ having compact support. From (86) we immediately have

Corollary 5.2. Let $\theta \in \Phi$ be fixed, then $\mathbf{u} = \lim_{\Delta_i \to 0} \mathbf{u}^{\Delta_i}$ is a weak solution of (81), (82) provided

$$\begin{aligned} &\mathbf{u}^{\Delta_i}(\cdot, 0) \to \mathbf{u}^0 \qquad \text{weakly}, \\ &\sum_{n=1}^{\infty} [\mathbf{u}^\Delta]_n \to 0 \qquad \text{weakly} \end{aligned}$$

as $\Delta_i \to 0$ by a subsequence as in Corollary 5.1.

That the first limit relation holds (indeed, in the sense of the convergence in L^1_{loc}) follows from (H1), the definition of $\mathbf{u}_0^\Delta$, and the fact that $\mathbf{u}^0(x)$ is continuous for a.e. $x \in \mathbb{R}$. It remains to prove the second, namely, that the sum of the jumps of the approximate solution across any finite number of horizontal lines $t = n\Delta t$ tends weakly to zero as the mesh step vanishes.

To this effect we fix a test function ψ, a sequence of random numbers $\theta \in \Phi$, the mesh step $\Delta > 0$, and let

$$F_n(\theta, \Delta, \psi) := \int_{\mathbb{R}} \psi(x, n\Delta t) \cdot [\mathbf{u}^{\Delta}]_n \, dx, \qquad F(\theta, \Delta, \psi) := \sum_{n=1}^{\infty} F_n(\theta, \Delta, \psi).$$

Note that F_n depends only on θ_i for $i \leq n$, and is linear in ψ.

Proposition 5.7. Let $\psi \in L_0^{\infty}(\mathbb{R} \times [0, \infty))$. Then there exist positive constants M, M' independent of θ, Δ, ψ such that

$$|F_n(\theta, \Delta, \psi)| \leq M\Delta \|\psi\|_{L^\infty}, \tag{87}$$

$$|F(\theta, \Delta, \psi)| \leq M' d \|\psi\|_{L^\infty}, \tag{88}$$

where d is the diameter of the support of $\psi = (\psi_1, \ldots, \psi_m)$, $d = \sup_i$ [diam supp ψ_i].

Proof. The second inequality follows from the first as, letting supp $\psi \subset [a, b] \times [0, T]$, F_n is different from zero only for $n < T/\Delta t \leq d/\Delta t$. Thus, by the CFL condition, there are at most $d/\Delta t = O(d/\Delta)$ nonzero terms in the series for F. It remains to prove (87). Using the fact that

$$\int_a^b |g(x) - g(x')| dx \leq (b - a)\, \mathrm{osc}(g) \leq (b - a) TV(g, \mathbb{R}),$$

we have

$$\begin{aligned} F_n(\theta, \Delta, \psi)| &\leq \sum_{\substack{m \in \mathbb{Z} \\ m \text{ odd}}} \int_{\mathbb{R}_m} \psi(x, n\Delta t) \cdot [\mathbf{u}^{\Delta}]_n \, dx \\ &\leq \|\psi\|_{L^\infty} \sum_{\substack{m \in \mathbb{Z} \\ m \text{ odd}}} \int_{\mathbb{R}_m} |\mathbf{u}^{\Delta}((m + \theta_n)\Delta x, n\Delta t - 0) - \mathbf{u}^{\Delta}(x, n\Delta t - 0)| dx \\ &\leq \|\psi\|_{L^\infty} c\Delta \sum_{\substack{m \in \mathbb{Z} \\ m \text{ odd}}} TV(\mathbf{u}^{\Delta}(\cdot, n\Delta t - 0), R_m) \leq c\Delta \|\psi\|_{L^\infty} TV(\mathbf{u}^0, \mathbb{R}), \end{aligned}$$

with constants $c > 0$ depending only on $m = \dim(\mathbf{u})$ and U [we have here used inequality (85)]. □

Proposition 5.8. Let ψ have compact support and be piecewise constant, i.e., constant on horizontal segments $R_m \times \{n\Delta t\}$. Then if $n \neq n'$, $F_n(\cdot, \Delta, \psi)$ is orthogonal to $F_{n'}(\cdot, \Delta, \psi)$ in the sense of the scalar product $\langle \cdot, \cdot \rangle$ in Φ.

Proof. The idea is that independent random variables with mean value zero are orthogonal. Suppose $n < n'$. Then F_n depends on θ_i for $i \leq n$ but not on $\theta_{n'}$, so that

$$\langle F_n, F_{n'} \rangle \equiv \int_\Phi F_n F_{n'} \, d\theta = \int F_n F_{n'} \prod_{i=1}^{n'} d\theta_i = \int F_n \left[\int F_{n'} \, d\theta_{n'} \right] \prod_{i<n'} d\theta_i.$$

As ψ is piecewise constant, letting $t' := n'\Delta t - 0$ we have

$$\begin{aligned} \int F_{n'} \, d\theta_{n'} &= \tfrac{1}{2} \int_{-1}^{1} dz \sum_{\substack{m \in \mathbb{Z} \\ m \text{ odd}}} \psi_m \cdot \int_{R_m} [\mathbf{u}_\theta^\Delta((m+z)\Delta, t') - \mathbf{u}_\theta^\Delta(x, t')] dx \\ &= \tfrac{1}{2} \sum_{\substack{m \in \mathbb{Z} \\ m \text{ odd}}} \psi_m \cdot \left\{ 2\Delta \int_{-1}^{1} \mathbf{u}_\theta^\Delta((m+z)\Delta, t') dz - 2 \int_{R_m} \mathbf{u}_\theta^\Delta(x, t')| dx \right\}, \end{aligned}$$

where z represents the variable $\theta_{n'}$ and the dependence of $\mathbf{u}_\theta^\Delta$ on the remaining θ_i does not influence the integral. By the change of variables $y = (m+z)\Delta$ we finally find $\int F_{n'} \, d\theta_{n'} = 0$. □

We consider from now on only Δ of the form $\bar{\Delta}_i = 2^{-i} (i = 1, 2, \ldots,)$ so that if ψ satisfies the assumptions of Proposition 5.8 for some Δ, it satisfies the assumptions for all smaller Δ.

Proposition 5.9. There exists a null set of $\mathcal{N} \subset \Phi$ and a sequence $\Delta_i \to 0$ such that for every $\theta \in \Phi \backslash \mathcal{N}$ and every $\psi \in C_0^1(\mathbb{R} \times [0, \infty))$, $F(\theta, \Delta_i, \psi) \to 0$ as $i \to \infty$.

Proof. Suppose first that ψ satisfies the assumptions of Proposition 5.8. Then by the orthogonality of $\{F_n\}$ we find

$$\|F(\cdot, \bar{\Delta}_i, \psi)\|_{L^2(\Phi)}^2 \equiv \int_\Phi |F|^2 \, d\theta = \int_\Phi \left| \sum_n F_n \right|^2 d\theta = \sum_n \|F_n\|_{L^2(\Phi)}^2 \leq \sum_n \|F_n\|_{L^\infty(\Phi)}^2$$

as Φ is a probability space. Inequality (87) yields

$$\sum_n \|F_n(\cdot, \bar{\Delta}_i, \psi)\|^2_{L^\infty(\Phi)} \leq \sum_n M^2 \|\psi\|^2_{L^\infty} \bar{\Delta}_i^2 \leq M^2 \|\psi\|^2_{L^\infty} \bar{\Delta}_i \bar{n}$$

as the sum is extended to a finite number $\bar{n}$ of terms (ψ has support with finite diameter d). Now $\bar{n} = O(d/\bar{\Delta}_i)$, hence

$$\|F(\cdot, \bar{\Delta}_i, \psi)\|^2_{L^2(\Phi)} \leq M^2 \|\psi\|^2_{L^\infty} d\bar{\Delta}_i \to 0 \qquad i \to \infty$$

(remember that $\bar{\Delta}_i = 2^{-i}$). For every piecewise constant ψ with compact support, $F(\theta, \bar{\Delta}_i, \psi) \to 0$ in the norm of $L^2(\Phi)$, hence there exists a subsequence $\Delta_{i'} \to 0$ such that $F(\theta, \Delta_{i'}, \psi) \to 0$ for a.e. $\theta \in \Phi$. On the other hand, (88) shows that, for every $\psi \in L_0^\infty$,

$$\|F(\cdot, \Delta, \psi)\|_{L^2(\Phi)} \leq \|F(\cdot, \Delta, \psi)\|_{L^\infty(\Phi)} \leq c\|\psi\|_{L^\infty(\Phi)}, \tag{89}$$

where $c = c(\mathrm{U}, \psi) > 0$. Let $\{\psi_\nu\}$ be a sequence of piecewise constant functions with compact support that uniformly approximate functions in C_0^1 (Exercise 5.5). By what was said above, for every ψ_ν there is a null set $\mathcal{N}_\nu$ and a sequence $\Delta_i^{(\nu)}$ such that

$$F(\theta, \Delta_{i'}^{(\nu)}, \psi_\nu) \to 0 \qquad \text{as} \quad i' \to \infty \qquad \forall \theta \in \Phi \backslash \mathcal{N}_\nu,$$

and we can choose $\Delta_{i'}^{(\nu)}$ to be a subsequence of $\Delta_{i'}^{(\nu')}$ for $\nu' < \nu$. Then, by the standard diagonal process, we can find a subsequence, call it Δ_i, such that

$$F(\theta, \Delta_i, \psi_\nu) \to 0 \qquad \text{as} \quad i \to \infty \qquad \forall \theta \in \Phi \backslash \mathcal{N} \qquad \text{and every} \quad \nu, \tag{90}$$

where $\mathcal{N} = \bigcup_\nu \mathcal{N}_\nu$ is still a null set. Finally, for an arbitrary test function $\psi \in C_0^1(\mathbb{R} \times [0, \infty))$ and $\theta \in \Phi \backslash \mathcal{N}$, from the linearity of F in ψ we obtain for $i \to \infty$,

$$|F(\theta, \Delta_i, \psi)| \leq |F(\theta, \Delta_i, \psi - \psi_\nu)| + |F(\theta, \Delta_i, \psi_\nu)| \leq c\|\psi - \psi_\nu\|_{L^\infty(\Phi)} + o(1)$$

[the first bound from (89), the second from (90)]. Choosing first ν large so that $\|\psi - \psi_\nu\|$ is small, and then i large so that $o(1)$ is small, proves the assertion. ∎

Remark 5.1. $F(\theta, \Delta_i, \psi) \to 0$ as $i \to \infty$ for every $\psi \in C_0^1(\mathbb{R} \times [0, \infty))$ is equivalent to the weak convergence $\sum_{n=1}^{\infty} [\mathbf{u}^{\Delta_i}]_n \to 0$.

The following corollary gives a precise formulation of Glimm's theorem.

Corollary 5.3. If $TV(\mathbf{u}^0, \mathbb{R})$ is sufficiently small, there exists a null set $\mathcal{N} \subset \Phi$ and a sequence $\Delta_i \to 0$ such that for every $\theta \in \Phi \backslash \mathcal{N}$ the limit in the sense of $L^1_{\text{loc}}(\mathbb{R} \times [0, \infty))$ (see Proposition 5.6),

$$\mathbf{u}_\theta = \lim_{\Delta_i \to 0} \mathbf{u}_\theta^{\Delta_i}$$

exists and defines a weak solution in $BV(\mathbb{R})$ of the Cauchy problem (81), (82).

Proof. Choose first a sequence $\bar{\Delta}_i = 2^{-i}$ so that $\mathbf{u}^{\bar{\Delta}_i}(\cdot, 0) \to \mathbf{u}^0$ as in Corollary 5.2. Then take a subsequence $\{\Delta_{i'}\}$ of $\{\bar{\Delta}_i\}$ such that for $\theta \in \Phi \backslash \mathcal{N}$, $\mathcal{N}$ a null set, we have $F(\theta, \Delta_{i'}, \psi) \to 0$ as $i' \to \infty$; this can be done by Proposition 5.9. Finally, let $\{\Delta_i\} \subset \{\Delta_{i'}\}$ be a further subsequence such that $\mathbf{u}_\theta^{\Delta_i}$ converges; this follows from Proposition 5.6. The assertion then is implied by Corollary 5.2. From (H2) it follows that Glimm's solution is $BV(\mathbb{R})$ for every $t \geq 0$. □

A natural question to ask is whether the solutions constructed by the Glimm scheme satisfy an entropy condition. We know from Lemma 4.1 that the geometric entropy condition and the Lax–Kruzhkov condition are equivalent under the assumptions of Glimm's theorem.

Proposition 5.10. Suppose the system admits an entropy–entropy flux pair (U, F), with U strictly convex. Then $\mathbf{u} = \lim_{\Delta_i \to 0} \mathbf{u}_\theta^{\Delta_i}$ satisfies the Lax–Kruzhkov entropy condition (LK).

Proof. We have seen that the $\mathbf{u}^\Delta$ are piecewise continuous in any strip $\mathbb{R} \times [n\Delta t, (n+1)\Delta t)$, and have only admissible shocks satisfying (GE) as discontinuities in each strip. From Lemma 4.1 it follows that $U^\Delta = U(\mathbf{u}^\Delta)$, $F^\Delta = F(\mathbf{u}^\Delta)$ satisfy the relation $U(\mathbf{u}^\Delta)_t + F(\mathbf{u}^\Delta)_x \leq 0$ in the weak sense in each strip, or

$$\int_{\mathbb{R}} \int_{n\Delta t}^{(n+1)\Delta t} (\psi_t U^\Delta + \psi_x F^\Delta) dx\, dt + \int_{\mathbb{R}} [\psi(x, \tau) U^\Delta(x, \tau)|_{\tau=(n+1)\Delta t - 0}^{\tau=n\Delta t+0}\, dx \geq 0,$$

$\forall\psi \in C_0^1(\mathbb{R} \times (0, \infty))$, $\psi \geq 0$. Summing over n we obtain

$$\sum_{n=1}^{\infty} \int_{\mathbb{R}} [\psi(x, n\Delta t)[U^{\Delta}(x, n\Delta t - 0) - U^{\Delta}(x, n\Delta t + 0)]dx$$

$$+ \int_{\mathbb{R}} \int_0^{\infty} (\psi_t U^{\Delta} + \psi_x F^{\Delta})dxdt \geq 0,$$

where the series is in reality a finite sum. By adapting the proof of Proposition 5.9, we see that the sum tends to zero for a suitable subsequence Δ_i tending to zero (and for almost every θ). As the approximants $\mathbf{u}_\theta^{\Delta_i}$ converge in such a way as to guarantee $F(\mathbf{u}_\theta^{\Delta_i}) \to F(\mathbf{u})$ and $U(\mathbf{u}_\theta^{\Delta_i}) \to U(\mathbf{u})$, cf. Corollary 5.1, passing to the limit in the last relation yields (LK). □

Remark 5.2. Concerning *uniqueness*, see Ref. 17. There are examples (e.g., Ref. 14) showing that weak solutions of (81), (82) may not be unique even if the entropy condition (LK) holds.

6. Appendix to Section 5

Proof of Proposition 5.1. If we write $R_j := \boldsymbol{r}_j \cdot \text{grad}_u$, then, as $\boldsymbol{r}_k = \boldsymbol{r}_k(\mathbf{u}(\varepsilon))$,

$$\frac{\partial \boldsymbol{r}_k}{\partial \varepsilon_j} = R_j \boldsymbol{r}_k.$$

Recall that $[\mathbf{u}_0, \ldots, \mathbf{u}_m;\ \varepsilon_1, \ldots, \varepsilon_m]$ is C^2 near $\varepsilon = 0$, C^3 for $\varepsilon \neq 0$, and

$$\left.\frac{\partial \mathbf{u}_i}{\partial \varepsilon_j}\right|_{\varepsilon_j=0} = \boldsymbol{r}_j(\mathbf{u}_{j-1}), \qquad \left.\frac{\partial^2 \mathbf{u}_i}{\partial \varepsilon_j^2}\right|_{\varepsilon_j=0} = R_j \boldsymbol{r}_j(\mathbf{u}_{j-1}),$$

so the Taylor formula yields for $(\mathbf{u}_c, \mathbf{u}_r)$

$$\mathbf{u}_r - \mathbf{u}_c = \sum_{j=1}^{m} (\mathbf{u}_j'' - \mathbf{u}_{j-1}'') = \sum_{j=1}^{m} \delta_j \boldsymbol{r}_j(\mathbf{u}_{j-1}'') + \tfrac{1}{2} \sum_{j=1}^{m} \delta_j^2 R_j \boldsymbol{r}_j(\mathbf{u}_{j-1}'') + O(|\delta|^3).$$

As $\boldsymbol{r}_j(\mathbf{u}_{j-1}'') = \boldsymbol{r}_j(\mathbf{u}_{j-2}'') + \delta_{j-1} R_{j-1} \boldsymbol{r}_j(\mathbf{u}_{j-2}'') + O(|\delta_{j-1}|^2)$, we find, recursively, that $\boldsymbol{r}_j(\mathbf{u}_{j-1}'') = \boldsymbol{r}_j(\mathbf{u}_c) + \sum_{i \leq j-1} \delta_i R_i \boldsymbol{r}_j(\mathbf{u}_c) + O(|\delta|^2)$, and

$$\mathbf{u}_r - \mathbf{u}_c = \sum_{j=1}^{m} \delta_j \boldsymbol{r}_j + \sum_{j=1}^{m} \sum_{i \leq j} \delta_i \delta_j R_i \boldsymbol{r}_j(\mathbf{u}_c)(1 - \tfrac{1}{2}\delta_{i,j}) + O(|\delta|^3),$$

all coefficients being evaluated at $\mathbf{u}_c$. Similarly, for $(\mathbf{u}_l, \mathbf{u}_c)$,

$$\mathbf{u}_l - \mathbf{u}_c = \sum_{j=1}^{m}(\mathbf{u}'_{j-1} - \mathbf{u}'_j) = -\sum_{j=1}^{m}\gamma_j \boldsymbol{r}_j(\mathbf{u}'_{j-1}) - \tfrac{1}{2}\sum_{j=1}^{m}\gamma_j^2 R_j \boldsymbol{r}_j(\mathbf{u}'_{j-1}) + O(|\gamma|^3),$$

and, as $\boldsymbol{r}_j(\mathbf{u}'_{j-1}) = \boldsymbol{r}_j(\mathbf{u}_c) - \sum_{j\le i}\gamma_i R_i \boldsymbol{r}_j(\mathbf{u}_c) + O(|\gamma|^2)$, we have

$$\mathbf{u}_l - \mathbf{u}_c = -\sum_{j=1}^{m}\gamma_j \boldsymbol{r}_j + \sum_{i=1}^{m}\sum_{j\le i}\gamma_i\gamma_j R_i \boldsymbol{r}_j(1 - \tfrac{1}{2}\delta_{i,j}) + O(|\gamma|^3),$$

where, again, all coefficients are evaluated at $\mathbf{u}_c$. Then

$$\begin{aligned}\mathbf{u}_r - \mathbf{u}_l = {} & \sum_{j=1}^{m}(\delta_j + \gamma_j)\boldsymbol{r}_j(\mathbf{u}_c) + \sum_{j=1}^{m}\sum_{i\le j}\delta_i\delta_j R_i \boldsymbol{r}_j(\mathbf{u}_c)(1 - \tfrac{1}{2}\delta_{i,j}) \\ & - \sum_{i=1}^{m}\sum_{j\le i}\gamma_i\gamma_j R_i \boldsymbol{r}_j(\mathbf{u}_c)(1 - \tfrac{1}{2}\delta_{i,j}) + O(|\gamma|^3 + |\delta|^3).\end{aligned} \tag{I}$$

If we pass directly from $\mathbf{u}_l$ to $\mathbf{u}_r$ via ε-waves, we obtain

$$\mathbf{u}_r - \mathbf{u}_l = \sum_{j=1}^{m}\varepsilon_j \boldsymbol{r}_j + \sum_{j=1}^{m}\sum_{i\le j}\varepsilon_i\varepsilon_j R_i \boldsymbol{r}_j(1 - \tfrac{1}{2}\delta_{i,j}) + O(|\varepsilon|^3), \tag{D}$$

all coefficients evaluated at $\mathbf{u}_l$. As $|\gamma|$, $|\delta| \to 0$, the solutions of the Riemann problems tend to the corresponding linearized solutions and

$$\varepsilon = O(|\gamma| + |\delta|).$$

Therefore, if use is made of

$$\boldsymbol{r}_j(\mathbf{u}_l) = \boldsymbol{r}_j(\mathbf{u}_c) - \sum_{i=1}^{m}\gamma_i R_i \boldsymbol{r}_j(\mathbf{u}_c) + O(|\gamma|^2), \tag{A1}$$

we obtain

$$\mathbf{u}_r - \mathbf{u}_l = \sum_{j=1}^{m}\varepsilon_j \boldsymbol{r}_j(\mathbf{u}_c) + O(|\gamma| + |\delta|)^2.$$

As $\boldsymbol{r}_j(\mathbf{u}_c)$ are linearly independent, this and (I) imply that $\varepsilon_j = \gamma_j + \delta_j + O(|\gamma| + |\delta|)^2$. On the other hand, direct substitution of (A1) in (D) yields

$$\mathbf{u}_r - \mathbf{u}_l = \sum_{j=1}^{m} \varepsilon_j \boldsymbol{r}_j(\mathbf{u}_c) - \sum_{j=1}^{m} \varepsilon_j \sum_{i=1}^{m} \gamma_i R_i \boldsymbol{r}_j(\mathbf{u}_c) + \sum_{i=1}^{m} \sum_{i \leq j} \varepsilon_i \varepsilon_j R_i \boldsymbol{r}_j(\mathbf{u}_c)(1 - \tfrac{1}{2}\delta_{i,j}) + O(|\gamma| + |\delta|)^3.$$

Comparison of this result with (I) implies then, up to terms $O(|\gamma|^3 + |\delta|^3)$,

$$\begin{aligned}\sum_{j=1}^{m}(\varepsilon_j - \gamma_j - \delta_j)\boldsymbol{r}_j(\mathbf{u}_c) &= \sum_{j=1}^{m}(\gamma_j + \delta_j) \sum_{i=1}^{m} \gamma_i R_i \boldsymbol{r}_j(\mathbf{u}_c) + \sum_{i=1}^{m}\sum_{i \leq j} \delta_i \delta_j R_i \boldsymbol{r}_j(\mathbf{u}_c)(1 - \tfrac{1}{2}\delta_{i,j}) \\ &\quad - \sum_{i=1}^{m}\sum_{i \leq j}(\gamma_i + \delta_i)(\gamma_j + \delta_j) R_i \boldsymbol{r}_j(\mathbf{u}_c)(1 - \tfrac{1}{2}\delta_{i,j}) - \sum_{i=1}^{m}\sum_{i \leq j} \gamma_i \gamma_j R_i \boldsymbol{r}_j(\mathbf{u}_c)(1 - \tfrac{1}{2}\delta_{i,j}) \\ &= \sum_{i=1}^{m}\sum_{j<i} \gamma_i \delta_j (R_i \boldsymbol{r}_j(\mathbf{u}_c) - R_j \boldsymbol{r}_i(\mathbf{u}_c)).\end{aligned}$$

Proposition 5.1 follows from this equation. □

Proof of Proposition 5.2. Consider first the case $D(\gamma, \delta) = 0$. Then there is a j such that $\mathbf{u}_i' = \mathbf{u}_c$ for $i \geq j$, $\mathbf{u}_k'' = \mathbf{u}_c$ for $k \leq j$, and neither of $\mathbf{u}_j'$, $\mathbf{u}_j''$ is a shock. The possibilities can be indicated schematically by

$$\mathbf{u}_l' \xrightarrow{\gamma_1} \mathbf{u}_1' \xrightarrow{\gamma_2} \mathbf{u}_2' \cdots \xrightarrow{\gamma_j} \mathbf{u}_j' = \mathbf{u}_c \xrightarrow{0} \mathbf{u}_c \cdots \xrightarrow{0} \mathbf{u}_c$$

and

$$\mathbf{u}_c \xrightarrow{0} \cdots \xrightarrow{0} \mathbf{u}_c \xrightarrow{\delta_{j+1}} \mathbf{u}_{j+1}'' \cdots \xrightarrow{\delta_m} \mathbf{u}_r \qquad (\delta_j = 0)$$

or

$$\mathbf{u}_c \xrightarrow{0} \cdots \xrightarrow{0} \mathbf{u}_c \xrightarrow{\delta_j} \mathbf{u}_j'' \cdots \xrightarrow{\delta_m} \mathbf{u}_r \qquad (\delta_j \neq 0).$$

If $\delta_j = 0$, the two sets of states merge together with

$$\varepsilon_i = \begin{cases} \gamma_i, & i \leq j, \\ \delta_i, & i > j. \end{cases}$$

If $\delta_j \neq 0$, both waves $(\mathbf{u}'_{j-1}, \mathbf{u}'_j) = (\mathbf{u}'_{j-1}, \mathbf{u}_c)$ and $(\mathbf{u}_c, \mathbf{u}''_j) = (\mathbf{u}''_{j-1}, \mathbf{u}''_j)$ have $\gamma_j > 0$ and $\delta_j > 0$, respectively, as $D(\gamma, \delta) = 0$, and they combine into a single centered wave $(\mathbf{u}'_{j-1}, \mathbf{u}''_j)$ with parameter $\gamma_j + \delta_j$. In each case $\varepsilon_i = \gamma_i + \delta_i$ for all i and (A2) holds when $D(\gamma, \delta) = 0$.

The proof in the general case $D(\gamma, \delta) > 0$ is accomplished by (finite) induction on the index p of the highest of nonnull wave in $\mathbf{u}''$. Let

$$\Delta = (\delta_1, \ldots, \delta_{p-1}, 0, \ldots, 0), \qquad \Delta_0 = (0, \ldots, 0, \delta_p, 0, \ldots, 0)$$

($\delta_{p-1} \neq 0$). We assume the induction hypothesis that the estimate holds for all δ of the form Δ, and we will show that it holds for all $\delta = \Delta + \Delta_0$. (At the initial induction step all $\delta_i = 0$ and the estimate is trivially true.) Let $\gamma_p \neq 0$ [otherwise $\gamma_p \delta_p = 0$ and $D(\gamma, \Delta) = D(\gamma, \delta)$]. We organize the proof into a sequence of four steps.

(To simplify things we use the same notation for the states.)

i. Define μ and ν by

$$\mu_i := \varepsilon_i(\gamma, \Delta, \mathbf{u}_c) - \nu_i, \qquad \nu_i := \begin{cases} \varepsilon_i(\gamma, \Delta, \mathbf{u}_c), & \text{if } \gamma_i \mathscr{A} \delta_p \\ 0, & \text{otherwise} \end{cases}$$

($i = 1, \ldots, m$). We have, if $\mu_p \neq 0$ (hence $\nu_p = 0$),

$$\mathbf{u}_l \xrightarrow{\mu_1} \mathbf{u}'_1 \cdots \xrightarrow{\mu_{p-1}} \mathbf{u}'_{p-1} \xrightarrow{\mu_p} \mathbf{u}'_p \xrightarrow{\nu_{p+1}} \mathbf{u}''_{p+1} \cdots \xrightarrow{\nu_m} \mathbf{u}_r,$$

and

$$\mathbf{u}_l \xrightarrow{\mu_1} \mathbf{u}'_1 \cdots \xrightarrow{\mu_{p-1}} \mathbf{u}'_{p-1} \xrightarrow{\nu_p} \mathbf{u}''_p \xrightarrow{\nu_{p+1}} \mathbf{u}''_{p+1} \cdots \xrightarrow{\nu_m} \mathbf{u}_r$$

if $\nu_p \neq 0$ (hence $\mu_p = 0$). Clearly, $|\mu| + |\nu| \to 0$ as $|\gamma| + |\delta| \to 0$.

By definition, $\mu_i \nu_j = 0$ for all $i \geq j$ so that $D(\mu, \nu) = 0$. The interaction of μ and ν leads by construction to the same (unique local) solution of Riemann's problem as the interaction of γ and Δ, so, by the first part of the proof, $\mu_i + \nu_i \to \varepsilon_i(\gamma, \Delta, \mathbf{u}_c)$. The induction hypothesis then implies

$$\mu_i + \nu_i = \varepsilon_i(\gamma, \Delta, \mathbf{u}_c) = \gamma_i + \Delta_i + c_i(\gamma, \Delta, \mathbf{u}_c) D(\gamma, \Delta) \tag{A3}$$

with $|c_i(\gamma, \Delta, \mathbf{u}_c)| \leq c|C_i(\mathbf{u}_c)|$. As $D(\gamma, \Delta) \leq D(\gamma, \delta)$, if we set

$$c_i := c_i(\gamma, \Delta, \mathbf{u}_c) D(\gamma, \Delta)/D(\gamma, \delta),$$

then $c_i(\gamma, \Delta, \mathbf{u}_c)D(\gamma, \Delta)$ can be replaced by $c_iD(\gamma, \delta)$ on the right-hand side of (A3). Letting $\tilde{\pi}_i = \nu_i + \delta_{ip}\delta_p$, δ_{ij} the Kronecker delta, we obtain

$$\mu_i + \tilde{\pi}_i = \gamma_i + \delta_i + c_iD(\gamma, \delta) \tag{A4}$$

with $|c_i| \leq c|C_i(\mathbf{u}_c)|$.

ii. We now consider ν interacting with Δ_0. Recall that $\nu_i = 0$ for $i < p$. The initial state in this interaction is $\mathbf{u}'_{p-1}$ if $\nu_p \neq 0$, or $\mathbf{u}'_p$ if $\nu_p = 0$, and the intermediate state is $\mathbf{u}''_{p-1}$. We write $\nu + \Delta_0 \to \pi = \varepsilon(\nu, \Delta_0, \mathbf{u}''_{p-1})$. This is indicated schematically as

$$\mathbf{u}'_{p-1} \xrightarrow{\nu_p} \mathbf{u}'_p \cdots \xrightarrow{\nu_m} \mathbf{u}'_m = \mathbf{u}''_{p-1} \xrightarrow{\delta_p} \mathbf{u}''_p.$$

As $(\Delta_0)_i = \delta_{ip}\delta_p$, Proposition 5.1 implies that, as $|\gamma| + |\delta| \to 0$,

$$\pi_i = \tilde{\pi}_i + C''_i|\nu||\delta_p|,$$

where $C''_i = C_i(\mathbf{u}''_{p-1})$ (+possible contributions from third-order terms), so that, by Lemma 5.1, C''_i differs from $C_i(\mathbf{u}_c)$ by first-order terms. The relationship between l_∞ and l_1 norms in $\mathbb{R}^m$ implies that there exists a fixed constant $C > 0$ such that

$$|\nu||\delta_p| = C|\nu_p||\delta_p| + C|\delta_p| \sum_{i>p} |\nu_i|.$$

We deal with the two terms separately. For the first, if $\nu_p \neq 0$, then $\nu_p = \varepsilon_p(\gamma, \Delta, \mathbf{u}_c) = \gamma_p + c_p(\gamma, \Delta, \mathbf{u}_c)D(\gamma, \Delta)$ from step (i). In this case $\gamma_p \mathscr{A} \delta_p$ by definition of ν_p, so by a suitable rescaling of c_p,

$$|\nu_p||\delta_p| = c_pD(\gamma, \delta).$$

If $\nu_p = 0$, this is trivially true.

For $i > p$, we have $\delta_i = \mu_i = 0$, so [from (A4)] $\nu_i = \gamma_i + c_iD(\gamma, \delta)$, and, as $\gamma_i \mathscr{A} \delta_p$, by a suitable rescaling of c_i we have

$$|\nu_i||\delta_p| = c_iD(\gamma, \delta).$$

Summarizing, we have shown that there are coefficients $c_i > 0$ such that

$$\pi_i = \tilde{\pi}_i + c_iD(\gamma, \delta), \tag{A5}$$

and $|c_i| \leq c|C_i(\mathbf{u}_c)|$ for some (rescaled) constant $c > 0$.

iii. Suppose μ interacts with $\tilde{\pi}$. Let $\tilde{\mathbf{u}}_c$ be the intermediate state connecting μ and ν, i.e., $\tilde{\mathbf{u}}_c = \mathbf{u}'_{p-1}$ if $\mu_p = 0$, $\tilde{\mathbf{u}}_c = \mathbf{u}'_p$ if $\mu_p \neq 0$. Then $\tilde{\mathbf{u}}_c$ is also the intermediate state between μ and $\tilde{\pi}$. This is indicated schematically by $\mu + \tilde{\pi} \to \varepsilon(\mu, \tilde{\pi}, \tilde{\mathbf{u}}_c)$, with

$$\mathbf{u}_l \xrightarrow{\mu_1} \mathbf{u}'_1 \cdots \xrightarrow{\mu_{p-1}} \mathbf{u}'_{p-1} \xrightarrow{\nu_p+\delta_p} \tilde{\mathbf{u}}''_p \xrightarrow{\nu_{p+1}} \tilde{\mathbf{u}}''_{p+1} \cdots \xrightarrow{\nu_m} \tilde{\mathbf{u}}''_m$$

for $\mu_p = 0$, and

$$\mathbf{u}_l \xrightarrow{\mu_1} \mathbf{u}'_1 \cdots \xrightarrow{\mu_p} \mathbf{u}'_p \xrightarrow{\delta_p} \tilde{\mathbf{u}}''_p \cdots \xrightarrow{\nu_m} \tilde{\mathbf{u}}''_m$$

for $\mu_p \neq 0$. (Note that this solves a different Riemann problem, and the states $\tilde{\mathbf{u}}''_i$ are "spurious.") We will show that $D(\mu, \tilde{\pi}) = 0$. This implies

$$\varepsilon_i(\mu, \tilde{\pi}, \tilde{\mathbf{u}}_c) = \mu_i + \tilde{\pi}_i \equiv \mu_i + \nu_i + \delta_{ip}\delta_p. \tag{A6}$$

As $\tilde{\pi}_i = 0$ for $i < p$, $\mu_i = 0$ for $i > p$, the only possible nonzero term in $D(\mu, \tilde{\pi})$ is $|\mu_p||\tilde{\pi}_p| = |\mu_p(\nu_p + \delta_p)|$. If $\mu_p \neq 0$, then $\nu_p = 0$ and γ_p, δ_p do not approach (each other), so that $\gamma_p > 0$, $\delta_p > 0$. It follows from (A4) that

$$\mu_p = \gamma_p + c_p D(\gamma, \delta),$$

which is positive for $|\gamma|$, $|\delta|$ small so that μ_p and δ_p do not approach. Thus, $D(\mu, \tilde{\pi}) = 0$ and (A6) follows. Taking (A4) into account, this yields

$$\varepsilon_i(\mu, \tilde{\pi}, \tilde{\mathbf{u}}_c) = \gamma_i + \delta_i + c_i D(\gamma, \delta). \tag{A7}$$

iv. Finally, suppose that μ interacts with π, as indicated schematically by

$$\mathbf{u}_l \xrightarrow{\mu_1} \mathbf{u}'_1 \cdots \xrightarrow{\mu_{p-1}} \mathbf{u}'_{p-1} \xrightarrow{\mu_p+\pi_p} \mathbf{u}''_p \xrightarrow{\pi_{p+1}} \mathbf{u}''_{p+1} \cdots \xrightarrow{\pi_m} \mathbf{u}_r$$

for $\mu_p \neq 0$, or

$$\mathbf{u}_l \xrightarrow{\mu_1} \mathbf{u}'_1 \cdots \xrightarrow{\mu_{p-1}} \mathbf{u}'_{p-1} \xrightarrow{\pi_p} \mathbf{u}''_p \xrightarrow{\pi_{p+1}} \mathbf{u}''_{p+1} \cdots \xrightarrow{\pi_m} \mathbf{u}_r$$

for $\mu_p = 0$. (Local) uniqueness of the solution of Riemann's problem implies that

$$\varepsilon_i(\gamma, \delta, \mathbf{u}_c) = \varepsilon_i(\mu, \pi, \tilde{\mathbf{u}}_c).$$

As ε is a C^2 function of its first two arguments (Exercise 5.1), there exist $L_i > 0$ such that

$$|\varepsilon_i(\mu, \pi, \tilde{\mathbf{u}}_c) - \varepsilon_i(\mu, \tilde{\pi}, \tilde{\mathbf{u}}_c)| = L_i|\pi - \tilde{\pi}|,$$

where, by a suitable rescaling of c, $L_i \leq c|C_i(\mathbf{u}_c)|$. The last two equations imply that

$$\varepsilon_i(\gamma, \delta, \mathbf{u}_c) = \varepsilon_i(\mu, \tilde{\pi}, \tilde{\mathbf{u}}_c) \pm L_i|\pi - \tilde{\pi}|.$$

Then by (A5) and (A7), Proposition 5.2 follows. □

Exercises

1.1. Show that a Riemann invariant z_j satisfies the relation $R_k z_j := \boldsymbol{r}_k \cdot \mathrm{grad}_u z_j = 0$ for $j \neq k$.

1.2. Show that for $m \geq 2$ there exist, for every $k = 1, \ldots, m$, $m - 1$ functions (Riemann invariants) $z_{j,k}(\mathbf{u})$ $(j = 1, \ldots, m - 1)$ satisfying $\boldsymbol{r}_k \cdot \mathrm{grad}_u\, z_{j,k} = 0$ in a neighborhood of every $\mathbf{u}_0 \in \mathbb{R}^m$, and such that the gradients $\mathrm{grad}_u z_{j,k}(\mathbf{u}_0)$ are linearly independent for each fixed k.

1.3. Prove that $z_2(\mathbf{u}) = \text{const.}$ defines a 1-simple wave and $z_1(\mathbf{u}) = \text{const.}$ defines a 2-simple wave for a 2×2 system. *Hint*: The proof follows from (4) and the relations $dz_j = d\mathbf{u} \cdot \mathrm{grad}_u z_j$, $R_k z_j = 0$ for $j \neq k$ (see Exercise 1.1).

1.4. Show that, for a barotropic gas, the relation $u = -2c/(\gamma - 1)$ characterizes a 1-simple wave, and the relation $u = 2c/(\gamma - 1)$ a 2-simple wave.

1.5. Show that $U = -\rho S \equiv -\rho c_v \ln(p/\rho^\gamma)$ is a convex function of ρ, ρu, ρW (i.e., the Hessian matrix of U is positive definite).

2.1. If $a_s = g_s(0; t, x)$ is the starting point (at $t = 0$) of the unique s-characteristic C_s: $x = g_s(t; 0, a_s)$ passing through the point (x, t), show (formally) that the group relation

$$g_s(\tau; t, g_s(t; t', x)) = g_s(\tau; t', x)$$

holds, and hence that $b_s = g_s(\tau; 0, a_s)$, $\mathbf{u}(b_s, \tau) = \mathbf{u}(g_s(\tau; 0, a_s), \tau)$.

2.2. Show that the solution $\bar{\mathbf{u}}(x, t)$ satisfies

$$\|\bar{\mathbf{u}}(x, t)\|_{S_1} \leq c\|\mathbf{u}^0\|_{S_1} \qquad 0 \leq t \leq a,$$

for some $c = c(\omega, \Lambda, Q, a)$, with c a fixed constant for diagonal systems and scalar conservation laws.

2.3. Adapt the proof of Theorem 2.1 to prove local a.e. existence for the system

$$\mathbf{u}_t + \mathbb{A}(\mathbf{u})\mathbf{u}_x = \boldsymbol{F}(\mathbf{u})$$

with source term $\boldsymbol{F} \in C^1(G)$. *Hint*: The quantities $\mathscr{F}_i = \sum_{j=1}^m \Lambda_{ij} F_j$ satisfy relations (29) with constants $C_7(\omega)$, $C_8(\omega)$ and the added terms $D_{s3} = \int_0^t \mathscr{F}_s(\mathbf{u}(b_s))d\tau$ in (TR) only modify the estimates for a.

3.1. Show that, if u_0 is differentiable, $x_0 = \alpha(x, t)$ satisfies $\alpha_t + f'(u_0(\alpha))\alpha_x = 0$, $\alpha(x, 0) = x$ and that equations (43) hold in the form

$$u_x(x, t) = \frac{u_0'(\alpha)}{(1 + tu_0'(\alpha)f''(u_0(\alpha)))}, \qquad u_t(x, t) = \frac{-u_0'(\alpha)f'(u_0(\alpha))}{(1 + tu_0'(\alpha)f''(u_0(\alpha)))}$$

If $f'' \neq 0$, show that $u(x, t) = b((x - x_0)/t)$, where b is the inverse function of f'.

3.2. Verify that the smooth solution of the equation $u_t + uu_x = 0$ with initial data $u(x, 0) = -\tanh(x/2\varepsilon)$ has finite life span $[0, T^*)$, with $T^* = 2\varepsilon$.

3.3. Prove that (W_1) and (W_2) imply (W). *Hint*: Choose a sequence of test functions of the form $\psi(x, 0)\xi_n(t)$ in (W_1), where $\xi_n(t)$ is a smooth approximation of the function

$$\Xi_n(t) = \begin{cases} 0, & 0 \leq t \leq 1/2n, \\ 2n(t - 1/2n), & 1/2n \leq t \leq 1/n, \\ 1, & 1/n \leq t, \end{cases}$$

and take the limit $n \to \infty$ after summing and subtracting $u_0(x)$. Verify that the same result holds if (bounded) a.e. convergence in (W_2) is replaced by weak convergence (in the sense of distributions).

3.4. Prove from (W) that every weak solution $u(x, \tau)$ converges weakly to $u_0(x)$ for a.e. $\tau \to 0$. *Hint*: Choose a sequence of test functions $\psi_n(x, t) = \theta_n(t)\phi(x)$, where $\phi \in C_0^1(\mathbb{R})$, $\theta_n(t)$ is a smooth approximation of the function

$$\Theta_n(t) = \begin{cases} 1, & 0 \leq t \leq \tau, \\ 1 - n(t - \tau), & \tau \leq t \leq \tau + 1/n, \\ 0, & \tau + 1/n \leq t, \end{cases}$$

and take the limit $n \to \infty$, and $\tau \to 0$ after applying the Lebesgue–Besicovitch theorem for integral means (Ref 18).

3.5. Solve the Burgers equation with $u_0(x) = 1$ for $x \leq 0$, 0 for $1 \leq x$, and $1 - x$ in the interval $0 \leq x \leq 1$. *Hint*: The method of characteristics can be useed to solve explicitly up to time $t = 1$ where a shock develops. Use the (RH) relation to determine the shock speed and the solution for $t > 1$.

3.6. (i) Show that, for the Burgers–Hopf equation, the following limits hold as $t \to 0$ for fixed μ:

$$\begin{aligned} \varphi(x, t) &\to \exp(-U_0(x)/2\mu), \\ 2\mu\varphi_x(x, t) &\to u_0(x)\exp(-U_0(x)/2\mu) \quad \text{if } u_0 \text{ is continuous at } x, \\ 2\mu\varphi_x(x, t) &\to u_0(x)\exp(-U_0(x)/2\mu) \quad \text{weakly (in the sense of distributions).} \end{aligned}$$

(ii) Deduce the convergence properties in (45).

3.7. Suppose $u_0(x) = \mathfrak{U}$ for $x < 0$, 0 for $x > 0$. Show that the solution of the Cauchy problem with this initial data is given by

$$\frac{\mathcal{U}}{u} = 1 + \exp\left(\frac{\mathcal{U}}{2\mu}(x - \tfrac{1}{2}\mathcal{U}t)\right)\frac{\operatorname{erfc}(-x/\sqrt{4\mu t})}{\operatorname{erfc}((x - \mathcal{U}t)/\sqrt{4\mu t})}$$

and discuss the nature of this solution for μ small. What happens when $\mu \to 0$?

3.8. Prove that inequality (46) implies the validity on admissible shocks of the (RH) relation and also of the local entropy condition (Oleinik's condition, Ref. 19),

$$s(u_r, u) \le s(u_r, u_l) \le s(u_l, u), \tag{E}$$

for all values of u between u_l and u_r. *Hint*: Suppose $u_l > u_r$. Taking $k < u_r < u_l$, $k > u_l > u_r$ in (46) yields (RH), while taking $u_r < k < u_l$ yields (E). Similarly for $u_l < u_r$.

Verify that (E) implies (GE) if $f''(u) > 0$.

3.9. Show that (70) holds.

3.10. Show, from the (RH) relation $s\Delta u = \Delta f$, that in the limit $\Delta u \to 0$ (hence for any Δu if f is affine) the shock line is a characteristic line (*infinitesimal shocks and "linear" shocks are characteristic*).

3.11. Use (64) and (70) to show that if the scalar law is *linear*, *all shocks are admissible provided they are characteristic, and the entropy condition becomes redundant*. *Hint*: If $f(u) = c + \lambda_0 u$, then $F = \lambda_0 U$ and, as $\mu(u_x^\mu)^2 \to 0$ weakly as $\mu \downarrow 0$, the (LK) entropy condition, with $U = u^2/2$, is always satisfied in the form $\int_{\mathbb{R}}\int_0^\infty(\psi_t U + \psi_x F)dxdt = \int_{\mathbb{R}}\int_0^\infty(\psi_t + \lambda_0\psi_x)U\,dxdt = 0$. On a shock line this implies $s = \lambda_0$.

3.12. Show that, if u is an N-wave, (i) $u,\ u^2 \in L^1_{\text{loc}}(\mathbb{R} \times (0, \infty))$, (ii) u satisfies (W_1), and (iii) u satisfies (GE). Moreover, verify that $u(x, t) = O(1/\sqrt{t})$ as $t \to +\infty$.

3.13. Carry out the details of computations in the cases (ii), (iii) of Section 3.4.

3.14. Show that a *rarefaction fan and a shock always collide*, by working out the case (ii) in Section 3.4 with $b > a = c$.

4.1. Verify the assertions in Example 4.1 and show that $[e] + \frac{1}{2}(p_l + p_r)[1/\rho] \equiv 0$ for $\gamma = -1$.

4.2. If (71) is such that $\mathbb{A} = \partial \mathbf{f}/\partial \mathbf{u}$ is a symmetric matrix for all $\mathbf{u} \in G$, then, if $g(\mathbf{u})$ is such that $\partial g/\partial u_i = f_i$, $U = \frac{1}{2}\sum u_i^2$, $F = \sum u_i f_i - g$ is an entropy–entropy flux pair.

4.3. The system $\mathbf{u}_t + \mathbb{A}(\mathbf{u})\mathbf{u}_x = 0$ is *symmetrizable* if there is a symmetric positive definite matrix $\mathbb{A}_0(\mathbf{u})$ such that $\tilde{\mathbb{A}} = \mathbb{A}_0\mathbb{A}$ is symmetric. The system can then be written as

$$\mathbb{A}_0\mathbf{u}_t + \tilde{\mathbb{A}}\mathbf{u}_x = 0.$$

Show that symmetrizable systems are hyperbolic.

4.4. Prove that if (71) admits a uniformly convex entropy $U(\mathbf{u})$, then it is symmetrizable with $\mathbb{A}_0(\mathbf{u})$ given by the Hessian matrix $\mathbb{H}(u)$ of U. *Hint*: Use (71), (73) and the symmetry of the Hessian.

4.5. Prove that using the *entropy variable* $\mathbf{v} = \text{grad}_u U(\mathbf{u})$, system (71) takes the *symmetric* form

$$\mathbf{g}(\mathbf{v})_t + \mathbf{G}(\mathbf{v})_x = 0 \qquad [\partial \mathbf{g}/\partial \mathbf{v} = \mathbb{H}(\mathbf{u})^{-1}, \quad \partial \mathbf{G}/\partial \mathbf{v} = \partial \mathbf{f}/\partial \mathbf{u}\mathbb{H}(\mathbf{u})^{-1}].$$

4.6. Perform the calculations leading to (78). *Hint*: Use the normalization $\boldsymbol{r}_k \cdot \text{grad}_u \lambda_k(\mathbf{u}) = 1$.

5.1. Show that $\varepsilon(\gamma, \delta, \mathbf{u}_c)$ is a C^2 function of γ and δ. *Hint*: $\mathbf{u}_r$ is a C^2 function of ε and δ for fixed $\mathbf{u}_l$ and $\mathbf{u}_c$, and we can invert to obtain $\varepsilon = \varepsilon(\mathbf{u}_r)$ for fixed δ, $\mathbf{u}_c$, $\mathbf{u}_l$; similarly, $\mathbf{u}_l$ is a C^2 function of ε and γ for fixed $\mathbf{u}_r$ and $\mathbf{u}_c, \ldots$.

5.2. If $\mathbf{u}_j' = (\mathbf{u}_l, \mathbf{u}_i)$, $\mathbf{u}_k'' = (\mathbf{u}_i, \mathbf{u}_r)$ are neighboring waves and $\mathbf{u}_j' \mathscr{A} \mathbf{u}_k''$, then $\mathbf{u}_j'$ and $\mathbf{u}_k''$ always interact. *Hint*: There are a number of cases according to whether $j = k$, $j > k$, and the individual wave is a shock or rarefaction wave. For example, if $j = k$, the result follows from the geometric entropy condition, $\lambda_k(\mathbf{u}_r) < s_k'' < \lambda_k(\mathbf{u}_i) < s_k' < \lambda_k(\mathbf{u}_l)$, which implies $s_k' > s_k''$ if both solutions are shock waves. (See Fig. 19.)

If $j > k$, the result is implied again by the geometric entropy conditions and the ordering of the eigenvalues, $s_k'' < \lambda_k(\mathbf{u}_i) < \lambda_j(\mathbf{u}_i) < s_j'$. See Fig. 20.

5.3. If $\mathbf{u}_k' = (\mathbf{u}_l, \mathbf{u}_i)$, $\mathbf{u}_k'' = (\mathbf{u}_i, \mathbf{u}_r)$ do not approach each other, then they do not interact. *Hint*: We have two centered waves such that the "tail" of the first travels with the same speed as the "head" of the second. See Fig. 21.

5.4. If $j < k$ and $|\mathbf{u}_j' - \mathbf{u}_k''|$ is sufficiently small, then $\mathbf{u}_j' = (\mathbf{u}_l, \mathbf{u}_i)$ and $\mathbf{u}_k'' = (\mathbf{u}_i, \mathbf{u}_r)$ do not interact. *Hint*: We have $s_k'' \simeq \lambda_k(\mathbf{u}_i) > \lambda_j(\mathbf{u}_i) \simeq s_j'$.

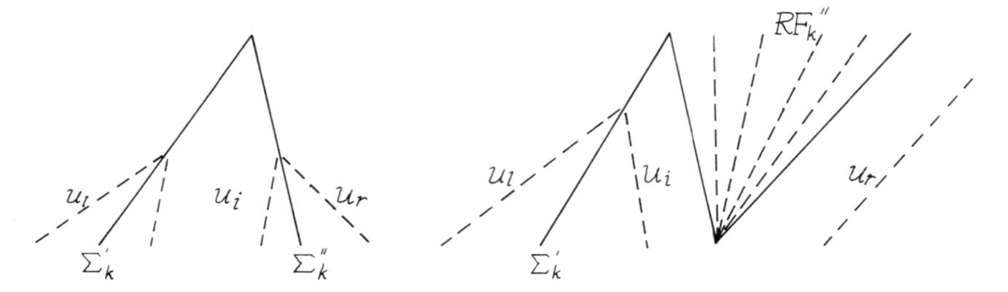

Fig. 19. Interaction when $j = k$.

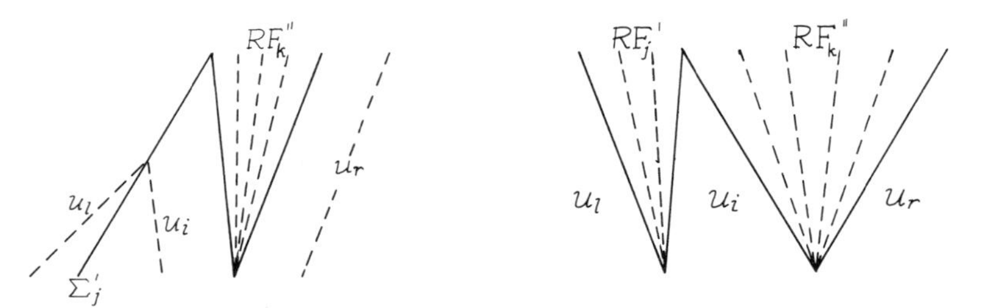

Fig. 20. Interaction when $j > k$.

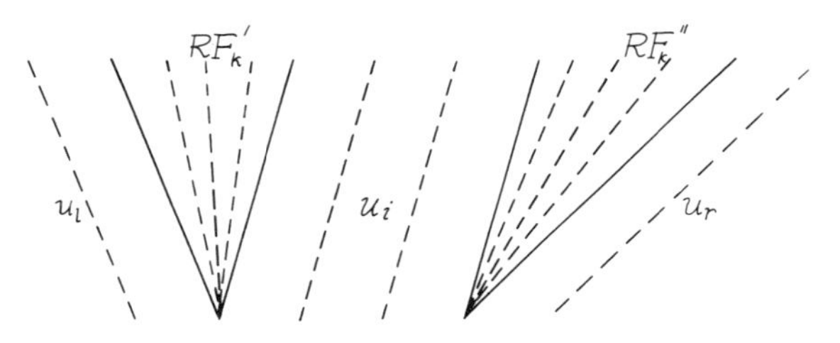

Fig. 21. Nonapproaching, noninteracting waves.

5.5. For a C_0^1 function $f(x)$ set $f_n(x) = f(m/2^n)$ for $m/2^n < x < (m+1)/2^n$ ($m = 0, 1, 2, \ldots$; $n = 1, 2, \ldots$). Show that $f_n \to f$ uniformly as $n \to \infty$.

References

1. NOBLE, B., *Applied Linear Algebra*, Prentice–Hall, Englewood Cliffs, New Jersey, 1969.
2. MAJDA, A., *Compressible Fluid Flow and Systems of Conservation Laws in Several Space Variables*, Applied Mathematical Sciences No. 53, Springer-Verlag, Berlin, Germany, 1984.

3. CHORIN, A.J., and MARSDEN, J.E., *A Mathematical Introduction to Fluid Mechanics*, Springer-Verlag, Berlin, Germany, 1990.
4. COURANT, R., and FRIEDRICHS, K.O., *Supersonic Flow and Shock Waves*, Interscience, New York, New York, 1948.
5. DOUGLIS, A., *Some Existence Theorems for Hyperbolic Systems of Partial Differential Equations in Two Independent Variables*. Communications on Pure and Applied Mathematics, Vol. V, pp. 119–154, 1952.
6. HARTMAN, P., and WINTER, A., *On Hyperbolic Differential Equations*, American Journal of Mathematics, Vol. 74, pp. 834–864, 1952.
7. CESARI, L., *A Boundary Value Problem for Quasilinear Hyperbolic Systems in the Schauder Canonic Form*, Annali Scuola Normale Superiore di Pisa (4) 1, pp. 311–358, 1974.
8. BASSANINI, P., and CESARI, L., *La duplicazione di frequenza nella radiazione laser*, Rendiconti Accademia Nazionale Lincei, Vol. LXIX, 3–4, pp. 166–173, 1980.
9. KRUZHKOV, S.N., *First Order Quasilinear Equations in Several Independent Variables*, Mathematics of the USSR Sbornik, Vol. 10, 2, pp. 217–243, 1970.
10. HOPF, E., *The Partial Differential Equation* $u_t + uu_x = \mu u_{xx}$, Communications on Pure and Applied Mathematics, Vol. III, pp. 201–230, 1950.
11. GODLEWSKI, E., and RAVIART, P.A., *Hyperbolic Systems of Conservation Laws*, SMAI No. 3/4, Paris, 1990–91.
12. LAX, P.D., *Hyperbolic Systems of Conservation Laws and the Mathematical Theory of Shock Waves*, Reg. Conference Series in Applied Mathematics, SIAM, Philadelphia, Pennsylvania, 1973.
13. SMOLLER, J., *Shock Waves and Reaction-Diffusion Equations*, Springer-Verlag, Berlin, Germany, 1983.
14. SEVER, M., *Uniqueness Failure for Entropy Solutions of Hyperbolic Systems of Conservation Laws*, Communications on Pure and Applied Mathematics, Vol. XLII, pp. 173–183, 1989.
15. GLIMM, J., *Solutions in the Large for Nonlinear Hyperbolic Systems of Equations*, Communications on Pure and Applied Mathematics, Vol. XVIII, pp. 697–715, 1965.
16. BILLINGSLEY, P., *Probability and Measure*, John Wiley, New York, New York, 1979.
17. LE FLOCH, P., and XIN, Z., *Uniqueness via the Adjoint Problems for Systems of Conservation Laws*, Communications on Pure and Applied Mathematics, Vol. XLVI, pp. 1499–1533, 1993.
18. EVANS, L.C., and GARIEPY, R.F., *Measure Theory and Fine Properties of Functions*, CRC Press, Boca Raton, Florida, 1992.
19. OLEINIK, O.A., *Uniqueness and Stability of the Generalized Solution of the Cauchy Problem for a Quasilinear Equation*, American Mathematical Society Translations, Series 2, Vol. 33, pp. 285–290, 1964.

Suggested Further Reading

BARDOS, C., *Introduction aux Problèmes Hyperboliques non Linéaires*, Lecture Notes, Corso CIME 40, Cortona, Italy, 1983.

DAFERMOS, C.M., *Characteristics in Hyperbolic Conservation Laws*. In: Nonlinear Analysis and Mechanics, Heriot-Watt Symposium, Vol. 1, Pitman, London, England, 1977.

FRIEDRICHS, K.O., and LAX, P.D., *Systems of Conservation Laws with a Convex Extension*, Proceedings of the National Academy of Sciences USA, Vol. 68, pp. 1686–1688, 1971.

LAX, P.D., *Hyperbolic Systems of Conservation Laws*, Communications on Pure and Applied Mathematics, Vol. X, pp. 537–566, 1957.

LAX, P.D., *Shock Waves and Entropy*. In: Proc. Symposium at the University of Wisconsin, 1971, Edited by E.H. Zarantonello, Academic Press, New York, New York, pp. 603–634, 1971.

LEVEQUE, R.J., *Numerical Methods for Conservation Laws*, Birkhäuser, Basel, Switzerland, 1990.

OLEINIK, O.A., *Discontinuous Solutions of Nonlinear Differential Equations*, American Mathematical Society Translations, Series 2, Vol. 26, pp. 95–172, 1957.

8

Distributions and Sobolev Spaces

This chapter contains a number of results that are used in the rest of the book. In particular, distribution theory and its natural progression into the theory of Sobolev spaces is presented. Enough material is given so that the discussion is coherent and not just a compendium of prerequisite theorems. Some of the proofs are sketched in the exercises, but several important theorems, for which the proofs are more involved, are not proven. The proofs are easily found in the references (Refs. 1–9). The acceptance of these results should not detract from following the main themes of the book.

1. Banach and Hilbert Spaces

If V is a real vector space, a *norm* is a real-valued function n defined on V such that

i. $n(x) \geq 0$ for $x \in V$ and $n(x) = 0$ if and only if $x = 0$;
ii. $n(\alpha x) = |\alpha| n(x)$ for all (real) scalars α;
iii. $n(x + y) \leq n(x) + n(y)$ for all $x, y \in V$.

We denote the norm of a vector x by $\|x\| = n(x)$. A metric on V is given by $\rho(x, y) = \|x - y\|$. If V is complete in this metric, i.e., Cauchy sequences converge to vectors in V, we call V with this norm a *Banach space*.

The following simple theorem is often useful in proving existence theorems for differential equations.

Theorem 1.1. Suppose that T is a mapping from a Banach space B into itself such that

$$\|T(x) - T(y)\| \leq k\|x - y\|,$$

where $0 < k < 1$, i.e., T is a contraction. Then there is a unique $x \in B$ such that $T(x) = x$.

Proof. Let $x_0 \in B$ and define the sequence $\{x_n\}$ by $x_{n+1} = T(x_n)$. Then, if $n > m$,

$$\|x_n - x_m\| \leq \sum_{j=m+1}^{n} \|x_j - x_{j-1}\| = \sum_{j=m+1}^{n} \|T^{j-1}(x_1) - T^{j-1}(x_0)\|$$
$$\leq \sum_{j=m+1}^{n} k^{j-1}\|x_1 - x_0\| \leq \frac{k^m}{1-k}\|x_1 - x_0\|. \qquad \square$$

A linear transformation T from a Banach space B_1 with norm $\|\cdot\|_1$ to a Banach space B_2 with norm $\|\cdot\|_2$ is said to be *bounded* if

$$\|T\| = \sup[\|Tx\|_2/\|x\|_1]$$

is finite, where the supremum is over nonzero vectors in B_1. It is easily shown that T is continuous if and only if T is bounded. We denote by $\mathscr{B}(B_1, B_2)$ the set of bounded linear transformations or *bounded linear operators* from B_1 into B_2. Another easy exercise shows that $\|T\|$ defines a norm on $\mathscr{B}(B_1, B_2)$.

Suppose that $B_1 = B_2 = B$. We say T is a bounded linear operator on B and write $T \in \mathscr{B}(B)$. The composition of operators in $\mathscr{B}(B)$ is again in $\mathscr{B}(B)$, and for $T, S \in \mathscr{B}(B)$,

$$\|TS\| \leq \|T\|\|S\|.$$

If T is one to one and onto and the inverse linear transformation is bounded, we say T is invertible and write T^{-1} for the inverse mapping. A simple sufficient condition for invertibility is given in the following result.

Theorem 1.2. Suppose that $\|S\| < 1$. Then $(I - S)^{-1}$ exists and

$$(I - S)^{-1} = \sum_{n=0}^{\infty} S^n.$$

Proof. $(I - S)(I + S + \cdots + S^n) = I - S^{n+1}$ and $\|S^n\| \leq \|S\|^n$ implies the result. $\square$

This result says a perturbation of the identity that is small enough in norm yields an invertible operator. We will next give similar results in which smallness is measured in a different way.

We say $T \in \mathscr{B}(B)$ is *compact* if T maps bounded sets into relatively compact sets. Equivalently, if $\|x_n\| \leq M$, $\{Tx_n\}$ has a convergent subsequence. If

$T(B)$ is finite-dimensional, we say T has *finite rank*. A finite rank operator is certainly compact. The following result is basic.

Theorem 1.3. A norm limit of compact operators is compact.

The proof is essentially a diagonalization argument and is sketched in Exercise 1.3. An immediate consequence is that the norm limit of finite rank operators is compact.

We mention also that a linear combination of compact operators is compact, and that T compact, $S \in \mathscr{B}(B)$ implies that TS and ST are compact. The following is a generalization to Banach spaces of a well-known alternative theorem for systems of linear algebraic equations.

Theorem 1.4 (Fredholm alternative). If $T \in \mathscr{B}(B)$ is compact, then either

i. $x - Tx = 0$ has a nonzero solution in B, or
ii. For each $y \in B, x - Tx = y$ has a unique solution in B, and $(I - T)^{-1} \in \mathscr{B}(B)$.

Proof. Let $S = I - T$ and $\mathscr{N} = S^{-1}(0)$. We claim that there is a $K > 0$ such that

$$\operatorname{dist}(x, \mathscr{N}) \leq K\|Sx\|.$$

If not, there is $\{x_n\}$ such that $\|Sx_n\| = 1$ and $d_n = \operatorname{dist}(x_n, \mathscr{N}) \to \infty$. Choose $y_n \in \mathscr{N}$ such that $d_n \leq \|x_n - y_n\| \leq 2d_n$, and let $z_n = (y_n - x_n)/\|y_n - x_n\|$. Then $\|z_n\| = 1$ and $\|Sz_n\| = \|Sx_n\|/\|y_n - x_n\| \leq d_n^{-1} \to 0$. By passing to a subsequence we may assume $Tz_n \to y_0 \in B$. Then $z_n = (S + T)z_n$ implies that $z_n \to y_0$, so that $y_0 \in \mathscr{N}$ (as S is continuous). But

$$\begin{aligned} \operatorname{dist}(z_n, \mathscr{N}) &= \|x_n - y_n\|^{-1} \inf_{\mathscr{N}} \|x_n - y_n - \|x_n - y_n\| y\| \\ &= \|x_n - y_n\|^{-1} \operatorname{dist}(x_n, \mathscr{N}) \geq \tfrac{1}{2}, \end{aligned}$$

so this cannot be true.

We claim now that $\mathscr{R} = S(B)$ is closed. Let $x_n \in B, Sx_n \to y \in B$. Let d_n and y_n be as above. Let $w_n = x_n - y_n$. As $\|w_n\| \leq 2d_n$ and $\{d_n\}$ is bounded, $\{w_n\}$ is a bounded sequence; also, $Sw_n \to y$. T is compact so that, after passing to a subsequence, $Tw_n \to w_0 \in B$. Then $w_n = (T + S)w_n \to y + w_0$, and continuity of S implies that $S(y + w_0) = y$. From this it follows that $\mathscr{R}$ is closed. □

Our next assertion is that if $\mathscr{N} = \{0\}$, then $\mathscr{R} \equiv S(B) = B$. This requires the following lemma. A proof is sketched in Exercise 1.4.

Lemma 1.1. If $\mathscr{M} \subset B$ is a closed linear manifold (or subspace), $\mathscr{M} \neq B$, then for any $\theta < 1$ there is an $x_\theta \in B$ such that $\|x_\theta\| = 1$ and $\operatorname{dist}(x_\theta, \mathscr{M}) \geq \theta$.

If $\mathscr{R}_j = S^j(B)$, our last result implies that they are closed subspaces of B and $\mathscr{R}_j \supset \mathscr{R}_{j+1}$. Suppose that each of these is a proper subspace of its predecessor. Then the lemma implies that we can choose $y_n \in \mathscr{R}_n$, $\|y_n\| = 1$ such that $\operatorname{dist}(y_n, \mathscr{R}_{n+1}) \geq 1/2$. Then, if $n > m$,

$$Ty_n - Ty_m = y_m + (-y_n - Sy_m + Sy_n) = y_m - y,$$

where $y \in \mathscr{R}_{m+1}$, and $\|Ty_n - Ty_m\| \geq 1/2$. This contradicts compactness of T so that $\mathscr{R}_j = \mathscr{R}_k, j \geq k$ for some k. For any $y \in B$, $S^k y = S^{k+1}x$ for some $x \in B$ as $\mathscr{R}_{k+1} = \mathscr{R}_k$, and $S^k(y - Sx) = 0$ implies $y = Sx$ as $S^{-k}(0) = S^{-1}(0) = \{0\}$. It follows that $\mathscr{R}_j = \mathscr{R} = B$ for all j.

Finally, if $\mathscr{R} = B$, then $\mathscr{N} = \{0\}$. To see this, consider $\mathscr{N}_j = S^{-j}(0)$, which are a sequence of closed subspaces with $\mathscr{N}_j \subset \mathscr{N}_{j+1}$. By applying Lemma 1.1 again we can show that $\mathscr{N}_j = \mathscr{N}_k$ for $j \geq k$ for some k (Exercise 1.5). $\mathscr{R} = B$ then implies that $S^k(B) = B$ and for $y \in \mathscr{N}_k$, $y = S^k x$. Then $S^{2k}x = S^k y = 0$, so that $x \in \mathscr{N}_{2k} = \mathscr{N}_k$, and $y = S^k x = 0$.

To see that $\mathscr{N} = \{0\}$ implies $(I - T)^{-1}$ is bounded we need only observe that our first assertion says $\|x\| \leq K\|(I - T)x\|$ in this case. Theorem 1.4 is proven.

A (real) number λ is called an *eigenvalue* of T if there is a nonzero vector x such that

$$Tx = \lambda x$$

and any such x is called a corresponding *eigenvector*. The dimension of the null space of $S_\lambda = (\lambda I - T)$ is called the multiplicity of λ. If $\lambda \neq 0$ is not an eigenvalue, Theorem 1.4 implies that $(\lambda I - T)^{-1} = R_\lambda$ is bounded.

Theorem 1.5. A compact operator has (at most) a countable set of eigenvalues having no limit points except possibly 0 and every nonzero eigenvalue has finite multiplicity.

The proof is sketched in Exercise 1.6.

If a vector space V has an *inner product*, i.e., a mapping q from $V \times V$ to $\mathbb{R}$ such that $(x, y \equiv q(x, y)$ satisfies for any (real) scalar λ, μ,

i. $(x, y) = (y, x)$,
ii. $(\lambda x + \mu y, z) = \lambda(x, z) + \mu(y, z)$,
iii. $(x, x) \geq 0$, $(x, x) > 0$ if $x \neq 0$,

V is said to be an inner product space. A first theorem on inner product spaces is the Cauchy–Schwarz inequality

$$|(x, y)|^2 \leq (x, x)(y, y).$$

It immediately follows that

$$(x + y,\ x + y) \leq [(x, x)^{1/2} + (y, y)^{1/2}]^2$$

and $\|x\| = (x, x)^{1/2}$ defines a norm. An additional property of this norm is

$$\|x + y\|^2 + \|x - y\|^2 = 2(\|x\|^2 + \|y\|^2).$$

This result is known as the parallelogram law. If all Cauchy sequences converge in this norm, we say V is a *Hilbert space*, and we denote it generically by H. A fundamental concept in Hilbert spaces is orthogonality. We say x and y are *orthogonal* if $(x, y) = 0$. If $\mathcal{M} \subset H$, $\mathcal{M}^\perp$ denotes the set of vectors orthogonal to all elements of $\mathcal{M}$. From the inequality $|(x_n - x, y)| \leq \|x_n - x\|\,\|y\|$, it follows that $\mathcal{M}^\perp$ is a closed linear manifold.

Theorem 1.6. If $\mathcal{M}$ is a closed subspace of H, then $x = y + z$, $y \in \mathcal{M}$, $z \in \mathcal{M}^\perp$ for any $x \in H$.

Proof. For $x \in \mathcal{M}$ set $z = 0$. Assuming $\mathcal{M} \neq H$ and $x \notin \mathcal{M}^\perp$, let $d := \text{dist}(x, \mathcal{M}) = \inf[\|y - x\| : y \in \mathcal{M}]$. Then there exists $y_n \in \mathcal{M}$ such that $\|y_n - x\| \to d > 0$. By the parallelogram law,

$$\|y_n - y_m\|^2 = 2\|x - y_m\|^2 + 2\|x - y_n\|^2 - 4\|x - \tfrac{1}{2}(y_n + y_m)\|^2.$$

As $\frac{1}{2}(y_n + y_m) \in \mathcal{M}$, this implies $\|y_n - y_m\| \to 0$, and as $\mathcal{M}$ is closed, $y = \lim_{n\to\infty} y_n \in \mathcal{M}$ satisfies $\|y - x\| = d$. Set $z = x - y$, then for all $y' \in \mathcal{M}$ and $\varepsilon > 0$,

$$d^2 \leq \|x - (y + \varepsilon y')\|^2 = \|z - \varepsilon y'\|^2 = d^2 - 2\varepsilon(y', z) + \varepsilon^2\|y'\|^2$$

as $\|z\| = d$. Letting $\varepsilon \to 0$ it follows that $(y', z) = 0$. $\square$

Theorem 1.6 also holds if $\mathcal{M}$ is a *closed convex* set in H (exercise 1.8). The element y is called the *projection* of x on $\mathcal{M}$.

In a Banach space B the operators $\mathscr{B}(B, \mathbb{R})$ are called bounded linear functionals, and this Banach space is called the *dual space* of B and is denoted by B'. If $f \in B'$, $x \in B$ we denote by $\langle f, x \rangle$ the (real) value taken by f at x.

Definition 1.1 (Weak convergence). A sequence $x_n \in B$ converges weakly to $x \in B$ if $\langle f, x_n \rangle \to \langle f, x \rangle$ for all $f \in B'$.

A subset $K \subset B$ is *weakly compact* if every sequence in K has a subsequence converging weakly to an element of K. An operator $A \in \mathscr{B}(B_1, B_2)$ is *weakly continuous* at x if Ax_n converges weakly to Ax in B_2 whenever x_n converges weakly to x in B_1.

The following theorem shows that the dual space of a Hilbert space has a very simple characterization.

Theorem 1.7 (Riesz representation theorem). If $f \in H'$, there is a uniquely determined $y = y_f \in H$ such that $f(x) = (x, y)$ for all $x \in H$. Further, $\| f \| = \|y\|$.

Proof. Let $\mathscr{N}$ be the null space of f. We may assume that $\mathscr{N} \neq H$. Choose $z \in H$, $z \neq 0$ such that $(x, z) = 0$ for all $x \in \mathscr{N}$. (This can be done by Theorem 1.6.) Then

$$f(xf(z) - zf(x)) = 0$$

for any $x \in H$, and $xf(z) - zf(x) \in \mathscr{N}$, so that $(xf(z) - zf(x), z) = 0$, that is,

$$f(x) = (x, y),$$

where $y = zf(z)/\|z\|^2$. In order to see that y is unique, we need only observe that $f(x) = (x, y) = (x, \tilde{y})$ implies that $y - \tilde{y} \in H^{\perp} = \{0\}$. As $|(x, y)| \leq \|x\| \|y\|$,

$$\|f\| = \sup_{x \neq 0} \frac{|(x, y)|}{\|x\|} \leq \| y\|$$

and

$$\|y\|^2 = (y, y) = f(y) \leq \| f\| \|y\|$$

so that $\|y\| \leq \|f\|$. □

The case we have in mind is infinite-dimensional Hilbert spaces, e.g., $H = L^2(\Omega)$ (the space of square integrable functions on an open set Ω, see Section 3). In view of the isometric isomorphism $y = Jf$ established in Theorem

1.7, one may identify H and H'. This identification is often convenient (but not always; e.g. Brézis, Ref. 1, p. 81).

A generalization of Theorem 1.7 is frequently useful in proving existence theorems for differential equations.

Theorem 1.8 (Lax–Milgram lemma). Suppose that $b : H \times H \to \mathbb{R}$ is linear in each variable and satisfies

i. $|b(x, y)| \leq K\|x\|\,\|y\|$, $x, y \in H (K > 0)$,
ii. $b(x, x) \geq \nu\|x\|^2$, $x \in H (\nu > 0)$

(in words, b is a *bounded* and *coercive* bilinear form defined on H). Then for every $f \in H'$ there is a unique $y \in H$ such that $b(x, y) = f(x)$.

Proof. For each $y \in H$, $b(x, y)$ defines an element of H' so there is a unique $z \in H$ such that $b(x, y) = (x, z)$. Letting $z = Ty$, we see that $T : H \to H$ is a linear mapping. Further, $|(x, Ty)| \leq K\|x\|\,\|y\|$ implies that $\|Ty\| \geq K\|y\|$, i.e., T is bounded, $\|T\| \geq K$. By (ii),

$$\nu\|y^2 \leq b(y, y) = (y, Ty) \leq \|y\|\,\|Ty\|,$$

so $\|Ty\| \geq \nu\|y\|$. This implies that $T(H)$ is closed and that T is one to one. We claim that T is onto. If not, there is $z \neq 0$ such that $(z, Ty) = 0$ for all $y \in H$, and choosing $y = z$ yields $(z, Tz) = b(z, z) = 0$, a contradiction. The inequality $\|Ty\| \geq \nu\|y\|$ shows that the linear mapping T^{-1} is bounded (with $\|T^{-1}\| \leq \nu^{-1}$). If $f(x) = (x, y_f)$, then the result follows if we set $y = T^{-1}y_f$. □

If $T \in \mathcal{B}(H)$ and $y \in H$, $f(x) = (Tx, y)$ defines a bounded linear functional on H so there is a unique $z \in H$ such that $(Tx, y) = (x, z)$; letting $z = T^*y$, then T^* is a linear mapping on H and $\|z\| = \|f\| \leq \|T\|\,\|y\|$. Hence, T^* is bounded, $\|T^*\| \leq \|T\|$. We can form $T^{**} := (T^*)^*$ and

$$(Tx, y) = (x, T^*y) = (T^*y, x) = (y, T^{**}x) = (T^{**}x, y)$$

implies $T^{**} = T$. As $\|T^{**}\| \leq \|T^*\|$, we see that $\|T^*\| = \|T\|$. T^* is called the *adjoint* of T. (For finite-dimensional spaces, T^* is the transpose of the matrix T.)

An operator T has *finite rank* if $T(H)$ is finite dimensional. Such an operator is certainly compact. The following result is an immediate consequence of Theorem 1.3.

Lemma 1.2. If $T \in \mathcal{B}(H)$ and there is a sequence of finite rank operators $\{T_n\}$ such that $\|T_n - T\| \to 0$, then T is compact.

This statement has a converse (at least for Hilbert spaces, as is the case here).

Lemma 1.3. If $T \in \mathscr{B}(H)$ is compact, there is a sequence of finite rank operators $\{T_n\}$ such that $\|T_n - T\| \to 0$.

The proof is left as an exercise.

The *Fredholm alternative* can be specialized in a useful way in the Hilbert space case.

Theorem 1.9. If $T \in \mathscr{B}(H)$ is compact, there is (at most) a countable set $\Lambda := \mathrm{sp}(T)$ of (real) eigenvalues having no nonzero limit points. Let $\lambda \neq 0$. Then the following alternative holds.

i. If $\lambda \notin \Lambda$, $(\lambda I - T)^{-1}$ and $(\lambda I - T^*)^{-1} \in \mathscr{B}(H)$.
ii. If $\lambda \in \Lambda$, $\lambda I - T$ and $\lambda I - T^*$ have finite-dimensional null spaces, having the same dimension. Further, the equation $\lambda x - Tx = y$ is solvable if and only if y is orthogonal to the null space of $\lambda I - T^*$, and the "adjoint" equation $\lambda x - T^* x = y$ is solvable if and only if y is orthogonal to the null space of $\lambda I - T$.

Proof. First we show that T^* is compact. If $\|x_n\| \leq M$, then

$$\|T^* x_n\|^2 = (x_n, TT^* x_n) \leq M\|T\|\|T^* x_n\|,$$

so $\{T^* x_n\}$ is bounded. Because $\{TT^* x_n\}$ converges,

$$\|T^*(x_n - x_m)\|^2 = (x_n - x_m, TT^*(x_n - x_m)) \leq 2m\|TT^*(x_n - x_m)\|,$$

so $\{T^* x_n\}$ converges.

If $\mathscr{R} = T(H)$, and $\mathscr{N}^*$ is the null space of T^*, we claim $\mathscr{R}^\perp = \mathscr{N}^*$. If $z \in \mathscr{N}^*$, and $y \in \mathscr{R}$, $y = Tx$, then

$$(y, z) = (Tx, z) = (x, T^* z) = 0,$$

so $\bar{\mathscr{R}} \subset \mathscr{N}^{*\perp}$. If $y \notin \bar{\mathscr{R}}$, let $y = y_1 + y_2$, $y_1 \in \bar{\mathscr{R}}$, $y_2 \in \bar{\mathscr{R}}^\perp \setminus \{0\}$. Then, as $0 = (y_2, Tx) = (T^* y_2, x)$ for all $x, y_2 \in \mathscr{N}^*$. As $(y_2, y) = \|y_2\|^2 \neq 0$, it follows $y \notin \mathscr{N}^{*\perp}$.

Finally, we need to show that the null spaces of $\lambda I - T$ and $\lambda I - T^*$ have the same dimensions. By Lemma 1.3 we can write $T = T_0 + T_1$ where T_0 has finite rank and $\|T_1\| < \lambda$. Then $T_2 := (\lambda I - T_1)^{-1} T_0$ has finite rank, where $(\lambda I - T_1)^{-1}$ is given by the norm-convergent series

$$(\lambda I - T_1)^{-1} = \sum_{n=0}^{\infty} \lambda^{-n-1} T_1^n.$$

A direct calculation shows that $Tx = \lambda x$ if and only if $T_2 x = x$ and $T^* x = \lambda x$ if and only if $T_2^* x = x$. These last are finite matrix equations for which the result is true. □

We will always consider *separable* Hilbert spaces, i.e., such that they contain a countable dense set. We have seen that a sequence $x_n \in H$ is said to be *weakly convergent* to $x \in H$ if $f(x_n) \to f(x)$ as $n \to \infty$ for every $f \in H'$. By Riesz's theorem, this implies $(y, x_n - x) \to 0$ as $n \to \infty$ for every $y \in H$. Separable Hilbert spaces are characterized by the existence of countable orthonormal bases $\{y_n\}$ with $(y_n, y_m) = \delta_{n,m}$ such that every $x \in H$ can be expanded in the (generalized) Fourier series

$$x = \sum_{n=0}^{\infty} (x, y_n) y_n \tag{FS}$$

convergent in the norm of H. It follows that $(x, y_n) = 0$ for every n implies $x = 0$. We say that $\{y_n\}$ is a *complete orthonormal set*.

Theorem 1.10. Every bounded sequence in a separable Hilbert space H has a weakly convergent subsequence.

Thus, every bounded set in a separable Hilbert space is "weakly compact." For example, as $\|x\|^2 = \sum_{n=1}^{\infty} (x, y_n)^2 < \infty$ for every $x \in H$ (Exercises 1.12, 1.13), the sequence y_n converges weakly to zero. On the other hand, $\|y_n - y_m\|^2 = 2$ for every $n \neq m$, hence the sequence is not norm convergent.

2. Theory of Distributions

We will denote by $\mathbf{x} = (x_1, \ldots, x_n)$ a point in $\mathbb{R}^n$. For any multindex $\alpha = (\alpha_1, \ldots, \alpha_n)$ of nonnegative integers, $|\alpha| := \alpha_1 + \cdots + \alpha_n$, $\alpha! := \alpha_1! \ldots \alpha_n!$, and $\alpha \leq \beta$ if $\alpha_j \leq \beta_j$, $j = 1, \ldots, n$. We can then write $\mathbf{x}^\alpha := x_1^{\alpha_1} \ldots x_n^{\alpha_n}$ for a monomial in $\mathbf{x}$, and

$$\partial^\alpha := \partial_1^{\alpha_1} \ldots \partial_n^{\alpha_n} \equiv \frac{\partial^{|\alpha|}}{\partial x_1^{\alpha_1} \ldots \partial x_n^{\alpha_n}}$$

for a general partial derivative of order $|\alpha|$. If $|\alpha| = k$ and $u \in C^k$, we can write $\partial^\alpha u$ without possible confusion as the result is independent of the order of differentiations. If $u \in C^k$, then Taylor's theorem can be written

$$u(\mathbf{x}+\mathbf{y}) = \sum_{|\alpha|<k} \frac{\mathbf{y}^\alpha}{\alpha!}\partial^\alpha u(\mathbf{x}) + \sum_{|\alpha|=k} k\frac{\mathbf{y}^\alpha}{\alpha!}\int_0^1 (1-t)^{k-1}\partial^\alpha u(\mathbf{x}+t\mathbf{y})dt$$

for $\mathbf{x}$, $\mathbf{y} \in \mathbb{R}^n$. The binomial coefficients $\binom{\alpha}{\beta}$ are defined by

$$\binom{\alpha}{\beta} = \frac{\alpha!}{\beta!(\alpha-\beta)!}$$

for $0 \le \beta \le \alpha$ and $\binom{\alpha}{\beta}$ otherwise. We then have the binomial formula

$$(\mathbf{x}+\mathbf{y})^\alpha = \sum_\beta \left\{\begin{matrix}\alpha\\ \beta\end{matrix}\right\} \mathbf{x}^\beta \mathbf{y}^{\alpha-\beta}$$

and Liebnitz's formula

$$\partial^\alpha(uv) = \sum_\beta \binom{\alpha}{\beta}\partial^\beta u \partial^{\alpha-\beta} v$$

for $u, v \in C^{|\alpha|}$. We will write $\mathbf{x}\cdot\mathbf{y}$ for the scalar product of $\mathbf{x}, \mathbf{y} \in \mathbb{R}^n$ and $|\mathbf{x}| = (\mathbf{x}\cdot\mathbf{x})^{1/2}$. We remark that the integrals

$$\int_{\mathbb{R}^n} (1+|\mathbf{x}|)^{-s} d\mathbf{x}, \qquad \int_{\mathbb{R}^n} (1+|\mathbf{x}|^2)^{-s/2} d\mathbf{x} \tag{1}$$

are finite if $s > n$, and $\int (1+|\mathbf{x}|^2)^{-n}\, d\mathbf{x} \le \pi^n$. These will be useful in estimating functions near infinity and zero.

A central role is placed here by the space

$$\mathscr{S} = \left\{\varphi \in C^\infty : \sup_{\mathbb{R}^n} |\mathbf{x}^\alpha \partial^\beta \varphi| < \infty \text{ for all } \alpha, \beta\right\}.$$

We denote

$$|\varphi|_k = \sup_{|\alpha+\beta| \le k} \sup_{\mathbb{R}^n} |\mathbf{x}^\alpha \partial^\beta \varphi|$$

for $\varphi \in \mathscr{S}$. Convergence in $\mathscr{S}$ is then defined by simultaneous convergence in all of the norms $|\cdot|_k \cdot \mathscr{S}$ is closed under differentiation and multiplication by C^∞

functions with polynomial growth at infinity. We denote this class of functions by $\mathscr{P}$,

$$\mathscr{P} = \{\psi : (1 + |\mathbf{x}|^2)^{-N} \partial^\alpha \psi(\mathbf{x}) \text{ is bounded for some } N = N(\alpha) \text{ for all } \alpha\}.$$

Then

$$|\partial^\alpha \varphi|_k \leq |\varphi|_{k+|\alpha|} \tag{2}$$

and, for $\psi \in \mathscr{P}$, there are constants C_k and N_k such that

$$|\psi\varphi|_k \leq C_k |\varphi|_{k+2N_k} \tag{3}$$

for $k \geq 0$ and $\varphi \in \mathscr{S}$. These inequalities show that differentiation and multiplication by functions in $\mathscr{P}$ are continuous operations on $\mathscr{S}$. We denote by C_0^∞ the subspace of $\mathscr{S}$ consisting of C^∞ functions with compact support. We recall that the support, supp φ, can be characterized by $\mathbf{x} \notin \text{supp } \varphi$ if $\varphi = 0$ in a neighborhood of $\mathbf{x}$. If Ω is an open set containing supp φ, $\varphi \in C_0^\infty$, we write $\varphi \in C_0^\infty(\Omega)$. The following lemma is useful in extending from local to global properties.

Lemma 2.1 (Partitions of unity). If K is a compact set in $\mathbb{R}^n$, $K \subset \bigcup_j \Omega_j$, Ω_j open, then there are a finite number of functions $\varphi_j \in C_0^\infty(\Omega_j)$, $\varphi \geq 0$, $\sum_j \varphi_j \leq 1$, and $\sum_j \varphi_j = 1$ in a neighborhood of K.

The proof makes use of a "unit test function" that is supported in the unit ball, nonnegative, and has integral 1 (Exercises 2.4, 2.5).

We denote by L^p the usual Banach spaces of (Equivalence classes equal a.e.) of Lebesgue measurable functions with norm

$$\|u\|_{L^p} = \left(\int_{\mathbb{R}}^n |u(\mathbf{x})|^p d\mathbf{x} \right)^{1/2}, \qquad 1 \leq p < \infty,$$

and

$$\|u\|_{L^\infty} = \inf\{M : |u(\mathbf{x})| \leq M \textit{ for a.e. } \mathbf{x}\}.$$

If $u\bar{v} \in L^1(\bar{v}$ the complex conjugate of $v)$, we write

$$(u, v) = \int_{\mathbb{R}^n} u(\mathbf{x})\overline{v(\mathbf{x})} d\mathbf{x}.$$

(Although we will often be interested in real-valued functions in the applications in this book, complex-valued functions are required for a convenient discussion of the Fourier transform. It is easy to pass from the real-valued to the complex-valued case using the complexification of a real vector space.) In particular, this defines the inner product on the Hilbert space L^2.

Proposition 2.1 $\mathscr{S} \subset L^p$ for all $1 \le p \le \infty$ and $\|\varphi\|_{L^p} \le (2\pi)^n |\varphi|_{2n}$ for $\varphi \in \mathscr{S}$. If $u \in L^p$, $1 \le p \le \infty$, and $\varphi \in \mathscr{S}$, then $u\bar{\varphi} \in L^1$ and

$$|(u, \varphi)| \le (2\pi)^n \|u\|_{L^p} |\varphi|_{2n}.$$

Further, if u is measurable and $u\bar{\varphi} \in L^1$, then $(u, \varphi) = 0$ for all $\varphi \in \mathscr{S}$ implies $u = 0$ a.e., and if U is a linear functional on $\mathscr{S}$ satisfying $|U(\varphi)| \le C\|\varphi\|_{L^2}$, there is a unique $u \in L^2$, $\|u\|_{L^2} \le C$, such that $U(\varphi) = (u, \varphi)$ for $\varphi \in \mathscr{S}$.

If $u \in L^1$, the *Fourier transform* $\hat{u}(\xi)$ is defined by

$$\hat{u}(\xi) \equiv \mathscr{F}[u(\mathbf{x})] := \int_{\mathbb{R}^n} e^{-i\mathbf{x} \cdot \xi} u(\mathbf{x}) d\mathbf{x}. \tag{4}$$

The dominated convergence theorem implies that $\hat{u}(\xi)$ is a bounded ($|u(\xi)| \le \|u\|_{L^1}$), continuous function, and the content of the Riemann–Lebesgue Lemma is that $\hat{u}(\xi)$ vanishes at infinity, i.e., $\lim_{|\xi| \to \infty} \hat{u}(\xi) = 0$, for all $u \in L^1$. We will need some properties of $\hat{u}(\xi)$ for $u \in \mathscr{S}$. We denote $u(-\mathbf{x})$ by $\check{\varphi}(\mathbf{x})$, and $D_j = -i\partial_j$. [It follows that $D^\alpha = (-i)^{|\alpha|}\partial^\alpha$.]

Proposition 2.2 If $\varphi \in \mathscr{S}$, then $\hat{\varphi} \in \mathscr{S}$ and $|\hat{\varphi}|_k \le (8\pi)^n (k+1)! |\hat{\varphi}|_{2n+k}$. Further, for any multindex α,

$$\mathscr{F}[D_x^\alpha \varphi(\mathbf{x})] = \xi^\alpha \hat{\varphi}(\xi), \qquad \mathscr{F}[x^\alpha \varphi] = -D_\xi^\alpha \hat{\varphi}(\xi). \tag{5}$$

This follows from the first inequality in Proposition 2.1 and differentiation under the integral.

Proposition 2.3. If $u \in L^1$, $\varphi \in \mathscr{S}$, then $(\hat{u}, \varphi) = (\check{\varphi}, \varphi)$.

Proposition 2.4. Suppose that $\varphi, \psi \in \mathscr{S}$. Then

$$\varphi(\mathbf{x}) = (2\pi)^{-n} \int_{\mathbb{R}^n} e^{i\mathbf{x} \cdot \xi} \hat{\varphi}(\xi) d\xi \qquad \text{(Inversion Theorem)}, \tag{6}$$

i.e., $\hat{\hat{\varphi}} \equiv \mathscr{F}^2[\varphi] = (2\pi)^n \check{\varphi}$, and

$$(\hat{\varphi}, \hat{\psi}) = (2\pi)^n (\varphi, \psi) \qquad \text{(Parseval's identity).} \tag{7}$$

Corollary 2.1. Suppose that $\varphi, \psi \in \mathscr{S}$. Then

$$\mathscr{F}\left[\int_{\mathbb{R}^n} \varphi(\mathbf{x} - \mathbf{y})\psi(\mathbf{y}) d_{\mathbf{y}}\right] = \hat{\varphi}\hat{\psi} \qquad \text{(Convolution theorem).}$$

The proof follows immediately from Fubini's theorem.

A linear functional U on $\mathscr{S}$ will be said to be continuous if there is a constant $C > 0$ and an integer N such that

$$|U(\varphi)| \leq C|\varphi|_N$$

for $\varphi \in \mathscr{S}$. Proposition 2.1 implies that $U(\varphi) = (u, \varphi)$, where $u \in L^p$, is a continuous linear functional on $\mathscr{S}$. We adopt the notation (u, φ) for $U(\varphi)$ whether or not U has a representation in terms of an L^p function. The set of continuous linear functionals on $\mathscr{S}$ are called the *temperate distributions* and are denoted by $\mathscr{S}'$. If $u \in \mathscr{S}'$, $\varphi \in \mathscr{S}$, then

$$(\check{u}, \varphi) = (u, \check{\varphi}) \quad \text{and} \quad (\hat{u}, \varphi) = (\check{u}, \hat{\varphi})$$

define $\check{\varphi}$ and $\hat{u}$ as elements of $\mathscr{S}'$.

By Proposition 2.4, if $\varphi \in \mathscr{S}$ and $\hat{\psi} = \varphi$, then $(\hat{\hat{u}}, \varphi) = (2\pi)^n (\check{u}, \varphi)$. Hence, if $u \in \mathscr{S}'$, then $\hat{\hat{u}} = (2\pi)^n \check{u}$.

Proposition 2.5 (Plancherel's theorem). If $u \in L^2$, then $\hat{u} \in L^2$,

$$(2\pi)^{-n/2} \|\hat{u}\|_{L^2} = \|u\|_{L^2}$$

and $(\hat{u}, \hat{v}) = (2\pi)^n (u, v)$.

A "direct" definition of the Fourier transform in L^2 is

$$\hat{u}(\xi) = \lim_{L \to \infty} \int_{Q_L} e^{-i\mathbf{x} \cdot \xi} u(\mathbf{x}) d\mathbf{x}, \qquad Q_L := \{x_i : |x_i| < L\},$$

the convergence being in the norm of L^2 (Fourier–Plancherel transform).

We can also define differentiation on $\mathscr{S}'$ by

$$(D^\alpha u, \varphi) = (u, D^\alpha \varphi) \qquad \text{for all } \varphi \in \mathscr{S}. \tag{8}$$

For a function $u \in C^{|\alpha|}$, integration by parts yields (8) for all $\varphi \in C_0^\infty$. If u and $D^\alpha u$ have behavior at infinity such that $u, D^\alpha u \in \mathscr{S}'$, the following proposition implies that the distribution given by the function $D^\alpha u$ and (8) are consistent.

Proposition 2.6. If u, v agree on C_0^∞, then $u = v$.

If $u \in \mathscr{S}'$ and $\psi \in \mathscr{P}$, we can define ψu by

$$(\psi u, \varphi) = (u, \bar{\psi}\varphi)$$

as $\bar{\psi}\varphi \in \mathscr{S}$. We can extend familiar properties of Fourier transforms and partial differentiation [in particular (5)] to temperate distributions.

Proposition 2.7. For $u \in \mathscr{S}'$ and $\psi \in \mathscr{P}$,

$$D^\alpha(\psi u) = \sum_\beta \binom{\alpha}{\beta} (D^\beta \psi)(D^{\alpha-\beta} u)$$

and

$$\mathscr{F}[D_x^\alpha u] = \xi^\alpha \hat{u}, \qquad \mathscr{F}[x^\alpha u] = -D_\xi^\alpha \hat{u}.$$

If we give up the use of Fourier transforms, we can extend the idea of distributions considerably. If $\Omega \subset \mathbb{R}^n$ is open, $\mathscr{D}'(\Omega)$ is the set of linear functionals on $C_0^\infty(\Omega)$ that satisfy: for each compact $K \subset \Omega$ there are nonnegative constants C_K and N_K (with N_K an integer) such that

$$|(u, \varphi)| \leq C_K |\varphi|_{N_K} \tag{9}$$

for $\varphi \in C_0^\infty(\Omega)$ with supp $\varphi \subset K$. Differentiation of $u \in \mathscr{D}'(\Omega)$ is defined as in $\mathscr{S}'$,

$$(D^\alpha u, \varphi) = (u, D^\alpha \varphi)$$

for $\varphi \in C_0^\infty(\Omega)$. If $u \in \mathscr{D}'(\Omega)$ and $\psi \in C^\infty(\Omega)$, $(\psi u, \varphi) = (u\bar{\psi}\varphi)$. Proposition 2.6 shows that $\mathscr{S}'$ can be identified with a linear manifold contained in $\mathscr{D}' = \mathscr{D}'(\mathbb{R}^n)$. More generally, if $u \in \mathscr{D}'(\Omega)$, and ω is an open subset of Ω, the restriction of u to ω is just u restricted to test functions φ in $C_0^\infty(\omega)$. This leads naturally to study the local behavior of a distribution. The following proposition shows that the local behavior determines a distribution completely.

Proposition 2.8. If u, $v \in \mathscr{D}'(\Omega)$ and $u = v$ in a neighborhood of every point of Ω, $u = v$.

The proof is accomplished using a partition of unity.

We can then define the support of a distribution, supp u for $u \in \mathscr{D}'(\Omega)$, by $\mathbf{x} \notin \text{supp } u$ if $u = 0$ in some neighborhood of $\mathbf{x}$. Similarly for the singular support, sing supp u, $\mathbf{x} \notin \text{sing supp } u$ if $u = \psi \in C^\infty(\omega)$ for some neighborhood ω of $\mathbf{x}$. Distributions with compact supports extend naturally to $\mathbb{R}^n$ as elements of $\mathscr{S}'$. [If $\psi \in C_0^\infty(\Omega)$, $\psi = 1$ on supp u and supp u compact, then $(u, \varphi) = (u, \psi\varphi)$ for all $\varphi \in C_0^\infty(\Omega)$ and this definition extends to $\mathscr{S}$ as $\psi\varphi \in C_0^\infty(\Omega)$ for $\varphi \in \mathscr{S}$. Further, $|(u, \varphi)| \leq C|\psi\varphi|_N \leq C_\psi C|\varphi|_N$. In fact, this definition extends to C^∞.]

We mention at this point several commonly occurring examples. If $u \in L^1_{\text{loc}}(\Omega)$, then u defines a distribution in $\mathscr{D}'(\Omega)$. The "δ-function" is the distribution $(\delta, \varphi) = \varphi(0)$ (unit measure concentrated at $\mathbf{x} = 0$.) As supp $\delta = \{0\}$, certainly $\delta \in \mathscr{S}'$. If u has discontinuities, the (distribution) derivative contains δ-functions: If $n = 1$ and $H(x)$ is the Heaviside function,

$$H(x) = \begin{cases} 1, & x \geq 0, \\ 0, & x < 0, \end{cases}$$

then a direct calculation shows that $H' = \delta$. More generally, if f is continuously differentiable for $x \neq 0$ where $f(0+) - f(0-) = \sigma \neq 0$, we get the integration by parts formula

$$(f', \varphi) = \int_{-\infty}^{\infty} \{f'(x)\}\varphi(x)dx + \sigma(\delta, \varphi) = -\int_{-\infty}^{\infty} f(x)\varphi'(x)dx.$$

In the integral $\{f'\}$ is the function equal to the derivative where it exists. This result generalizes to functions defined and differentiable except for jumps across smooth (hyper) surfaces in $\mathbb{R}^n$. If $\mathscr{S}$ is such a hypersurface, then

$$(\partial^i f, \varphi) = -\int_{\mathbb{R}^n} f(\mathbf{x})\partial^i\varphi(\mathbf{x})d\mathbf{x} = \int_{\mathbb{R}^n} \{\partial^i f(\mathbf{x})\}\varphi(\mathbf{x})d\mathbf{x} + \int_{\mathscr{S}} \sigma\varphi dS, \tag{10}$$

where σ is the jump of f across S, and $\{\partial^i f(\mathbf{x})\}$ is the function equal to the ith partial derivative where it exists. We may describe the last term as arising from a δ measure supported by $\mathscr{S}$.

The principal value integral

$$PV \int_{\mathbb{R}} \frac{\varphi(x)}{x}\, dx = \lim_{\varepsilon \to 0} \int_{|x| > \varepsilon} \frac{\varphi(x)}{x}\, dx$$

exists if $\varphi \in C_0^\infty(\mathbb{R})$. This defines a distribution u that we denote by $PV(1/x)$.

The Fourier transform of δ can be calculated directly to be 1. More generally, for any distribution u with compact support we can show that $\hat{u}$ is given by the function

$$\hat{u}(\xi) = (u, e^{-i\xi \cdot \mathbf{x}}).$$

[$\hat{u}$ can be extended to $C^\infty(\mathbb{R}^n)$.] If we write, for $\varsigma \in \mathbb{C}^n$, $\xi = \text{Re}\ (\varsigma)$, then $\hat{u}$ extends to $\mathbb{C}^n$ as an analytic function. The celebrated Paley–Wiener–Schwartz theorem gives the following further information.

Proposition 2.9. If u is a distribution with compact support contained in the ball B_r, then $\hat{u}(\xi)$ is the restriction of an analytic function $\hat{u}(\varsigma)$ defined on $\mathbb{C}^n$ such that

$$|\hat{u}(\varsigma)| \le C(1 + |\varsigma|)^N e^{r|\text{IM}(\varsigma)|}$$

for some positive constants C and N. This condition is also sufficient that $\hat{u}(\xi)$ be the Fourier transform of a distribution with support contained in B_r.

The following proposition gives a useful characterization of distributions. We need to define formally convergence in $\mathscr{D}'(\Omega)$:

$$u_n \to u \qquad \text{in } \mathscr{D}'(\Omega)$$

if $(u_n, \varphi) \to (u, \varphi)$ for all $\varphi \in C_0^\infty(\Omega)$.

Proposition 2.10. For any $u \in \mathscr{D}'(\Omega)$ there is a sequence of functions $u_n \in C_0^\infty(\Omega)$ such that $u_n \to u$ in $\mathscr{D}'(\Omega)$.

This proposition shows that a distribution can be envisaged as a weakly convergent sequence of test functions. The result extends to $\mathscr{S}'$. In particular, if $u = \delta$, we call such a sequence a δ-*approximate sequence*. A set of sufficient conditions of a sequence are

i. $u_n \ge 0$,
ii. $\int u_n = 1$,
iii. $u_n \to 0$ as $n \to \infty$ uniformly for $|\mathbf{x}| \ge \eta$ for each $\eta > 0$.

In fact, we have seen in Chapter 3 that the Gaussian kernel for the heat equation satisfies these hypotheses for any sequence $t_n \to 0$.
Finally, we consider distributions as possible solutions to differential equations. If $f \in L^1_{\text{loc}}(\mathbb{R})$, then $u(x, t) = f(x - ct)$ might be thought a reasonable candidate to be a solution of the wave equation $u_{tt} = c^2 u_{xx}$. As u defines a distribution in $\mathbb{R}^2$, we need only find $c^2 u_{xx} - u_{tt}$ in $\mathscr{D}'$. With $(u_{tt}, \varphi) = (u, \varphi_{tt})$ and $(u_{xx}, \varphi) = (u, \varphi_{xx})$, we have

$$(c^2 u_{xx} - u_{tt}, \varphi) = (u, c^2 \varphi_{xx} - \varphi_{tt}) = \int\int f(x - ct)(c^2 \varphi_{xx} - \varphi_{tt}) dx\, dt.$$

We introduce the change of variables $y = x - ct$, $z = x + ct$ in this integral obtaining

$$4c^2 \int\int f(y)\varphi_{yz}\, dzdy,$$

and this integral certainly vanishes by virtue of the compact support of φ. We see now that d'Alembert's solution with discontinuous data makes perfectly good sense in $\mathscr{D}'$.

As another example, suppose $n = 3$, and $u = 1/|\mathbf{x}|$. Then, as $u \in L^1_{\text{loc}}(\mathbb{R}^3)$, u defines a distribution. A calculation using Stokes' theorem shows that

$$\Delta u = -4\pi\delta.$$

More generally, if $u = E(\mathbf{x} - \mathbf{y})$ is the fundamental solution of the Laplace operator in $\mathbb{R}^n$, then the Green theorem (R) in Section 2.1 of Chapter 4 and (2.8) shows that

$$\Delta_x E(\mathbf{x} - \mathbf{y}) - \delta(\mathbf{x} - \mathbf{y}). \tag{11}$$

3. Sobolev Spaces

We begin with the special case in which functions are defined in all of $\mathbb{R}^n$ and the derivatives are square integrable. We define

$$\lambda^s(\xi) = (\lambda(\xi))^s = (1 + |\xi|^2)^{s/2}$$

for $s \in \mathbb{R}$. Then

$$H^s := \{u \in \mathscr{S}' : \lambda^s \hat{u} \in L^2\}$$

and $\|\ \|_s$ is defined by

$$\|u\|_s^2 := (2\pi)^{-n} \int_{\mathbb{R}^n} (1 + |\xi|^2)^s |\hat{u}(\xi)|^2 d\xi.$$

As $s \geq t$ implies $H^s \subset H^t$, we can define $H^\infty = \cap H^s, H^{-\infty} = \cup H^s$, and $\mathscr{S} \subset H^\infty \subset H^{-\infty} \subset \mathscr{S}'$. One has $u \in H^{s+1}$ if and only if u, $D_1 u, \ldots,$ $D_n u \in H^s$ as

$$|\lambda^{s+1}\hat{u}|^2 = \lambda^2 |\lambda^s \hat{u}|^2 = |\lambda^s \hat{u}|^2 + \sum_j |\lambda^s \xi_j \hat{u}|^2$$

and $\xi_j \hat{u} = \mathscr{F}[D_j u]$. For s a positive integer we obtain, by induction, $u \in H^k$ if and only if $D^\alpha u \in L^2$ for $|\alpha| \leq k$. If $u \in H^s, s > n/2$, then, as $\lambda^{-s} \in L^2, \hat{u} = \lambda^{-s}(\lambda^s \hat{u}) \in L^1$ so we know that u is bounded and continuous. Also, if $s = n$,

$$|u|_0 \geq (2\pi)^{-n} \|\hat{u}\|_{L^1} \leq (2\pi)^{-n/2} \|\lambda^{-n}\|_0 (2\pi)^{-n/2} \|\lambda^n u\|_0 \leq 2^{-n/2} \|u\|_n.$$

More generally, for $s > n/2$, $|u|_o \leq C_s \|u\|_s$ if $u \in H^s$. We will return to results of this type shortly.

We can characterize H^s as the space of continuous linear functionals on H^{-s}. We remark that if $u \in H^s$, $s > 0$, then, as $\hat{u} \in L^2$, $u \in L^2$.

Proposition 3.1. If $u \in H^s$, $\varphi \in \mathscr{S}$, then $|(u, \varphi)| \leq \|u\|_s \|\varphi\|_{-s}$. If $u \in \mathscr{S}'$ and $|(u, \varphi)| \leq C\|\varphi\|_{-s}$ for all $\varphi \in \mathscr{S}$, then $u \in H^s$ and $\|u\|_s \leq C$.

Because

$$|(u, \varphi)| = (2\pi)^{-n} |(\hat{u}, \hat{\varphi}| = (2\pi)^{-n} |(\lambda^s \hat{u}, \lambda^{-s} \hat{\varphi})| \leq \|u\|_s \|\varphi\|_{-s},$$

the first conclusion is immediate. On the other hand, if $u \in \mathscr{S}'$ satisfies $|(u, \varphi)| \leq C\|\varphi\|_{-s}$,

$$\begin{aligned} |(\lambda^s \hat{u}, \varphi)| &= |(\hat{u}, \lambda^s \varphi)| = |(\check{u}, \mathscr{F}[\lambda^s \varphi])| \leq C \|\mathscr{F}[\lambda^s \varphi]\|_{-s} \\ &= C(2\pi)^{-n/2} \|\lambda^{-s} \mathscr{F}^2[\lambda^s \varphi]\|_0 = C(2\pi)^{n/2} \|\varphi\|_0, \end{aligned}$$

so the Riesz representation theorem implies that $\lambda^s \hat{u} \in L^2$ and $\|\lambda^s \hat{u}\|_0 \leq C(2\pi)^{n/2}$.

We can get better insight into the relationship between $\mathscr{S}'$ and H^s from the following

Proposition 3.2. If $u \in \mathscr{S}'$, there is an N such that $\psi u \in H^{-N}$ if $\psi \in \mathscr{S}$.

The spaces H^s are called *Sobolev spaces*. We have need of other classes of functions that also carry this name. The definitions and initial propositions will be given independently, and the connection with H^s and these new spaces will be made afterwards.

Ω will denote an open subset of $\mathbb{R}^n$. $C_0^1(\Omega)$ is the set of continuously differentiable functions with compact support in Ω. If $f \in L^1_{\text{loc}}(\Omega)$, $g_i \in L^1_{\text{loc}}$, and

$$\int_\Omega f \partial^i \varphi d\mathbf{x} = -\int_\Omega g_i \varphi d\mathbf{x}$$

for all $\varphi \in C_0^1(\Omega)$, we say g_i is the weak partial derivative with respect to x_i. A weak derivative is automatically a distribution derivative so there is no ambiguity in writing $g_i = \partial^i f$. We define $W^{1,p}(\Omega)$, $1 \le p \le \infty$, to be the set of functions in $L^2(\Omega)$ for which the weak derivatives $\partial^i f$ exist and $\partial^i f \in L^p(\Omega)$, $i = 1, \dots, n$. For $1 \le p < \infty$, the norm in $W^{1,p}(\Omega)$ is given by

$$\|f\|_{1,p} = \left(\int_\Omega (|f|^p + |Df|^p) d\mathbf{x}\right)^{1/p},$$

where $Df = (\partial^1 f, \dots, \partial^n f)$. In $W^{1,\infty}(\Omega)$ the norm is given by $\|f\|_{L^\infty} + \|Df\|_{L^\infty}$. $f \in W_{\text{loc}}^{1,p}(\Omega)$ if $f \in W^{1,p}(V)$ for all open sets V such that $\bar{V} \subset \Omega$. We say f_k converges to f in $W_{\text{loc}}^{1,p}(\Omega)$ if f_k converges to f in $W^{1,p}(V)$ for all such V. Let $\eta_\varepsilon(\mathbf{x})$ be the function defined in Exercises 2.4 and 2.5, and, for $f \in L_{\text{loc}}^1(\Omega)$, define

$$f_\varepsilon(\mathbf{x}) = \int_\Omega \eta_\varepsilon(\mathbf{x} - \mathbf{y}) f(\mathbf{y}) d\mathbf{y} := \eta_\varepsilon * f \tag{12}$$

(f_ε is called the mollification of f, η_ε the *mollifier*). Then we have the following result on approximation of f by f_ε as ε goes to zero.

Proposition 3.3

i. $f_\varepsilon \in C^\infty(\Omega_\varepsilon)$, where $\Omega_\varepsilon = \{\mathbf{x} \in \Omega : \text{dist}(\mathbf{x}, \partial\Omega) > \varepsilon\}$.
ii. If $f \in C(\Omega)$, then f_ε converges uniformly to f on compact subsets of Ω.
iii. If $f \in L_{\text{loc}}^p(\Omega)$, $1 \le p < \infty$, then f_ε converges to f in $L_{\text{loc}}^p(\Omega)$.
iv. If $f \in W_{\text{loc}}^{1,p}(\Omega)$, $1 \le p < \infty$, then $\partial^i f_\varepsilon = \eta_\varepsilon * \partial^i f$.
v. If $f \in W_{\text{loc}}^{1,p}(\Omega)$, $1 \le p < \infty$, then f_ε converges to f in $W_{\text{loc}}^{1,p}(\Omega)$.

Statements (i), (ii), and (iv) can be used to show that distribution derivatives are weak derivatives for functions in $W^{1,p}(\Omega)$. Equivalently, $f \in W^{1,p}(\Omega)$ if and only if $f \in L^p(\Omega)$ and $\partial^i f$ (distribution derivative) $\in L^p(\Omega)$, $i = 1, \dots, n$.

We recall that a *Lebesgue point* of a measurable function is a point $\mathbf{x}$ for which

$$\lim_{r\to 0} \frac{1}{|B(\mathbf{x}, r)|} \int_{B(\mathbf{x},r)} |f - f(\mathbf{x})| \equiv \lim_{r \to 0} \frac{1}{v_n r^n} \int_{|\mathbf{y}-\mathbf{x}| \le r} |f(\mathbf{y}) - f(\mathbf{x})| \, d\mathbf{y} = 0$$

[where $v_n = \omega_n/n$, $\omega_n = 2\pi^{n/2}/\Gamma(n/2)$ the n-dimensional solid angle]. A basic result in real analysis (Ref. 3) says that if $f \in L_{\text{loc}}^1$, then almost all points are Lebesgue points.

Proposition 3.4. If $f \in L^1_{\text{loc}}(\Omega)$ and $\mathbf{x}$ is a Lebesgue point of f, then $\lim_{\varepsilon\to 0} f(\mathbf{x}) = f(\mathbf{x})$.

The relationship between smooth functions and functions in Sobolev spaces is made clearer by the following two propositions.

Proposition 3.5. If $f \in W^{1,p}(\Omega)$, $1 \leq p < \infty$, there is a sequence $f_k \in W^{1,p}(\Omega) \cap C^\infty(\Omega)$ converging to f in $W^{1,p}(\Omega)$.

We have made no hypothesis on the regularity of the boundary of Ω in this proposition. In order to go further we have to make a hypothesis. A domain Ω is said to be Lipschitz if for each point of $\partial\Omega$ there is a neighborhood in which $\partial\Omega$ is a graph of a Lipschitz continuous function of $n-1$ variables in a suitably rotated coordinate system.

Proposition 3.6. Assume Ω is Lipschitz and $f \in W^{1,p}(\Omega)$, $1 \leq p < \infty$. Then there is a sequence $f_k \in C^\infty(\bar{\Omega})$ converging to f in $W^{1,p}(\Omega)$.

Note that a Lipschitz domain necessarily lies only one side of its boundary. It is easy to see that $C^\infty(\bar{\Omega})$ is not dense in $W^{1,p}(\Omega)$ if this property is not satisfied (see Exercise 3.18).

Proposition 3.7. If Ω is a bounded domain, $1 \leq p < \infty$, Ω is Lipschitz, and Ω' is a domain, $\bar{\Omega} \subset \Omega'$, then there is an extension operator from $W^{1,p}(\Omega)$ to $W^{1,p}(\mathbb{R}^n)$, i.e., there is a linear transformation $\mathbb{E}$,

$$\mathbb{E}\colon W^{1,p}(\Omega) \to W^{1,p}(\mathbb{R}^n),$$

Such that $\mathbb{E}f = f$ on Ω, supp $\mathbb{E}f \subset \Omega'$ for $f \in W^{1,p}(\Omega)$, and

$$\|\mathbb{E}f\|_{W^{1,p}} \leq C\|f\|_{1,p},$$

where $C = C(p, \Omega, \Omega')$.

Suppose, in particular, that $p = 2$. Then the properties of H^1 given previously show that $H^1 = W^{1,2}(\mathbb{R}^n)$. A function in H^1 then automatically defines one in $W^{1,2}(\Omega)$, Ω an open set, $\Omega \subset \mathbb{R}^n$, by restriction. Conversely, if $\Omega \subset \mathbb{R}^n$ is a bounded, Lipschitz domain, Proposition 3.7 implies that $W^{1,2}(\Omega)$ can be embedded in H^1.

Higher-order weak derivatives can be defined exactly as first-order ones. We say $g = \partial^\alpha f$ where $f, g \in L^1_{\text{loc}}(\Omega)$ if

$$\int_\Omega f \partial^\alpha \varphi \, d\mathbf{x} = (-1)^{|\alpha|} \int_\Omega g \varphi \, d\mathbf{x}$$

for all $\varphi \in C_0^{|\alpha|}(\Omega)$. The spaces $W^{m,p}(\Omega)$ are defined by $f \in W^{m,p}(\Omega)$ if f, $\partial^\alpha f \in L^p(\Omega)$ for $|\alpha| \le m$. The norm is defined by

$$\|u\|_{m,p} = \left(\int_\Omega \sum_{|\alpha| \le m} |\partial^\alpha u|^p \, d\mathbf{x} \right)^{1/p}$$

Then $W^{m,p}(\Omega)$ is a Banach space (a Hilbert space for $p = 2$), and the closure of $C_0^m(\Omega)$ in this space is denoted by $W_0^{m,p}(\Omega)$. If Ω is all of $\mathbb{R}^n$, $W_0^{m,p}(\mathbb{R}^n) = W^{m,p}(\mathbb{R}^n)$. A statement analogous to Proposition 3.5, namely, $W^{m,p}(\Omega) \cap C^\infty(\Omega)$ is dense in $W^{m,p}(\Omega)$, is also true. These results imply that "strong derivatives," i.e., limits in L^p of smooth functions, are equal to "weak derivatives," i.e., those given by the definition we have adopted. The product rule for differentiation follows immediately by approximation. More generally we can say that for u, v that have weak first derivatives in Ω, and for which $u\partial^i v + v\partial^i u \in L^1_{\text{loc}}(\Omega)$, the product formula

$$\partial^i(uv) = u\partial^i v + v\partial^i u \tag{13}$$

holds.

For $p = 2$, we have $W^{m,2}(\mathbb{R}^n) = H^m$ (the two definitions of the norm are equal when $m = 0$ or $m = 1$, otherwise they are equivalent). If $\partial\Omega$ satisfies certain regularity conditions, extension operators exist (Adams, Ref. 2, Chapter 4). Because we have no need of these in this book, explicit statements will not be given. We will denote (see Ref. 1)

$$H^m(\Omega) := W^{m,2}(\Omega), \qquad H_0^n(\Omega) := W_0^{m,2}(\Omega).$$

The following result has been repeatedly used in the book.

Lemma 3.1 (Poincaré's inequality). Let Ω be a bounded domain in $\mathbb{R}^n$. There exists a constant $\sigma = \sigma(\Omega) > 0$ such that

$$\|u\|_{L^2(\Omega)} \le \sigma \|\text{grad}\, u\|_{L^2(\Omega)} \tag{PI}$$

for every $u \in C_0^1(\Omega)$ and, by density, for every $u \in H_0^1(\Omega)$.

Proof. Suppose, without loss of generality, that $x_i > 0$ for all $\mathbf{x}$ in Ω. If $u \in C_0^1(\Omega)$, then

$$|u(\mathbf{x})|^2 = \left| \int_0^{x_i} u_{x_i}(t)dt \right|^2 \leq x_i \int_0^{x_i} |u_{x_i}|^2 dt \leq x_i \int_0^d |\text{grad } u|^2 dx_i,$$

where d is the diameter of Ω, and by Fubini's theorem we have

$$\int_\Omega |u(\mathbf{x})|^2 d\mathbf{x} \leq cd^2 \int_\Omega |\text{grad } u(\mathbf{x})|^2 d\mathbf{x}, \tag{14}$$

where $c > 0$ is another constant, depending on Ω. As $C_0^1(\Omega)$ is dense in $H_0^1(\Omega)$, the result follows. □

We frequently have use of the following

Proposition 3.8. If $u \in W^{1,p}(\Omega)$, $1 \leq p < \infty$, and $f \in C^1(\mathbb{R})$, $\sup|f'| < \infty$, then $f(u) \in W^{1,p}(\Omega)$ and $\partial^i(f(u)) = f'(u)\partial^i u$ a.e.

This can be used to prove, by approximation using Proposition 3.3, the following two results. We recall that $u^+ := \max\{u, 0\}$, $u^- := \max\{-u, 0\}$, and that $u = u^+ - u^-$, $|u| = u^+ + u^-$.

Proposition 3.9. If $u \in W^{1,p}(\Omega)$, $1 \leq p < \infty$, then u^+, u^-, $u \in W^{1,p}(\Omega)$ and for almost all $\mathbf{x}$,

$$Du^+ = \begin{cases} Du, & \{u > 0\}, \\ 0, & \{u \leq 0\} \end{cases}, \qquad Du^- = \begin{cases} 0, & \{u \geq 0\}, \\ Du, & \{u < 0\}, \end{cases}$$

and

$$D|u| = \begin{cases} \text{sgn}(u)Du, & \{u \neq 0\}, \\ 0, & \{u = 0\}. \end{cases}$$

We can also deduce the useful result that $Du = 0$ a.e. on $\{u = 0\}$.

Using Proposition 3.9 we can give the following generalization of Proposition 3.8.

Proposition 3.10. Suppose that $f \in C(\mathbb{R})$, f' exists except at a finite number of points at which right- and left-hand limits of f' exist, and $\sup|f'| < \infty$. Then if $u \in W^{1,p}(\Omega)$, $1 \leq p < \infty$, $f(u) \in W^{1,p}(\Omega)$ and $\partial^i(f(u)) = f'(u)\partial^i u$ a.e.

A similar result in which f is a general Lipschitz function can also be proven. We do not need this generalization in this book, so we refer the reader to the references for its statement and proof. We will mention the following result that is of historical significance in this subject: A function is weakly differentiable if and only if it is equal a.e. to a function that is absolutely continuous on almost all line segments parallel to the coordinate axes and whose partial derivatives are locally integrable.

For $1 \leq p < \infty$, we define $p^* := np/(n-p)$, the Sobolev conjugate of p. In particular, $p^* = 6$ for $n = 3$, $p = 2$.

Proposition 3.11. If Ω is an arbitrary domain in $\mathbb{R}^n$ and $u \in W_0^{1,p}(\Omega)$, $1 \leq p < n$, then $u \in L^{p^*}(\Omega)$. Further,

$$\|u\|_{L^{p^*}(\Omega)} \leq C(n, p)\|Du\|_{L^p(\Omega)}. \tag{15}$$

If $p = n$, then $u \in L^q(\Omega)$ for $p \leq q < \infty$.

Proposition 3.12. If Ω is a bounded domain and $u \in W_0^{1,p}(\Omega)$, $p > n$, then $u \in C(\bar{\Omega})$ and

$$\|u\|_{L^\infty(\Omega)} \leq C(n, p)|\Omega|^{(1/n)-(1/p)}\|Du\|_{L^p(\Omega)}. \tag{16}$$

The inequalities (15), (16) are called *Sobolev inequalities*, and the inclusions are called Sobolev *embedding theorems*. Proposition 3.12 can be refined to show that $u \in C^\gamma(\bar{\Omega})$ where $\gamma = 1 - n/p$.

We also need a result on compact embedding of Sobolev spaces.

Proposition 3.13. If Ω is a bounded domain, $W_0^{1,p}(\Omega)$ is compactly embedded in $L^q(\Omega)$ if $1 \leq 1 < np/(n-p)$ when $p < n$, $1 \leq 1 < +\infty$ if $p = n$.

The particular case $p = 2$, $q = 2$ with $W_0^{1,2}(\Omega) = H_0^1(\Omega)$ is known as *Rellich's lemma*. We need the following classic theorem of Rademacher (Refs. 10, 11).

Proposition 3.14. If f is a locally Lipschitz function defined on an open set in $\mathbb{R}^n$, then f is differentiable a.e.

We can complete our characterization of Sobolev spaces with the following result.

Proposition 3.15. f is locally Lipschitz on an open set Ω if an only if $f \in W^{1,\infty}_{\text{loc}}(\Omega)$.

We need to use a special case of "Hausdorff measure" for sets in $\mathbb{R}^n$. Suppose that A is a set in $\mathbb{R}^3$. We define (Evans and Gariepy, Ref. 3)

$$\mathscr{H}^2_\delta(A) = \inf\left\{\sum_j \frac{\pi}{2}\left(\frac{\text{diam}C_j}{2}\right)^2 : A \subset \cup C_j, \text{diam } C_j \leq \delta\right\},$$

where C_j are arbitrary sets with finite diameter, and

$$\mathscr{H}^2(A) = \lim_{\delta \to 0} \mathscr{H}^2_\delta(A).$$

This set function extends our intuitive idea of surface area. In particular, if $z = f(x, y)$ is a Lipschitz function defined over an open set $\mathcal{O}$, G is the graph of f, and $W = (1 + z_x^2 + z_y^2)^{1/2}$, then we have the following result, originally proven by Geöcze.

Proposition 3.16. $\mathscr{H}^2(G) = \int\int_{\mathcal{O}} W dx dy := A(z, \mathcal{O})$.

The following proposition gives a generalization of the Gauss lemma that we occasionally find useful. We denote the surface measure on $\partial\Omega$ by $\mathscr{H}^{n-1}$. Note that Proposition 3.14 implies that a Lipschitz domain has a normal vector $\mathscr{H}^{n-1}$–a.e.

Proposition 3.17. Suppose that Ω is a bounded, Lipschitz domain and $1 \leq p < \infty$. Then there is a bounded linear operator T (the *trace*),

$$T : W^{1,p}(\Omega) \to L^p(\partial\Omega),$$

such that $Tf = f$ if $f \in W^{1,p}(\Omega) \cap C(\bar{\Omega})$. If $\varphi \in (C^1_0(\mathbb{R}^n))^n$, then

$$\int_\Omega f \, \text{div}\, \varphi \, d\mathbf{x} = -\int_\Omega Df \cdot \varphi \, d\mathbf{x} + \int_{\partial\Omega} (\varphi \cdot \mathbf{n}) Tf \, d\mathscr{H}^{n-1},$$

where $\mathbf{n}$ is the unit outer normal (defined $\mathscr{H}^{n-1}$–a.e.) to $\partial\Omega$. By a standard density argument, this formula of integration by parts holds for $\varphi \in (\mathbf{W}^{1,p'}(\Omega))^n$, $p' = p/(p-1)$, with $\varphi|_{\partial\Omega}$ replaced by $T\varphi$.

A function $f \in L^1(\mathcal{O})$, $\mathcal{O}$ an open subset of $\mathbb{R}^n$, has *bounded variation* in $\mathcal{O}$ if

$$\sup \int_{\mathcal{O}} f \operatorname{div} \varphi \, d\mathbf{x} < \infty, \tag{17}$$

where the supremum is over $\varphi \in C_0^1(\mathcal{O}, \mathbb{R}^n)$ with $|\varphi| \leq 1$. The space of such functions is denoted by $BV(\mathcal{O})$. We define $TV(f) \equiv TV(f, \mathcal{O})$ to be the left-hand side in (17). It is shown in the references that the distribution derivatives of f are measures and $TV(f)$ is the total variation of the vector-valued measure Df. We have $W^{1,1}(\mathcal{O}) \subset BV(\mathcal{O})$ and, for $f \in W^{1,1}(\mathcal{O})$,

$$TV(f) = \int_{\mathcal{O}} |Df| d\mathbf{x}.$$

In general, the measures Df have singular parts, as the simple example of the Heaviside function in one dimension shows. The basic compactness theorem for $BV(\mathcal{O})$ is given in the following proposition.

Proposition 3.18. Suppose that $\mathcal{O}$ is a bounded Lipschitz domain and $\{f_k\}$ is a sequence of functions in BV($\mathcal{O}$) with $\{TV(f_k)\}$ bounded. Then there is a subsequence $\{f_{k_j}\}$ and $f \in BV(\mathcal{O})$ such that

$$\|f_{k_j} - f\|_{L^1(\mathcal{O})} \to 0.$$

In the usual course in advanced calculus the space $BV([a, b])$ where $-\infty \leq a < b \leq +\infty$ is defined to be the set of functions for which

$$\sup \sum_{j=1}^{m} |f(t_{j+1}) - f(t_j)| < \infty \tag{18}$$

for all partitions $a < t_1 < \cdots < t_{m+1} < b$. We need to modify this definition in order to fit the general framework introduced above. In particular, if f is modified on a set of measure zero, we should have the same "function." For a real-valued function f we say l is the *approximate limit* of f as $y \to x$,

$$\operatorname{ap}\lim_{y \to x} f(y) = l,$$

if, for each $\varepsilon > 0$,

$$\lim_{r \to 0} \frac{|\{x : |f(x) - l| \geq \varepsilon\} \cap [x - r, x + r]|}{2r} = 0.$$

The *essential variation* of f over $[a, b]$ is given by (18) with the additional requirement that each t_j be a point of *approximate continuity*, i.e., ap $\lim_{y \to t_j}$

$f(y) = f(t_j)$. A basic result in real analysis says that for a measurable function f almost all points are points of approximate continuity (Ref. 3). It follows from this that for $f \in L^1(a, b)$, $g = f$ a.e., the essential variations are the same.

Proposition 3.19. For $f \in L^1(a, b)$, $TV(f) =$ the essential variation of f over $[a, b]$.

Therefore, using the essential variation in the definition of $BV[a, b]$ gives the same result as using (17). For a function in which (18) is satisfied for all partitions we know that the discontinuities form a countable set, and our expanded definition just identifies all functions that are modifications on a set of measure zero. In our use of BV functions in Chapter 7 we need to use the characterization of TV given in (17) and this certainly holds for any f satisfying (18).

Finally, we give two propositions about difference quotients and weak derivatives that are used in the regularity theory for elliptic equations (Chapter 5).

Proposition 3.20. If $u \in W^{1,p}(\Omega)$, then $\Delta^h u \in L^p(\Omega')$ for any $\Omega' \subset \Omega$ with $h < \text{dist}(\Omega', \partial\Omega)$ and

$$\|\Delta^h u\| L^p(\Omega') \leq \|D_i u\|_{L^p(\Omega)}.$$

Proof. We prove the result for $u \in C^1(\Omega) \cap W^{1,p}(\Omega)$. Then

$$\Delta^h u(\mathbf{x}) = \frac{1}{h} \int_0^h D_i u(\mathbf{x} + \mathbf{y}) dy,$$

where $\mathbf{y} = (0, \ldots, y \ldots, 0)$, and Hölders inequality implies

$$|\Delta^h u(\mathbf{x})|^p \leq \frac{1}{h} \int_0^h |D_i u(\mathbf{x} + \mathbf{y})|^p dy.$$

Integration over Ω' implies the result. □

Proposition 3.21. If $u \in W^{1,p}(\Omega)$, $1 < p < \infty$, and there is a constant K such that $\Delta^h u \in L^p(\Omega')$ and $\|\Delta^h u\|_{L^p(\Omega')} \leq K$ for all $\overline{\Omega'} \subset \Omega$ and $h \leq \text{dist}(\Omega', \partial\Omega)$, then $D_i u$ exists and $\|D_i u\|_{L^p(\Omega)} \leq K$.

Proof. Choose $h_m \to 0$ and $v \in L^p(\Omega)$ such that $\|v\|_{L^p(\Omega)} \leq K$ and

$$\int_\Omega \varphi \Delta^{h_m} u \, dx \to \int_\Omega \varphi v \, dx$$

for all $\varphi \in C_0^1(\Omega)$. When $h_m < \operatorname{dist}(\operatorname{supp} \varphi, \partial\Omega)$,

$$\int_\Omega \varphi \Delta^{h_m} u \, dx = -\int_\Omega u \Delta^{-h+m} \varphi \, dx \to -\int_\Omega u \, D_i \varphi \, dx.$$

It follows that $v = D_i u$ □

4. Vector-Valued Functions

In the study of partial differential equations with a distinguished variable t, usually associated with time, we have need of functions of a real variable that take values in a Banach space. If J is an interval on the real axis and f is a function defined in J taking values in a Banach space $\mathscr{B}$, then we can define continuity and differentiability of f in the usual way in terms of limits in $\mathscr{B}$. (We mean limits in the sense described by the norm in $\mathscr{B}$. We make no explicit use of weak topologies here.) This leads, in a natural way, to spaces of functions such as $C(0, 1, \mathscr{B})$, the functions valued in $\mathscr{B}$ that are continuous on $[0, 1]$. This is a Banach space with norm

$$\sup_{0 \leq t \leq 1} \|f(t)\|_{\mathscr{B}},$$

where $\| \|_{\mathscr{B}}$ is the norm in $\mathscr{B}$. In a similar way we can define $C^k(0, 1, \mathscr{B})$ in analogy to the scalar-valued case.

For f defined on J taking values in $\mathscr{B}$ we say f is *measurable* if for every compact subset K of J and $\varepsilon > 0$, there is a compact set $K' \subset K$ such that $|K \setminus K'| < \varepsilon$ and f is continuous on K'. (A proof that this property is equivalent to the usual definition of measurability when $\mathscr{B} = \mathbb{R}$ can be found in Wheeden and Zygmund, Ref. 4.) If f is measurable and $\|f(t)\|_{\mathscr{B}} \in L^p(J)$, $1 \leq p \leq \infty$, and we identify functions $f_1(t)$, $f_2(t)$ for which $f_1(t) = f_2(t)$ a.e. in t, we say that $f \in L^p(J, \mathscr{B})$ and we introduce the norm

$$\left(\int_J \|f(t)\|_{\mathscr{B}}^p dt \right)^{1/p}$$

for $1 \leq p < \infty$, or

$$\sup_J \|f(t)\|_{\mathscr{B}}$$

for $p = \infty$. The spaces $L^p(J, \mathscr{B})$ are Banach spaces, and if $p = 2$ and $\mathscr{B} = \mathscr{H}$ is a Hilbert space, with inner product $\langle , \rangle_{\mathscr{H}}$, $L^2(J, \mathscr{H})$ is a Hilbert space with inner product

$$(f, g) = \int_J \langle f(t), g(t) \rangle_{\mathscr{H}} dt.$$

We need the space

$$W(0, T) = \{u \in L^2(0, T, H^1) : u_t \in L^2(0, T, H^{-1})\}$$

for the study of the heat equation and related equations. Here and in what follows H^s are the spaces defined in Section 1, and the norm in $W(0, T)$ is given by

$$\|u\|^2 = \int_0^T (\|u(t)\|_1^2 + \|u_t(t)\|_{-1}^2) dt,$$

Proposition 4.1. $C^\infty(0, t, H^1)$ is dense in $W(0, T)$.

The proof of this fundamental fact requires some sophisticated machinery concerning the integration of vector-valued functions and is sketched as a "research exercise" at the end of the section.

Proposition 4.2. The natural injection of $C^\infty(0, T, H^1)$ into $C(0, T, L^2)$ can be extended to a continuous injection of $W(0, T)$ into $C(0, T, L^2)$.

Proof. Suppose that $u(t) \in C^\infty(0, T, H^1)$. We can extend u to $[-T, T]$ as an even function, and then multiply by a C^∞ function that vanishes for $t < -T$ and is identically 1 for $t \geq 0$. We denote the extended function by $\tilde{u}$. Note that for this function

$$(\tilde{u}'(t), \tilde{u}(t)) \leq \|\tilde{u}'(t)\|_{-1} \|\tilde{u}(t)\|_1 \leq \tfrac{1}{2} \|\tilde{u}'(t)\|_{-1}^2 + \tfrac{1}{2} \|\tilde{u}(t)\|_1^2$$

for each t. Then

$$\|\tilde{u}(t)\|_0^2 = \int_{-T}^{t} \frac{d}{d\tau}(\|\tilde{u}(\tau)\|_0^2)d\tau = 2\int_{-T}^{t} (\tilde{u}'(\tau), \tilde{u}(\tau))d\tau$$
$$\leq \int_{-T}^{T} (\|\tilde{u}'(\tau)\|_{-1}^2 + \|\tilde{u}(\tau)\|_1^2)d\tau \leq C\int_0^T (\|\tilde{u}'(\tau)\|_{-1}^2 + \|\tilde{u}(\tau)\|_1^2)d\tau.$$

In particular, for $t \in [0, T]$ we obtain

$$\|u(t)\|_0 \leq C\|u\|_{W(0,T)},$$

thus the injection of $C^\infty(0, T, H^1)$ into $C(0, T, L^2)$ is a bounded linear operator if $C^\infty(0, T, H^1)$ is given the norm of $W(0, T)$. The theorem follows from Proposition 4.1. □

If we replace H^1 by $W_0^{1,2}(\Omega) = H_0^1(\Omega)$ and H^{-1} by $(H_0^1(\Omega))' = H^{-1}(\Omega)$, Propositions 4.1 and 4.2 are still true.

Proposition 4.3. If $v \in W(0, T)$, then $v_+ \in L^2(0, T, H^1) \cap C(0, T, L^2)$ and, for $t_1, t_2 \in [0, T]$,

$$2\int_{t_1}^{t_2} (v'(\tau), v_+(\tau))d\tau = \|v_+(t_2)\|_0^2 - \|v_+(t_1)\|_0^2.$$

Proof. If $f(u) = \max\{u, 0\}$, and $v \in W(0, T)$, then Proposition 3.10 with $u = v$ implies that $v_+(t) \in H^1$ for a.e. t, and $|f'(u)| \leq 1$ implies that

$$\|v_+(t)\|_1 \leq \|v(t)\|_1$$

a.e. in $[0, T]$. It follows that $v_+ \in L^2(0, T, H^1)$, and

$$\|v_+(t)\|_{L^2(0,T,H^1)} \leq \|v(t)\|_{L^2(0,T,H^1)}.$$

For functions $C^\infty(0, T, H^1)$ the identity follows from $v_+ \partial v_+/\partial t = v_+ \partial v/\partial t$, which follows in turn from Proposition 2.10. Suppose $v_n \in C^\infty(0, T, H^1)$ and

$v_n \to v \in W(0, T)$ in the norm of $W(0, T)$. As $(v_n)_+(t) \to v_+(t)$ in L^2 for each fixed t we can pass to the limit on the right. We have

$$\left| \int_{t_1}^{t_2} (v_t, v_+) d\tau - \int_{t_1}^{t_2} ((v_n)_t, (v_n)_+) d\tau \right|^2$$
$$\leq \|v\|^2 \int_{t_1}^{t_2} \|v_+ - (v_n)_+\|_1^2 d\tau + \|v - v_n\|^2 \|v_n\|^2,$$

where $\|\|$ is the norm in $W(0, T)$. We need only show that the integral in the first term on the right tends to zero. As $v_+ = (v + |v|)/2$, $v \to v_+$ is continuous on H^1 and the integrand tends to zero for each t. The integrand can be bounded in terms of $\|v\|$ so the dominated convergence theorem implies the result. □

The *integration of vector-valued functions* will now be discussed in more detail. For the basic facts we follow Yoshida (Ref. 5) and Edwards (Ref. 6). We will work only in the generality needed for the results of this section, and this is considerably less than that of the references. The domain of our functions will be an interval J on the real line, and the values will be taken on in a real, separable Hilbert space $\mathscr{H}$. A *simple function* is one of the form

$$\sum_i v_i I_{B_i},$$

where I_{B_i} is the characteristic function of a measurable set on the real line with finite (Lebesgue) measure, $v_i \in \mathscr{H}$, and the sum is finite. It can be proven that a function is measurable according to our definition if and only if it is the a.e. $\mathscr{H}$-norm limit of a sequence of simple functions (Ref. 6, Theorem 8.15.1). A function $f(t)$ is (*Bochner*) *integrable* over J if there is a sequence of simple functions $f_n(t)$ converging to $f(t)$ a.e. on J such that

$$\lim_{n \to \infty} \int_J \|f(t) - f_n(t)\| dt = 0, \tag{19}$$

where $\|\|$ is the norm in $\mathscr{H}$. We define

$$\int_J f(t) dt := \lim_{n \to \infty} \int_J f_n(t) dt.$$

In order for this definition to make sense we need to invoke a theorem of Pettis that implies that $f(t)$ is measurable if and only if $\|f(t)\|$ is, and to observe that (19) implies that $\{\int_J f_n(t) dt\}$ is a Cauchy sequence whose limit is independent of the sequence $\{f_n\}$. Then a theorem of Bochner can be proven that says $f(t)$ is integrable if and only if $\|f(t)\|$ is.

The inequality

$$\left\| \int_J f(t)dt \right\| \leq \int_J \|f(t)\|dt \tag{20}$$

also holds.

Another fundamental result is that $C(0, T, \mathscr{H})$ is dense in $L^2(0, T, \mathscr{H})$. Suppose that $f \in L^2(0, T, \mathscr{H})$ and $\varepsilon > 0$. For any $M > 0$ we can define

$$f_M(t) = \begin{cases} f(t), & \|f(t)\| \leq M, \\ 0, & \|f(t)\| > M. \end{cases}$$

We can show that $f_M \to f$ as $M \to \infty$ in $L^2(0, T, \mathscr{H})$, using (20) and well-known properties of the Lebesgue integral. Choose M so that $\|f - f_M\|_{L^2(0,T,\mathscr{H})} < \varepsilon/2$. By virtue of the definition of measurability we can choose a compact set $K \subset [0, T]$ such that

$$|[0, T] \setminus K| < (\varepsilon/4M)^2$$

and f_M is continuous on K. It can then be proven that there exists an extension $\tilde{f}_M$ of f_M to $\mathbb{R}$ that is continuous and satisfies

$$\sup\|\tilde{f}_M(t)\| = \sup\|f_M(t)\|.$$

(This theorem is proven in Ref. 3, Chapter 1, for functions taking values in $\mathbb{R}^n$. The proof carries over verbatim to the present situation.) As

$$\|f_M - \tilde{f}_M\|_{L^2(0,T,\mathscr{H})} \leq \sup\|f_M(t) - \tilde{f}_M(t)\|(|[0, T] \setminus K|)^{1/2},$$

we can deduce

$$\|f_M - \tilde{f}_M\|_{L^2(0,T,\mathscr{H})} < \varepsilon/2$$

and the result follows.

All of this sets the context for our assertion in Proposition 4.1 that $C^\infty(0, T, H^1)$ is dense in $W(0, T)$. Let $u \in W(0, T)$. We define an extension $\tilde{u}$ by first extending u to $[-T, T]$ by $u(-t) = u(t)$ and then letting $\tilde{u} = \alpha(t)u$ where $\alpha \in C_0^\infty(\mathbb{R})$, $\alpha = 0$ for $t < -T$, and $\alpha = 1$ for $t \in [0, T]$. We let

$$\tilde{u}_m(t) = \int_{\mathbb{R}} \tilde{u}(\xi)\eta_m(t-\xi)d\xi = (\eta_m * \tilde{u})(t),$$

where η_m is the usual mollifer. We claim that $\tilde{u}_m$ is an infinitely differentiable H^1-valued function. This can be proven by showing differentiation under the integral can be justified using (20) and the dominated convergence theorem.

We claim now that $u_m \to u$ in $L^2(0, T, H^1)$ and $u'_m \to u'$ in $L^2(0, T, H^{-1})$. Observe that integration by parts shows that

$$\tilde{u}'_m(t) = \int_{\mathbb{R}} \tilde{u}'(\xi)\eta_m(t-\xi)d\xi = (\eta_m * \tilde{u}')(t).$$

It suffices then to show that if $v \in L^2(0, T, \mathscr{H})$, $\mathscr{H}$ a Hilbert space, then $\eta_m * v \to v$ in $L^2(0, T, \mathscr{H})$ as $m \to \infty$. We may extend v to $\tilde{v}$ as above. Observe first that for $f \in L^2(\mathbb{R}, \mathscr{H})$ the Cauchy–Schwarz inequality implies

$$\|(\eta_m * f)(t)\|_{\mathscr{H}} \leq \int_{\mathbb{R}} \eta_m(t-\xi)\|f(\xi)\|^2_{\mathscr{H}} d\xi,$$

and Tonelli's theorem then implies

$$\int_{\mathbb{R}} \|\eta_m * f\|^2_{\mathscr{H}} dt \leq \int_{\mathbb{R}} \|f(\xi)\|^2_{\mathscr{H}} \int_{\mathbb{R}} \eta_m(t-\xi)dt\, d\xi = \int_{\mathbb{R}} \|f(\xi)\|^2_{\mathscr{H}}\, d\xi. \tag{21}$$

Now let $\varepsilon > 0$ and choose $w \in C(0, T, \mathscr{H})$ such that $\|v - w\|_{L^2(0,T,\mathscr{H})} < \varepsilon$. Then

$$\|\eta * v - v\|_{L^2} \leq \|\eta_m * (v - w)\|_{L^2} + \|\eta_m * w - w\|_{L^2} + \|w - v\|_{L^2},$$

where $L^2 = L^2(0, T, \mathscr{H})$. The inequality (21) implies that the first term is bounded above by $C\|v - w\|_{L^2}$ where C depends only on α. It suffices then to show that the middle term can be made small. We have

$$\begin{aligned}\|\eta_m * w(t) - w(t)\|_{\mathscr{H}} &= \left\| \int_{\mathbb{R}} \eta_m(t-\xi)(\tilde{w}(\xi) - \tilde{w}(t))d\xi \right\|_{\mathscr{H}} \\ &\leq \sup_{|\xi - t| < \frac{1}{m}} \|\tilde{w}(\xi) - \tilde{w}(t)\|_{\mathscr{H}}.\end{aligned}$$

As $\tilde{w}$ is uniformly continuous on its (fixed) support, the right-hand side goes to zero uniformly with respect to m.

Exercises

1.1. Prove that a (linear) operator T is bounded if and only if T is continuous, and is a contraction if and only if $\|T\| < 1$.

1.2. Prove that $\|T\| = \sup[\|Tx\|_2/\|x\|_1]$ defines a norm in $\mathscr{B}(B_1, B_2)$ and $\mathscr{B}(B_1, B_2)$ is a Banach space with this norm.

1.3. For the proof of Theorem 1.3, consider T_m compact with $\|T_m - T\| \to 0$, and a bounded sequence $\{x_n\}$. By proceeding inductively we can choose nested subsequences such that $\{T_m x_{n_j}\}$ converges. The "diagonal" subsequence $\{y_j\}$ is such that $\{T_m y_j\}$ converges for all m. Because

$$Ty_j - Ty_k = (T - T_m)y_j + T_m y_j - T_m y_k + (T_m - T)y_k,$$

we can shown that $\{Ty_j\}$ is a Cauchy sequence.

1.4 (Proof of Lemma 1.1). For any $x \in B \setminus \mathscr{M}$, $d = \text{dist}(x, \mathscr{M}) > 0$ and there is $y \in \mathscr{M}$ such that $\|x - y\| \leq d/\theta$. Choosing $x_\theta = (x - y_\theta)/\|x - y_\theta\|$ works.

1.5 (From the proof of Theorem 1.4). Prove that $\mathscr{N}_j$ a proper subset of $\mathscr{N}_{j+1}$ for all j is impossible.

1.6. In order to prove Theorem 1.5, suppose that λ_n is a sequence of eigenvalues with corresponding eigenvectors x_n such that $\{x_n\}$ are linearly independent and $\lambda_n \to \lambda \neq 0$. If $\mathscr{M}_n$ is the span of $\{x_1, \ldots, x_n\}$, Lemma 1.1 implies existence of $y_n \in \mathscr{M}_n$, $\|y_n\| = 1$, such that $\text{dist}(y_n, \mathscr{M}_{n-1}) \geq \frac{1}{2}$ for $n \geq 2$. If $n > m$, show that

$$\lambda_n^{-1} Ty_n - \lambda_m^{-1} Ty_m = y_n - z, \qquad z \in \mathscr{M}_{n-1},$$

and that

$$\|\lambda_n^{-1} Ty_n - \lambda_m^{-1} Ty_m\| \geq \tfrac{1}{2}.$$

1.7. If V is a *complex* inner product vector space, $q : V \times V \to \mathbb{C}$, and the definition of inner product is modified by replacing (i) with $(x, y) = \overline{(y, x)}$ (complex conjugate). The basic facts are the same. Check this.

1.8. Prove that Theorem 1.6 also holds if $\mathscr{M}$ is a closed *convex* set, i.e., $\mathscr{M}$ is closed and $\lambda x + \mu y \in \mathscr{M}$ whenever $x, y \in \mathscr{M}$ and $\lambda + \mu = 1$, $\lambda \geq 0$, $\mu \geq 0$.

1.9. Show that the projection onto a closed linear subspace $\mathscr{M} \neq \{0\}$ of H is a bounded linear operator P satisfying $P^2 = P$ and $\|P\| = 1$.

1.10 (*Lemma 1.3*). Show that a compact operator T in a Hilbert space is the norm limit of finite rank operators. *Hint*: Recall that a set S has a compact closure if and only if it is totally bounded, i.e., for each $\in > 0$ there is a finite set of elements x_j, $j = 1, \ldots, N$, such that

$$\sup_{x \in S} \inf_j \|x - x_j\| < \varepsilon.$$

Apply this to $T(B)$ where B is the unit ball in H. Let P_ε be the projection onto the span of $\{x_1, \ldots, x_N\}$ and set $T_\varepsilon = P_\varepsilon T$. T_ε has finite rank and is uniformly close to T.]

1.11. Suppose $x_1, \ldots, x_N$ is a finite orthonormal set in H, i.e., $(x_i, x_j) = \delta_{ij}$, and define $T \in \mathscr{B}(H)$ by

$$Tx := \sum_{i=1}^{n} \lambda_i (x, x_i) x_i, \qquad \lambda_i \neq 0.$$

Show that T is a finite rank operator, T has nonzero eigenvalues λ_i with corresponding eigenvectors x_i, and, letting $\mathscr{M}$ be the span of $\{x_1, \ldots, x_N\}$, the null space of T is $\mathscr{M}^\perp$. Hence, $\lambda = 0$ is an eigenvalue of T with (possibly infinite) multiplicity equal to dim $\mathscr{M}^\perp$.

1.12. Prove that if $\{y_n\}$ is a complete orthonormal set, a series of the form $\sum_{n=1}^{\infty} a_n y_n$ converges in the norm of H if and only if $\sum_{n=1}^{\infty} |a_n|^2 < \infty$, and in this case $a_n = (x, y_n)$, where x is the element of H defined by the series, $x = \sum_{n=1}^{\infty} a_n y_n$.

1.13. (Parseval's equality). Prove that if $\{y_n\}$ is a complete orthonormal set, for any two vectors $x, z \in H$ has $(x, z) = \sum_{n=1}^{\infty} (x, y_n)(z, y_n)$, hence $\|x\|^2 = \sum_{n=1}^{\infty} |(x, y_n)|^2$.

1.14. If H is a complex Hilbert space, $(\alpha A)^* = \bar{\alpha} A^*$ if $A \in \mathscr{B}(H)$ and $\alpha \in \mathbb{C}$. In particular, $(\lambda I - T)^* = \bar{\lambda} I - T^*$, and in the Fredholm alternative an eigenvalue λ of T corresponds to an eigenvalue $\bar{\lambda}$ of T^*.

1.15. If $\mathscr{B}$ is a Banach space and B' its dual space, define linear functionals on B' by $x(y) = y(x)$ for $x \in B$, $y \in B'$. If every element of the second dual $(B')'$ arises in this way, B is called *reflexive*. Show that $L^p(\Omega)$, $1 < p < \infty$, is reflexive.

2.1. Use the relation $\binom{a+1}{b} = \binom{a}{b-1} + \binom{a}{b}$ for a, b positive integers and an inductive argument to prove correctness of the binomial formula and Leibniz's formula. (Assume they are true for α, show they must hold for $\alpha + \gamma$ where $\gamma_i = 1$ in the jth place, 0 otherwise.)

2.2. Use $(1 + |\mathbf{x}|^2)^{-s} \leq (1 + |\mathbf{x}|^2)^{-s/2} \leq (1 + x_1^2)^{-s/2n} \cdots (1 + x_n^2)^{-s/2n}$ to show that (1) hold.

2.3. Prove (3) using Leibniz's rule and a choice of C_k, N_k such that

$$|\partial^\gamma \psi(\mathbf{x})| \le \frac{C_k}{2^k(n+1)^{N_k}}(1+|\mathbf{x}|^2)^{N_k}, \qquad |\gamma| \le k.$$

2.4. Show that $\varphi(\mathbf{x}) = f(|(\mathbf{x})|^2 - 1)$, where

$$f(t) = \begin{cases} e^{1/t}, & t \le 0, \\ 0, & t \ge 0, \end{cases}$$

is a function in C_0^∞ with support contained in $B_1 = \{|\mathbf{x}| \le \mathbf{1}\}$. Let $\eta := \varphi / \int \varphi\, dx$. Then $\eta_\varepsilon(\mathbf{x}) = \varepsilon^{-n}\eta(\mathbf{x}/\varepsilon)$ has support in B_ε.

2.5. A proof of the existence of a partition of unity can be given as follows. If $K \subset \Omega_1$ and K_δ is the set of points whose distance from K is $\le \delta$, let $\varepsilon = \frac{1}{4}$ dist$(K, \partial\Omega_1)$, η_ε as in Exercise 2.4, and $\psi(\mathbf{x}) = \int_{K_{2\varepsilon}} \eta_\varepsilon(\mathbf{x}-\mathbf{y})d\mathbf{y}$. Then $\psi = 1$ on K_ε, supp $\psi \subset K_{3\varepsilon} \subset \Omega_1$, and $\psi \in C^\infty$. In general, use compactness of K to complete the proof.

2.6 (Proposition 2.1). If $1 \le p < \infty$ and $\varphi \in \mathscr{S}$,

$$|\varphi(\mathbf{x})|^p \le (2^n|\varphi|_{2n})^p \prod_1^n (1+x_j^2)^{-1}.$$

Use (1) and Holder's inequality. To prove the second statement, approximate the characteristic function of measurable sets by that of an open set, say χ_E, and use the construction in Exercise 2.5 to find a sequence $\varphi_j \in C_0^\infty$, $\varphi_j = 1$ on a compact subset of E, such that $\varphi_j \to \chi_E$ pointwise.

Because $\mathscr{S}$ is dense in L^2, U extends uniquely as a linear functional in L^2 and the Riesz representation theorem (Theorem 1.7) implies existence of $u \in L^2$ such that $U(\varphi) = (u, \varphi)$ and $\|U\| = \|u\|_{L^2} \le C$.

2.7. (Proposition 2.3). For $u \in L^1$, $\varphi \in \mathscr{S}$, $u(\mathbf{x})\overline{\varphi(\xi)} \in L^1(\mathbb{R}^{2n})$ and Fubini's theorem applied to $(\hat{u}, \varphi)$ implies the result,

2.8. If $\varphi(\mathbf{x}) = e^{-|\mathbf{x}|^2/2}$, then $\hat{\varphi}(\xi) = (2\pi)^{n/2}e^{-|\xi|^2/2}$ (hence φ is an eigenfunction of the Fourier transform). To show this it suffices to consider the one-dimensional case. We can write

$$\hat{\varphi}(\xi) = e^{-\xi^2/2}\int_{-\infty}^{+\infty} e^{-(x+i\xi)^2/2}\, dx.$$

For $\xi > 0$ a contour integral around a rectangular path with base $[-A, A]$ and top $[-A + i\xi, A + i\xi]$ can be used to show that

$$\int_{\mathbb{R}} r^{-(x+i\xi)^2}\, dx = \int_{\mathbb{R}} e^{-x^2/2}\, dx = \sqrt{2\pi}.$$

The Fourier transform of an even function is even.

2.9 (Proposition 2.4). Let $\psi(\xi) = e^{-|\xi|^2/2}$, and $\varepsilon > 0$. A change of variable $\mathbf{x} - \mathbf{y} = \varepsilon\mathbf{z}$, $\varepsilon\xi = \eta$ shows that

$$\int_{\mathbb{R}^n} \psi(\varepsilon\xi)\hat{\varphi}(\xi)e^{i\mathbf{x}\cdot\xi}\, d\xi = \int_{\mathbb{R}^n}\int_{\mathbb{R}^n} \psi(\varepsilon\xi)\varphi(\mathbf{y})e^{i(\mathbf{x}-\mathbf{y})\cdot\xi}\, d\xi\, d\mathbf{y}$$
$$= \int_{\mathbb{R}^n} \hat{\psi}(\mathbf{z})\varphi(\mathbf{x} + \varepsilon\mathbf{z})d\mathbf{z}.$$

The dominated convergence theorem implies that the first and last term tend, as $\varepsilon \to 0$, to

$$\psi(0)\int_{\mathbb{R}^n} \hat{\varphi}(\xi)e^{i\mathbf{x}\cdot\xi}\, d\xi$$

and

$$\varphi(\mathbf{x})\int_{\mathbb{R}^n} \hat{\psi}(\mathbf{z})d\mathbf{z},$$

respectively. Exercise 2.8 implies (6). Equation (7) follows from (6) and Proposition 2.3.

2.10 (Proposition 2.5). If $\varphi \in \mathscr{S}$ and $\hat{\psi} = \varphi$, then $(\hat{\hat{u}}, \varphi) = (2\pi)^n(\check{u}, \varphi)$ by Proposition 2.4. Further, if $u \in L^2$,

$$|(\hat{u}, \varphi)| = |(\check{u}, \hat{\varphi})| \le \|u\|_{L^2}\|\hat{\varphi}\|_{L^2} = (2\pi)^{n/2}\|u\|_{L^2}\|\varphi\|_{L^2},$$

so $|U(\varphi)| = |(\hat{u}, \varphi)| \le C\|\varphi\|_{L^2}$ with $C = (2\pi)^{n/2}\|u\|_{L^2}$ and Proposition 2.1 implies $\hat{u} \in L^2$ with $\|\hat{u}\|_{L^2} \le (2\pi)^{n/2}\|u\|_{L^2}$. As $\hat{\hat{u}} = (2\pi)^n\check{\varphi}$,

$$(2\pi)^{n/2}\|u\|_{L^2} = (2\pi)^{-n/2}\|\hat{\hat{u}}\|_{L^2} \le \|\hat{u}\|_{L^2} \le (2\pi)^{n/2}\|u\|_{L^2}.$$

Finally, use

$$(u, v) = \tfrac{1}{4}(\|u + v\|^2 - \|u - v\|^2 + i\|u + iv\|^2 - i\|u = iv\|^2),$$

which holds in any Hilbert space.

2.11 (Proposition 2.6). If $\psi \in C_0^\infty$, $0 \le \psi \le 1$, $\psi = 1$ on B_1 and $\psi_\varepsilon(\mathbf{x}) = \psi(\varepsilon\mathbf{x})$, set $\varphi_\varepsilon = \psi_\varepsilon(\mathbf{x})\varphi$ for $\varphi \in \mathscr{S}$ and show that, for $k \ge 0$,

$$|\varphi - \varphi_\epsilon|_k \le n\varepsilon^2|\varphi|_{k+2} + C_k\varepsilon|\varphi|_k.$$

[Write $\mathbf{x}^\alpha\partial^\beta(\varphi - \varphi_\varepsilon) = (1 - \psi_\varepsilon)\mathbf{x}^\alpha\partial^\beta\varphi + \ldots$ and use $0 \le 1 - \psi_\varepsilon \le 1 \le \varepsilon^2|\mathbf{x}|^2$ on $\mathrm{supp}(1 - \psi_\varepsilon)$.] If $|(u - v, \varphi)| \le C|\varphi|_N$ for $\varphi \in \mathscr{S}$, then $(u - v, \varphi_\epsilon) = 0$ implies

$$|(u - v, \varphi)| = |(u - v, \varphi - \varphi_\epsilon)| \le C|\varphi - \varphi_\epsilon|_n \le C_1\varepsilon|\varphi|_{N+2}$$

for $\varepsilon < 1$.

2.12. Prove Proposition 2.7.

2.13. Prove that $(\ln(|\mathbf{x}|))' = PV(1/x)$.

2.14. Under what circumstances does $u \in L^1_{\mathrm{loc}}(\mathbb{R}^n)$ define a distribution in $\mathscr{S}'$?

2.15. Prove (11).

2.16. Deduce from (11) that $\mathscr{F}[1/|\mathbf{x}|] = 4\pi/|\xi|^2$. Compute the two-dimensional Fourier transform of $1/|\mathbf{x}|$ restricted to the plane $x_3 = 0$, i.e., of $(x_1^2 + x_2^2)^{-1/2}$. *Hint:* $2\pi(\xi_1^2 + \xi_2^2)^{-1/2}$.

2.17. Prove that $xf(x) = 1$ has solutions $f = PV(1/x) + C\delta$, C an arbitary constant. Find the solutions of $x^nf(x) = 0$ (n an interger).

2.18. Verify that all possible eigenvalues of the Fourier transform $\mathscr{F}$ are given by $\pm\,(2\pi)^{n/2}$, $\pm\, i(2\pi)^{n/2}$. *Hint*: $\mathscr{F}^4[f] = (2\pi)^{2n}f$.

2.19. Prove that the Poisson kernel (see Section 2 of Chapter 4) satisfies $K_a(\mathbf{x}, \mathbf{y}) \to \delta_y$ in $\mathscr{D}'$ as $|\mathbf{x}| \to a$, where δ_y denotes the "shifted δ-function," $(\delta_y, \varphi) = \varphi(\mathbf{y})$.

3.1. Show that $|e^{i\mathbf{x}\cdot\xi} - e^{i\mathbf{y}\cdot\xi}| \le 2^{1-\varepsilon}|\mathbf{x} - \mathbf{y}|^\varepsilon\lambda^\varepsilon(\xi)$ for $\mathbf{x}, \mathbf{y}, \xi \in \mathbb{R}^n$, $0 < \varepsilon < 1$. Use this inequality to show that $u \in H^s$ with $s > \frac{1}{2}n + \varepsilon$ implies that

$$|u(\mathbf{x}) - u(\mathbf{y})| \le C|\mathbf{x} - \mathbf{y}|^\varepsilon$$

for some $C > 0$ and all $\mathbf{x}, \mathbf{y}, \in \mathbb{R}^n$.

3.2 (Proposition 3.2). As $u \in \mathscr{S}'$ there is a k such that $|(u, \varphi)| \le C|\varphi|_k$ for all $\varphi \in \mathscr{S}$. Use this to show that

$$|(\psi u, \varphi)| \le C2^k|\bar{\psi}|_k \max_{|\beta|\le k} |\partial^\beta\varphi|_0.$$

Then show that $|\partial^\beta\varphi|_0 \le \|\varphi\|_{n+|\beta|}$. Using Proposition 3.1, show $N = n + k$ works. In a similar way, $\psi(\mathbf{x}) = (1 + |\mathbf{x}|^2)^{-N}$, $N \ge k/2$ implies $\psi u \in H^{-N}$.

3.3. Prove that $\delta \in H^{-(n/2)-\varepsilon}(\mathbb{R}^n)$ for any $\varepsilon > 0$. Hence, show that, for $n = 1$, $H(x) \in H^{(1/2)-\varepsilon}_{\mathrm{loc}}(\mathbb{R})$ [functions in $H^{1/2}(\mathbb{R})$ can be discontinuous, e.g., logarithimically, but cannot have jump discontinuities; e.g., see Example 5.2 of Chapter 4 and Ref. 7).

3.4. Prove (13) for $u, v \in W^{1,p}(\Omega) \cap L^\infty(\Omega)$ using u_ε, v_ε [equation (12)] and Proposition 3.3.

3.5 (Proposition 3.8). Choose u_ε as in Exercise 3.4 and apply Proposition 3.3.

3.6 (Proposition 3.9). Let

$$f_\varepsilon(r) = \begin{cases} (r^2 + \varepsilon^2)^{1/2} - \varepsilon, & r \geq 0, \\ 0, & r < 0. \end{cases}$$

Apply Proposition 3.8 to get

$$\int_\Omega u^+ \partial^i \varphi \, d\mathbf{x} = - \int_{\Omega+} \varphi \partial^i f \, d\mathbf{x},$$

where $\Omega^+ = \Omega \cap \{u > 0\}$. Use the $u^- = (-u)^+$, $|u| = u^+ + u^-$ to get the other assertions.

3.7. Establish the equivalence of the norms in $W^{m,2}(\mathbb{R}^n)$ and in H^m.

3.8. For all $\mathbb{R}$ and $\xi, \eta \in \mathbb{R}^n$, $\lambda^s(\xi) \leq 2^{|s|}\lambda^{|s|}(\xi - \eta)\lambda^s(\eta)$ (Peetre's inequality). As

$$\begin{aligned} &(1 + |\xi|) \leq (1 + |(\xi - \eta)|)(1 + |\eta|), \\ &\lambda^2(\xi) \leq (1 + |\xi|^2) \leq (1 + |(\xi - \eta)|)^2(1 + |\eta|)^2, \end{aligned}$$

and $(1 + |\eta|)^2 \leq (1 + |\eta|)^2 + (1 - |\eta|)^2 = 2\lambda^2(\eta)$, so $\lambda^2(\xi) \leq 2^2\lambda^2(\xi - \eta)\lambda^2(\eta)$. Raise this to the power s to obtain the result for $s \geq 0$. For $s < 0$ exchange the roles of ξ and η.

3.9. $\mathscr{F}[uv](\xi) = (2\pi)^{-n} \int \hat{u}(\xi - \eta)\hat{v}(\eta)d\eta = (2\pi)^{-n} \int \hat{u}(\xi)\hat{v}(\eta - \xi)d\eta$ for any $u \in H^\infty$, $v \in H^{-\infty}$.

3.10. If $a(\mathbf{x}, D) = \sum_{|\alpha| \leq m} a_\alpha(\mathbf{x})D^\alpha$ is a linear partial differential operator of order m with coefficients $a_\alpha \in H^\infty$, $a(\mathbf{x}, D)$ maps H^s continuously into H^{s-m} for any $s \in \mathbb{R}$. It suffices to consider φu for $\varphi \in H^\infty$, $u \in H^m$. Apply Exercises 3.8 and 3.9 to show that $\|\varphi u\|_s \leq 2^{|s|-(n/2)}\|\varphi\|_{s+n}\|u\|_s$. *Hint:*

$$\begin{aligned} |\lambda^s(\xi)\mathscr{F}[\varphi u](\xi)|^2 &= |(2\pi)^{-n} \int \lambda^s(\xi)\hat{\varphi}(\xi - \eta)\hat{u}(\eta)d\eta|^2 \\ &\leq (2\pi)^{-n}\|\varphi\|_{s+n} \int \lambda^{2s}(\xi) \cdot |\hat{u}(\eta)|^2\lambda^{-2|s|-2n}(\xi - \eta)d\eta. \end{aligned}$$

3.11. Prove that $W_0^{m,p}(\mathbb{R}^n) = W^{m,p}(\mathbb{R}^n)$. *Hint:* Let η be as in Exercise 2.4, and for $u \in W^{m,p}(\Omega)$, let $u_0 = \eta(\varepsilon\mathbf{x})u$ where $0 < \varepsilon \leq 1$. Show that

$$|D^\alpha u_0| \leq M \sum_{\beta \leq \alpha} \binom{\alpha}{\beta} |D^\beta u|$$

where M depends only on m (through bounds on $D^\gamma \eta$, $|\gamma| \leq M$). Then, if $\Omega_\varepsilon = \{\mathbf{x} : |\mathbf{x}| > 1/\varepsilon\}$,

$$\|u - u_o\|_{W^{m,p}(\mathbb{R}^n)} = \|u - u_o\|_{W^{m,p}(\Omega_\varepsilon)} \leq C\|u\|_{W^{m,p}(\Omega_\varepsilon)},$$

where C is independent of ε. Show that the right-hand side tends to zero as $\varepsilon \to 0$.

3.12. If $N = \sum_{|\alpha| \leq m} 1$, then $H^m(\Omega) \equiv W^{m,2}(\Omega)$ may be thought of as a subspace of $L_N^2(\Omega) = \prod_{j=1}^{N} L^2(\Omega)$ using the mapping $Pu = \{D^\alpha u\}_{|\alpha| \leq m}$.

3.13. Using Exercises 3.11 and 3.12, show that a bounded linear functional on $H^m(\Omega)$, $U \in (H^m)\Omega))'$, can be represented as

$$U(u) = \sum_{|\alpha| \leq m} (D^\alpha u, v^\alpha)$$

for $\{v^\alpha\} \in L_N^2(\Omega)$ and v^α can be chosen so that $\|U\|^2 = \sum_{|\alpha| \leq m} \int_\Omega |D^\alpha v|^2 d\mathbf{x}$.
Hint: The Hahn–Banach theorem implies that U can be extended to $(L_N^2(\Omega))' = L_N^2(\Omega)$ without increasing $\|U\|$.

3.14. As Exercise 3.11 shows that $\mathscr{D}$ is dense in $W^{1,2}(\mathbb{R}^n) = H^1$, we can deduce from Proposition 3.1 that $(H^1)' = H^{-1}$. Deduce that for $f \in H^{-1}$ there are $f_0, f_1, \ldots, f_n \in L^2$ such that $f = f_0 + \sum_1^n (f_i)_{x_i}$, as distributions and $\|f\|_{-1}^2 = \|f_0\|_0^2 + \sum_1^n \|f_i\|_0$.

3.15. Prove a result analogous to that in Exercise 3.14 for $(H_0^1(\Omega))' = H^{-1}(\Omega)$.

3.16 Recall from Proposition 3.9 that $|v| \in H^1$ whenever $v \in H^1$. We can deduce from Proposition 3.10 that $\operatorname{grad}|v| = \operatorname{sgn}(v)\ \operatorname{grad} v$, and hence $\|v\|_1 = \||v|\|_1$. Show that the map $v \to |v|$ is continuous on H^1. *Hint:* Suppose that $v_n \to v$ in H^1. As bounded sets in H^1 are weakly precompact, we can choose a subsequence $|v_{n_i}|$ of $|v_n|$ that is weakly convergent in H^1. But $|v_n|$ converges to $|v|$ in L^2, hence this weak limit is necessarily $|v|$. As any weakly convergent subsequence of $|v_n|$ must converge to $|v|$, $|v_n|$ converges weakly to $|v|$ in H^1. Moreover, $\||v_n|\|_1 \to \||v|\|_1$. As a weakly convergent sequence in a Hilbert space whose norms also converge is norm convergent, it follows that $|v_{n_i}| \to |v|$ in H^1.

3.17. Prove that $f \in W^{1,p}(\mathbb{R}^n)$, $1 \le p < n$, implies that $\|f\|_{L^{p*}} \le C\|\operatorname{grad} f\|_{L^p}$, $p^* = np/(n-p)$. *Hint:* It suffices to consider $f \in C_0^1$. The fundamental theorem of calculus implies immediately that

$$|f(\mathbf{x})|^{n/n-p} \le \prod_{i=1}^{n} \left(\int_{\mathbb{R}} |\partial^i f(x_1, \ldots, t_i, \ldots, x_n)| dt_i \right)^{1/n-p}$$

and integration with respect to $x_1, x_2, \ldots, x_n$ successively together with Hölder's inequality yields

$$\int_{\mathbb{R}} |f|^{p*} \, d\mathbf{x} \le C \left(\int_{\mathbb{R}} |\operatorname{grad} f|^p dt_i \right)^{p*/p},$$

whence the assertion.

3.13. Let $n = 2$, $\Omega = \{(x, y) \in \mathbb{R}^2;\ 0 < |x| < 1, 0 < |y| < 1\}$. Show that Proposition 3.6 is false for this domain. *Hint:* Consider $u(x, y) = \operatorname{sgn}(x)$ in Ω.

3.19. (a) Extend the result of Exercise 3.13 to obtain a representation of $T \in (W_0^{m,p}(\Omega))'$ as

$$T = \sum_{|\alpha| \le m} (-1)^{|\alpha|} D^{\alpha} v_{\alpha},$$

where $v_\alpha \in L^{p'}(\Omega)$, $1/p + 1/p' = 1, 1 \le p < \infty$, and $D^\alpha v_\alpha$ denote the distributional derivatives (Ref. 2).

(b) Show that $T \in \mathscr{D}'(')$ of this form extends uniquely to an element of $(W_0^{m,p}(\Omega))'$. This set of distributions is denoted by $W^{-m,p'}(\Omega)$.

4.1. Show that $u \in L^2(0, T, L^2)$ implies that $u \in L^2([0, T] \times \mathbb{R})$. *Hint*: The sticky point is measurability. After this, finiteness of the integral follows immediately from Tonelli's theorem.

References

1. BRÉZIS, H., *Analyse Fonctionnelle*, Masson, Paris, France, 1983.
2. ADAMS, R. A., *Sobolev Spaces*, Academic Press, New York, New York, 1975.
3. EVANS, L. C., and GARIEPY, R. F., *Measure Theory and Fine Properties of Functions*, CRC Press, Boca Raton, Florida, 1992.
4. WHEEDEN, R. L., and ZYGMUND, A., *Measure and Integral*, Marcel Dekker, New York, New York, 1977.
5. YOSIDA, K., *Functional Analysis*, 2nd Edition, Springer-Verlag, Berlin, Germany, 1968.

6. EDWARDS, R. E., *Functional Ayalysis*, Holt, Rinehart & Winston, New York, New York, 1965.
7. BREZZI, F., and GILARDI, G., *Fundamentals of PDE's for Numerical Analysis*, Report No. 446, IAN-CNR, Pavia, Italy, 1984.
8. SAINT RAYMOND, X., *Elementary Introduction to the Theory of Psueodifferential Operators*, CRC Press, Boca Raton, Florida, 1991.
9. TREVES, F., *Basic Linear Partial Differential Equations*, Academic Press, New York, New York, 1975.
10. RADEMACHER, H., *Ueber partielle und totale Differenzierbarkeit von Funktionen mehrerer Variablen. I*, Mathematische Annalen. Vol. 79, pp. 340–359, 1919,
11. RADEMACHER, H., *Ueber partielle und totale Differenzierbarkeit von Funktionen mehrere Variablen. II*, Mathematische Annalen Vol. 81, pp. 52–63, 1920.

Index

Complete Series Listing

Below is a complete listing of the volumes in the *Mathematical Concepts and Methods in Science and Engineering* series.

1 **INTRODUCTION TO VECTORS AND TENSORS, Volume 1: Linear and Multilinear Algebra**
• *Ray M. Bowen and C.-C. Wang*

2 **INTRODUCTION TO VECTORS AND TENSORS, Volume 2: Vector and Tensor Analysis**
• *Ray M. Bowen and C.-C. Wang*

3 **MULTICRITERIA DECISION MAKING AND DIFFERENTIAL GAMES**
• *Edited by George Leitmann*

4 **ANALYTICAL DYNAMICS OF DISCRETE SYSTEMS**
• *Reinhardt M. Rosenberg*

5 **TOPOLOGY AND MAPS**
• *Taqdir Husain*

6 **REAL AND FUNCTIONAL ANALYSIS**
• *A. Mukherjea and K. Pothoven*

7 **PRINCIPLES OF OPTIMAL CONTROL THEORY**
• *R. V. Gamkrelidze*

8 **INTRODUCTION TO THE LAPLACE TRANSFORM**
• *Peter K. F. Kuhfittig*

9 **MATHEMATICAL LOGIC: An Introduction to Model Theory**
• *A. H. Lightstone*

10 **SINGULAR OPTIMAL CONTROLS**
• *R. Gabasov and F. M. Kirillova*

11 **INTEGRAL TRANSFORMS IN SCIENCE AND ENGINEERING**
• *Kurt Bernardo Wolf*

12 **APPLIED MATHEMATICS: An Intellectual Orientation**
• *Francis J. Murray*

13 **DIFFERENTIAL EQUATIONS WITH SMALL PARAMETERS AND RELAXATION OSCILLATIONS**
• *E. F. Mishchenko and N. Kh. Rozov*

14 **PRINCIPLES AND PROCEDURES OF NUMERICAL ANALYSIS**
- *Ferenc Szidarovszky and Sidney Yakowitz*

16 **MATHEMATICAL PRINCIPLES IN MECHANICS AND ELECTROMAGNETISM**
Part A: Analytical and Continuum Mechanics
- *C.-C. Wang*

17 **MATHEMATICAL PRINCIPLES IN MECHANICS AND ELECTROMAGNETISM**
Part B: Electromagnetism and Gravitation
- *C.-C. Wang*

18 **SOLUTION METHODS FOR INTEGRAL EQUATIONS**
- *Edited by Michael A. Golberg*

19 **DYNAMIC OPTIMIZATION AND MATHEMATICAL ECONOMICS**
- *Edited by Pan-Tai Liu*

20 **DYNAMICAL SYSTEMS AND EVOLUTION EQUATIONS**
- *J. A. Walker*

21 **ADVANCES IN GEOMETRIC PROGRAMMING**
- *Edited by Mordecai Avriel*

22 **APPLICATIONS OF FUNCTIONAL ANALYSIS IN ENGINEERING**
- *J. L. Nowinski*

23 **APPLIED PROBABILITY**
- *Frank A. Haight*

24 **THE CALCULUS OF VARIATIONS AND OPTIMAL CONTROL**
- *George Leitmann*

25 **CONTROL, IDENTIFICATION, AND INPUT OPTIMIZATION**
- *Robert Kalaba and Karl Spingarn*

26 **PROBLEMS AND METHODS OF OPTIMAL STRUCTURAL DESIGN**
- *N. V. Banichuk*

27 **REAL AND FUNCTIONAL ANALYSIS, Second Edition**
Part A: Real Analysis
- *A. Mukherjea and K. Pothoven*

28 **REAL AND FUNCTIONAL ANALYSIS, Second Edition**
Part B: Functional Analysis
- *A. Mukherjea and K. Pothoven*

29 **AN INTRODUCTION TO PROBABILITY THEORY WITH STATISTICAL APPLICATIONS**
• *Michael A. Golberg*

30 **MULTIPLE-CRITERIA DECISION MAKING**
• *Po-Lung Yu*

31 **NUMERICAL DERIVATIVES AND NONLINEAR ANALYSIS**
• *Harriet Kagiwada, Robert Kalaba, Nima Rasakhoo, and Karl Spingarn*

32 **PRINCIPLES OF ENGINEERING MECHANICS**
Volume 1: Kinematics—The Geometry of Motion
• *M. F. Beatty, Jr.*

33 **PRINCIPLES OF ENGINEERING MECHANICS**
Volume 2: Dynamics—The Analysis of Motion
• *M. F. Beatty, Jr.*

34 **STRUCTURAL OPTIMIZATION**
Volume 1: Optimality Criteria
• *Edited by M. Save and W. Prager*

35 **OPTIMAL CONTROL APPLICATIONS IN ELECTRIC POWER SYSTEMS**
• *Edited by G. S. Christensen, M. E. El-Hawary, and S. A. Soliman*

36 **GENERALIZED CONCAVITY**
• *Mordecai Avriel, Walter E. Diewert, Siegfried Schaible, and Israel Zang*

37 **MULTICRITERIA OPTIMIZATION IN ENGINEERING AND IN THE SCIENCES**
• *Edited by Wolfram Stadler*

38 **OPTIMAL LONG-TERM OPERATION OF ELECTRIC POWER SYSTEMS**
• *G. S. Christensen and S. A. Soliman*

39 **INTRODUCTION TO CONTINUUM MECHANICS FOR ENGINEERS**
• *Ray M. Bowen*

40 **STRUCTURAL OPTIMIZATION**
Volume 2: Mathematical Programming
• *Edited by M. Save and W. Prager*

41 **OPTIMAL CONTROL OF DISTRIBUTED NUCLEAR REACTORS**
• *G. S. Christensen, S. A. Soliman, and R. Nieva*

42 **NUMERICAL SOLUTION OF INTEGRAL EQUATIONS**
• *Edited by Michael A. Golberg*

43 **APPLIED OPTIMAL CONTROL THEORY OF DISTRIBUTED SYSTEMS**
• *K. A. Lurie*

44 **APPLIED MATHEMATICS IN AEROSPACE SCIENCE AND ENGINEERING**
• *Edited by Angelo Miele and Attilio Salvetti*

45 **NONLINEAR EFFECTS IN FLUIDS AND SOLIDS**
• *Edited by Michael M. Carroll and Michael A. Hayes*

46 **THEORY AND APPLICATIONS OF PARTIAL DIFFERENTIAL EQUATIONS**
• *Piero Bassanini and Alan R. Elcrat*